Plant Biochemistry: Concepts and Applications

Plant Biochemistry: Concepts and Applications

Editor: Timothy Granger

www.callistoreference.com

Callisto Reference,
118-35 Queens Blvd., Suite 400,
Forest Hills, NY 11375, USA

Visit us on the World Wide Web at:
www.callistoreference.com

ISBN: 978-1-63239-987-8 (Hardback)

Trademark Notice: Registered trademark of products or corporate names are used only for explanation and identification without intent to infringe.

Cataloging-in-Publication Data

Plant biochemistry : concepts and applications / edited by Timothy Granger.
   p. cm.
Includes bibliographical references and index.
ISBN 978-1-63239-987-8
1. Botanical chemistry. 2. Plant biochemical genetics. 3. Phytochemicals. I. Granger, Timothy.
QK861 .P53 2018
581.192--dc23

Table of Contents

Preface

The world is advancing at a fast pace like never before. Therefore, the need is to keep up with the latest developments. This book was an idea that came to fruition when the specialists in the area realized the need to coordinate together and document essential themes in the subject. That's when I was requested to be the editor. Editing this book has been an honour as it brings together diverse authors researching on different streams of the field. The book collates essential materials contributed by veterans in the area which can be utilized by students and researchers alike.

Plant biochemistry studies the chemical mechanisms of plants. It is an important field as it helps in explaining the molecular functions of a plant. Crop modification techniques like mutagenesis, protoplast fusion, and traditional breeding help in promoting desired traits in plants. This book explores all the important aspects of plant biochemistry in the present day scenario. The various studies that are constantly contributing towards advancing technologies and evolution of this field are examined in detail. For all readers who are interested in plant biochemistry, the case studies included in this book will serve as an excellent guide to develop a comprehensive understanding.

Each chapter is a sole-standing publication that reflects each author's interpretation. Thus, the book displays a multi-facetted picture of our current understanding of application, resources and aspects of the field. I would like to thank the contributors of this book and my family for their endless support.

Editor

Antioxidant and Anti-Inflammatory Activities of Extracts from *Cassia alata, Eleusine indica, Eremomastax speciosa, Carica papaya* and *Polyscias fulva* Medicinal Plants Collected in Cameroon

Bertrand Sagnia[1]*, Donatella Fedeli[2], Rita Casetti[3], Carla Montesano[4], Giancarlo Falcioni[2], Vittorio Colizzi[3,4]

1 Laboratory of Microbiology and Immunology, Chantal BIYA International Reference Centre for Research on Prevention and Management of HIV/AIDS (CIRCB), Yaounde, Cameroon, 2 School of Pharmacy, University of Camerino, Camerino (MC), Italy, 3 Laboratory of Cellular Immunology, National Institute for Infectious Diseases "Lazzaro Spallanzani", Rome, Italy, 4 Department of Biology, University of Rome Tor Vergata, Rome, Italy

Abstract

Background: The vast majority of the population around the world has always used medicinal plants as first source of health care to fight infectious and non infectious diseases. Most of these medicinal plants may have scientific evidence to be considered in general practice.

Objective: The aim of this work was to investigate the antioxidant capacities and anti-inflammatory activities of ethanol extracts of leaves of *Cassia alata, Eleusine indica, Carica papaya, Eremomastax speciosa* and the stem bark of *Polyscias fulva*, collected in Cameroon.

Methods: Chemiluminescence was used to analyze the antioxidant activities of plant extracts against hydrogen peroxide or superoxide anion. Comet assays were used to analyze the protection against antioxidant-induced DNA damage induced in white blood cells after treating with hydrogen peroxide. Flow cytometry was used to measure $\gamma\delta$ T cells proliferation and anti-inflammatory activity of $\gamma\delta$ T cells and of immature dendritic cells (imDC) in the presence of different concentrations of plant extracts.

Results: Ethanol extracts showed strong antioxidant properties against both hydrogen peroxide and superoxide anion. *Cassia alata* showed the highest antioxidant activity. The effect of plant extracts on $\gamma\delta$ T cells and imDC was evidenced by the dose dependent reduction in TNF-α production in the presence of *Cassia alata, Carica papaya, Eremomastax speciosa Eleusine indica,* and *Polyscias fulva*. $\gamma\delta$ T cells proliferation was affected to the greatest extent by *Polyscias fulva*.

Conclusion: These results clearly show the antioxidant capacity and anti-inflammatory activities of plant extracts collected in Cameroon. These properties of leaves and stem bark extracts may contribute to the value for these plants in traditional medicine and in general medical practice.

Editor: Mario D. Cordero, University of Sevilla, Spain

Funding: This work was supported by the School of Advanced Studies, University of Camerino, Italy, and the University of Rome Tor Vergata, Italy. The funders had no role in study design, data collection and analysis, decision to publish, or preparation of the manuscript.

Competing Interests: The authors have declared that no competing interests exist.

* Email: bertrandsagnia@yahoo.fr

Introduction

Medicinal plants have been used in traditional health care systems since prehistoric times and are still the most important health care source for the most of the world's population. The World Health Organization (WHO) has estimated that more than 75% of the world's total population depends on herbal drugs for their primary healthcare needs [1]. Therefore, there is a major research emphasis on discovering plants with antioxidant and anti-inflammatory potential that may be treat various kinds of injuries or protect against diseases [2]. Cellular and tissue damage caused by oxidative stress is defined by the elevated levels of free radicals or other reactive oxygen species (ROS) that can elicit direct or indirect damage to the body. The generation and subsequent involvement of free radicals contributes to a large number of diseases including AIDS [3,4], carcinogenesis and liver damage, inflammatory diseases, cataract formation and Alzheimer's disease, are recognized [5,6,7,8]. Intracellular protective mechanisms against inflammatory stresses involve antioxidant enzymes, including superoxide dismutase (SOD), catalase (CAT) and glutathione peroxidase (GPx) in tissues. It has been shown that faulty cellular antioxidant systems cause organisms to develop a series of inflammatory or malignant diseases [9]. However, it appears that the various roles of enzymatic antioxidants help to protect organisms from excessive generation of oxidative stress during inflammation process, which leads to studies focusing on the role for natural products in suppressing the production of oxidation by increasing enzymatic antioxidants in tissues [10].

Under normal conditions, ROS levels are controlled by the body's complex antioxidant defense system and there is an equilibrium between ROS formation and degradation. Overproduction of ROS and/or inadequate antioxidant defense disturbs this equilibrium in favor of a ROS upsurge that results in oxidative stress. A deficiency in the body's natural antioxidant defense mechanisms has been implicated as the etiological or pathological factor in several clinical disorders. This has led to scientific research focusing on identifying safe and effective sources of antioxidants. Plant extracts and plant-derived antioxidant compounds may potentiate the body's antioxidant and anti-inflammatory defense mechanisms or act as antioxidants.

Inflammation is a normal response to tissue injury but, if uncontrolled, leads to additional complications. At the injury site, an increase in blood vessel wall permeability followed by migration of immune cells can cause edema formation during inflammation. Excessive inflammation contributes to many acute and chronic human diseases [11].

The immune system continuously monitors resident microbiota and utilizes constitutive antimicrobial mechanisms to maintain immune homeostasis. For example, lipopolysaccharide (LPS) is an endotoxin and a constituent of the outer membrane of gram-negative bacteria. LPS stimulates innate immunity, by regulating the productions of inflammatory mediators, like, Nitric Oxide (NO), TNF-α (Tumor Necrosis Factor-alpha), Interleukin-6 (IL-6), prostanoids, and leukotrienes. Dendritic cells (DCs), the main antigen presenting cells, play a pivotal role in priming T cells upon immune challenge [12]. DCs respond to innate immune activation by upregulating costimulatory molecules, lymphoid-homing chemokines and producing proinflammatory cytokines, such as TNF-α. There is a reciprocal activation between DCs and Vγ9Vδ2 T cells, which are the major population of peripheral blood γδ T cells [13,14]. Vγ9Vδ2 T cells, when activated by phosphoantigen compounds after viral or bacterial infection, produce some cytokines and chemokines that recruit other immune cells to the site of infection. TNF-α is one of the most pro-inflammatory cytokine produce by Vγ9Vδ2 T cells when activated [15] and if this reaction is not controlled, it may lead to chronic, destructive inflammation. Thus, antioxidant and anti-inflammatory activities are very important for controlling oxidative stress and inflammatory processes generated during the response to infectious diseases.

In this study, we used plant extracts of Cassia alata Linn (Caesalpinaceae), Carica papaya Linn (Caricaceae), Eremomastax speciosa Hochst (Acanthaceae), Eleusine indica Linn Gaertn (Poaceae) and the stem bark of Polyscias fulva Hiern HARMS (Araliaceae) to evaluate their antioxidant activities compared to the antioxidant enzyme present in humans, the superoxide dismutase (SOD).

Leaves of Carica papaya, L. [16], have been used to treat malaria, yellow fever, oral candidosis [17], dengue, as a diuretic and in case of anemia [18,19,20,21]. Leaves of Eremomastax speciosa Hochst, are used as anti-diarrheal [22,23]. Leaves of Cassia alata are used against yellow fever or malaria, and as anti asthmatics, or anti diabetics [24,25,26]. Eleusine indica Gaertn or Wiregrass (grass family Poaceae), is used in traditional medicine as a diuretic, anti-helminthic, febrifuge and for treating cough [27], [28]. Eleusine indica was evaluate for hepatoprotective effects and its mechanism of action was studied [29]. Polyscias fulva, Harms,

is used in traditional medicine to fight against venereal infections and against dermatoses [30,31]. Since these plants are widely used in Cameroon as traditional folk medicines there is a need for chemical and pharmacological studies that may help to corroborate the scientific bases of the use of these plant products and standardize their use in general medical practice. The E. coli plasmid, the DNA of lymphocytes and the liposomes were used as a model to evaluate the antioxidant properties of plant extracts against anion superoxide and hydrogen peroxide. γδ T cells and imDC were used as a model to evaluate the anti-inflammatory activity. Our results showed that certain medicinal plants are promising sources of potential antioxidants and anti-inflammatory and may be effective as preventive agents in the pathogenesis of some diseases.

Materials and Methods

Plant Material

Fresh aerial parts of Carica papaya (leaves), Eremomastax speciosa (leaves), Eleusine indica (leaves), Cassia alata (leaves), and Polyscias fulva (stem bark), were collected in the Centre Region of Cameroon, in January–June, 2002. These plants were identified at the Cameroon National Herbarium, where the voucher specimens were kept under the reference numbers.

Ethics Statement

For the collection of plants, no specific permits were required for the described field studies. For any locations/activities, no specific permissions were required. All locations where the plants were collected were not privately-owned or protected in any way and the field studies did not involve endangered or protected species. This study was approved by the University of Camerino's institutional review board.

Extraction

Air-dried and powdered samples from each plant were macerated separately in an ethanol for 48 h at room temperature with shaking. The extract was filtered through a Whatman no. 1 filter paper. The filtrate was then evaporated to dryness at 50°C for ethanol under reduced pressure using a rotary evaporator. This produced a residue which constituted the crude extract. The extraction yield was calculated and the crude extract was kept at + 4°C until further use.

Chemiluminescence Measurement of Plant Extracts Antioxidant Activity

Chemiluminescence measurements (CL) to evaluate the antioxidant activity of plant extracts were performed using an Autolumat Berthold LB 953 (Berthold Co., Wilbard, Germany). In order to measure the scavenger activity of these compounds against hydrogen peroxide, a reaction mixture containing different concentrations of plant extracts, 100 μM luminol in 50 mM phosphate buffer pH 7.0 were prepared. The reaction was initiated by injecting 0.05 ml of H_2O_2 to a final concentration of 50 mM. To assess the scavenger activity toward superoxide anion, a reaction mixture containing 0.9 U/ml Xanthine oxidase, 150 μM lucigenin in 50 mM phosphate buffer pH 7.0, and different concentrations of plant extracts were used. The reaction was started by injecting Xanthine at a final concentration of 50 μM. The data were reported as the percentage (%) of inhibition (I) of the CL (chemiluminescence) signal and calculated as follows: where **CL sample** is the chemiluminescence signal obtained for a sample in the presence of plant extracts and **CL blank** is the

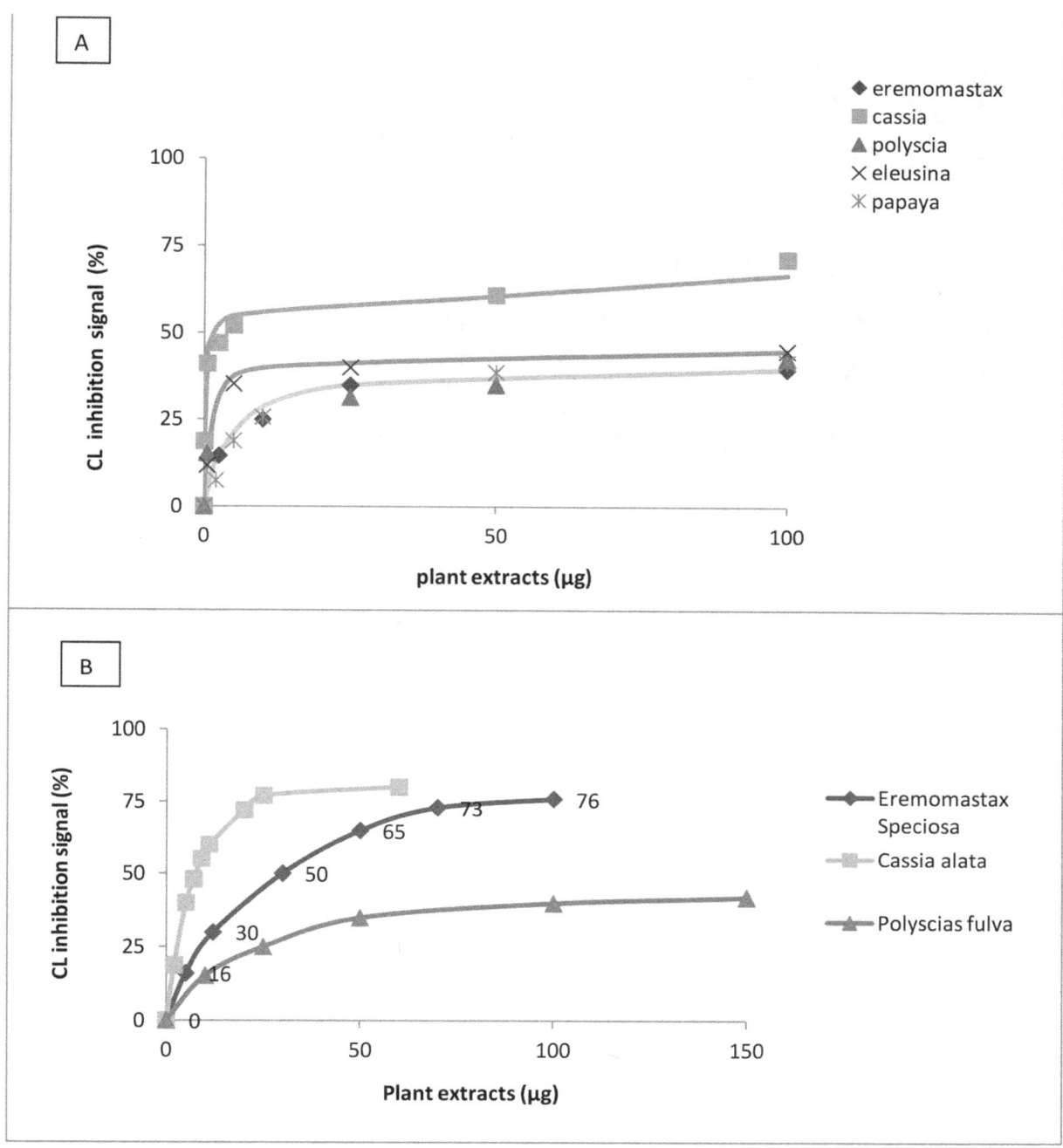

Figure 1. Evaluation of the chemiluminescence (CL) signal inhibition in presence of different concentrations of plant extracts versus: A) Hydrogen peroxide (H_2O_2) B) Anion superoxide (O_2) generated by Xanthine – Xanthine oxidase system.

chemiluminescence signal obtained in the sample without plant extracts.

E. coli DNA damage

The antioxidant capacity of plant extracts was evaluated on *E. coli* plasmid DNA pBR322 in the presence of oxidant agents. 100 ηg DNA +10 μM Fe^{++} +100 μM H_2O_2+12.5 μg plant extracts. Samples were incubated for 1 h at 35°C. The agarose gel (1.2%) was prepared by dissolving the solid agarose powder in the electrophoresis buffer, the Tris-AceticAcid-EDTA ("TAE"). (A commonly used stock solution for TAE is 50 times concentrated

("50xTAE"); the standard procedure for preparation of 50xTAE is: mix deionised water with solid Tris powder, a certain amount of an EDTA stock solution, usually 0.5 M EDTA pH 8.0, and concentrated acetic acid to adjust the pH to 7.6). The agarose gel was dissolved at 100°C for a few minutes and the dissolved solution was then poured into an electrophoretic plate which contained combs. Cooling down to room temperature results in the slow formation of a solid gel when the concentration of agarose is between 0.5 and 2% (weight/volume). Ethidium bromide (Et-Br) was added to the gel solution for visualization of DNA bands. The Et-Br agarose gel was then submerged in a horizontal electropho-

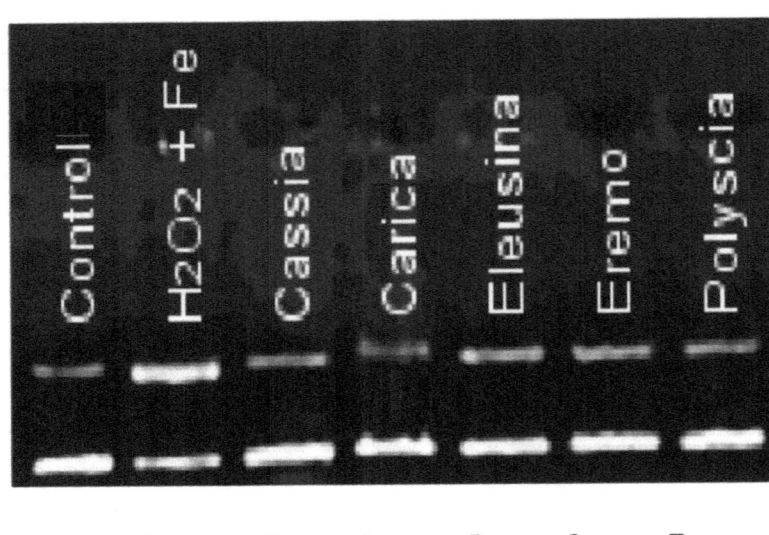

1. DNA;
2. DNA+ Fe⁺⁺/H2O2;
3. DNA+ Fe⁺⁺/H2O2+ 12µg *Cassia alata*;
4. DNA+ Fe⁺⁺/H2O2+ 12µg Carica papaia;
5. DNA+ Fe⁺⁺/H2O2+ 12µg Eleusine indica
6. DNA+ Fe⁺⁺/H2O2+ 12µg Eremomastax speciosa
7. DNA+ Fe⁺⁺/H2O2+ 12µg Poliscia fulva

Figure 2. Protective effect of 12 µg plant extracts from Fe^{2+} and H_2O_2 induced *E. coli* **plasmid damage.** *Ctr: control.*

resis apparatus. Loading buffer (Bromo Phenol Blue) was used to fill the samples into the wells and to serve as the migration tracking dye. Migration was done at about 85 Volts for 45 min at room temperature. After electrophoresis, the gel was placed under UV light and standard or digital photograph of the fluorescent DNA bands were saved.

Comet Assay

DNA damage in lymphocytes was evaluated using the alkaline single-cell microgel electrophoresis ('comet' assay) [32,33]. The comet assay was carried out under yellow light. About 2×10^5 cells were mixed with 65 µl of 0.7% low melting agarose (LMA) in Ca^{2+} and Mg^{2+} free PBS to form a cell suspension. The cell

suspension was rapidly spread over a pre-cleaned microscope slide previously conditioned by spreading a 1 ml aliquot of 1% LMA in Ca^{2+} and Mg^{2+} free PBS. After solidification, the cells were protected with a top layer of 75 µl of 0.7% LMA. To lyse the embedded cells and to permit DNA unfolding, the slides were immersed in freshly prepared ice-cold lysis solution (1% sodium *N*lauroyl- sarcosinate, 2.5 M NaCl, 100 mM Na₂EDTA, 10 mM Tris–HCl, pH 10, with 1% Triton X-100 and 10% DMSO added just before use) for 1 h at +4°C in the dark. After the lysis, the slides were placed on a horizontal electrophoresis box. The unit was filled with freshly made alkaline buffer (300 mM NaOH, 1 mM Na₂EDTA, pH>13) and, to allow DNA unwinding and expression of alkali labile damage, the embedded cells were left in

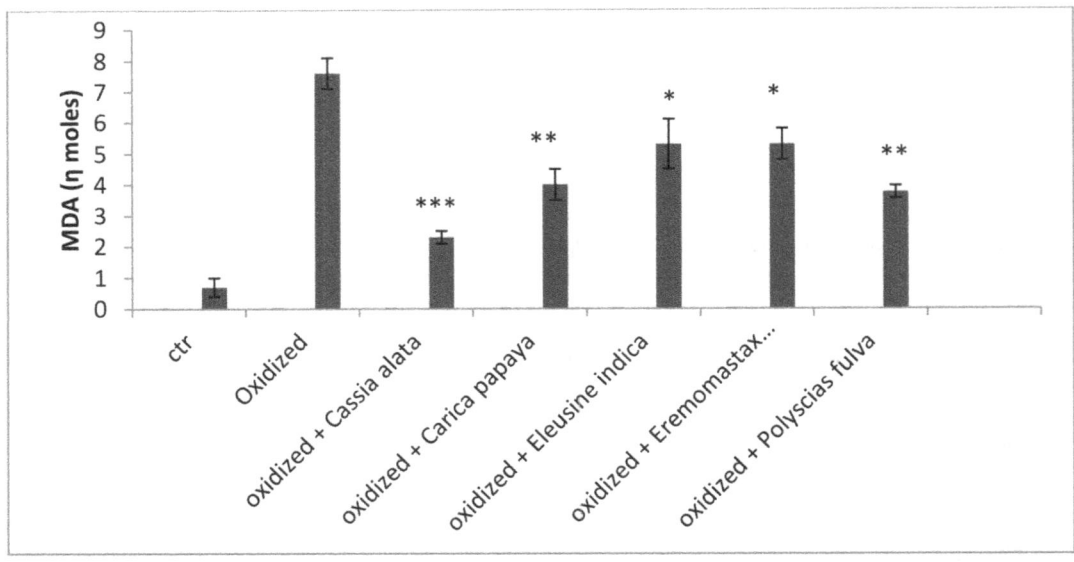

Figure 3. Reduced amount of malonil dialdeide (MDA) by lipid peroxydation in presence of a lucegenin probe and the xanthine/ xanthine oxydase system with 150 µg of different plant extracts. *Ctr: Control, ***$p < 0.001$, **$p < 0.01$, *$P < 0.05$.*

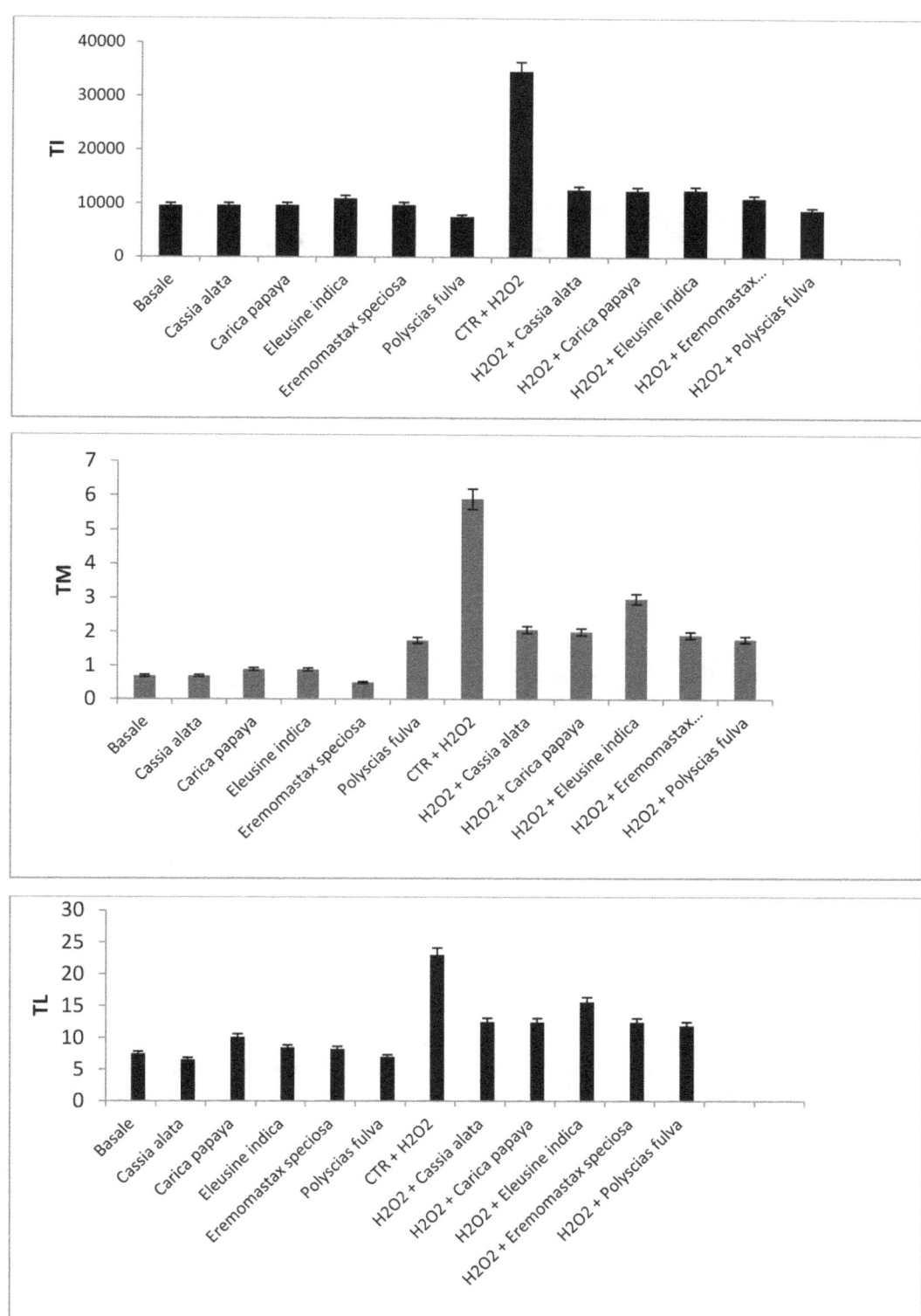

Figure 4. Distributions of: a) comet tail moment (TM); b) tail length (TL); c) tail intensity (TI) in white blood cells from healthy human donors treated with hydrogen peroxide and 100 μg of plant extracts. Data (at least 150 scores/sample) are mean values +/− SEM. $p < 0.05$ is considered significant.

the solution for 20 min. Electrophoresis was performed for 20 min by applying an electric field of 25 V and adjusting the current to 200 mA. After the electrophoresis, the slides were washed gently with 0.4 M Tris- HCl buffer pH 7.5 to neutralize the excess alkali and remove detergents. Slides were stained by adding 20 μl of ethidium bromide (2 μg/ml) and the rate of DNA damage was

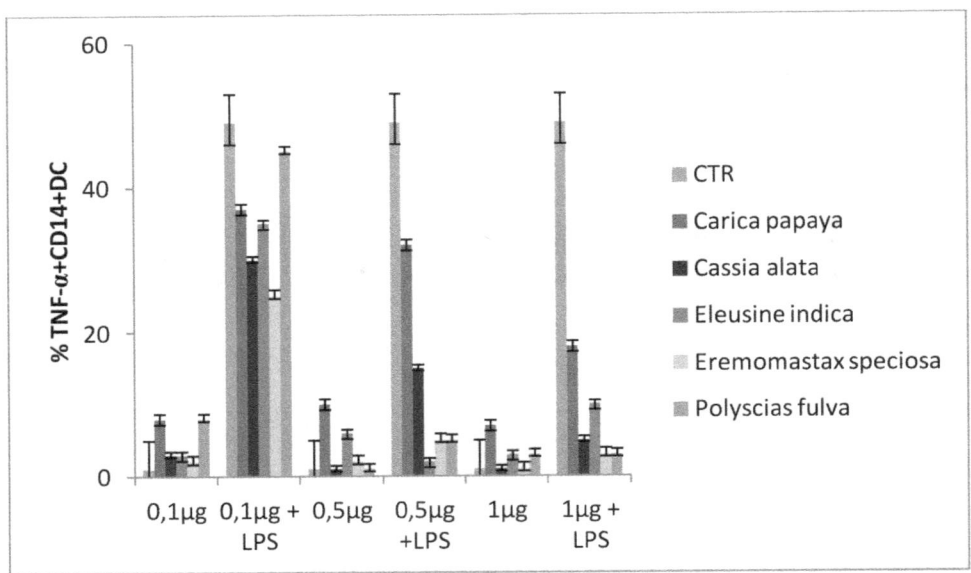

Figure 5. Plant extracts modulate LPS –induced TNF-α production by imDC. imDC were generated from monocytes positively separated from PMBC by anti-CD14 magnetic beads and cultured for 5 days in the presence of GM-CSF and IL-4. Different concentrations of plant extracts were added to imDC in the presence of LPS and TNF-α production was assessed by Flow Cytometry. Values of $p < 0.05$ were considered statistically significant. *CTR: Control; imDC: immature Dendritic Cells; LPS: Lipopolysaccharide; TNF-α: Tumor necrosis factor alpha.*

evaluated by analyzing at a magnification of ×20 the images of 150 randomly selected cells (50 cells from each of three replicate slides) by using an Axioskop 2 plus epi-fluorescence microscope (Carl Zeiss, Germany) equipped with an excitation filter of 515–560 nm. Imaging was performed using a specialized analysis system ('Metasystem' Altlussheim, Germany) to determine tail length, tail intensity and tail moment.

Inhibition of lipid peroxidation assay

Fe^{2+} induced lipid peroxidation is one of the established system for assessing antioxidant action of different plant extracts. A modified thiobarbituric acid-reactive species (TBARS) assay [34,35] was used to measure the lipid peroxide formed using liposomes homogenate as lipid rich media. Malondialdehyde (MDA), a secondary end product of the oxidation of polyunsaturated fatty acids, reacts with two molecules of TBA yielding a pinkish red chromogen, homogenate was centrifuged at 800 g for 15 min at 4°C and the supernatant was used in thiobarbituric acid assays. Different concentrations of plant extracts (40–400 μg/mL) were fixed with the liposome preparation and incubated at room temperature for 10 min. Then, 50 μL Fenton's reagents (10 mM $FeCl3$; 10 μL of 2.5 mM H_2O_2) in phosphate buffer (0.2 M, pH 7.4) were added, and the volume was made to 1 mL. The tubes were then incubated for 30–45 min at 37°C to induce lipid peroxidation. Thereafter, 2 mL of ice-cold HCl (0.25 N) containing 15% trichloroacetic acid, 0.5% thiobarbituric acid and 0.5% butylated hydroxytoluene (BHT) were added in each sample that were boiled for 15 min. The mixture was then centrifuged at 1000 rpm for 10 min and the extent of lipid peroxidation was subsequently monitored by formation of thiobarbituric acid reactive substances (TBARS) as pink chromogen in presence or absence of extracts. The absorbance of the supernatant was measured spectrophotometrically at 532 nm and decline in formation of pink chromogen in pre-treated reactions was considered as inhibition of lipid peroxidation.

Cell isolation and culture conditions

PBMC from healthy donors were isolated from fresh buffy coats by Lympholyte (Cedarlane) density gradient centrifugation. Cells were cultured at a concentration of 4×10^6/ml in the presence of IL-2 (6.5 U/ml), IL-15 (10 ηg/ml) (both from Sigma-Aldrich), TGF-β (1.7 g/ml; Calbiochem) and isopentenyl pyrophosphate (IPP; 20 μg/ml; Sigma). On days 3, 6, and 9, half of the supernatant volume was discarded and replaced with fresh medium containing cytokines. Aliquots of PBMC and bead-sorted γδ T cells (Miltenyi Biotec) were used fresh or were frozen for use at later time points.

Monocyte purification, imDC generation

PBMCs were isolated from buffy coats of healthy donors by density gradient centrifugation using Lympholyte-H (Cederlane Laboratories). In this study we used anonymous residual samples (buffy coats) and approval was not necessary. These residual samples were collected in the Azienda Ospedaliera Universitaria Policlinico Umberto I - Centro Trasfusionale e Immunoematologia. URL: www.policlinicoumberto1.it.

Monocytes were positively separated by anti-CD14 magnetic beads (MACS; Miltenyi Biotec) according to the manufacturer's instructions. The cells were then resuspended in RPMI 1640 (Euroclone) supplemented with 10% FCS (HyClone, Invitrogen Life Technologies), L-glutamine (2 mM), HEPES buffer (10 mM), and gentamicin (10 μg/ml) (Sigma-Aldrich), and cultured for 5 days in the presence of GM-CSF (200 U/ml) and IL-4 (10 ηg/ml; Euroclone) to generate immature DC (imDC). Then, imDC were put in culture with different concentration of plant extracts in the presence of 200 ηg of LPS.

γδ T cells purification and proliferation

γδ T cells were separated from autologous PBMCs by positive selection using anti-γδ-magnetic beads (MACS; Miltenyi Biotec) according to the manufacturer's instructions. Purified cell populations contained 98% of viable γδ T cells as assessed by flow

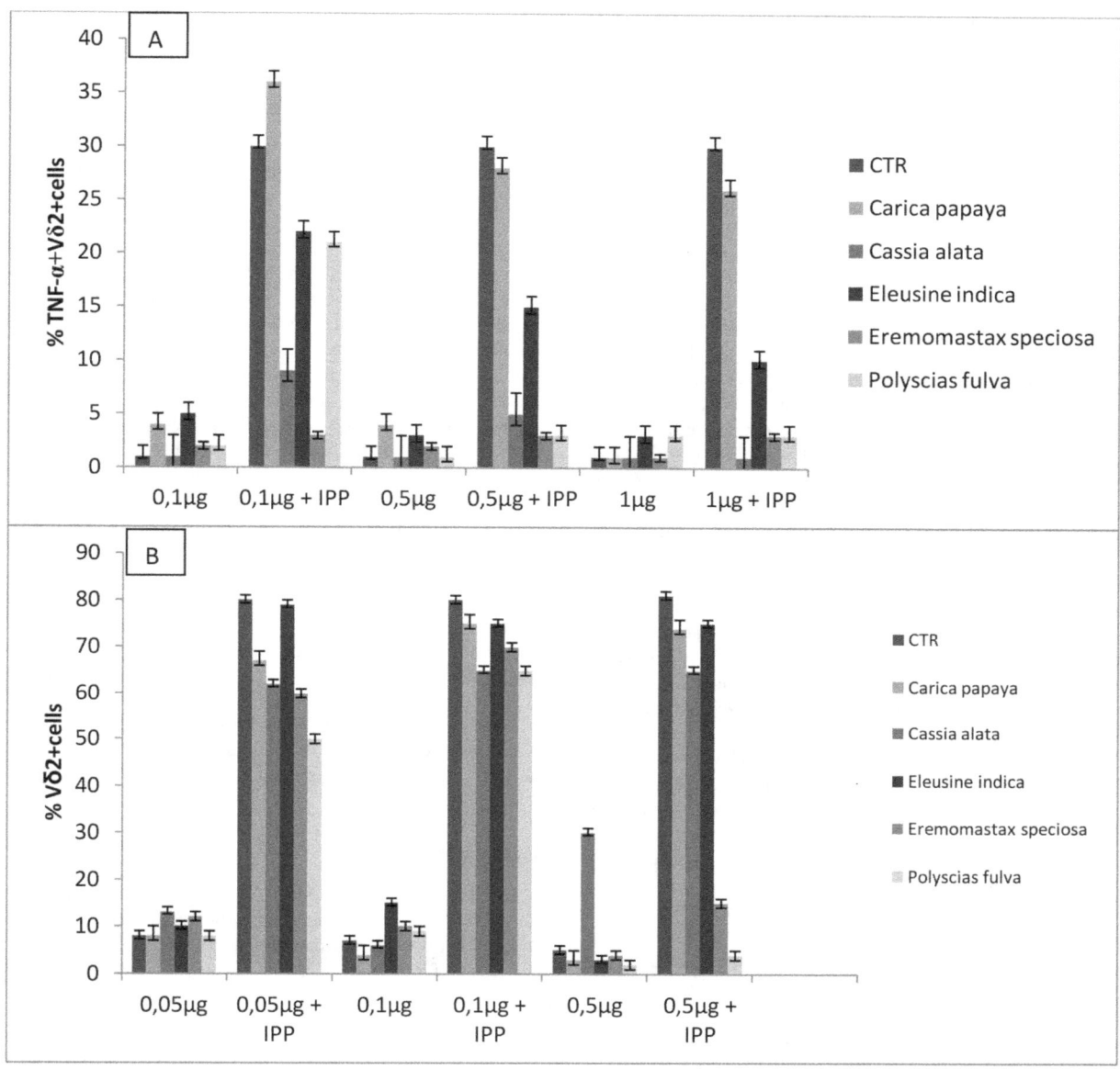

Figure 6. Plant extracts modulate γδ T cell cytokine production and proliferation. Purified-γδ T cells were incubated with IPP and IL-2 in the presence of different concentrations of plant extracts induced TNF-α production (A) and on γδ T cell proliferation (B) were assessed by Flow Cytometry. Values of p<0.05 were considered statistically significant. *CTR: Control; TNF-a: Tumor necrosis Factor-alpha; IPP: isopentenyl pyrophosphate.*

cytometry. Purified-γδ T cells were incubated for 10 days with 3 µg of IPP and 100 U/mL of Interleukine-2 (IL-2) (Euroclone) and in the presence of different concentrations of plant extracts.

FACS analysis

The following FITC-, PE-, PerCP-, or APC- conjugated Abs: CD25, CD14, CD27, CD45RA, CD62L, TNF-α, IFN-γ, CD3, HLA-DR, and Vδ2 (BD Biosciences) were used for direct immunofluorescence staining to characterize the phenotype of γδ T cells. Isotype-matched mAbs (BD Biosciences) were used in all experiments as controls. In brief, the cells were washed twice in PBS, 1% BSA, and 0.1% sodium azide, and were stained with the mAbs for 15 min at 4°C. The cells were then washed and analyzed using a FACSCalibur instrument with CellQuest software (BD Biosciences). The following PE and allophycocyanin-conjugated

anti-TNF-α and anti-IFN-γ mAbs (BD Pharmingen) were used for intracellular immunostaining to characterize Vδ2 T cells and imDC producing cytokines.

Statistical analysis

Statistical analysis was determined using a Mann-Whitney *U* test. Values of p<0.05 were considered statistically significant for flow cytometry. Data are expressed as mean values SEM except for chemiluminescence results, for which data are presented as mean values SD.

Results

In this study we evaluated both antioxidant and anti-inflammatory activity of *Cassia alata* Linn (Caesalpinaceae), *Carica papaya* Linn (Caricaceae), *Eremomastax speciosa* Hochst (Acanthaceae), *Eleusine indica* Linn Gaertn (Poaceae) and the stem bark of *Polyscias fulva* Hiern HARMS (Araliaceae) by using chemiluminescence, gel electrophoresis and flow cytometry respectively. First, we showed the presence of phenol in all extracts by the spectrophotometric analysis.

Antioxidant scavenging activities

In order to evaluate the scavenger activities of plant extracts against hydrogen peroxide or superoxide anion, luminol and lucigenin amplified chemiluminescence inhibition respectively was measured. As showed in Fig. 1A, Luminol amplified chemiluminescence signal reduced of 67%, 45.5%, 42.9%, 41.8% and 39.7% for *Cassia alata*, *Eleusine indica*, *Polyscias fulva*, *Carica papaya* and *Eremomastax speciosa*, respectively by using less than 12.5 μg of plant extracts. Concerning the scavenger activity versus superoxide anion generated by xanthine-xanthine oxidase system, lucigenin amplified chemiluminescence was reduced of 65%, 38% and 10% in presence of *Cassia alata*, *Eremomastax speciosa* or *Polyscias fulva*, respectively Fig. 1B. For both systems, *Cassia alata* resulted to have the best scavenger activity. Moreover, the antioxidant capacity of plant extracts was evaluated by measuring the reduction of Fe^{3+} ferricyanide complex to the ferrous form, Fe^{2+} due to the presence of a reductant, such as the antioxidant substances in plant extracts. As shown in Fig. 2, 12 μg of plant extracts, protect the E. coli plasmid from the oxidation with hydrogen peroxide by maintaining the DNA plasmid supercoiled. The relevant information is conversion of supercoiled DNA to the slower running nicked circular DNA. The protective effects of plant extracts are seen by the reduction in nicked circular DNA.

In Fig. 3, the plant extracts were tested on chicken liposomes about the reduction of MDA (Malonil dialdeid) in presence of the anion superoxide produced by Xanthine – Xanthine oxidase and amplified by lucegenin as a probe. In this system, using 150 μg of plant extracts, the highest reducing power activity was observed on *Cassia alata*, followed by *Polyscias fulva*, *Carica papaya*, *Eremomastax* and *Eleusine indica*, if compared with the positive control. The ferric reducing power may be attributed to the phenolic and flavonoid contents of the extracts. The ability to reduce Fe (III) may be attributed to the hydrogen donation from phenolic compound, which is related to the presence of a reducing agent. In addition, the number and position of hydroxyl group of phenolic compounds also govern their antioxidant activity [36].

For evaluating DNA damage in lymphocytes following hydrogen peroxide treatment and to explore the potential protective effect of plant extracts the comet assay was used. Comet assay parameters – tail moment, tail length, and tail intensity – have been used widely all over the world for determination of DNA damage. As the amount of the damage increases in a cell, more DNA migrates into the tail intensity and is quantified in terms of increased amount of determined fluorescence in the tail region and tail length. The percentage of DNA in the tail region (tail intensity) has been in use for quantifying DNA strand breakage and is the most advised parameter to use [37,38]. A major advantage of using tail moment as the index of DNA damage is that both the amount of damaged DNA and the distance of migration of the genetic material in the tail are represented by a single number. In this study, data analysis was performed using tail moment, tail length, and tail intensity. In our study, the main outcome of the comet assay was the detection of DNA strand breaks caused by exposure to hydrogen peroxide. Fig. 4 shows the data obtained after incubating cells with 100 μg of plant extracts in the presence or absence of hydrogen peroxide. Our results show that plant extracts protect the DNA of lymphocytes from damage because

there was a significant decreased of Tail Intensity (Fig. 4A), Tail Moment (Fig. 4B), and Tail length (Fig. 4C) in the Comet assays.

Anti-inflammatory activity

As known, Lipopolysaccharide (LPS) induce dendritic cell maturation producing a great amount of TNF-α. To assess the anti-inflammatory activity of plant extracts, TNF-α production by imDC was evaluated after culture of DC with different concentration of plant extracts in the presence of LPS.

As shown in Fig. 5, all the considered plant extracts inhibited in a dose dependent manner the LPS-induced TNF-α production by imDC. The most inhibition was obtained with 1 μg of plants extracts by *Eremomastax speciosa Polyscias fulva* and *Cassia alata* followed by *Eleusine indica* and *Carica papaya*.

In response to phosphoantigen stimulation $\gamma\delta$ T cells produce high levels of cytokines including TNF-α and begin to proliferate. In our experiment, we evaluated the effect of plant extracts on both the cytokine production and proliferation of $\gamma\delta$ T lymphocytes. IPP-induced TNF-α production was inhibited by plant extracts in a dose dependent manner. The greatest inhibition was obtained by 1 μg of *Cassia alata*, followed in the same extent by *Eremomastax speciosa* and *Polyscias fulva*, and finally by *Eleusine indica* and *Carica papaya* ($p<0.05$) (Fig. 6A).

As regards the proliferative activities, PBMCs of healthy donors were incubated for 10 days with IPP and different concentrations of plant extracts in the presence of IL-2. As shown in Fig. 6B, only *Eremomastax speciosa* and *Polyscias fulva* at 0.5 μg inhibited $\gamma\delta$ T cells proliferation.

Taken together these results show that *Cassia alata*, but also *Eremomastax speciosa* and *Polyscias fulva* possess the best antioxidant and scavenger activities as well as anti-inflammatory activity.

Discussion

In this study, we used plants extracts to evaluate their antioxidant activities compared to the antioxidant enzyme present in humans, the superoxide dismutase (SOD). The *E. coli* plasmid, the DNA of lymphocytes and the liposomes were used as a model to evaluate the antioxidant properties of plant extracts against anion superoxide and hydrogen peroxide. $\gamma\delta$ T cells and imDC were used as a model to evaluate the anti-inflammatory activity. The antioxidant and anti-inflammatory activities are very important for humans because the oxidative stress and the inflammatory process were generated during different infectious diseases. In this study, we observed that extracts from leaves of *Cassia alata* as well as *Eremomastax speciosa*, and *Polyscias fulva* have better antioxidant activities than *Carica papaya* or *Eleusine Indica*. This could be explained in part because the spectrophotometric analysis show the presence of phenols in all extracts and these extracts protect from hemolysis except for the *Polyscias fulva* stem bark extract. The antioxidant and anti-inflammatory activities could also been explained by the chemical analysis of *Cassia alata*, showing the presence of the Chrysarobin, the tannin, Kaempferol, Isochrysophanol, Chrysophanol glycoside Chrysarobin 1,5 [24]. Leaves of *Eremomastax speciosa* show the presence of alkaloids, of iridoids, and arthraquinone, flavonols like kaempferol and trace of quercetin [39]. Chemical analyses of the sterm bark of *Polyscias fulva* present the tri terpène glycoside drammarane-type, 2 types of saponin (α -hederin and 3-O- α -L-rhamnopyranosyl-(1→2)- α -L-arabinopyranosyl-hederagenin-28-O-α-L-rhamnopyranosyl-(1→4)-β-.glucopyranosyl-(1→6)- β-D-glucopyranosyl ester and the quercetin 3-O-β-D-glucopyranoside [31]. Leaves of *Carica papaya* are rich in anthraquinone and in alkaloids like carpaine, the flavonols, the vitamin C and E. These molecules have a wide range of biological activities. The popularity of herbs in traditional medicine has been linked to their higher likelihood of containing pharmacologically active compounds compared to woody plant forms [40]. Previous studies showed chemical constituents like alkaloids, volatile and essential oils, phenolic compounds, triterpenoids, saponins, phytosterols, tannins, flavanoids all possess, anti-inflammatory and antioxidant activity [41]. Leaves of plants have been reported to accumulate inulins, tannins and other alkaloids which may be responsible for their medicinal properties. Other studies reported that leaves are the most widely used parts of plant [42]. This may be because some plants contain many secondary metabolites which could have different pharmacological activities and consequently treat different diseases [43,44,45] or that leaves can be identified clearly and labeled for commercial trade. Our studies begin to address the complexity of pharmacologic activities present in traditional plant medicines used in Cameroon and help to further our understanding of mechanisms for action and why specific plants are used to treat individual diseases. According to World Health Organization, medicinal plants may be the best source for a variety of drugs. More than 75% of the total population in developing countries relies on traditional medicines based on plant products [1]. Cameroon is rich in the variety of medicinal plants that could help to fight disease like the HIV/AIDS, the malaria and tuberculosis. These diseases are concentrated in lower income countries where the health care access is difficult. The use of medicinal plants could constitute a reservoir of new molecules important for antifungal, antibacterial, antiviral, antioxidant and anti-inflammatory substances therapies and research on composition and mechanism of action will create better treatment standards and improve the value of traditional plants as sources of new medications.

Conclusion

The findings of this study support the view that medicinal plants are promising sources of potential antioxidants and anti-inflammatory agents that may be effective for therapy of human diseases. Studying the ethanol soluble molecules revealed a rich variety of antioxidant and anti-inflammatory molecules mainly present in leaves. In the context of HIV/AIDS and other infections, research and development of novel and effective treatments from safe herbal medicines will improve the quality of life for persons in need. The results presented here should encourage the use of these plants for medicinal health and nutraceutical applications, due to their antioxidant and anti-inflammatory properties.

Acknowledgments

We dedicate this work in memoriam to our dearest friend Angelo Martino, who helps in the anti-inflammatory and proliferative assays in this work at Spallanzani Hospital in Rome. It was a great privilege and an immense pleasure for us to have been associated with Angelo. We will miss forever his friendship and infectious enthusiasm for immunological investigations.

Author Contributions

Conceived and designed the experiments: GF BS VC. Performed the experiments: BS RC CM. Analyzed the data: BS DF RC. Contributed reagents/materials/analysis tools: GF VC DF. Wrote the paper: BS RC GF.

References

1. Organization WH (2002) WHO traditional medicine strategy 2002–2005 Geneva: World Health Organization; 2002.
2. Huang SS, Chiu CS, Lin TH, Lee MM, Lee CY, et al. (2013) Antioxidant and anti-inflammatory activities of aqueous extract of Centipeda minima. J Ethnopharmacol 147: 395–405.
3. Morris MR, Khurasany M, Nguyen T, Kim J, Guilford F, et al. (2013) Glutathione and infection. Biochim Biophys Acta 1830: 3329–3349.
4. Mandas A, Iorio EL, Congiu MG, Balestrieri C, Mereu A, et al. (2009) Oxidative imbalance in HIV-1 infected patients treated with antiretroviral therapy. J Biomed Biotechnol 2009: 749575.
5. Kumar H, Koppula S, Kim IS, More SV, Kim BW, et al. (2012) Nuclear factor erythroid 2 - related factor 2 signaling in Parkinson disease: a promising multi therapeutic target against oxidative stress, neuroinflammation and cell death. CNS Neurol Disord Drug Targets 11: 1015–1029.
6. Valko M, Leibfritz D, Moncol J, Cronin MT, Mazur M, et al. (2007) Free radicals and antioxidants in normal physiological functions and human disease. Int J Biochem Cell Biol 39: 44–84.
7. Finkel T (2003) Oxidant signals and oxidative stress. Curr Opin Cell Biol 15: 247–254.
8. Floyd RA (1990) Role of oxygen free radicals in carcinogenesis and brain ischemia. FASEB J 4: 2587–2597.
9. Valko M, Rhodes CJ, Moncol J, Izakovic M, Mazur M (2006) Free radicals, metals and antioxidants in oxidative stress-induced cancer. Chem Biol Interact 160: 1–40.
10. Huang L, Guan T, Qian Y, Huang M, Tang X, et al. (2011) Anti-inflammatory effects of maslinic acid, a natural triterpene, in cultured cortical astrocytes via suppression of nuclear factor-kappa B. Eur J Pharmacol 672: 169–174.
11. Rao CV, Verma AR, Gupta PK, Vijayakumar M (2007) Anti-inflammatory and anti-nociceptive activities of Fumaria indica whole plant extract in experimental animals. Acta Pharm 57: 491–498.
12. Mellman I, Steinman RM (2001) Dendritic cells: specialized and regulated antigen processing machines. Cell 106: 255–258.
13. Nussbaumer O, Gruenbacher G, Gander H, Thurnher M (2011) DC-like cell-dependent activation of human natural killer cells by the bisphosphonate zoledronic acid is regulated by gammadelta T lymphocytes. Blood 118: 2743–2751.
14. Cunningham AL, Harman A, Kim M, Nasr N, Lai J (2013) Immunobiology of dendritic cells and the influence of HIV infection. Adv Exp Med Biol 762: 1–44.
15. Dunne MR, Mangan BA, Madrigal-Estebas L, Doherty DG (2010) Preferential Th1 cytokine profile of phosphoantigen-stimulated human Vgamma9Vdelta2 T cells. Mediators Inflamm 2010: 704941.
16. Afolabi IS, Osikoya IO, Fajimi OD, Usoro PI, Ogunleye DO, et al. (2012) Solenostemon monostachyus, Ipomoea involucrata and Carica papaya seed oil versus Glutathione, or Vernonia amygdalina: methanolic extracts of novel plants for the management of sickle cell anemia disease. BMC Complement Altern Med 12: 262.
17. Kisangau DP, Hosea KM, Joseph CC, Lyaruu HV (2007) In vitro antimicrobial assay of plants used in traditional medicine in Bukoba Rural district, Tanzania. Afr J Tradit Complement Altern Med 4: 510–523.
18. Ghoti H, Fibach E, Dana M, Abu Shaban M, Jeadi H, et al. (2011) Oxidative stress contributes to hemolysis in patients with hereditary spherocytosis and can be ameliorated by fermented papaya preparation. Ann Hematol 90: 509–513.
19. Aruoma OI, Hayashi Y, Marotta F, Mantello P, Rachmilewitz E, et al. (2010) Applications and bioefficacy of the functional food supplement fermented papaya preparation. Toxicology 278: 6–16.
20. Iweala EE, Uhegbu FO, Ogu GN (2010) Preliminary in vitro antisickilng properties of crude juice extracts of Persia Americana, Citrus sinensis, Carica papaya and Ciklavit(R). Afr J Tradit Complement Altern Med 7: 113–117.
21. Sripanidkulchai B, Wongpanich V, Laupattarakasem P, Suwansaksri J, Jirakulsomchok D (2001) Diuretic effects of selected Thai indigenous medicinal plants in rats. J Ethnopharmacol 75: 185–190.
22. Tan PV, Nditafon NG, Yewah MP, Dimo T, Ayafor FJ (1996) Eremomastax speciosa: effects of leaf aqueous extract on ulcer formation and gastric secretion in rats. J Ethnopharmacol 54: 139–142.
23. Momo CE, Oben JE, Tazoo D, Dongo E (2006) Antidiabetic and hypolipidaemic effects of a methanol/methylene-chloride extract of Laportea ovalifolia (Urticaceae), measured in rats with alloxan-induced diabetes. Ann Trop Med Parasitol 100: 69–74.
24. Varghese GK, Bose LV, Habtemariam S (2013) Antidiabetic components of Cassia alata leaves: identification through alpha-glucosidase inhibition studies. Pharm Biol 51: 345–349.
25. Ouedraogo M, Da FL, Fabre A, Konate K, Dibala CI, et al. (2013) Evaluation of the Bronchorelaxant, Genotoxic, and Antigenotoxic Effects of Cassia alata L. Evid Based Complement Alternat Med 2013: 162651.
26. Khan M, Reddy CN, Ravindra G, Reddy KV, Dubey PK (2012) Development and validation of a stability indicating HPLC method for simultaneous determination of four novel fluoroquinolone dimers as potential antibacterial agents. J Pharm Biomed Anal 59: 162–166.
27. Ettebong EO, Nwafor PA, Okokon JE (2012) In vivo antiplasmodial activities of ethanolic extract and fractions of Eleucine indica. Asian Pac J Trop Med 5: 673–676.

28. De Melo GO, Muzitano MF, Legora-Machado A, Almeida TA, De Oliveira DB, et al. (2005) C-glycosylflavones from the aerial parts of Eleusine indica inhibit LPS-induced mouse lung inflammation. Planta Med 71: 362–363.

29. Iqbal M, Gnanaraj C (2012) Eleusine indica L. possesses antioxidant activity and precludes carbon tetrachloride (CCl(4))-mediated oxidative hepatic damage in rats. Environ Health Prev Med 17: 307–315.

30. Njateng GS, Gatsing D, Mouokeu RS, Lunga PK, Kuiate JR (2013) In vitro and in vivo antidermatophytic activity of the dichloromethane-methanol (1: 1 v/v) extract from the stem bark of Polyscias fulva Hiern (Araliaceae). BMC Complement Altern Med 13: 95.

31. Bedir E, Toyang NJ, Khan IA, Walker LA, Clark AM (2001) A new dammarane-type triterpene glycoside from Polyscias fulva. J Nat Prod 64: 95–97.

32. Gabbianelli R, Falcioni G, Lupidi G, Greci L, Damiani E (2004) Fluorescence study on rat epithelial cells and liposomes exposed to aromatic nitroxides. Comp Biochem Physiol C Toxicol Pharmacol 137: 355–362.

33. Gabbianelli R, Lupidi G, Villarini M, Falcioni G (2003) DNA damage induced by copper on erythrocytes of gilthead sea bream Sparus aurata and mollusk Scapharca inaequivalvis. Arch Environ Contam Toxicol 45: 350–356.

34. Ohkawa H, Ohishi N, Yagi K (1979) Assay for lipid peroxides in animal tissues by thiobarbituric acid reaction. Anal Biochem 95: 351–358.

35. Kirkpatrick DT, Guth DJ, Mavis RD (1986) Detection of in vivo lipid peroxidation using the thiobarbituric acid assay for lipid hydroperoxides. J Biochem Toxicol 1: 93–104.

36. Rice-Evans CA, Miller NJ, Bolwell PG, Bramley PM, Pridham JB (1995) The relative antioxidant activities of plant-derived polyphenolic flavonoids. Free Radic Res 22: 375–383.

37. Mitchelmore CL, Chipman JK (1998) DNA strand breakage in aquatic organisms and the potential value of the comet assay in environmental monitoring. Mutat Res 399: 135–147.

38. Mitchelmore CL, Birmelin C, Livingstone DR, Chipman JK (1998) Detection of DNA strand breaks in isolated mussel (Mytilus edulis L.) digestive gland cells using the "Comet" assay. Ecotoxicol Environ Saf 41: 51–58.

39. Birdi T, Daswani P, Brijesh S, Tetali P, Natu A, et al. (2010) Newer insights into the mechanism of action of Psidium guajava L. leaves in infectious diarrhoea. BMC Complement Altern Med 10: 33.

40. Thomas E, Vandebroek I, Sanca S, Van Damme P (2009) Cultural significance of medicinal plant families and species among Quechua farmers in Apillapampa, Bolivia. J Ethnopharmacol 122: 60–67.

41. Sen TA, Kundak AA, Guraksin O, Demir T, Narci A (2010) Acute rheumatic carditis associated with Schoenlein-Henoch vasculitis. Anadolu Kardiyol Derg 10: 465–466.

42. Focho DA, Newu MC, Anjah MG, Nwana FA, Ambo FB (2009) Ethnobotanical survey of trees in Fundong, Northwest Region, Cameroon. J Ethnobiol Ethnomed 5: 17.

43. Cheynier V, Comte G, Davies KM, Lattanzio V, Martens S (2013) Plant phenolics: Recent advances on their biosynthesis, genetics, and ecophysiology. Plant Physiol Biochem.

44. Cesari I, Hoerle M, Simoes-Pires C, Grisoli P, Queiroz EF, et al. (2013) Anti-inflammatory, antimicrobial and antioxidant activities of Diospyros bipindensis (Gurke) extracts and its main constituents. J Ethnopharmacol 146: 264–270.

45. Brusotti G, Cesari I, Dentamaro A, Caccialanza G, Massolini G (2013) Isolation and characterization of bioactive compounds from plant resources: The role of analysis in the ethnopharmacological approach. J Pharm Biomed Anal.

2

Metallothionein 2 (*SaMT2*) from *Sedum alfredii* Hance Confers Increased Cd Tolerance and Accumulation in Yeast and Tobacco

Jie Zhang, Min Zhang, Shengke Tian, Lingli Lu, M. J. I. Shohag, Xiaoe Yang*

MOE Key Laboratory of Environment Remediation and Ecosystem Health, College of Environmental and Resource Sciences, Zhejiang University, Hangzhou, China

Abstract

Metallothioneins are cysteine-rich metal-binding proteins. In the present study, *SaMT2*, a type 2 metallothionein gene, was isolated from Cd/Zn co-hyperaccumulator *Sedum alfredii* Hance. *SaMT2* encodes a putative peptide of 79 amino acid residues including two cysteine-rich domains. The transcript level of SaMT2 was higher in shoots than in roots of *S. alfredii*, and was significantly induced by Cd and Zn treatments. Yeast expression assay showed *SaMT2* significantly enhanced Cd tolerance and accumulation in yeast. Ectopic expression of *SaMT2* in tobacco enhanced Cd and Zn tolerance and accumulation in both shoots and roots of the transgenic plants. The transgenic plants had higher antioxidant enzyme activities and accumulated less H_2O_2 than wild-type plants under Cd and Zn treatment. Thus, *SaMT2* could significantly enhance Cd and Zn tolerance and accumulation in transgenic tobacco plants by chelating metals and improving antioxidant system.

Editor: Wagner L. Araujo, Universidade Federal de Vicosa, Brazil

Funding: The present study was supported by the Fundamental Research Funds for the Central Universities (No. 2013FZA6005), National Natural Science Foundation of China (No. 31372128; 21177107). The funders had no role in study design, data collection and analysis, decision to publish, or preparation of the manuscript.

Competing Interests: The authors have declared that no competing interests exist.

* Email: xyang@zju.edu.cn

Introduction

Heavy metals are known to cause toxic effects and inhibition of plant growth. However, rare plant species, which can accumulate and tolerate extremely high concentrations of heavy metals in their shoots without toxicity effects, have been defined as "hyperaccumulators" [1]. The elucidation of the mechanisms underlying metal hyperaccumulation may enable the phytoremediation of metal-contaminated soils and the biofortification of trace elements in food crops [2–4].

Higher plants have evolved various defense mechanisms to detoxify excess metals. These mechanisms contain compartmentalization in inactive tissues, chelation by metal ligands and detoxification by antioxidants [5]. Metal chelators such as organic acids, amino acids, phytochelatins and metallothioneins play important roles in metal detoxification [6]. Metallothioneins (MTs) are low-molecular-mass, cysteine-rich proteins which are broadly distributed in microorganisms, plants and animals [7]. Plant MTs can be divided into four subfamilies based on the distribution of cysteine residues in their amino- and carboxyl-terminal regions [8]. Several MT genes have been isolated and characterized from plants. There are some evidence indicating that plant MTs are involved in metal homeostasis, detoxification and reactive oxygen species (ROS) scavenging [8–11].

Hyperaccumulating ecotype (HE) of *Sedum alfredii* Hance is a Zn/Cd hyperaccumulator discovered from an old Pb/Zn mining area of China [12,13]. It can accumulate up to 9000 $\mu g\ g^{-1}$ Cd and 29000 $\mu g\ g^{-1}$ Zn in its shoots without toxicity symptoms [12,13]. This large amount of metals in plant cells needs a powerful detoxification system to protect plants from the deleterious effects of the metals. Earlier studies have demonstrated that the hyperaccumulating ecotype of *Sedum alfredii* has a more effective antioxidant enzyme system than non-hyperaccumulating ecotype (NHE) [14,15]. However, the mechanism of hypertolerance of metals in this species has not been fully understood. In the present study, a metallothionein gene from hyperaccumulating ecotype of *Sedum alfredii* Hance, named *SaMT2*, was isolated and cloned The expression pattern of this gene was studied by Real Time-PCR. To analysis the function of *SaMT2*, its full length cDNA was cloned and expressed in yeast and tobacco. The transgenic yeast and tobacco plants were analyzed to evaluate whether *SaMT2* protein played a role in Cd or Zn tolerance and accumulation.

Materials and Methods

Ethics statement

These field studies did not involve any protected species. No specific permits were required for the collection of samples in the study location.

Plant growth

The hyperaccumulating ecotype of *S. alfredii* Hance was collected from an old Pb/Zn mining site in Zhejiang Province, P. R. China. Plants were grown in non-polluted soils for several generations to minimize internal metal concentrations. Similar size shoot branches were cut and cultured hydroponically. After two weeks, rooted seedlings were then subjected to 4 days exposure of one-fourth, half and full strength nutrient solutions containing 2 mM $Ca(NO_3)_2$, 0.7 mM K_2SO_4, 0.1 mM KH_2PO_4, 0.1 mM KCl, 0.5 mM $MgSO_4$, 10 μM H_3BO_3, 0.5 μM $MnSO_4$, 5 μM $ZnSO_4$, 0.2 μM $CuSO_4$, 0.01 μM $(NH_4)_6 \cdot Mo_7O_{24}$, and 20 μM Fe-EDTA. Nutrient solution pH was adjusted daily to 5.8 with 0.1 M NaOH or HCl. Plants were grown under glasshouse conditions with natural light (day/night of 16/8 h), day/night temperature of 26/20°C and day/night humidity of 70/85%. The nutrient solution was aerated continuously and renewed every 3 d. To compare the expression of *SaMT2*, the precultured seedlings were treated with 100 μM $CdCl_2$ and 500 μM $ZnSO_4$ for 8 days.

Cloning of *SaMT2* cDNA and sequence analysis

The cDNA fragment of *SaMT2* was isolated from RNA-Seq data of *Sedum alfredii* Hance (Gao et al. 2013). The full length of *SaMT2* was isolated using 3′ and 5′ RACE methods as described by the supplier (Smart RACE cDNA amplification kit; Clontech Laboratories, Inc. CA, USA). PCR was performed with the following primers: 5′-CTGGGCGTGGCTCCGAAGCAAGT-GTA-3′ for 3′ RACE and 5′-CGCAACCACAGTTTCCACCA-CAGCA-3′ for 5′ RACE. Alignment of *SaMT2* was performed by ClustalW on the internet (http://clustalw.ddbj.nig.ac.jp/). The phylogenetic tree was constructed using the neighbor-joining algorithm by MEGA 5 software (released from http://www.megasoftware.net/) after ClustalW alignment with 1000 bootstrap trials.

Real time RT-PCR analysis

The total RNA was extracted from various tissues by RNAiso plus (Takara Bio, Inc. Shiga, Japan), and then converted to cDNA using Primescript™ RT regent kit with gDNA eraser (Takara Bio,

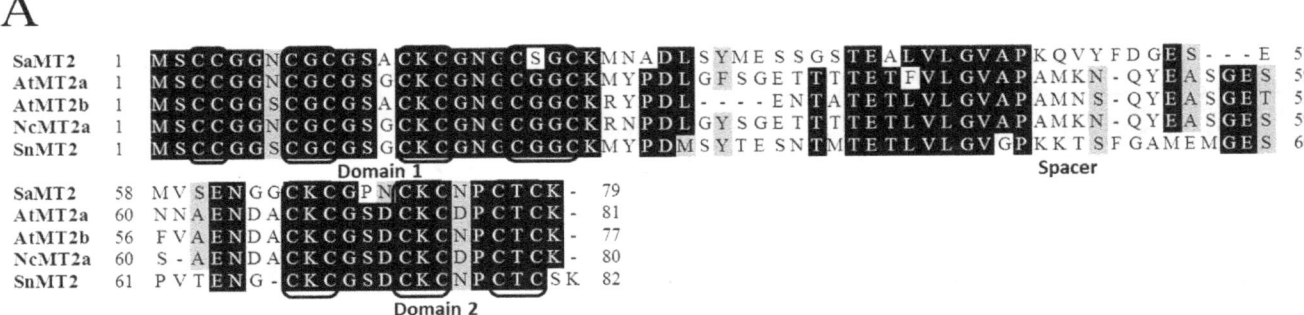

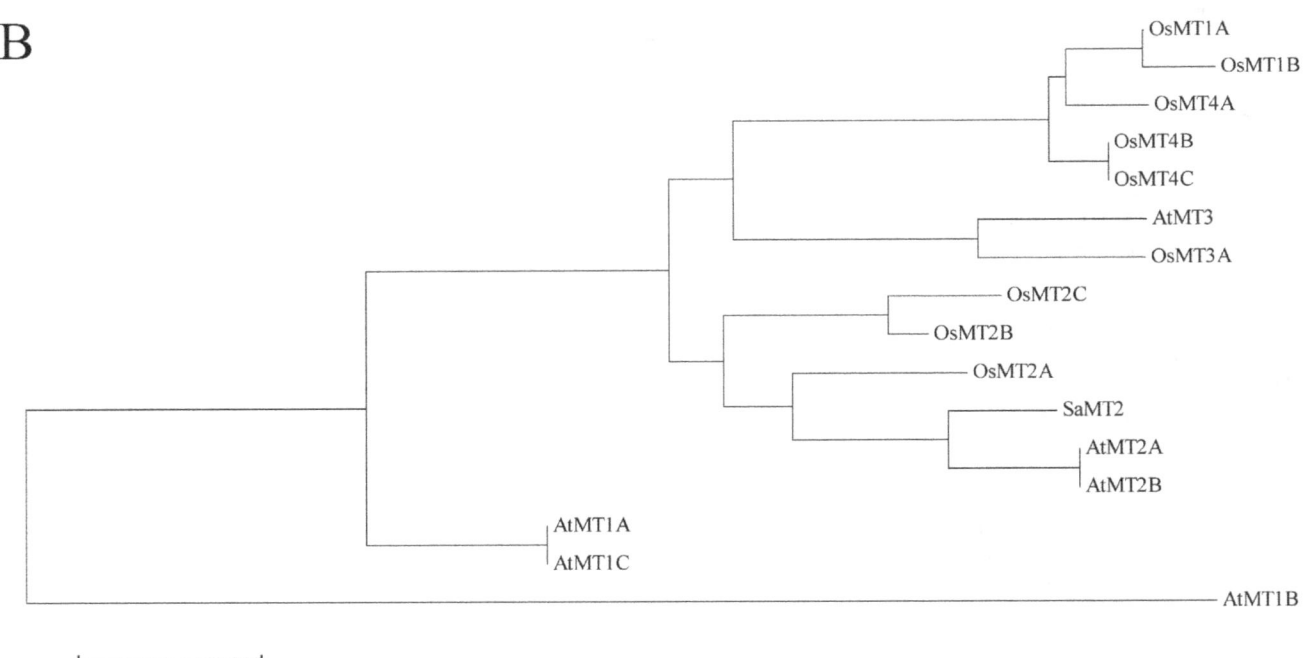

Figure 1. Sequence alignment and phylogenic analysis of *SaMT2* with other MTs. (A) The deduced amino acid sequences encoded by *SaMT2* were aligned with MTs from *Arabidopsis thaliana*, *Noccaea caerulescens* and *Solanum nigrum*. The cysteine-rich domains are boxed. (B) The phylogenic tree of *SaMT2* and MTs from Arabidopsis and rice.

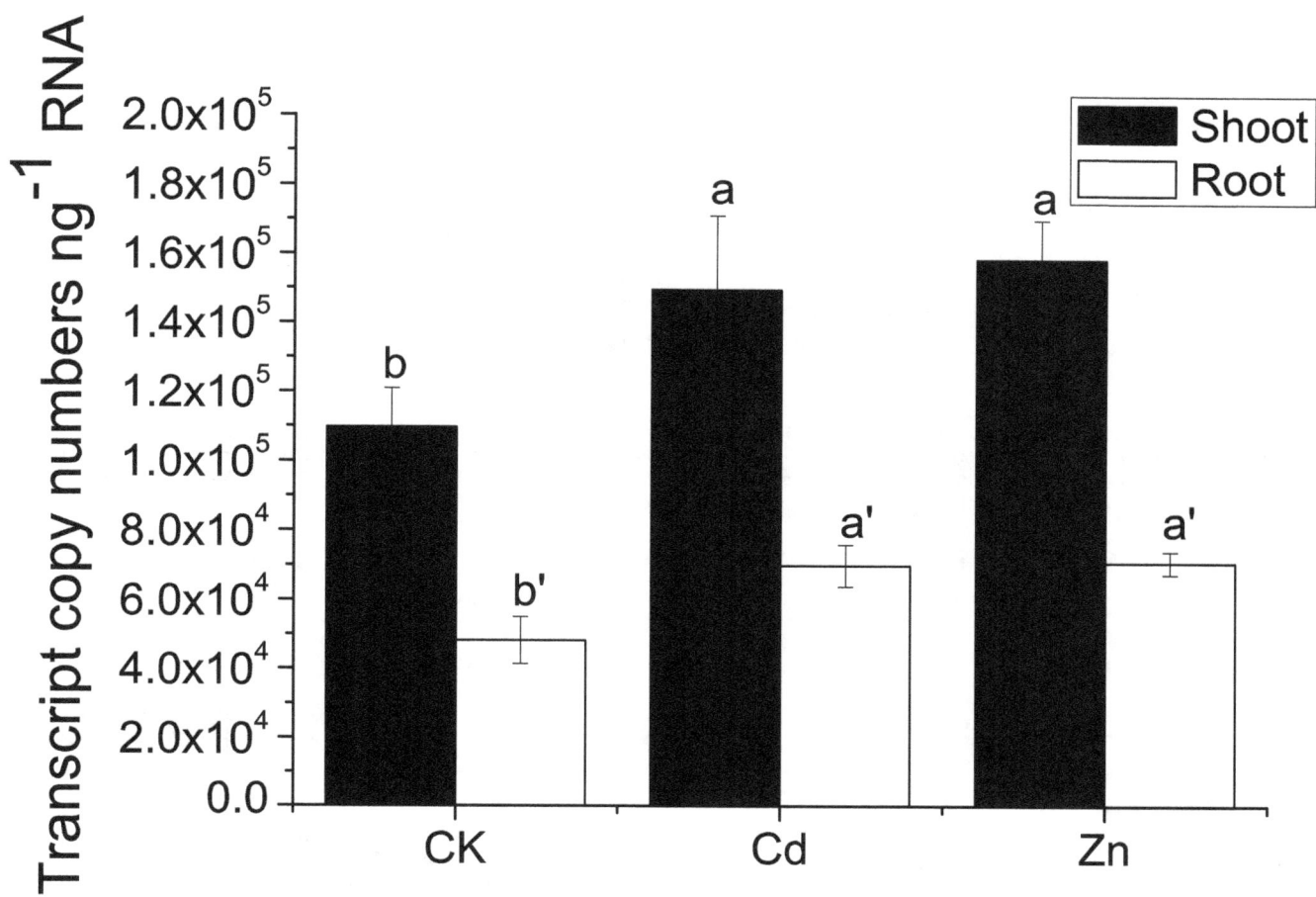

Figure 2. The expression level of *SaMT2* **in** *Sedum alfredii.* The transcript level of *SaMT2* induced by Cd and Zn treatments. The different letters above the columns indicate the significant difference between the treatments (p<0.05, Tukey's test). CK represents the control group.

Inc. Shiga, Japan). Expression of the *SaMT2* was determined by quantitative RT-PCR with the SYBR Green I reagent (SYBR Premix Ex Taq II; Takara Bio, Inc. Shiga, Japan) on an Eppendorf Mastercycler Epgradient Realplex^2 (Eppendorf AG, Hamburg, Germany). A portion (10 ng) of cDNA was used for the template. The primers used for *SaMT2* were forward 5′-CTGTGGTTGCGGATCTGCTT-3′ and reverse 5′- TCCATT-CTCCGACACCATCT-3′. To generate standard curves for the absolute quantification for *SaMT2* copy number, a series of dilutions (from 1×10^{-1} to 1×10^{-6} ng) of plasmids were made and then subjected to real-time PCR.

Plasmids construction

To express *SaMT2* in *Saccharomyces cerevisiae*, the full ORF of *SaMT2* was amplified from the cDNA of *S. alfredii* using primers: 5′-AG<u>CTCGAG</u>ATGTCTTGCTGTGGTGGA-3′ contains an XhoI site and 5′-GA<u>GGATCC</u>TCATTTGCAAGTGCAGGG-3′ contains a BamHI site. The PCR products were then cloned into a pEASY Blunt simple vector (Transgen, Beijing, China) and its sequence confirmed. This vector was double digested with XhoI and BamHI, and the obtained fragment was cloned into pDR195 between XhoI and BamHI sites.

To construct the plant overexpression vector, the full ORF of *SaMT2* was amplified using primers: 5′-AA<u>AGATCT</u>GATGTC-TTGCTGTGGTGGA-3′ and 5′-AA<u>GGTGACC</u>TCATTTG-CAAGTGCAGG-3′, which contained BglII and BstpI restriction

sites, respectively. The obtained fragment was restricted with BglII and BstpI, and then cloned into the BglII and BstpI sites of pCAMBIA 1302 vector and its sequence was confirmed.

Yeast complementation assay

The *S. cerevisiae* strains BY4741 (wild type, *MATα; his2Δ0; met15Δ0; ura3Δ0*), *Δycf1* (*MATa; his3Δ1; leu2Δ0; lys2Δ0; ura3Δ0; YDR135c::kanMX4*) and *Δzrc1* (*MATα; his3Δ1; leu2Δ0; met15Δ0; ura3Δ0; YMR243c::kanMX4*) mutants were used to investigate the role of *SaMT2* in Cd and Zn tolerance. The yeast transformation was conducted using LiAc/PEG/ssDNA methods, as described by Gietz and Schiestl [16]. To obtain cells for transformation: Inoculate a single colony of the yeast strain with a sterile inoculation loop from a fresh SD (synthetic medium plus dextrose, 0.67% yeast nitrogen base, 2% D-glucose, and amino acids) plate into 5 ml of YPD medium (2% peptone, 1% yeast extracts, 2% D-glucose) and incubate overnight at 30°C. Add 2.5×10^8 cells to 50 ml of YPD medium in a culture flask and incubate until the cell titer is at least 2×10^7 cells ml^{-1}. Cells were harvested by centrifugation at 3,000 g for 5 min and washed twice with sterilized water. Then the cells were re-suspended in 1.0 ml of sterile deionized water and pelleted by centrifugation (13 000×g for 30 sec). The supernatant was discarded and the transformation mixture {containing 240 µl PEG 3350 (50% w/v), 36 µl 1.0 M lithium acetate, 10 µl single-stranded carrier DNA (10 mg ml^{-1}) and plasmid DNA (0.5–1 µg), and sufficient sterile deionized water

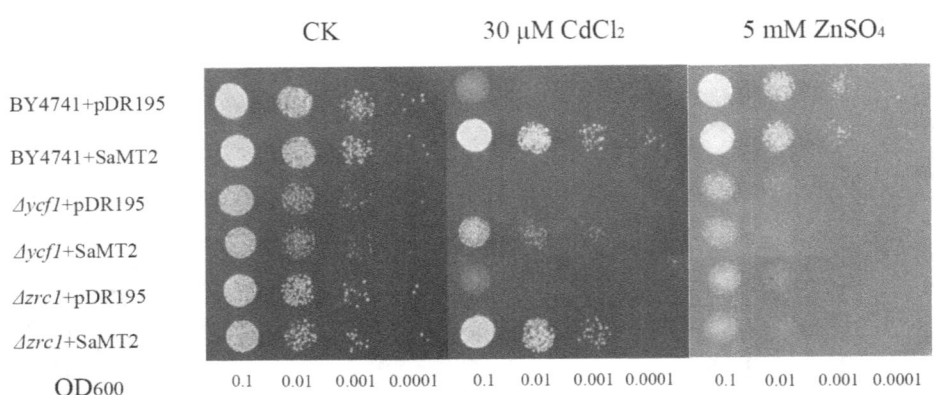

Figure 3. Cd and Zn tolerance of yeast cells expressing *SaMT2*. The *Saccharomyces cerevisiae* BY4741, *Δycf1* and *Δzrc1* yeast cells harboring pDR195 (vector control) or pDR195-*SaMT2* were grown in liquid SD selective medium. Cultures were adjusted to OD_{600nm} of 0.1 and serially 10-fold diluted in water. 10 μl aliquots of each dilution were spotted either on SD selective plates or on plates with 30 μM $CdCl_2$ or 5 mM $ZnSO_4$. After 3 days of incubation at 30°C, plates were photographed. CK represents the control group.

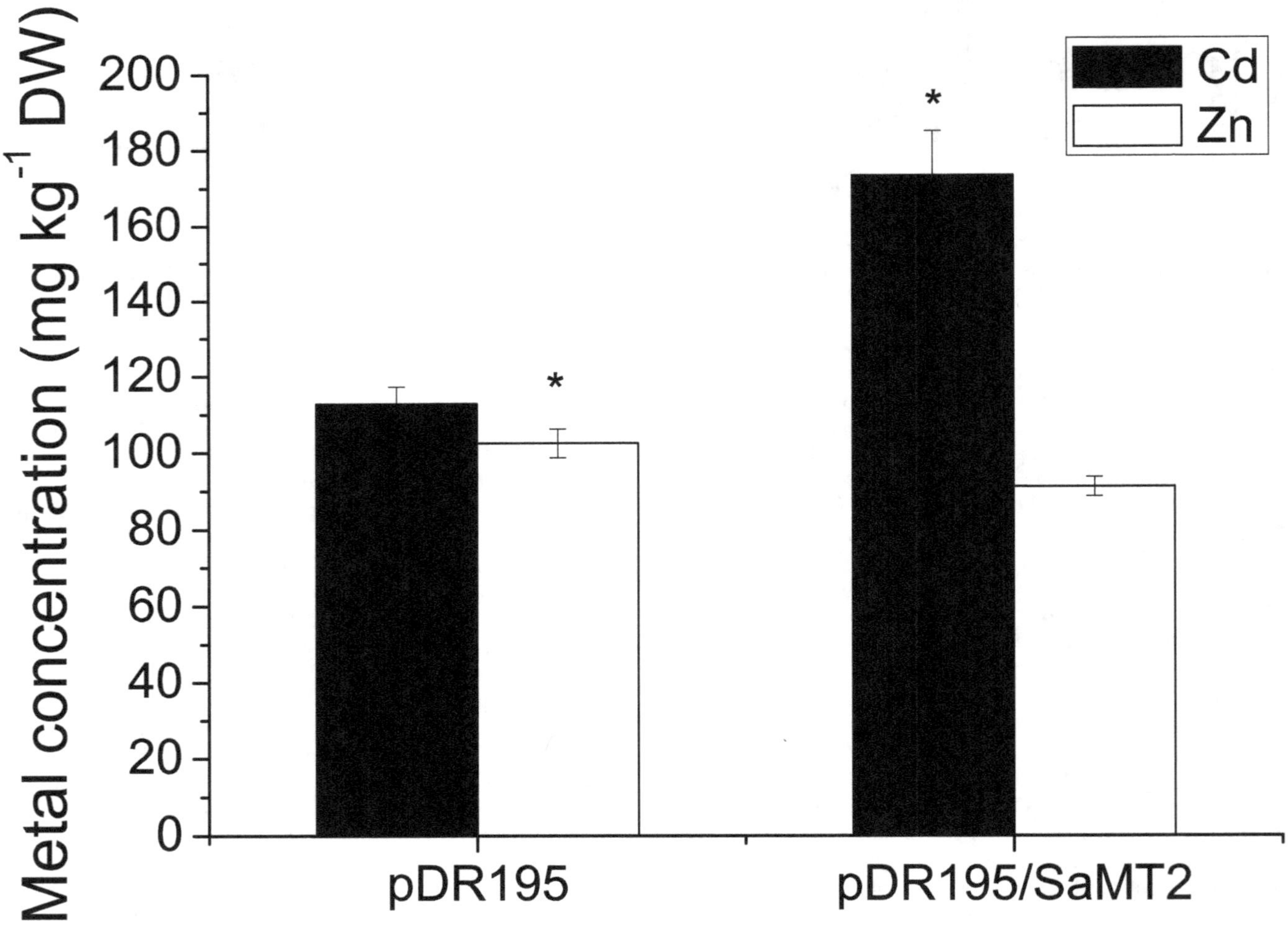

Figure 4. Cd and Zn concentration in *Δycf1* and *Δzrc1* yeast cells expressing *SaMT2*. The yeast transformants containing pDR195 or pDR195-*SaMT2* were grown in liquid SD selective medium with 30 μM $CdCl_2$ and 100 μM $ZnSO_4$ for *Δycf1* and *Δzrc1*, respectively. Cells were incubated at 30°C for 48 h and metal contents were measured by ICP-MS. Results are averages (±S.E.) from three independent experiments done with four different colonies. The '*' symbol indicates the mean values were significantly different at $p < 0.05$ (Tukey's test).

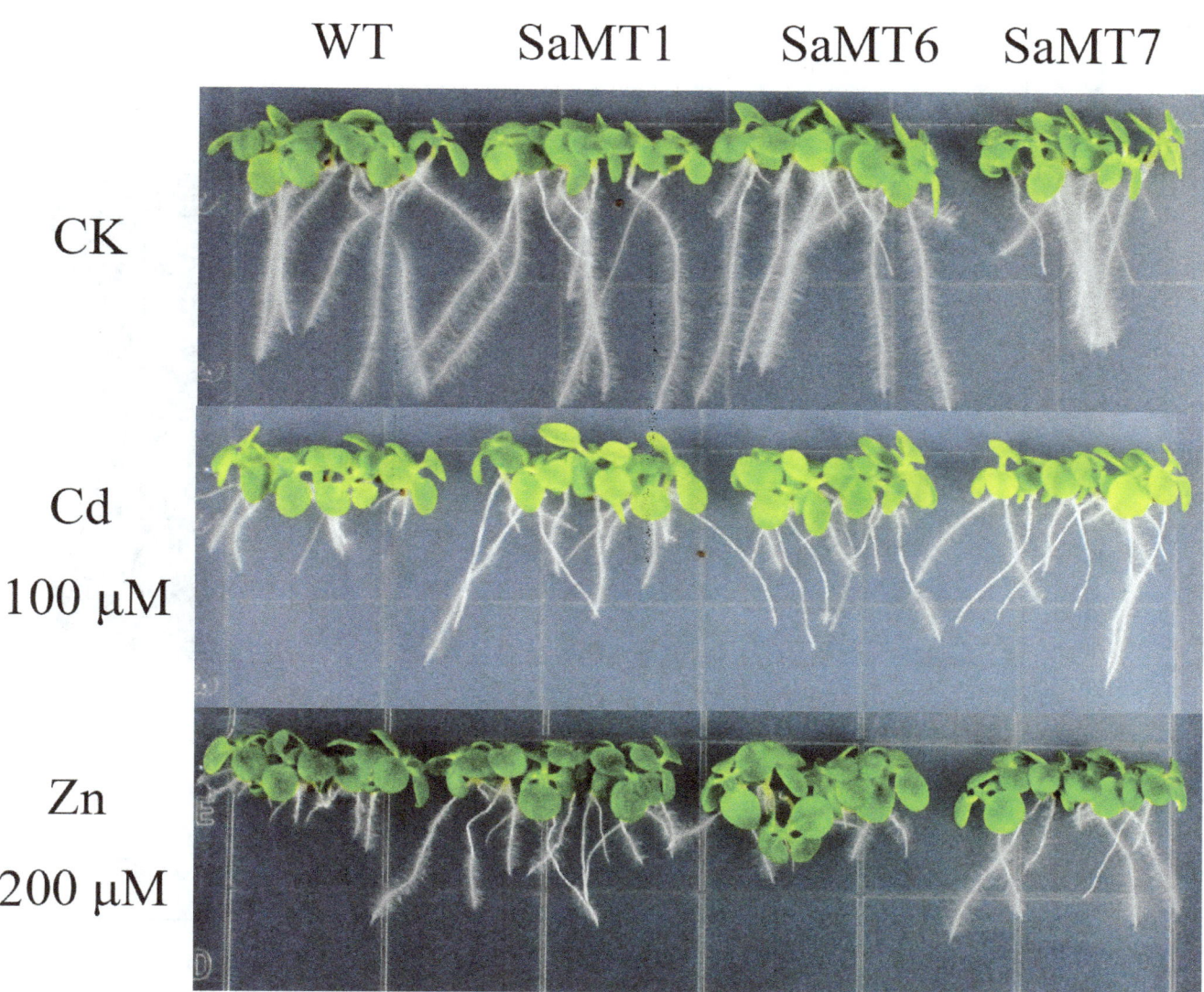

Figure 5. Metal tolerance analysis of transgenic tobacco plants over-expressing *SaMT2*. The figure shows the effect of 200 µM ZnSO$_4$ or 100 µM CdCl$_2$ on the growth of WT and transgenic plants on B5 medium. CK represents the control group.

to provide a final volume of 360 µl} were layered over the pellet. The mixture was vortex vigorously for 1 min and subjected to a heat shock at 42°C for 40 min. The transformation mixture was then centrifuged at 13 000×g for 30 sec to pellet the cells. After the supernatant was decanted, the cells were resuspended in 1.0 ml of sterile deionized water. Aliquots of the resuspended cells were plated onto SD-URA media. Plates were incubated for 2–3 days at 30°C until transformants were observed. Single colonies were picked from each transformant plate and established on fresh SD-URA plates.

For the metal tolerance assay, single colonies from SD-URA plates were cultured in liquid SD-URA medium until OD$_{600}$ reached 1.0. After serial dilutions (OD$_{600}$ = 0.1, 0.01, 0.001, 0.0001, respectively) were prepared, each dilution was spotted onto SD-URA medium with or without 5 mM ZnSO$_4$ or 30 µM CdCl$_2$. Plates were photographed after incubation at 30°C for 3 d.

For determination of metal concentration in yeast, transformants were grown in liquid SD-URA medium overnight. Then, cells were adjusted to OD$_{600}$ = 0.2 in the presence of 10 µM

CdCl$_2$ or 100 µM ZnSO$_4$ for Zn determination. After incubation for 48 h, the cells were harvested and washed with distilled water, 20 mM Na$_2$EDTA and distilled water, respectively. Dry weight was determined after 3 days at 60°C. Cells were digested using 5 ml concentrated HNO$_3$, incubated at 95°C for 2 h. The Zn and Cd concentrations were determined by using ICP-MS (Inductively Coupled Plasma Mass Spectrometry, Agilent 7500a, CA, USA).

Heterogeneous expression of *SaMT2* in tobacco

The transformation of tobacco was constructed using the leaf disk method according to Horsch *et al* [17]. Surface-sterilized T$_1$ seeds of two transgenic tobacco lines were germinated on Murashige Skoog (MS) plates containing 40 mg/L hygromycin to select hygromycin-resistance seedlings. For Zn/Cd tolerance analysis, the wild type (WT) and transgenic plants were transferred to MS plates containing 100 µM CdCl$_2$ or 200 µM ZnSO$_4$ for 14 d. To determine the Zn/Cd concentrations in plants, both WT and transgenic plants were transferred to hydroponic culture. One-month old plants were then treated with 50 µM CdCl$_2$ or

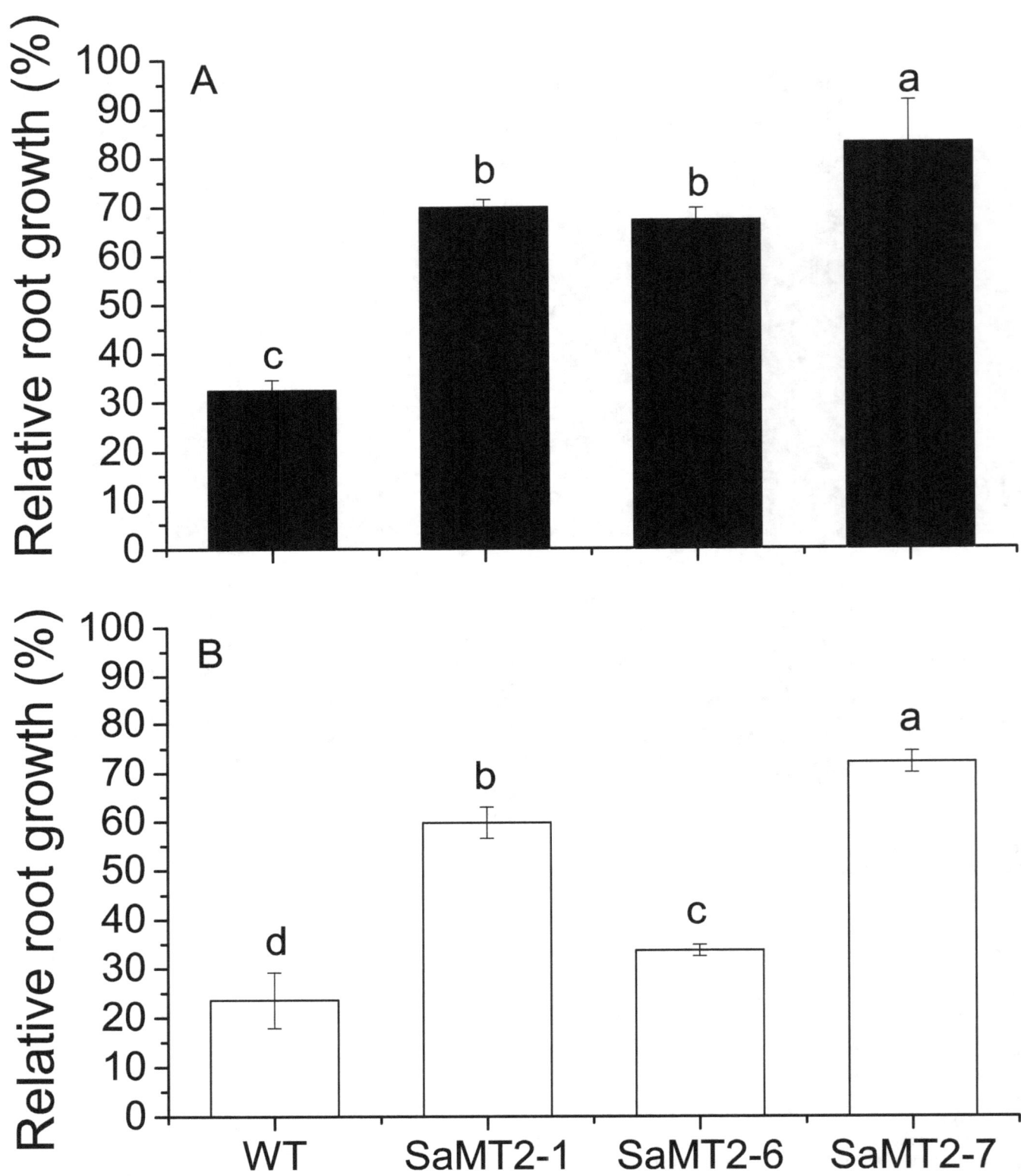

Figure 6. Relative root growth of transgenic tobacco plants. The relative root growth of WT and transgenic tobacco plants under Cd (A) and Zn (B) treatments. Different letters above the columns indicate a significant difference among different plant lines (p<0.05, Tukey's test).

100 µM ZnSO₄ for one week. Cadmium and zinc concentrations in plant tissues were measured using ICP-MS as described by Yang *et al.* [13].

Determination of SOD, POD, CAT and H₂O₂

For antioxidant enzyme activity determination, a 0.5-g aliquot of plant sample was homogenized in 5 ml potassium phosphate

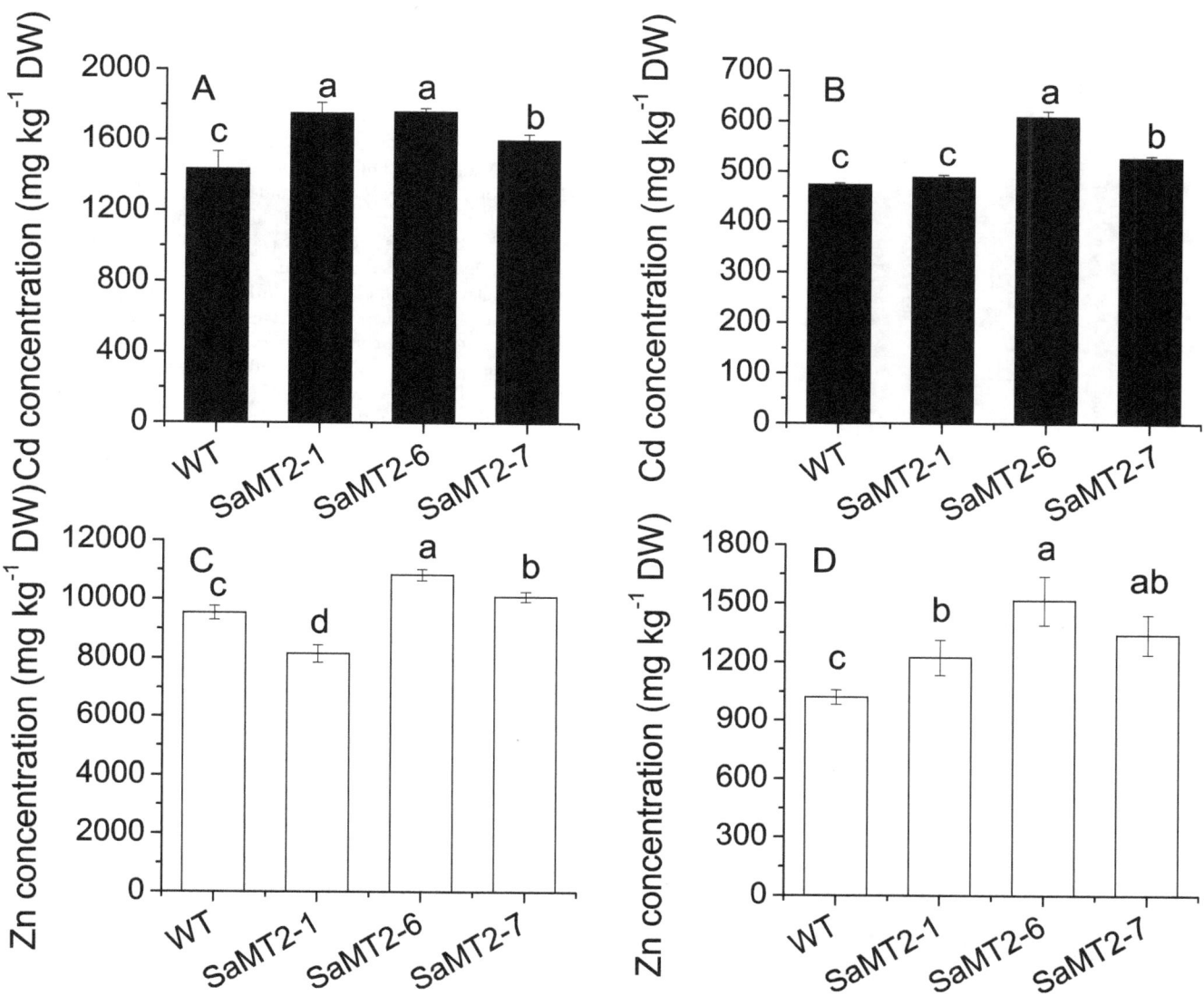

Figure 7. Cd and Zn concentrations in wild type amd transgenic tobacco lines overexpressing *SaMT2*. Three independent *SaMT2* over-expressing lines and wild-type tobacco were grown in nutrient solution containing 50 μM CdCl$_2$, 100 μM ZnSO$_4$ for 1 week. A: Cd concentration in roots, B: Cd concentration in shoots, C: Zn concentration in roots, D: Zn concentration in shoots. Results are means ± S.E. (n = 3). Different letter indicate the mean values were significantly different from WT tobacco determined by Tukey's test (p<0.05).

buffer (50 mM, pH 7.8). The homogenates were then centrifuged at 12000×g for 20 min at 4 °C. The supernatants were used for the analysis of enzyme activity. Superoxide dismutase (SOD) activity was determined by the photochemical method described by Giannopotitis and Ries [18]. One unit of the enzyme activity was defined as the amount of enzyme required to result in a 50% inhibition of the rate of nitro blue tetrazolium reduction measured at 560 nm. Catalase (CAT) activity was estimated according to Cakmak *et al.* [19]. The reaction mixture in a total volume of 2 ml contained 25 mM sodium phosphate buffer (pH 7.0), 10 mM H$_2$O$_2$. The reaction was initiated by the addition of 100 μl of enzyme extract and activity was determined by measuring the initial rate of disappearance of H$_2$O$_2$ at 240 nm (E = 39.4 mM^{-1} cm^{-1}) for 30 s. Peroxidase (POD) activity was measured as the increase of absorbance due to guaiacol oxidation [20]. The reaction mixture contained 25 mM phosphate buffer (pH 7.0), 10 mM H$_2$O$_2$, 0.05% guaiacol and 100 μl of enzyme

extract. The reaction was initiated by the addition of H$_2$O$_2$. The oxidation of guaiacol was measured at 470 nm (E = 26.6 mM^{-1} cm^{-1}).

H$_2$O$_2$ was determined according to Loreto & Velikova [21]. Leaf tissues (0.07 g) were homogenized in an ice bath with 5 ml of 0.1% (w/v) trichloroacetic acid (TCA). The homogenate was centrifuged at 12,000×g for 15 min and 0.5 ml of the supernatant was added to 0.5 ml of 10 mM potassium phosphate buffer (pH 7.0) and 1 ml of 1 M KI. The absorbance of the supernatant was measured at 390 nm. The content of H$_2$O$_2$ was calculated by comparison with a standard calibration curve previously made by using different concentrations of H$_2$O$_2$.

Statistical analysis of data

All data were statistically analyzed by using the SPSS package (version 20.0). All values were performed as means of three replicates. Data was tested at significant levels of p<0.05 using

Table 1. The activities of SOD, POD CAT and the content of H_2O_2 in the shoot and roots of wild-type and transgenic tobacco plants.

Metal treatment		SOD (U mg⁻¹ Protein)		POD (nanokatals mg⁻¹ Protein)		CAT (nanokatals mg⁻¹ Protein)		H_2O_2 (µg g⁻¹ FW)	
		WT	Transgenic	WT	Transgenic	WT	Transgenic	WT	Transgenic
Shoot	CK	20.1±0.4d	21.4±1.5d	20190.7±205.0d	20045.6±175.0d	36.7±5.0b	35.0±3.3b	154.4±10.0d	156.1±12.1d
	50 µM Cd	23.5±0.3c	28.6±1.2a	29267.5±720.1b	33821.7±688.4a	28.3±3.3c	45.0±1.7a	411.2±20.7a	167.9±10.1d
	100 µM Zn	24.3±0.5c	25.5±0.4b	26777.0±855.1c	30396.0±373.4b	25.0±1.7c	35.0±3.3b	267.3±9.7b	215.2±2.0c
Root	CK	24.2±0.4e	25.7±1.3e	30199.3±338.4e	31156.2±1171.9e	5.0±1.7b	3.3±1.7b	131.6±6.4d	132.0±7.4d
	50 µM Cd	30.2±0.6c	41.6±1.4a	46077.5±606.7b	54069.1±825.1a	5.0±1.7b	11.7±1.7a	400.5±23.0a	210.8±7.7c
	100 µM Zn	28.3±1.0d	34.2±0.7b	34723.6±270.0d	38681.0±1073.5c	6.7±1.7ab	10.0±1.7a	310.1±3.9b	198.6±10.0c

The data presented are mean ± SD of three replicates. Different letters indicate significant differences (p<0.05) among the different treatments and different plant lines.

one-way ANOVA (analysis of variance). The graphical works were made by using Origin software.

Results

Clone and sequence analysis of *SaMT2*

A cDNA fragment of metallothionein like gene was obtained from RNA-seq of *S. alfredii*. RT-PCR and RACE techniques were used to obtain the full length cDNA of this gene, whose sequence was identified by BLAST search (www.ncbi.nlm.nih.gov/BLAST). The obtained cDNA encoded a 79 amino acids protein, which showed certain similarity to the cDNA of *AtMT2a* or *AtMT2b*. According to the amino acid sequences, it belonged to the Type 2 MTs, and was named *SaMT2* (GeneBank accession number: KJ862538).

Multiple sequence alignment of the deduced amino acid sequences of *SaMT2* with *AtMT2a* (*Arabidopsis thaliana*, NP_187550.1), *AtMT2b* (NP_195858.1), *NcMT2a* (*Noccaea caerulescens*, ACR46966.1) and *SnMT2* (*Solanum nigrum*, ACF10396.1) were conducted (Figure 1A). *SaMT2* shared 62%, 59%, 59% and 66% similarities with *AtMT2a*, *AtMT2b*, *NcMT2a* and *SnMT2*, respectively. Similar to other plant MT proteins, *SaMT2* contained two cysteine-rich domain separated by a large cysteine-free domain [22]. The cysteine-rich domains in the N-terminal region is CCxxxCGCxxxCKCxxxCxGC, which was highly conserved, and that in the C-terminal region contained three CxC motifs. The spacer region between the two terminal regions contained approximately 40 amino acids.

The phylogenetic tree of *SaMT2* and MTs from Arabidopsis and rice was conducted using MEGA software (Figure 1B). These MTs were divided into several groups and *SaMT2* was closely clustered with *AtMT2a* and *AtMT2b*.

Expression analysis of *SaMT2* in *Sedum alfredii* Hance

The expression of *SaMT2* was investigated using absolute quantitative RT-PCR. To investigate whether Cd or Zn were involved in the regulation of *SaMT2*, *S. alfredii* seedlings were treated with 100 µM $CdCl_2$ or 500 µM $ZnSO_4$ and were subjected to determine the transcript level of *SaMT2*. The expression level of *SaMT2* in roots was higher than that in shoots. The expression of *SaMT2* was significantly (p<0.05) increased in both roots and shoots treated with Cd and Zn (Figure 2).

SaMT2 enhanced cadmium but not zinc tolerance in yeast mutants

The *Δycf1* and *Δzrc1* yeast mutants and the wild type strain BY4741 were used to test the Cd and Zn tolerance ability of *SaMT2*. When grown in the control medium, yeasts containing either pDR195 or pDR195-*SaMT2* could grow well. When grown in a medium containing 30 µM $CdCl_2$, the growth of yeast contain pDR195 was significantly inhibited; however, the expression of *SaMT2* could markedly mitigate this growth defect (Figure 3). However, the growth of yeasts in a Zn containing medium was not affected whether *SaMT2* was expressed or not (Figure 3).

Similar trends were found for Cd and Zn concentrations in yeast mutants. Expression of *SaMT2* significantly increased the Cd concentration in yeast *Δycf1* mutant; however, it decreased the concentration of Zn in yeast *Δzrc1* mutant significantly (p<0.05, Figure 4).

Overexpression of *SaMT2* in tobacco enhanced Cd and Zn tolerance and accumulation

To evaluate the functions of *SaMT2* in plants, transgenic tobacco plants were generated, ectopically expressing *SaMT2* under the control of CaMV 35S promoter. Three independent transgenic tobacco lines over-expressing *SaMT2* were selected for Cd and Zn tolerance analysis. The wild plants were used as control.

There was no difference in growth between wild type and transgenic plants under control condition. Exposure of the plants to 100 μM CdCl$_2$ or 200 μM ZnSO$_4$ significantly decreased root elongation and plant growth of both wild type and transgenic plants; however, the growth deficiency was less pronounced in transgenic plants (Figure 5). Under 100 μM CdCl$_2$ or 200 μM ZnSO$_4$ treatments, the root growth of wild type plants was decreased by 68% or 76%, compared to the control, respectively. However, the transgenic plants showed a significantly higher resistance to Cd and Zn. Compared to the control, the root growth of the transgenic plants was only decreased by 17%–33% under Cd treatment, and decreased by 28%–66% under Zn treatment (Figure 6).

Over-expression of *SaMT2* gene significantly increased both Cd and Zn concentration in transgenic tobacco plants (p<0.05) (Figure 7). Compared to the wild type plants, the Cd concentration was increased by 11–22% in the roots and by 3–28% in the shoots of the transgenic plants, respectively (Figure 7A, B). Except for the SaMT2-1 line, the Zn concentration was increased by 6–14% in roots and 20–48% in shoots of transgenic plants, respectively (Figure 7C, D). The SaMT2-6 line accumulated the highest amount of Cd and Zn among the three transgenic lines.

Based on the tolerance and accumulation of Cd and Zn in transgenic tobacco lines, the SaMT2-7 line - having similar Cd and Zn tolerance and accumulation level- was selected, to evaluate the reason of elevated tolerance to Cd and Zn of the transgenic plants. The plants treated with different metals were used to determine the activities of SOD, POD, CAT and the content of H$_2$O$_2$. The transgenic plant accumulated significantly less H$_2$O$_2$ in both roots and shoots than the WT plants under Cd and Zn treatments. The activities of SOD and POD were significantly increased in both roots and shoots of transgenic plants compared to that of wild type plants under Cd and Zn treatments (Table 1). For CAT activity, however, no significant difference was observed between WT and transgenic plants.

Discussion

Metallothioneins (MTs) are cysteine-rich proteins involved in metal tolerance of diverse living organisms. Plant metallothioneins can be divided into four subfamilies based on their sequence similarities and phylogenetic relationships [7,8]. In the present study, the MT gene cloned from *S. alfredii* encoded a protein with two Cys-rich regions, showing high identity with the N- and C-terminal regions of type 2 MTs of other plants; therefore, this MT gene was named as *SaMT2*. *S. alfredii* is a Cd/Zn co-hyperaccumulator, which shows extremely high tolerance to Cd and Zn [12,13]. Thus, It was hypothesized that *SaMT2* cloned from *S. alfredii* might be involved in Cd or Zn tolerance.

It has been reported that different MT genes have distinct tissue specific expression patterns in plants [8]. Generally, MT1s are predominantly expressed in roots, MT2s and MT3s in shoots [22,23]. In the present study, *SaMT2* was more highly expressed in shoots of *S. alfredii* than in roots. Similar results have also been found in Arabidopsis, rice and other plants [22–25]. The expression of MT genes in plants is regulated by many factors, including metal ions, oxidative stress, and stresses such as heat, salt, wounding and so on [11]. Here, the expression of *SaMT2* was significantly increased in both roots and shoots of *S. alfredii* treated with Cd or Zn; in contrast, it has been reported that Cd

and Zn do not induce the expression of *TcMT2* and *TcMT3* in *Thlaspi caerulescens* (now *Noccaea caerulescens*) another Cd/Zn hyperaccumulator [29].

The plant MTs are suggested to be involved in metal homeostasis or tolerance, such as Cu, Cd and Zn. When expressed in yeast or *E. coli*, the plant MTs are able to restore Cu, Cd and Zn tolerance [9,11,26,27]. In the present study, Cd and Zn induced the expression of *SaMT2*, suggesting its possible involvement in Cd and Zn tolerance. This was confirmed in yeast and tobacco plant overexpressing *SaMT2*, which exhibited increased Cd and Zn tolerance and accumulation. Previous studies also reported enhanced tolerance and accumulation of Cd or other heavy metals by over-expressing plant MT genes. For example, the over-expression of *Cajanus cajan* MT1 enhances Cd and Cu tolerance in *E. coli* and Arabidopsis [27]. The expression of *Colocasia esculenta* CeMT2b increases Cd tolerance and accumulation in *E. coli* and tobacco [28]. However, Hassinen et al. [29] have observed that MT expression and Cd accumulation are not correlated among *T. caerulescens* accessions. Furthermore, the overexpression of *TcMT2* and *TcMT3* do not increase Cd accumulation in Arabidopsis shoots. On the other hand, Lv et al. [30] have observed that the ectopic expression of either *BcMT1* or *BcMT2* does increase Cd tolerance, but not the Cd accumulation in Arabidopsis shoots and roots. Thus, the MT genes may have different specific functions, depending on plant species.

The plant MTs are thought to function as metal chelators or ROS scavengers in heavy metal stress [8]. On one hand, plant MT proteins are supposed to have binding activities to heavy metals, such as Cd, Zn and Cu [7,8]. In the present study, the ectopic expression of *SaMT2* in tobacco enhanced Cd tolerance and accumulation, which might be due to reduced activities of free Cd ions in the cytoplasm, by the binding of overexpressed MT protein and Cd. On the other hand, MTs can also function as ROS scavengers which can reduce the ROS induced by Cd or other metals [8]. The plants exposure to heavy metals, such as Cd, can produce ROS and oxidative stress. The present study demonstrated that overexpression of *SaMT2* could significantly reduce H$_2$O$_2$ in tobacco exposure to excess Cd. Several studies have demonstrated that MTs can effectively scavenge ROS in plants. Over-expression of *BcMT1*, *BcMT2* [30], *EhMT1* [31], *pCeMT* [9] reduces ROS production in transgenic plants. Using recombinant GhMT3a protein, Xue *et al.* [32] have demonstrated that GhMT3a can scavenge ROS *in vitro*. Plants themselves have developed various antioxidant defense mechanisms to protect from deleterious effects of ROS. One of them is the enzymatic system, which includes SOD, APX, POD, and CAT. Plants overexpressing MT genes show higher antioxidant enzyme activities [8]. The present study demonstrated that tobacco plants overexpressing *SaMT2* showed higher SOD and POD activities than wild type plants, indicating that *SaMT2* might also act as an activator of antioxidant enzyme system.

It has been demonstrated that MTs are not related with Cd or Zn tolerance and accumulation in hyperaccumulator *T. caerulescens*, even though the expression of MT genes varies among *T. caerulescens* accessions [29]. However, in the present study, ectopic expression study in yeast and tobacco revealed that *SaMT2* might play certain roles in Cd and Zn tolerance and accumulation. It is not certain whether *SaMT2* is directly involved in Cd or Zn tolerance in *Sedum alfredii*. There are clear evidence that MTs are not directly related in Zn or Cd tolerance in *T. caerulescens* [29]. Data from the present study demonstrated that *SaMT2* might be involved in the Cd or Zn induced antioxidant stress in *Sedum alfredii*. However, the exact role of *SaMT2* in metal tolerance and accumulation in *Sedum alfredii* needs to be examined by further study in the future.

In conclusion, *SaMT2* is a metallothionein gene cloned from Cd/Zn hyperaccumulator *Sedum alfredii* Hance. Overexpression of this gene could significantly enhance Cd tolerance and accumulation in yeasts and tobacco plants. The mechanism of the elevated Cd tolerance and accumulation by overexpressing of *SaMT2* includes binding of *SaMT2* with Cd and improving the antioxidant system.

Acknowledgments

The *Δzrc1* and *Δycf1* yeast mutant strains and the wildtype strain BY4741 were kindly supplied by Prof. Eide, University of Wisconsin-Madison, USA.

Author Contributions

Conceived and designed the experiments: JZ SKT LLL XY. Performed the experiments: JZ MZ. Analyzed the data: JZ MZ XY. Contributed reagents/materials/analysis tools: JZ MZ XY. Contributed to the writing of the manuscript: JZ MZ MJIS XY.

References

1. Brooks R (1998) Geobotany and hyperaccumulators. In: Robert R. Brooks editor. Plants that Hyperaccumulate Heavy Metals. New York: CAB International. pp. 55–94.
2. Kramer U (2010) Metal hyperaccumulation in plants. Annual Review of Plant Biology 61: 517–534.
3. McGrath SP, Zhao FJ (2003) Phytoextraction of metals and metalloids from contaminated soils. Current Opinion in Biotechnology 14: 277–282.
4. Zhao FJ, McGrath SP (2009) Biofortification and phytoremediation. Current Opinion in Plant Biology 12: 373–380.
5. Verbruggen N, Hermans C, Schat H (2009) Molecular mechanisms of metal hyperaccumulation in plants. New Phytologist 181: 759–776.
6. Hall JL (2002) Cellular mechanisms for heavy metal detoxification and tolerance. Journal of Experimental Botany 53: 1–11.
7. Cobbett C, Goldsbrough P (2002) Phytochelatins and metallothioneins: Roles in heavy metal detoxification and homeostasis. Annual Review of Plant Biology 53: 159–182.
8. Hassinen VH, Tervahauta AI, Schat H, Karenlampi SO (2011) Plant metallothioneins - metal chelators with ROS scavenging activity? Plant Biology 13: 225–232.
9. Kim YO, Jung S, Kim K, Bae HJ (2013) Role of pCeMT, a putative metallothionein from *Colocasia esculenta*, in response to metal stress. Plant Physiology and Biochemistry 64: 25–32.
10. Mir G, Domenech J, Huguet G, Guo WJ, Goldsbrough P, et al. (2004) A plant type 2 metallothionein (MT) from cork tissue responds to oxidative stress. Journal of Experimental Botany 55: 2483–2493.
11. Xia Y, Lv Y, Yuan Y, Wang G, Chen Y, et al. (2012) Cloning and characterization of a type 1 metallothionein gene from the copper-tolerant plant *Elsholtzia haichowensis*. Acta Physiologiae Plantarum 34: 1819–1826.
12. Yang X, Long XX, Ni WZ, Fu CX (2002) *Sedum alfredii* H: A new Zn hyperaccumulating plant first found in China. Chinese Science Bulletin 47: 1634–1637.
13. Yang XE, Long XX, Ye HB, He ZL, Calvert DV, et al. (2004) Cadmium tolerance and hyperaccumulation in a new Zn-hyperaccumulating plant species (*Sedum alfredii* Hance). Plant and Soil 259: 181–189.
14. Jin XF, Yang X, Mahmood Q, Islam E, Liu D, et al. (2008) Response of antioxidant enzymes, ascorbate and glutathione metabolism towards cadmium in hyperaccumulator and nonhyperaccumulator ecotypes of *Sedum alfredii* H. Environmental Toxicology 23: 517–529.
15. Jin XF, Yang XO, Islam E, Liu D, Mahmood Q (2008) Effects of cadmium on ultrastructure and antioxidative defense system in hyperaccumulator and non-hyperaccumulator ecotypes of *Sedum alfredii* Hance. Journal of Hazardous Materials 156: 387–397.
16. Gietz RD, Schiestl RH (2007) High-efficiency yeast transformation using the LiAc/SS carrier DNA/PEG method. Nature Protocols 2: 31–34.
17. Horsch RB, Fry JE, Hoffmann NL, Eichholtz D, Rogers SG, et al. (1985) A simple and general method for transferring genes into plants. Science 227: 1229–1231.
18. Giannopolitis CN, Ries SK (1977) Superoxide dismutases: I. Occurrence in higher plants. Plant Physiology 59: 309–314.
19. Cakmak I, Strbac D, Marschner H (1993) Activities of hydrogen peroxide-scavenging enzymes in germinating wheat seeds. Journal of Experimental Botany 44: 127–132.
20. Zheng X, van Huystee RB (1992) Peroxidase-regulated elongation of segments from peanut hypocotyls. Plant Science 81: 47–56.
21. Loreto F, Velikova V (2001) Isoprene produced by leaves protects the photosynthetic apparatus against ozone damage, quenches ozone products, and reduces lipid peroxidation of cellular membranes. Plant Physiology 127: 1781–1787.
22. Zhou JM, Goldsbrough PB (1995) Structure, organization and expression of the metallothionein gene family in Arabidopsis. Molecular & General Genetics 248: 318–328.
23. Guo WJ, Bundithya W, Goldsbrough PB (2003) Characterization of the Arabidopsis metallothionein gene family: tissue-specific expression and induction during senescence and in response to copper. New Phytologist 159: 369–381.
24. Hsieh HM, Liu WK, Chang A, Huang PC (1996) RNA expression patterns of a type 2 metallothionein-like gene from rice. Plant Molecular Biology 32: 525–529.
25. Hsieh HM, Liu WK, Huang PC (1995) A novel stress-inducible metallothionein-like gene from rice. Plant Molecular Biology 28: 381–389.
26. Roosens NH, Bernard C, Leplae R, Verbruggen N (2004) Evidence for copper homeostasis function metallothionein of metallothionein (MT3) in the hyperaccumulator *Thlaspi caerulescens*. FEBS Letters 577: 9–16.
27. Sekhar K, Priyanka B, Reddy VD, Rao KV (2011) Metallothionein 1 (CcMT1) of pigeonpea (*Cajanus cajan*, L.) confers enhanced tolerance to copper and cadmium in *Escherichia coli* and *Arabidopsis thaliana*. Environmental and Experimental Botany 72: 131–139.
28. Kim YO, Patel DH, Lee DS, Song Y, Bae HJ (2011) High cadmium-binding ability of a novel *Colocasia esculenta* metallothionein increases cadmium tolerance in *Escherichia coli* and tobacco. Bioscience Biotechnology and Biochemistry 75: 1912–1920.
29. Hassinen VH, Tuomainen M, Peraniemi S, Schat H, Karenlampi SO, et al. (2009) Metallothioneins 2 and 3 contribute to the metal-adapted phenotype but are not directly linked to Zn accumulation in the metal hyperaccumulator, *Thlaspi caerulescens*. Journal of Experimental Botany 60: 187–196.
30. Lv YY, Deng XP, Quan LT, Xia Y, Shen ZG (2013) Metallothioneins BcMT1 and BcMT2 from *Brassica campestris* enhance tolerance to cadmium and copper and decrease production of reactive oxygen species in *Arabidopsis thaliana*. Plant and Soil 367: 507–519.
31. Xia Y, Qi Y, Yuan YX, Wang GP, Cui J, et al. (2012) Overexpression of *Elsholtzia haichowensis* metallothionein 1 (EhMT1) in tobacco plants enhances copper tolerance and accumulation in root cytoplasm and decreases hydrogen peroxide production. Journal of Hazardous Materials 233: 65–71.
32. Xue TT, Li XZ, Zhu W, Wu CG, Yang GG, et al. (2009) Cotton metallothionein GhMT3a, a reactive oxygen species scavenger, increased tolerance against abiotic stress in transgenic tobacco and yeast. Journal of Experimental Botany 60: 339–349.

3

Exogenous 5-Aminolevulenic Acid Promotes Seed Germination in *Elymus nutans* against Oxidative Damage Induced by Cold Stress

Juanjuan Fu, Yongfang Sun, Xitong Chu, Yuefei Xu*, Tianming Hu*

Department of Grassland Science, College of Animal Science and Technology, Northwest A&F University, Yangling, Shaanxi Province, P. R. China

Abstract

The protective effects of 5-aminolevulenic acid (ALA) on germination of *Elymus nutans* Griseb. seeds under cold stress were investigated. Seeds of *E. nutans* (Damxung, DX and Zhengdao, ZD) were pre-soaked with various concentrations (0, 0.1, 0.5, 1, 5, 10 and 25 mg l^{-1}) of ALA for 24 h before germination under cold stress (5°C). Seeds of ZD were more susceptible to cold stress than DX seeds. Both seeds treated with ALA at low concentrations (0.1–1 mg l^{-1}) had higher final germination percentage (FGP) and dry weight at 5°C than non-ALA-treated seeds, whereas exposure to higher ALA concentrations (5–25 mg l^{-1}) brought about a dose dependent decrease. The highest FGP and dry weight of germinating seeds were obtained from seeds pre-soaked with 1 mg l^{-1} ALA. After 5 d of cold stress, pretreatment with ALA provided significant protection against cold stress in the germinating seeds, significantly enhancing seed respiration rate and ATP synthesis. ALA pre-treatment also increased reduced glutathione (GSH), ascorbic acid (AsA), total glutathione, and total ascorbate concentrations, and the activities of superoxide dismutase (SOD), catalase (CAT), ascorbate peroxidase (APX) and glutathione reductase (GR), whereas decreased the contents of malondialdehyde (MDA) and hydrogen peroxide (H_2O_2), and superoxide radical ($O_2^{\cdot-}$) release in both germinating seeds under cold stress. In addition, application of ALA increased H^+-ATPase activity and endogenous ALA concentration compared with cold stress alone. Results indicate that ALA considered as an endogenous plant growth regulator could effectively protect *E. nutans* seeds from cold-induced oxidative damage during germination without any adverse effect.

Editor: Zhulong Chan, Chinese Academy of Sciences, China

Funding: This work was supported by Northwest A & F University Fundamental Research Special Fund (No. QN2011100) and Key Projects in the National Science & Technology Pillar Program in the Twelfth Five-year Plan Period (No. 2011BAD17B05). The funders had no role in study design, data collection and analysis, decision to publish, or preparation of the manuscript.

Competing Interests: The authors have declared that no competing interests exist.

* Email: xuyfgrass@gmail.com (YX); hutianming@126.com (TH)

Introduction

Cold stress is commonly defined as the low temperature range that is adequate to alter growth without stopping cellular processes [1]. Cold greatly influences seed germination, and consequently induces a reduction in germination rate and a delay in the initiation of the germination and seedling establishment [2]. Thus, it is worthwhile to clarify the physiological mechanisms of poor seed germination caused by cold stress and to develop reasonable strategies to alleviate the adverse effects of cold on seed germination thereby plants establishment on low temperature environment, especially at high altitude.

Cold is one of severe environmental stresses that disrupts the metabolic balance of cells, resulting in membrane damage [3], reduction of cellular respiration [4], and production of reactive oxygen species (ROS) [5]. In plants, the antioxidant enzymes are important defense systems to detoxify ROS [6]. ROS scavenging enzymes in plants include superoxide dismutase (SOD), peroxidases (POD), catalase (CAT), guaiacol peroxidase (GPX), ascorbate peroxidase (APX), dehydroascorbate reductase (DHAR) and glutathione reductase (GR) [7,8]. A large body of evidence has demonstrated that the antioxidant systems play important roles in

protecting plants against oxidative damage induced by cold stress [3,9].

The 5-aminolevulenic acid (ALA) is a key precursor in the biosynthesis of all porphyrins compounds such as chlorophyll, heme and phytochrome [10]. A number of reports show that exogenous ALA improves the growth and yield of a number of plants by enhancing chlorophyll contents and the rate of photosynthesis [11,12]. It is also known that ALA in low concentration regulates key physiological processes associated with plant growth under various abiotic and biotic stresses, including low or high temperature [7,13], salinity [14], drought [15] and heavy metals [16]. In contrast, high levels of ALA can promote enhanced production of ROS, thereby enhancing oxidative stress in plants [17]. These results suggest that ALA has a great application potential in agricultural production as a new non-toxic endogenous substance [18].

Elymus nutans Griseb., a perennial cool-season grass, is distributed in the north, northwest and southwest regions, especially on the Qinghai-Tibetan Plateau from 3,000 to 5,000 m in China [19]. *E. nutans* has been traditionally used as typical native forage and has often been collected and dried as long

cool season [20]. Recently, it has been widely planted in cultivated pastures in alpine areas, owing to its high adaptability, good nutrition, high yield and good resistance to cold, drought and biotic stress [19]. Thus, an investigation of seed germination in low temperature is important to wild *E. nutans* establishment at high altitude in Qinghai-Tibetan plateau. Chen and Jia [20] reported *E. nutans* also plays a pivotal role in animal husbandry and environmental sustenance in China. However, to date, no specific information is available regarding the effects of ALA on cold stress resistance of *E. nutans* seeds. Moreover, further studies are required to elucidate the mechanism of how ALA application could regulate specific metabolic reactions to achieve enhanced resistance in seeds to temperature stress. Therefore, this study provides the first investigation ALA effects *E. nutans* on cold stress. Our specific objectives were: (1) to investigate whether ALA could improve *E. nutans* seed germination under low temperature and (2) to further explore the mechanism of exogenous ALA pre-soaking improving seed germination via determining antioxidant enzyme activities, lipid peroxidation, seed respiration rate, H^+-ATPase activity and endogenous ALA concentration in *E. nutans* seeds under cold stress.

Materials and Methods

Plant Material and Treatments

Elymus nutans seeds were obtained from two sources: seeds of Damxung (DX) were collected in September 2012, from wild plants growing in Damxung County (30°28.535′N, 91°06.246′E, altitude 4678 m), located in the middle of Tibet, China. Agriculture and Animal Husbandry Bureau in Tibet responses for Damxung County. *E. nutans* occurs naturally and abundantly at altitudes between 3,000 and 5,000 m in the Qinghai-Tibetan Plateau, the field studies did not involve endangered or protected species. And Zhengdao (ZD) seeds were obtained in September 2012, from Beijing Rytway Ecotechnology Co., Ltd., located in Changping District (40°06.595′N, 116°24.383′E, altitude 550 m), Beijing, China. Seeds were cleaned and stored at 4°C in paper bags until the start of the experiments. In a preliminary experiment, two sources of *E. nutans*, DX and ZD, were found that ZD was more susceptible to cold stress than DX.

Seeds were surface sterilized in 1% (w/v) sodium hypochlorite for 10 min and rinsed several times with distilled water. Seeds were placed on double layers of filter paper wetted with 5 ml of 0, 0.1, 0.5, 1, 5, 10 or 25 mg l^{-1} ALA (Sigma Aldrich, St. Louis, MO, USA) solution in Petri dishes of 9 cm diameter. Seeds were kept at 25°C in the dark for 24 h [9]. Soaked seeds were then washed for 1 min under running water.

Preliminary investigation demonstrated inhibition of germination of *E. nutans* at 5°C compared to higher temperatures (10–30°C). After ALA application, germination tests were carried out in a plant growth chamber (Percival E-36L, Percival Scientific. Inc., USA) at a day/night temperature 5°C/5°C, a relative humidity of 70%, a day/night regime of 14 h/10 h and a photosynthetic photon flux density (PPFD) of 100 μmol m^{-2} s^{-1}. The lighting system is lit by (16) 17 W cool white fluorescent lamps and (2) 40 W incandescent lamps properly spaced for uniform light intensity. Fifty seeds were placed on two layers of filter paper moistened with 5 ml of distilled water in covered 9 cm Petri dishes. To prevent fungal contamination, 1 ml of 0.5% Captan was added. Petri dishes were arranged in completely randomized design with five replications. Seed germination was defined as root or shoot emergence [21]. Germination was recorded daily until the numbers stabilized (for 15 days). Dry weights of seeds and

chlorophyll (Chl) contents of seedlings were measured following germination determination.

To further explore possible cold adaptation mechanisms of seed germination in *E. nutans* promoted by ALA treatment, a second batch from both sources remained untreated or was imbibed in ALA solution using previously described conditions at either 5°C or 25°C for 24 h. All seeds were germinated at 5°C for 5 d in the growth chamber using previously described temperature and light conditions. After 5 d of cold stress [22], the germinating seeds were used for further bio-chemical and physiological measurements. Seed treatments at 25°C was used as the control (and identical the rest of growth conditions).

Assay of Dry Weight

Seeds (mainly endosperms and pericarps) were isolated and dried in oven at 80°C for 72 h and their dry weights were determined.

Determination of Chlorophyll (Chl) Concentration

The concentration of Chl were determined spectrophotometrically using 80% acetone as a solvent [23]. Extract absorbance was measured at 645 and 663 nm with Optizen 5100 UV spectrophotometer (Shanghai, China).

Analysis of Lipid Peroxidation

Membrane lipid peroxidation was measured as the concentration of malondialdehyde (MDA) produced using 10% (w/v) trichloroacetic acid (TCA), according to Dhindsa et al. [24]. The absorbance of the supernatant was measured at 450, 532, and 600 nm.

Measurement of Hydrogen Peroxide and Superoxide Radical

Hydrogen peroxide concentration was measured according to Veljovic-Jovanovic et al. [25]. Seeds (0.5 g) were ground in liquid nitrogen and the powder was extracted in 2 ml 1 M $HClO_4$ in the presence of 5% polyvinylpyrrolidone (PVP). The absorbance was read at 590 nm. Hydrogen peroxide concentration was calculated from a standard curve prepared in a similar way and expressed as nmol g^{-1} FW.

Superoxide radical production rate was determined by the modified method according to Elstner and Heupel [26]. Seeds (1.0 g) were homogenized in 3 ml 50 mM potassium phosphate buffer (pH 7.8) and centrifuged at 12,000×g for 20 min. The final solution was mixed with an equal volume of ethyl ether, and the absorbance of the pink phase was read at 530 nm.

Quantification of Non-enzymatic Antioxidant Concentrations

Reduced glutathione (GSH) and oxidized glutathione (GSSG) concentrations were determined according to according to Law et al. [27] with some modifications. The germinating seeds (0.3 g) were homogenized with 5 ml of 10% (w/v) TCA and homogenate was centrifuged at 15,000×g for 15 min. To assay total glutathione, 150 ml supernatant was added to 100 ml of 6 mM 5,5′-dithiobis-(2-nitrobenzoic acid) (DTNB), 50 ml of glutathione reductase (10 units ml^{-1}), and 700 ml 0.3 mM nicotinamide adenine dinucleotide phosphate (NADPH). The total glutathione content was calculated from the standard curve. All the reagents were prepared in 125 mM NaH_2PO_4 buffer, containing 6.3 mM ethylene diamine tetraacetic acid (EDTA), at pH 7.5. To measure GSSG, 120 ml of supernatant was added to 10 ml of 2-vinylpyridine followed by 20 ml of 50% (v/v) triethanolamine.

The solution was vortex-mixed for 30 s and incubated at 25°C for 25 min. The mixture was assayed as mentioned above. Calibration curve was developed by using GSSG samples treated exactly as above and GSH was determined by subtracting GSSG from the total glutathione content.

After 0.2 g of germinating seeds was suspended in 3 ml of 6% TCA and was centrifuged at 4°C and 15,000×g for 20 min, the contents of AsA and total ascorbate were assayed at 525 nm [28]. The difference between the levels of total ascorbate and AsA was used for estimating the content of oxidized ascorbate.

Assay of Antioxidant Enzymes

The germinating seeds (0.5 g) were homogenized with a mortar and pestle at 4°C in 5 ml 50 mM phosphate buffer (pH 7.8) containing 1 mM EDTA and 2% PVP. Homogenate was centrifuged at 12,000×g for 20 min at 4°C and the supernatant was used for enzyme activity assays. Protein content in the supernatant was determined according to the method of Bradford [29] with bovine serum albumin (BSA) as standard.

The assay for ascorbate peroxidase (APX) activity was measured in a reaction mixture of 3 ml containing 100 mM phosphate (pH 7), 0.1 mM EDTA-Na$_2$, 0.3 mM ascorbic acid, 0.06 mM H$_2$O$_2$ and 100 µl enzyme extract. Change in absorption was observed at 290 nm 30 s after addition of H$_2$O$_2$ [30]. One unit of APX forms 1 µM of ascorbate oxidized per minute under assay conditions. Activity of catalase (CAT) was measured by following the consumption of H$_2$O$_2$ at 240 nm according to Cakmak and Marschner [31]. The decrease in the absorption was followed for 3 min and a breakdown of 1.0 µM H$_2$O$_2$ ml^{-1} min^{-1} was defined as 1 Unit of CAT activity. Glutathione reductase (GR) activity was measured by following the decrease in absorbance at 340 nm due to NADPH oxidation. The reaction mixture contained tissue extract, 1 mM EDTA, 0.5 mM GSSG, 0.15 mM NADPH and 50 mM Tris–HCl buffer (pH 7.5) and 3 mM MgCl$_2$ [32]. The reaction was started by using NADPH. Activity of superoxide dismutase (SOD) was determined according to Beauchamp and Fridovich [33] by following the photo-reduction of nitroblue tetrazolium (NBT) at 560 nm. One Unit of SOD activity was defined as the amount of enzyme required to cause a 50% inhibition of NBT reduction.

Determination of Seed Respiration Rates

Respiration rates were measured according to Zheng et al. [34]. A closed gas collecting system was used to measure CO$_2$ production during seed germination. After 5 d of cold stress, germinating seeds were sealed in an internally ventilated chamber with a volume of 0.2 L. The chamber was coupled with a GXH-3010F IRGA (infra-red gas analyzer, Huayun Instrument Research Institute Co., Beijing, China). Respiration rate of seed was calculated according to the slope of CO$_2$ increase in the chamber.

Assay of Seed Adenosine Triphosphate (ATP) content

The ATP content of seeds was determined by spectrofluorometry as described by Zheng et al. [34]. One g of germinating seeds were finely sliced and put into 5 ml acetone, and placed in boiling water bath for 5 min until the acetone fully evaporated. Three ml of 20 mM Tris–HCl buffer (pH 7.6) were added to the sample and heated in a boiled water bath for 10 min, and then immediately cooled down in an ice bath. The tubers was centrifuged at 3,000×g for 10 min, and the supernatant was collected. Bioluminescence produced by adding the ATP extract was measured with the ATP Bioluminescent Assay Kit (luciferin-luciferase reagent, the Detect Technical Institute, Shenzhen, China) using a SHG-D Bioluminescence and Chemiluminescence Meter (The Detect Technical Institute, Shenzhen, China).

Assay of Plasma Membrane (PM) H$^+$-ATPase Activity

Plasma membrane vesicles were isolated from germinating seeds by phase partitioning according to the procedure by Palmgren et al. [35]. Samples were ground in ice cold homogenization buffer containing 50 mM 3-(N-morpholino) propanesulphonic acid- Bistris Propane (MOPs–BTP) (pH 7.5), 330 mM sucrose, 5 mM EDTA, 5 mM dithiothreitol (DTT), 0.5 mM phenylmethanesulphonylfluoride, 0.2% (w/v) casein, 0.2% bovine serum albumin, and 0.5% PVP-40. Homogenate was filtered through four layers of cheesecloth and the filtrate was centrifuged at 10,000×g for 15 min at 4°C. Supernatant was collected and centrifuged at 80,000×g for 45 min, and the resulting precipitate was resuspended in buffer consisting of 330 mM sucrose, 5 mM potassium phosphate (pH 7.8), 5 mM KCl, 0.1 mM EDTA, and 1 mM DTT. Homogenate was loaded onto a two-phase system containing 6.5% Dextran T-500 (Sigma–Aldrich, USA), 6.5% (w/w) polyethylene glycol (PEG)-3350 (Sigma–Aldrich), 250 mM sucrose, 5 mM KH$_2$PO$_4$ (pH 7.8), 4 mM KCl, and sterile distilled water. After the batch procedure, the resulting upper phase was mixed with a dilution buffer consisting of 5 mM MOPs–BTP (pH 7.5), 330 mM sucrose, and 5 mM KCl, and was centrifuged at 100,000×g for 60 min. PM vesicles obtained were either used immediately or stored at −80°C, pending analysis.

PM H$^+$-ATPase activity was measured according to the procedure of Ahn et al. [36]. PM H$^+$-ATPase activity was measured with 5 µg protein in 0.5 ml of reaction solution that contained 30 mM MOPs–BTP (pH 6.5), 3 mM MgSO$_4$, 50 mM KCl, 1.5 mM ATP, and 0.05% Triton-X100. After 30 min at 37°C, the reaction was stopped by adding 500 µl of 5% trichloroacetic acid, 2 ml of 100 mM sodium acetate, 300 µl of 1% ascorbic acid, 60 µl of 10 µM CuSO$_4$, and 300 µl of 1% ammonium molybdate in 0.025 mM H$_2$SO$_4$. Following additional 10 min at 30°C, absorbance at 720 nm was measured with an Optizen 5100 UV spectrophotometer (Shanghai, China). Difference between samples with and without 0.1 mM vanadate, which is a specific PM H$^+$-ATPase inhibitor, was expressed as the PM H$^+$-ATPase activity. A standard curve of phosphate in the reaction mixture was included for each assay.

Determination of ALA Concentration

Germinating seeds (0.1 g) were homogenized in 5 ml of 1 M sodium acetate buffer (pH 4.6) and centrifuged at 12,000×g for 10 min. The assay mixture consisted of 0.1 ml of supernatant, 0.4 ml of distilled water, and 25 µl of acetylacetone. The assay medium was mixed and heated in a boiling water bath for 10 min. The extract was then cooled at room temperature, and an equal volume of modified Ehrlich's reagent was added and vortexed for 2 min. After 10 min of incubation, absorbance of the extract was measured at 555 nm and ALA concentration was determined from the standard curve of ALA [37].

Statistical Analysis

Each experiment was repeated three times. All values were expressed as means ± SD. Statistical analyses were performed by analysis of variance (ANOVA) using SPSS-17 statistical software (SPSS Inc., Chicago, IL, USA). Means were separated using Duncan's least significance difference test at $P < 0.05$.

Results

Effect of ALA on Seed Germination, Dry Weight and Chl Content

ALA concentrations ranging from 0 to 25 mg l^{-1} were applied to *E. nutans* seeds to investigate the response for cold resistance. The ratio of germination percentage (GP) was improved for seeds from both sources when pre-treated with 0.5, 1 or 5 mg l^{-1} ALA. However, ALA concentrations above 1 mg l^{-1} caused the ratio of GP reduction in a dose dependent manner (Fig. 1A). On the other hand, 1 mg l^{-1} ALA alone did not have any effect on the ratio of GP compared to the unsoaked (data not shown).

The ratio of dry weight of DX and ZD increased with the concentrations of ALA, peaking at 1 mg l^{-1} ALA prior to cold stress, and decreased at 5 mg l^{-1} ALA. These concentrations alone did not alter dry weight compared to controls (data not shown). Treatment with 10 or 25 mg l^{-1} ALA resulted in a major seed dry weight loss compared to the unsoaked (Fig. 1B).

Pre-treatment with low ALA concentrations (0.1–5 mg l^{-1}) prevented the loss of Chl content in both sources of *E. nutans* seed, whereas exposure to higher ALA concentrations (10 and 25 mg l^{-1}) brought about a dose dependent decrease, reaching a minimum in DX and ZD pre-treated with 25 mg l^{-1} ALA compared to the untreated (Fig. 1C).

ALA was effective in enhancing cold resistance in *E. nutans* seeds and seedlings up to 5 mg l^{-1}. The best results were obtained in seeds pre-treated with 1 mg l^{-1} ALA. As a result, 1 mg l^{-1} ALA was applied in subsequent experiments.

Effect of ALA on MDA, H$_2$O$_2$ Concentrations and O$_2^{\bullet -}$ Level

Contents of MDA and H$_2$O$_2$, and release rate of O$_2^{\bullet -}$ in both sources of *E. nutans* seed increased after exposure to cold stress for 5 d (Fig. 2). Cold treatment increased ($P<0.05$) MDA contents by 48.6% and 120.0% in seeds of DX and ZD, respectively. DX seeds exhibited a 76.5% and 119.7% increase in levels of H$_2$O$_2$ and O$_2^{\bullet -}$, while a 82.4% and 120.9% increase was observed in ZD under cold stress. Exogenous ALA treatment alleviated ($P<0.05$) the cold induced accumulation of MDA and H$_2$O$_2$, and decreased ($P<0.05$) the release rate of O$_2^{\bullet -}$ in germinating *E. nutans* seeds. Under normal condition, pretreatment of seeds with 1 mg l^{-1} ALA did not significantly change MDA, H$_2$O$_2$ concentrations and O$_2^{\bullet -}$ generation in seed from either source.

Effect of ALA on Concentrations of Non-enzymatic Antioxidants

The data regarding increased glutathione (GSH), ascorbic acid (AsA), total glutathione, total ascorbate concentrations, and the ratios of reduced/oxidized glutathione (GSH/GSSG) and reduced/oxidized ascorbate (AsA/oxidized ascorbate) in DX seeds under cold stress had been shown in Fig. 3, while the ratio of GSH/GSSG and AsA/oxidized ascorbate decreased in ZD ($P<0.05$). ALA at lower concentration 1 mg l^{-1} showed improvement in GSH, AsA, total glutathione and ascorbate concentrations, and the ratio of GSH/GSSG and AsA/oxidized ascorbate in germinating *E. nutans* seeds under cold stress. Antioxidants showed no significant changes when seeds were treated with ALA alone.

Effect of ALA on Activities of Antioxidant Enzymes

Activities of SOD, CAT, APX and GR increased ($P<0.05$) by 34.3%, 76.5%, 75.4%, 63.3% in DX seeds, but decreased ($P<0.05$) by 22.1%, 12.6%, 11.9%, 12.1% in ZD seeds subjected only

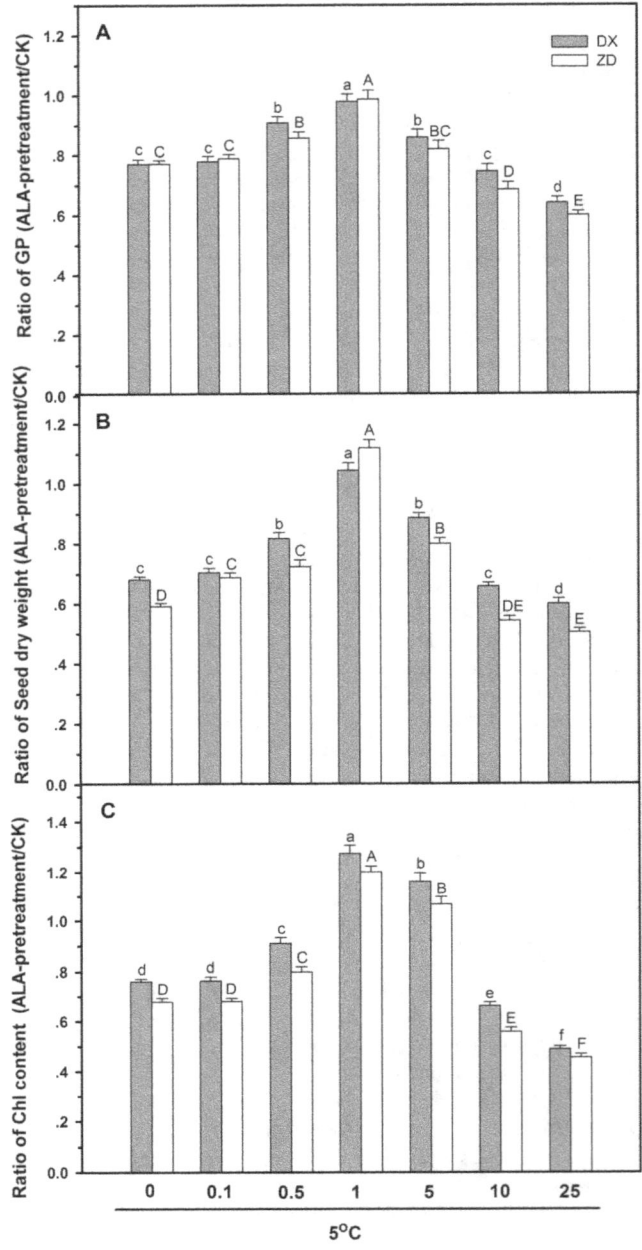

Figure 1. Effect of ALA applications on the ratio of germination percentage (GP) (A), dry weights of seed (B), and chlorophyll (Chl) concentration (C) under cold stress in *E. nutans* (DX, Damxung and ZD, Zhengdao). ALA pretreatment at different concentrations (0, 0.1, 0.5, 1, 5, 10, and 25 mg l^{-1}) were carried out prior to cold stress (5°C). Bars represent the mean ± SD (n = 3). Bars with different letters are significantly different at the 5% level.

to cold treatment, respectively (Fig. 4). Pretreatment with ALA increased SOD, CAT, APX and GR activities ($P<0.05$), especially in DX seeds. Under control conditions, activities of all four antioxidant enzymes were not significantly influenced by exogenous ALA.

Effect of ALA on Seed Respiration Rate and ATP Content

After 5 d of cold stress, seed respiration rate sharply decreased in seed from *E. nutans* sources (Fig. 5A). Compared to control,

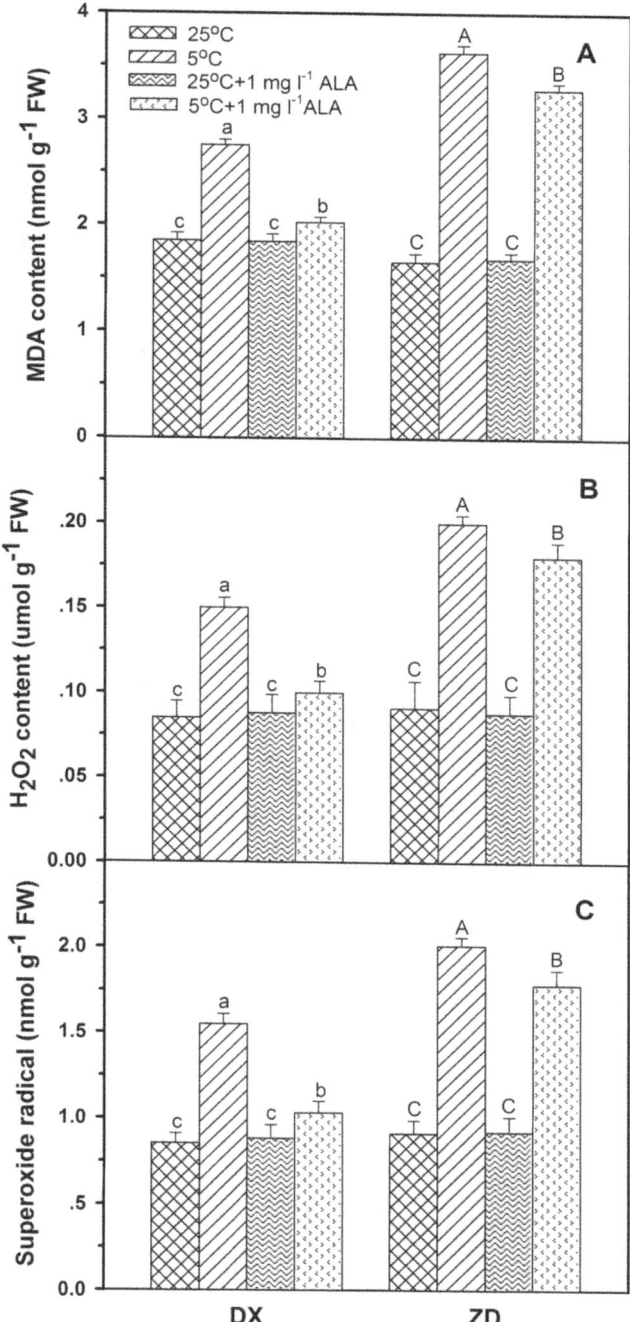

Figure 2. Effect of ALA applications on MDA (A), H₂O₂ concentrations (B) and the rate of O₂·⁻ generation (C) under cold stress in E. nutans (DX, Damxung and ZD, Zhengdao). Bars represent the mean ± SD (n = 3). Bars with different letters are significantly different at the 5% level.

exogenous ALA treatment increased the respiration rate in both seed sources of *E. nutans*. Treatment with ALA, under normal conditions, had no effect on seed respiration rate compared with control. A similar pattern of changes in germinating seed ATP content in response to exogenous ALA was observed in *E. nutans* under cold stress (Fig. 5B). Exogenous ALA treatment increased seed ATP content in seeds. ATP content in ALA soaked DX and

ZD seeds increased 83.1% and 61.7% after 5 d stress compared to non-ALA treatment.

Effect of ALA on Activities of PM H⁺-ATPase

Cold stress resulted in a 73.3% and 53.8% increase ($P<0.05$) in H⁺-ATPase activity in DX and ZD, compared with control (Fig. 6). Exogenous ALA treatment prior to cold stress further enhanced ($P<0.05$) the activities of H⁺-ATPase in both seeds. In contrast, enhancement of the activities of H⁺-ATPase did not occur in seeds pretreated with ALA alone.

Endogenous ALA Production

To verify the protective effect of exogenous ALA applied to seeds under cold stress, endogenous ALA release rates were measured. Endogenous ALA release rates decreased ($P<0.05$) after 5 d of cold stress. Under cold stress, pretreatment with 1 mg l⁻¹ ALA increased ($P<0.05$) endogenous ALA release in both seeds, especially in DX throughout the stress period. Application of 1 mg l⁻¹ ALA alone showed no change in endogenous ALA in both seeds (Fig. 7).

Discussion

Cold stress can lead to biochemical and physiological changes in plant tissues. Inhibitory effects of cold stress on *E. nutans* seed germination is consistent with earlier reports of low temperature stress in pepper seed [2]. In the present study, cold stress significantly reduced seed GP and dry weight for both seed sources. Under cold stress, cold-resistant DX showed significantly greater GP and seed dry weight than cold-sensitive ZD. Observed reduction of GP and dry weight in both seeds might be due to oxidative stress induced by cold [38]. Treatment with ALA significantly enhanced GP and dry weight for sources of both *E. nutans* seeds. Pre-soaking with ALA may have potential to enhance stress resistance by decreasing the lipid peroxidation [39] by activating the heme-based antioxidant enzyme systems to scavenge ROS like H₂O₂ [40].

Membranes are most susceptible to damage resulting from low temperature [41]. MDA is considered a final decomposition product of polyunsaturated fatty acids and it is used to determine oxidative damage [42]. In this study, MDA contents were increased after 5 d of cold stress. ALA alone had no effect on MDA level, but application of ALA under cold stress decreased MDA contents in germinating seeds, paralleling findings of Naeem et al. [43] who observed that ALA reduced MDA content under salinity stress in *Brassica napus*. Oxidative stress is increased by increases in ROS in cells under low temperature [7]. Observed increases in ROS level due to cold stress are similar to the findings of Zhang et al. [13]. In our study, ALA significantly reduced production of ROS under cold stress, suggesting that ALA can improve plant resistance to oxidative stress. Ali et al. [16] reported that application of ALA at 25 mg l⁻¹ concentration facilitated Cd stressed plants to detoxify the ROS using the antioxidant enzyme in *B. napus*.

ALA alleviates the membrane peroxidation resulting from ROS produced under stress conditions through different metabolism, and antioxidant capacity modulation was reported to be one of important pathways in many investigations [15,16,44]. GSH is an important component of the antioxidant system that scavenges ROS either directly or indirectly by participating in the ascorbate–glutathione cycle [45]. The key role of GSH in the antioxidant defense system is due to its ability to regenerate ascorbate (AsA) through reduction of dehydroascorbate via the ascorbate–gluta-thione cycle [46]. The high concentrations of AsA and GSH play

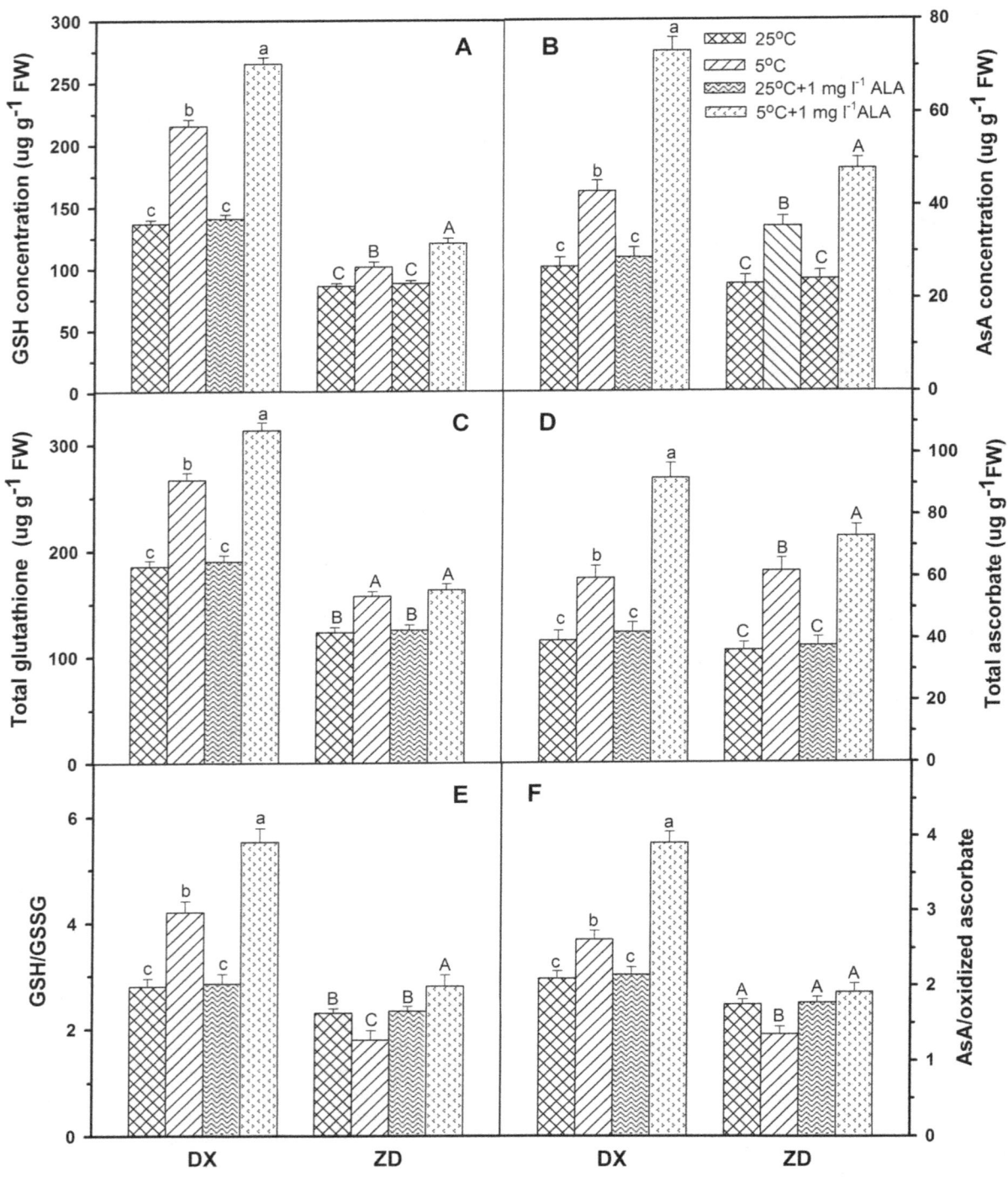

Figure 3. Effect of ALA applications on GSH (A), AsA (B), total glutathione (C), total ascorbate concentrations (D), and the ratios of GSH/GSSG (E) and AsA/oxidized ascorbate (F) under cold stress in *E. nutans* (DX, Damxung and ZD, Zhengdao). Bars represent the mean ± SD (n = 3). Bars with different letters are significantly different at the 5% level.

roles in alleviating the injury caused by ROS [47]. ALA application at low concentrations can enhance the GSH contents in the roots of *B. napus* [16]. In the presence of cold stress, GSH, AsA, total glutathione, total ascorbate concentrations, and the ratios of GSH/GSSG and AsA/oxidized ascorbate increased significantly when applied with ALA. Similarly, Liu et al. [48] reported that PEG treatment increases the concentrations of GSH and AsA and the ratios of AsA/oxidized ascorbate and GSH/GSSG, as well as decreases ROS level. ALA pretreatment can induce the synthesis of heme-based molecules [49]. ROS induction under cold stress may be due to the reason that ALA is a precursor of heme biosynthesis, so it can boast up the activities of heme-based molecules and can help in scavenging the ROS under cold conditions [16].

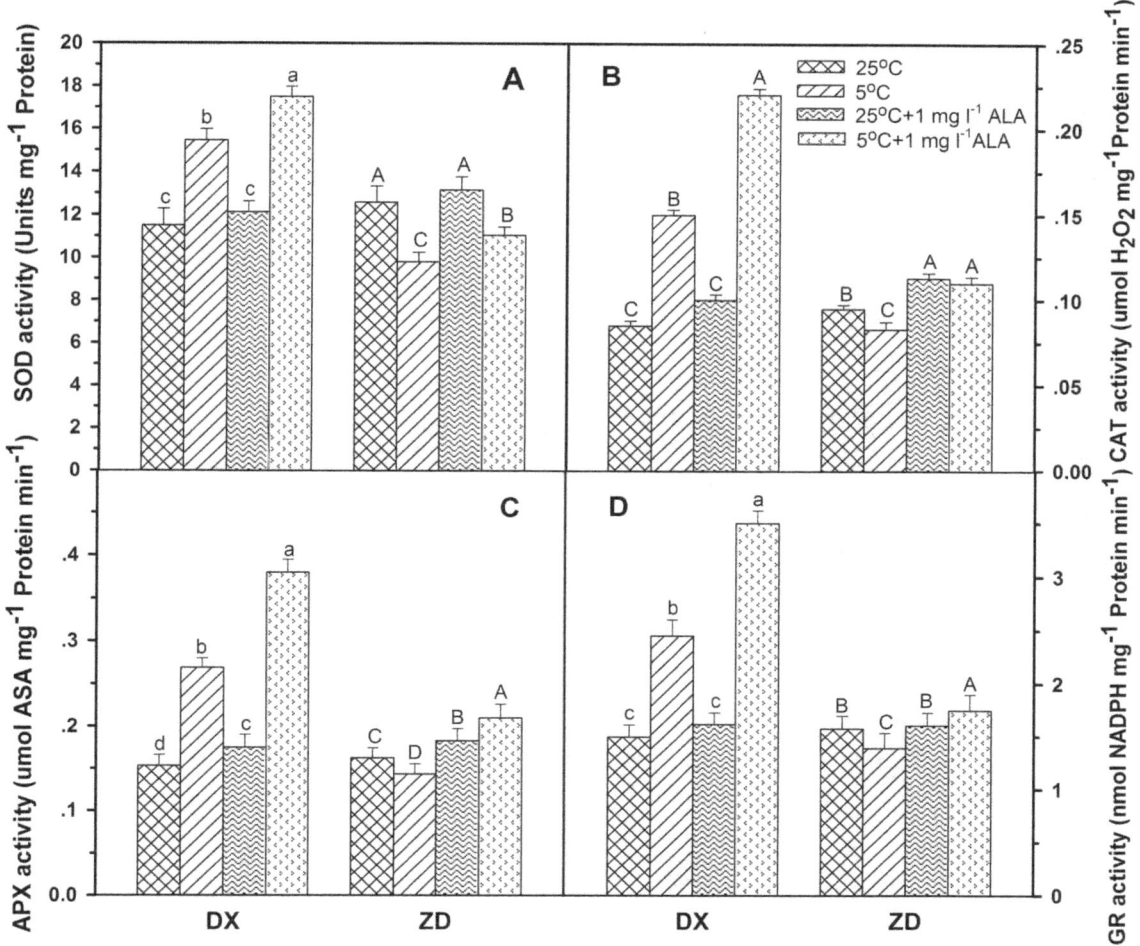

Figure 4. Effect of ALA applications on the activities of SOD (A), CAT (B), APX (C), and GR (D) under cold stress in *E. nutans* (DX, Damxung and ZD, Zhengdao). Bars represent the mean ± SD (n = 3). Bars with different letters are significantly different at the 5% level.

Antioxidant enzymes are considered to be the most efficient mechanisms against oxidative stress. When exposed to oxidative stress, the synthesis and activity of antioxidant enzymes are increased [50]. Among these enzymes, SOD is a major scavenger of $O_2^{\bullet-}$, catalyzing the dismutation of superoxide radicals to H_2O_2 and O_2. CAT directly scavenge H_2O_2, while APX and GR are involved in the AsA–GSH cycle, a non-enzymatic pathway that removes $O_2^{\bullet-}$ and H_2O_2 [44,51]. The enhancement of the activities of antioxidant enzymes suggests that ROS induced these changes in different cellular compartments [52]. In this study, DX is more resistant to cold because antioxidant enzyme activities to remove newly produced ROS are higher. Under cold stress, a greater increase in antioxidant enzyme activities and lower levels of H_2O_2 and $O_2^{\bullet-}$ were found in DX than in ZD. ALA has been reported to stimulate the activities of antioxidative enzymes under stress conditions [3]. Our results showed treatment with ALA further enhanced those antioxidant enzyme activities under cold stress. Similar observation were made by Zhang et al. [13] in cucumber. Thus, ALA contributed to reduce oxidative stress via higher antioxidant concentrations and antioxidant enzyme activities in germinating *E. nutans* seeds, thereby improving germination percentages under cold stress.

Seed treatment with ALA is known to improve GP and physiological processes under various stress conditions [10].

Similar observation had shown that ALA as a pre-soaking seed treatment improved the low-temperature resistance of pepper (*Capsicum annuum*) by enhancing final GP and germination rate [2]. Our results indicated that the treatment with ALA significantly enhanced GP for both sources of *E. nutans* seeds. Likewise, application of ALA treatment increased the seed germination of pakchoi (*Brassica campestris*), which was due to the improved seed respiration rate under salt stress [53]. The results were consistent with our finding that ALA treatment increased seed respiration rate in both sources of *E. nutans* seeds. Seed priming techniques have been used to increase germination and improve activities antioxidant enzyme by plant hormones/regulators under different various stress conditions [54,55]. In pepper, a remarkable enhancement in GP was observed through seed priming with ALA under cold stress [2]. The improved GP observed in our study is most likely due to the enhanced antioxidant enzymes activities just like ALA improved GP in pepper [2,3]. Therefore, ALA may be employed as effective approach to improving seed germination and plant growth under stress conditions.

In the plant mitochondria, electron transfer along the respiration chain is coupled to the formation of ATP [56], and the redundant electron leads to the formation of ROS if ATP synthesis is blocked [57]. ALA is the first precursor in the biosynthesis of porphyrin compounds such as chlorophyll and heme, a key

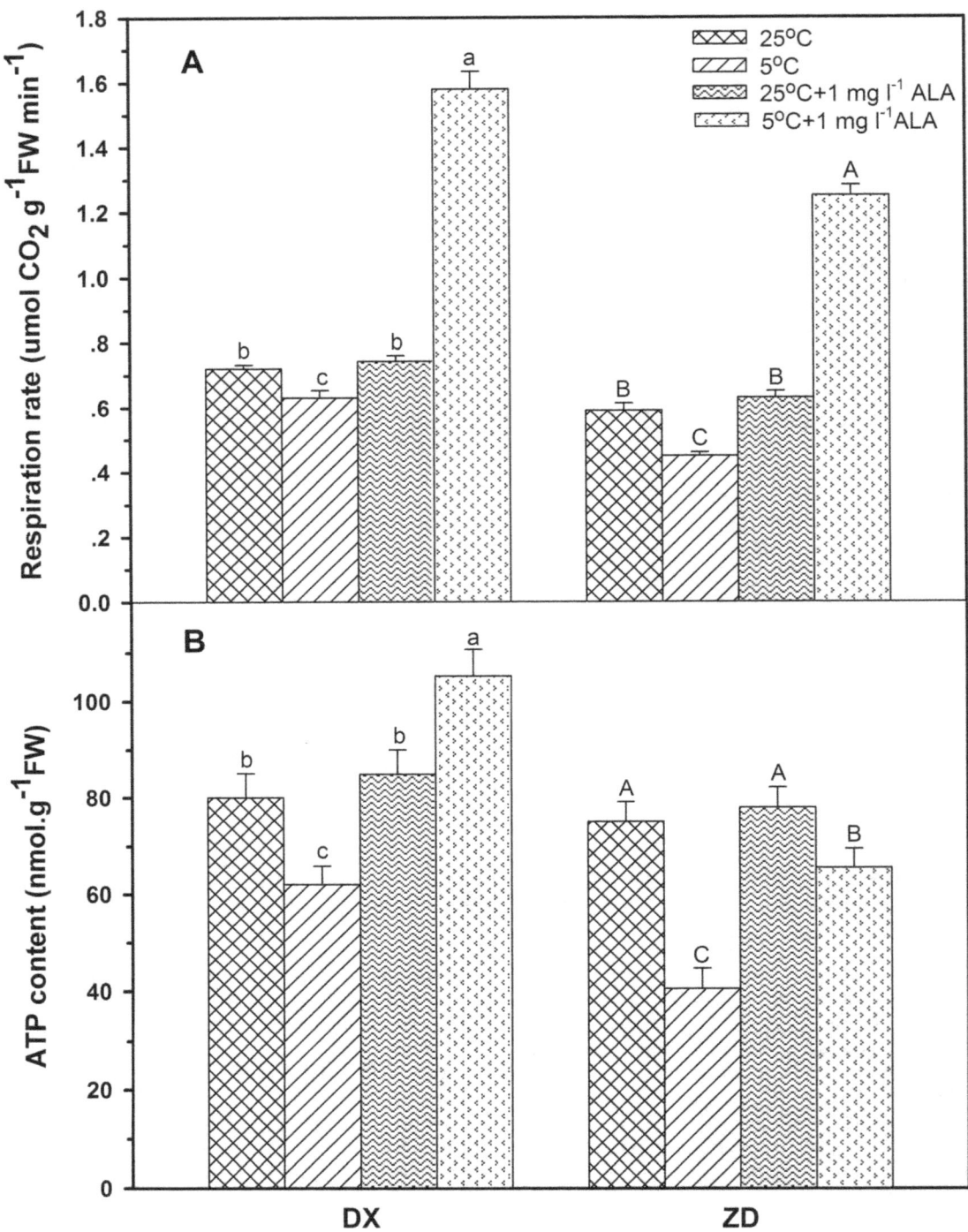

Figure 5. Effect of ALA applications on seed respiration rate (A) and ATP content (B) under cold stress in *E. nutans* (DX, Damxung and ZD, Zhengdao). Bars represent the mean ± SD (n = 3). Bars with different letters are significantly different at the 5% level.

element required for cytochrome c activity in the respiration chain of the mitochondrion [10]. Respiration, a temperature-dependent and heme-requiring process, increases during germination in order to provide necessary energy. Under cold stress, decreased respiration rates and ATP contents were observed in germinating seeds from both sources, while respiration was enhanced in seeds treated with ALA. Thus, it is suggested that exogenous ALA may promote ATP synthesis and enhance seed activity (respiration rate), both having a positive effect on seed germination under cold stress. A similar antioxidant stress effect of exogenous ALA was observed in salt-stressed pakchoi seeds [53]. In addition, cold stress reduced Chl concentration and endogenous ALA level while application of exogenous ALA increased Chl concentration and endogenous ALA release in germinating seeds, suggesting application of exogenous ALA prior to cold stress could mitigate inadequate biosynthesis problem [3].

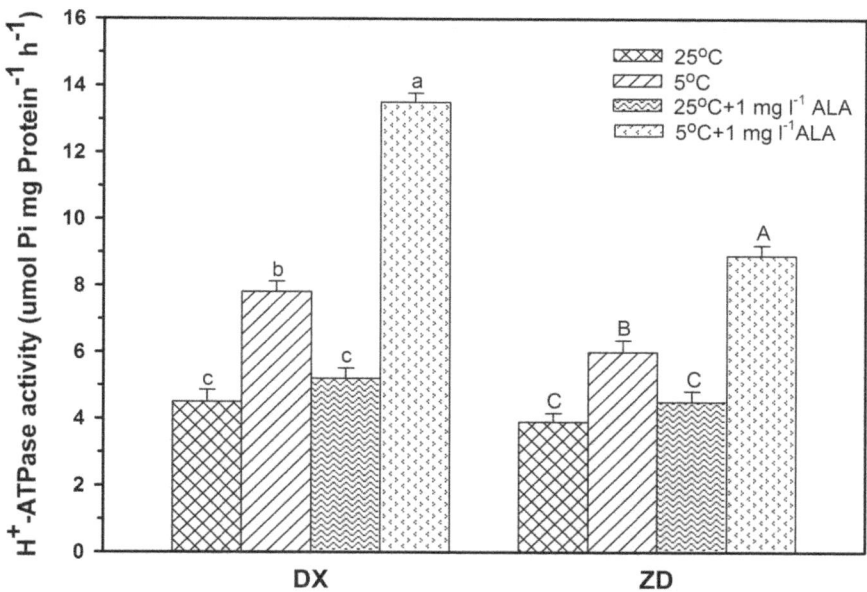

Figure 6. Effect of ALA applications on the activities of PM H⁺-ATPase under cold stress in *E. nutans* (DX, Damxung and ZD, Zhengdao). Bars represent the mean ± SD (n = 3). Bars with different letters are significantly different at the 5% level.

PM H⁺-ATPase plays a role in the adaptation of plants to stress conditions. An increase in permeability related to membrane damage and a change in its viscosity and fluidity are observed in plants that have been subjected to low temperature [41]. Increased generation of a proton gradient by PM H⁺-ATPase is required to maintain ionic balance and replenish the loss of organic compounds [58]. In this study, the activity of PM H⁺-ATPase increased after 5 d of cold stress in germinating seeds, agreeing with reports of Kim et al. [59] in cold treated camelina and rapeseed. Pretreatment with ALA further elevated H⁺-ATPase activity compared with cold stress alone, which might indicate

ALA acts as a signaling molecule, inducing increases in the activities of H⁺-ATPase.

In conclusion, results of our study revealed that pre-soaking with 1 mg l⁻¹ ALA improved *E. nutans* germination compared with non-ALA treated seeds. Protective effects of ALA on germinating seeds result from stimulation of activities of heme-based non-enzymatic antioxidants and antioxidant enzymes which help in scavenging the ROS under cold stress, especially in cold resistant DX seeds. In addition, ALA might act as a signaling molecule, inducing increased H⁺-ATPase activity and promoting ATP synthesis. Our finding that ALA could be used as a seed treatment

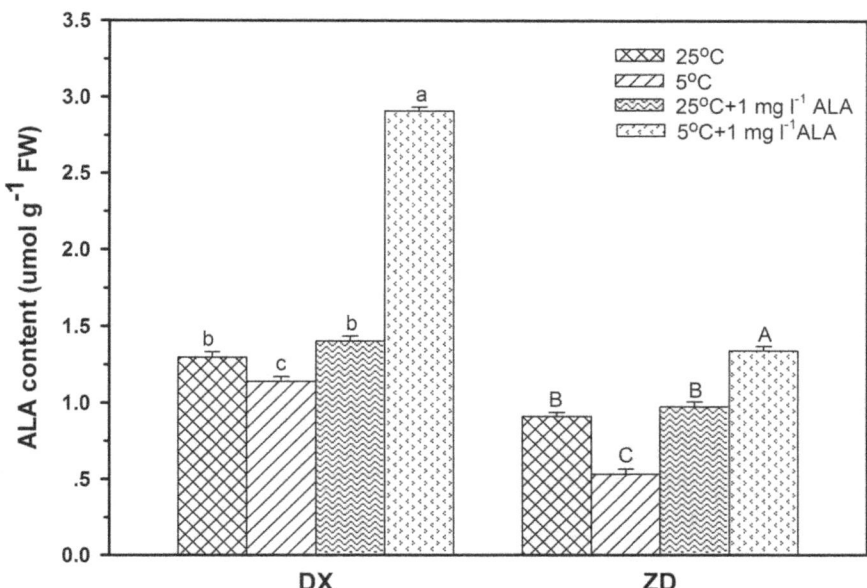

Figure 7. Effect of ALA applications on endogenous ALA concentrations under cold stress in *E. nutans* (DX, Damxung and ZD, Zhengdao). Bars represent the mean ± SD (n = 3). Bars with different letters are significantly different at the 5% level.

to enhance GP in *E. nutans* under cold stress may be useful in helping to solve serious problems occurring on a global scale due to low temperatures.

Acknowledgments

We are grateful to Associate Professor Yanjun Miao (Tibet University, Linzhi, P R China) for his donating *E. nutans* seed materials. Professor Roger N. Gates (South Dakota State University, USA) helps to revise

language. We would also like to thank anonymous reviewers for careful revision and critical comments on a earlier version of the manuscript.

Author Contributions

Conceived and designed the experiments: YFX TMH. Performed the experiments: JJF YFS XTC. Analyzed the data: JJF. Contributed reagents/materials/analysis tools: YFX TMH. Contributed to the writing of the manuscript: JJF.

References

1. Sanghera GS, Wani SH, Hussain W, Singh NB (2011) Engineering cold stress tolerance in crop plants. Curr Genomics 12: 30–43.
2. Korkmaz A, Korkmaz Y (2009) Promotion by 5-aminolevulenic acid of pepper seed germination and seedling emergence under low temperature stress. Sci Hortic-Amsterdam 119: 98–102.
3. Korkmaz A, Korkmaz Y, Demirkıran AR (2010) Enhancing chilling stress tolerance of pepper seedlings by exogenous application of 5-aminolevulinic Acid. Environ Exp Bot 67: 495–501.
4. Sugie A, Naydenov N, Mizuno N, Nakamura C, Takumi S (2006) Overexpression of wheat alternative oxidase gene Waox1a alters respiration capacity and response to reactive oxygen species under low temperature in transgenic Arabidopsis. Genes Genet Syst 81: 349–354.
5. Suzuki N, Mittler R (2006) Reactive oxygen species and temperature stresses: a delicate balance between signaling and destruction. Physiol Plant 126: 45–51.
6. Ashraf M (2009) Biotechnological approach of improving plant salt tolerance using antioxidants as markers. Biotechnol Adv 27: 84–93.
7. Balestrasse KB, Tomaro ML, Batlle A, Noriega GO (2010) The role of 5-aminolevulinic acid in the response to cold stress in soybean plants. Phytochemistry 71: 2038–2045.
8. Duan M, Ma NN, Li D, Deng YS, Kong FY, et al. (2012) Antisense-mediated suppression of tomato thylakoidal ascorbate peroxidase influences anti-oxidant network during chilling stress. Plant Physiol Bioch 58: 37–45.
9. Imahori Y, Takemura M, Bai J (2008) Chilling-induced oxidative stress and antioxidant responses in mume (*Prunus mume*) fruit during low temperature storage. Postharvest Biol Tec 49: 54–60.
10. Akram NA, Ashraf M (2013) Regulation in plant stress tolerance by a potential plant growth regulator, 5-aminolevulinic acid. J Plant Growth Regul 32: 663–679.
11. Awad MA (2008) Promotive effects of a 5-aminolevulinic acid-based fertilizer on growth of tissue culture-derived date palm plants (*Phoenix dactylifera* L.) during acclimatization. Sci Hortic-Amsterdam 118: 48–52.
12. Maruyama-Nakashita A, Hira MY, Funada S, Fuek S (2010) Exogenous application of 5-aminolevulinic acid increases the transcript levels of sulfur transport and assimilatory genes, sulfate uptake, and cysteine and glutathione contents in Arabidopsis thaliana. Soil Sci Plant Nutr 56: 281–288.
13. Zhang J, Li DM, Gao Y, Yu B, Xia CX, et al. (2012) Pretreatment with 5-aminolevulinic acid mitigates heat stress of cucumber leaves. Biol Plantarum 56: 780–784.
14. Naeem MS, Warusawitharana H, Liu H, Liu D, Ahmad R, et al. (2012) 5-Aminolevulinic acid alleviates the salinity-induced changes in *Brassica napus* as revealed by the ultrastructural study of chloroplast. Plant Physiol Bioch 57: 84–92.
15. Li DM, Zhang J, Sun WJ, Li Q, Dai AH, et al. (2011) 5-Aminolevulinic acid pretreatment mitigates drought stress of cucumber leaves through altering antioxidant enzyme activity. Sci Hortic-Amsterdam 130: 820–828.
16. Ali B, Tao QJ, Zhou YF, Gill RA, Ali S, et al. (2013) 5-Aminolevolinic acid mitigates the cadmium-induced changes in *Brassica napus* as revealed by the biochemical and ultra-structural evaluation of roots. Ecotox Environ Safe 92: 271–280.
17. Pattanayak GK, Tripathy BC (2011) Overexpression of protochlorophyllide oxidoreductase C regulates oxidative stress in Arabidopsis. PloS ONE 6: e26532.
18. Wang LJ, Jiang WB, Zhang Z, Yao QH, Matsui H, et al. (2003) 5-Aminolevinilic acid and its potential application in agriculture. Plant Physiol Commun 39: 185–192.
19. Dong YC, Liu X (2007) Crops and their wild relatives in China. China Agriculture Press, Beijing, China.
20. Chen MJ, Jia SX (2002) China feeding plant. China Agriculture Press, Beijing, China.
21. Fernández-Torquemada Y, Sánchez-Lizaso JL (2013) Effects of salinity on seed germination and early seedling growth of the mediterranean seagrass *Posidonia oceanica* (L.) delile. Estuar Coast Shelf S 119: 64–70.
22. Aroca R, Irigoyen JJ, Sánchez-Dıaz M (2001) Photosynthetic characteristics and protective mechanisms against oxidative stress during chilling and subsequent recovery in two maize varieties differing in chilling sensitivity. Plant Sci 161: 719–726.
23. Lichtenthaler HK (1987) Chlorophylls and carotenoids: pigments of photosynthetic biomembranes. Method Enzymol 148: 350–382.
24. Dhindsa RS, Plumb-dhindsa P, Thorpe TA (1981) Leaf senescence: correlated with increased levels of membrane permeability and lipid peroxidation and decreased levels of superoxide dismutase and catalase. J Exp Bot 32: 93–101.
25. Veljovic-Jovanovic S, Noctor G, Foyer CH (2002) Are leaf hydrogen peroxide concentrations commonly overestimated? the potential influence of artefactual interference by tissue phenolics and ascorbate. Plant Physiol Bioch 40: 501–507.
26. Elstner EF, Heupel A (1976) Inhibition of nitrite formation from hydroxylammoniumchloride: a simple assay for superoxide dismutase. Anal Biochem 70: 616–620.
27. Law MY, Charles SA, Halliwell B (1983) Glutathione and ascorbic-acid in spinach (*Spinacia oleracea*) chloroplasts-the effect of hydrogen-peroxide and of paraquat. Biochem J 210: 899–903.
28. Kampfenkel K, Van Montagu M, Inzé D (1995) Extraction and determination of ascorbate and dehydroascorbate from plant tissue. Anal Biochem 225: 165–167.
29. Bradford MM (1976) A rapid and sensitive method for the quantitation of microgram quantities of protein utilizing the principle of protein-dye binding. Anal Biochem 72: 248–254.
30. Nakano Y, Asada K (1981) Hydrogen peroxide is scavenged by ascorbate-specific peroxidase in spinach chloroplast. Plant Cell Physiol 22: 867–880.
31. Cakmak I, Marschner H (1992) Magnesium deficiency and high light intensity enhance activities of superoxide dismutase, ascorbate peroxidase, and glutathione reductase in bean leaves. Plant Physiol 98: 1222–1227.
32. Shaedle M, Bassham JA (1977) Chloroplast glutathione reductase. Plant Physiol. 59: 1011–1012.
33. Beauchamp C, Fridovich I (1971) Superoxide dismutase: Improved assays and an assay applicable to acrylamide gels. Anal Biochem 44: 276–287.
34. Zheng CF, Jiang D, Liu FL, Dai TB, Liu WC, et al. (2009) Exogenous nitric oxide improves seed germination in wheat against mitochondrial oxidative damage induced by high salinity. Environ Exp Bot 67: 222–227.
35. Palmgren MG, Askerlund P, Fredrikson K, Widell S, Sommarin M, et al. (1990) Sealed inside-out and right-side-out plasma membrane vesicles: optimal conditions for formation and separation. Plant Physiol 92: 871–880.
36. Ahn SJ, Im YJ, Chung GC, Cho BH (1999) Inducible expression of plasma membrane H⁺-ATPase in the roots of figleaf gourd plants under chilling root temperature. Physiol Plant 106: 35–40.
37. Harel E, Klein S (1972) Light dependent formation of 5-aminolevulinic acid in etiolated leaves of higher plants. Biochem Biophys Res Commun 49: 364–370.
38. Mittler R (2002) Oxidative stress, antioxidants and stress tolerance. Trends Plant Sci 7: 405–410.
39. Youssef T, Awad MA (2008) Mechanisms of enhancing photosynthetic gas exchange in date palm seedlings (*Phoenix dactylifera* L.) under salinity stress by a 5-aminolevulinic acid-based fertilizer. J Plant Growth Regul 27: 1–9.
40. Nishihara E, Kondo K, Parvez MM, Takahashi K, Watanabe K, et al. (2003) Role of 5-aminolevulinic acid (ALA) on active oxygen-scavenging system in NaCl-treated spinach (*Spinacia oleracea*). Plant Physiol 160: 1085–1091.
41. Janicka-Russak M, Kabała K, Wdowikowska A, Kłobus G (2012) Response of plasma membrane H⁺-ATPase to low temperature in cucumber roots. J Plant Res 125: 291–300.
42. Gunes A, Inal A, Bagci EG, Coban S, Sahin O (2007) Silicon increases boron tolerance and reduces oxidative damage of wheat grown in soil with excess boron. Biol Plant 51: 571–574.
43. Naeem MS, Rasheed M, Liu D, Jin ZL, Ming DF, et al. (2011) 5-Aminolevulinic acid ameliorates salinity-induced metabolic, water-related and biochemical changes in *Brassica napus* L. Acta Physiol Plant 33: 517–528.
44. Liu D, Wu LT, Naeem MS, Liu HB, Deng XQ, et al. (2013) 5-Aminolevulinic acid enhances photosynthetic gas exchange, chlorophyll fluorescence and antioxidant system in oilseed rape under drought stress. Acta Physiol Plant 35: 2747–2759.
45. Smirnoff N (2005) Antioxidants and reactive oxygen species in plants. Blackwell Publishing, Oxford, UK.
46. Foyer CH, Theodoulou FL, Delrot S (2001) The functions of inter- and intracellular glutathione transport systems in plants. Trends Plant Sci 6: 486–492.
47. Schonhof I, Kläring HP, Krumbein A, Claußen W, Schreiner M (2007) Effect of temperature increase under low radiation conditions on phytochemicals and ascorbic acid in greenhouse grown broccoli. Agr Ecosyst Environ 119: 103–111.
48. Liu ZJ, Zhang XL, Bai JG, Suo BX, Xu PL, et al. (2009) Exogenous paraquat changes antioxidant enzyme activities and lipid peroxidation in droughtstressed cucumber leaves. Sci Hortic-Amsterdam 121: 138–143.
49. Bhaya D, Castelfranco PA (1985) Chlorophyll biosynthesis and assembly into chlorophyll protein complexes in isolated developing chloroplasts. Proc Natl Acad Sci USA 82: 5370–5374.

50. Farooq M, Wahid A, Kobayashi N, Fujita D, Basra SMA (2009) Plant drought stress: effects, mechanisms and management. Agron Sustain Dev 29: 185–212.

51. Xu YF, Sun XL, Jin JW, Zhou H (2010) Protective effect of nitric oxide on light-induced oxidative damage in leaves of tall fescue. J Plant Physiol 167: 512–518.

52. Logan BA, Demmig-Adams B, Adams III WW, Grace SC (1998) Antioxidants and xanthophyll cycle-dependent energy dissipation in *Cucurbitapepo* L. and *Vinca* major acclimated to four growth PPFDs in the field. J Exp Bot 49: 1869–1879.

53. Wang JJ, Jiang WB, Liu H, Liu WQ, Kang L, et al. (2005) Promotion by 5-aminolevulinic acid of germination of pakchoi (*Brassica campestris* ssp. *Chinensis* var. *communis* Tsen et Lee) seeds under salt stress. J Integ Plant Biol 47: 1084–1091.

54. Jisha KC, Vijayakumari K, Puthur Jos T (2013) Seed priming for abiotic stress tolerance: an overview. Acta Physiol Plant 35: 1381–1396.

55. Eisvand HR, Tavakkol-Afshari R, Sharifzadeh F, Maddah Arefi H, Hesamzadeh Hejazi SM (2010) Effects of hormonal priming and drought stress on activity and isozyme profiles of antioxidant enzymes in deteriorated seed of tall wheatgrass (*Agropyron elongatum* Host). Seed Sci Technol 38: 280–297.

56. Affourtit C, Krab K, Moore AL (2001) Control of plant mitochondrial respiration. Biochim Biophys Acta 1504: 58–69.

57. Petrussa E, Casolo V, Peresson C, Krajňáková J, Macri F, et al. (2008) Activity of a K_{ATP}^+ channel in *Arum* spadix mitochondria during thermogenesis. J Plant Physiol 165: 1360–1369.

58. Palmgren MG (2001) Plant plasma membrane H^+-ATPases: powerhouses for nutrient uptake. Annu Rev Plant Physiol Plant Mol Biol 52: 817–845.

59. Kim HS, Oh JM, Luan S, Carlson JE, Ahn SJ (2013) Cold stress causes rapid but differential changes in properties of plasma membrane H^+-ATPase of camelina and rapeseed. J Plant Physiol 170: 828–837.

4

Inhibition of Catalase by Tea Catechins in Free and Cellular State: A Biophysical Approach

Sandip Pal, Subrata Kumar Dey, Chabita Saha*

Department of Biotechnology, West Bengal University of Technology, Kolkata, West Bengal, India

Abstract

Tea flavonoids bind to variety of enzymes and inhibit their activities. In the present study, binding and inhibition of catalase activity by catechins with respect to their structure-affinity relationship has been elucidated. Fluorimetrically determined binding constants for (−)-epigallocatechin gallate (EGCG) and (−)-epicatechin gallate (ECG) with catalase were observed to be 2.27×10^6 M^{-1} and 1.66×10^6 M^{-1}, respectively. Thermodynamic parameters evidence exothermic and spontaneous interaction between catechins and catalase. Major forces of interaction are suggested to be through hydrogen bonding along with electrostatic contributions and conformational changes. Distinct loss of α-helical structure of catalase by interaction with EGCG was captured in circular dichroism (CD) spectra. Gallated catechins demonstrated higher binding constants and inhibition efficacy than non-gallated catechins. EGCG exhibited maximum inhibition of pure catalase. It also inhibited cellular catalase in K562 cancer cells with significant increase in cellular ROS and suppression of cell viability (IC$_{50}$ 54.5 μM). These results decipher the molecular mechanism by which tea catechins interact with catalase and highlight the potential of gallated catechin like EGCG as an anticancer drug. EGCG may have other non-specific targets in the cell, but its anticancer property is mainly defined by ROS accumulation due to catalase inhibition.

Editor: Heidar-Ali Tajmir-Riahi, University of Quebect at Trois-Rivieres, Canada

Funding: Mr. Sandip Pal is financially supported by Senior Research Fellowship (SRF) Grant of Indian Council of Medical Research (ICMR). Dr. Chabita Saha is thankful to Department of Science and Technology (DST), Govt. of India for the financial support. The funders had no role in study design, data collection and analysis, decision to publish, or preparation of the manuscript.

Competing Interests: The authors have declared that no competing interests exist.

* Email: k.chabita@gmail.com

Introduction

Green tea polyphenols have received wide attention for their beneficial health effects. Catechins have been effective for cancer prevention studies [1–3]. The major catechins copiously present in tea extract, especially in green tea, are (−)-epicatechin (EC), (−)-epigallocatechin (EGC), (−)-epicatechin gallate (ECG), and (−)-epigallocatechin gallate (EGCG) as illustrated in Figure 1. The major anticancer activities of tea catechins are as antioxidants, pro-oxidants and enzyme inhibitors [4–8]. Antioxidant activity of tea polyphenols has found wide application in radioprotection and chemoprevention by scavenging reactive oxygen species (ROS) [9–13]. Galloylated catechins, especially EGCG, are known to inhibit growth of cancer cells and induce apoptosis in various types of tumor cells due to their pro-oxidant activity [14–17]. Various mechanisms may be associated with the pro-oxidant behavior of flavonoids in cancer cells, of which enzyme inhibition is a major process.

The ability of flavonoids to bind and inhibit some vital cellular enzymes leading to suppression of cell proliferation is being investigated widely [18–22]. EGCG is known to bind to proteins like salivary proline-rich proteins, fibronectin, fibrinogen and histidine rich glycoproteins [23]. Caseins and lactoglobulins are milk proteins which rendered the antioxidant activity of tea polyphenols upon binding with catechins [24–27]. EGCG binds to Bcl-2 proteins with inhibition constant (K_i) 0.33–0.49 μM and to vimentin and G3BP1 with dissociation constant (K_d) 3.3 nM and 0.4 μM, respectively [28–30]. In the present study, binding of catechins to catalase – an enzyme which maintains the cellular ROS levels is followed fluorimetrically and thermodynamically characterized.

Catalase is an antioxidant oligomeric enzyme (MW 2,40,000) with four identical subunits arranged tetrahedrally [31–33]. Each subunit consists of a single polypeptide chain that associates with a prosthetic group, ferric protoporphyrin IX [34]. It is a very important enzyme in protecting cells from oxidative damage by ROS. Drugs bind to enzymes and elicit enzyme inhibition; one such example is catalase inhibition by wogonin led to H$_2$O$_2$ accumulation and cytotoxicity in cancer cells through H$_2$O$_2$-mediated NF-κB suppression and apoptosis activation [35]. Similarly, EGCG at concentrations between 5–20 μM also inhibited phosphorylation of JNK, JUN, MEK1, MEK2 in JB6 epidermal cell lines [36]. In MCF7 breast cancer cell lines, cyclin dependent kinase 2 (CDK2) and CDK4 were reported to be inhibited by 30 μM EGCG [37]. In KYSE 510 human esophageal cancer cells, EGCG inhibited DNA methyltransferase with K_i 7 μM [38]. All these enzymes have relevance in cancer prevention. Inhibition of human cancer cell growth by tea polyphenols has been observed in H1299, H661, HT-29, H441 and breast cancer cell lines at IC$_{50}$ values between (20–75 μg/ml) [39]. Catalase is another cellular enzyme whose interaction with microsystin, a cyanotoxin drug, decreases its enzymatic action [40]. Here its

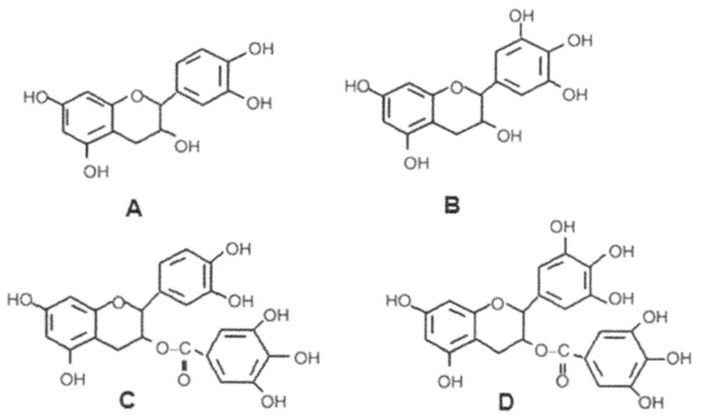

Figure 1. Chemical structures of four catechins: (A) EC, (B) EGC, (C) ECG, and (D) EGCG.

inhibition in K562 cancer cells is reported and its relevance in cancer is demonstrated by the suppression of cell viability. The study also highlights the influence of structure of the catechins on the inhibition of catalase activity by catechins.

As endogenously formed ROS are important in promoting carcinogenesis, tea polyphenols may have important role in quenching them and also tea polyphenols being redox active can under go auto-oxidation and produce ROS in media and in mitochondria [41,42]. This makes activity of tea polyphenols in cancer prevention very complex which needs active investigation. Changes in ROS population are expected manifestation of catalase inhibition and have been followed to monitor the enzyme activity on EGCG treatment.

Materials and Methods

Chemicals

Catalase (from bovine liver), EC, ECG, EGC, EGCG, dimethyl sulfoxide (DMSO), 2′,7′-dichloroflurescein diacetate (DCFDA) were obtained from Sigma-Aldrich, St. Louis, MO, USA. 3-amino-1,2,4-triazole (3-AT), 3-(4,5-dimethylthiazol-2-yl)-2, 5-diphenyltetrazolium bromide (MTT) were procured from Himedia, Mumbai, India. RPMI 1640 media, fetal bovine serum (FBS), penicillin-streptomycin were obtained from Gibco, Grand Island, NY. All the experiments were carried out in 50 mM phosphate buffer of pH 7.4.

Cell culture

K562 (human chronic myeloid leukemia cell line) was procured from NCCS, Pune and grown in RPMI 1640 supplemented with 10% FBS, 100 U/ml of penicillin and 100 μg/ml of streptomycin in 25 cm^2 T-flasks (Nunc, Roskilde, Denmark) with vented caps and incubated at 37°C in a humidified atmosphere of 5% CO_2 in air.

Fluorescence spectroscopy

Fluorescence spectra were recorded using a Perkin Elmer fluorescence spectrophotometer (LS-55) equipped with 150 W Xenon flash lamp and using fluorescence-free quartz cell of 1 cm path length. The widths of excitation and emission slits were set to 5 nm, and scan speed (100 nm/min), excitation voltage were kept constant for each data set. Quantitative analysis of the potential interaction between catechins and catalase was performed by the fluorimetric titration technique. Briefly, solution of catalase (5×10^{-7} M, as calculated using molar extinction coefficient of

3.24×10^5 M^{-1} s^{-1} at 405 nm for catalase) in 50 mM phosphate buffer of pH 7.4 was titrated in cuvette by successive additions of individual catechin solution aliquots from a stock of 5×10^{-7} M. Fluorescence emission spectra were recorded in the wavelength range of 290–450 nm upon excitation at 280 nm when catalase samples were titrated with catechins. All experiments were carried out at room temperature. Fluorescence intensity was measured at 340 nm of protein emission spectra. Fluorescence spectra of individual catechins in buffer were recorded as blanks under the same experimental conditions and subtracted from the corresponding sample to correct the fluorescence background.

Isothermal titration calorimetry (ITC)

The energetics of the binding of catechins to catalase was studied by ITC using a MicroCal ITC_{200}, (Northampton, MA, USA). All solutions were degassed under vacuum (140 mbar, 10 min) on the MicroCal's Thermovac unit to eliminate air bubble formation inside the calorimeter cell. Briefly, the calorimeter syringe was filled with a concentrated solution of catechins (10 μM each). Successive injections of this solution into 1 μM solution of catalase in the calorimeter cell were effected from the rotating syringe with constant stirring of the solution. The titration and analysis were performed through Origin 7 software provided with the unit.

Circular dichroism (CD) spectroscopy

All the CD experiments were carried out on a Jasco-815 automatic spectropolarimeter (Jasco International Co., Ltd. Hachioji, Japan) equipped with a peltier cuvette holder and temperature controller PFD425 L/15. The catalase concentration and path length of the cuvette used were 1 μM and 0.1 cm, respectively. The instrument parameters were set at scanning speed of 50 nm/min, bandwidth of 1.0 nm and sensitivity of 100 milli degree. The molar ellipticity values are expressed in terms of mean residue molar ellipticity, in units of deg. cm^2 $dmol^{-1}$.

Inhibition of catalase activity

Absorption spectrophotometer (Varian, CARY 100 Bio, USA) was used to determine pure catalase activity (cell free). Briefly, 900 μl of H_2O_2 (0.036%) was taken in quartz cell of 1 cm path length and its absorbance was recorded at 240 nm. To the solution, 100 μl of catalase (50 U/ml) was added and the decrease in absorbance was recorded at 10 s interval. This experiment was performed in presence 50 μM each of four catechins and 3-AT separately to evaluate the inhibitory effect of catechins on catalase activity, after 1 h incubation.

Estimation of cellular catalase activity was carried out by suspending K562 cells (5×10^4 for each sample) in 1 ml of PBS along with four catechins and 3-AT (50 μM each) separately and incubated at 37°C for 2 hours. Later, each sample was centrifuged and washed twice with PBS. The final pellet was suspended in 200 μl lysis buffer and kept on ice for 30 min. The lysate was then used in the same protocol used for evaluation of pure catalase activity.

Analysis of cell viability

Cell viability of K562 cells was quantified by MTT based colorimetric assay as described elsewhere [43]. Briefly, cells were rinsed with PBS and 0.5 mg/ml MTT was added into each sample. The mixture was incubated at 37°C for an additional 3 h. At the end of incubation period, the medium containing MTT was removed and the pellet was dissolved in 200 μl DMSO.

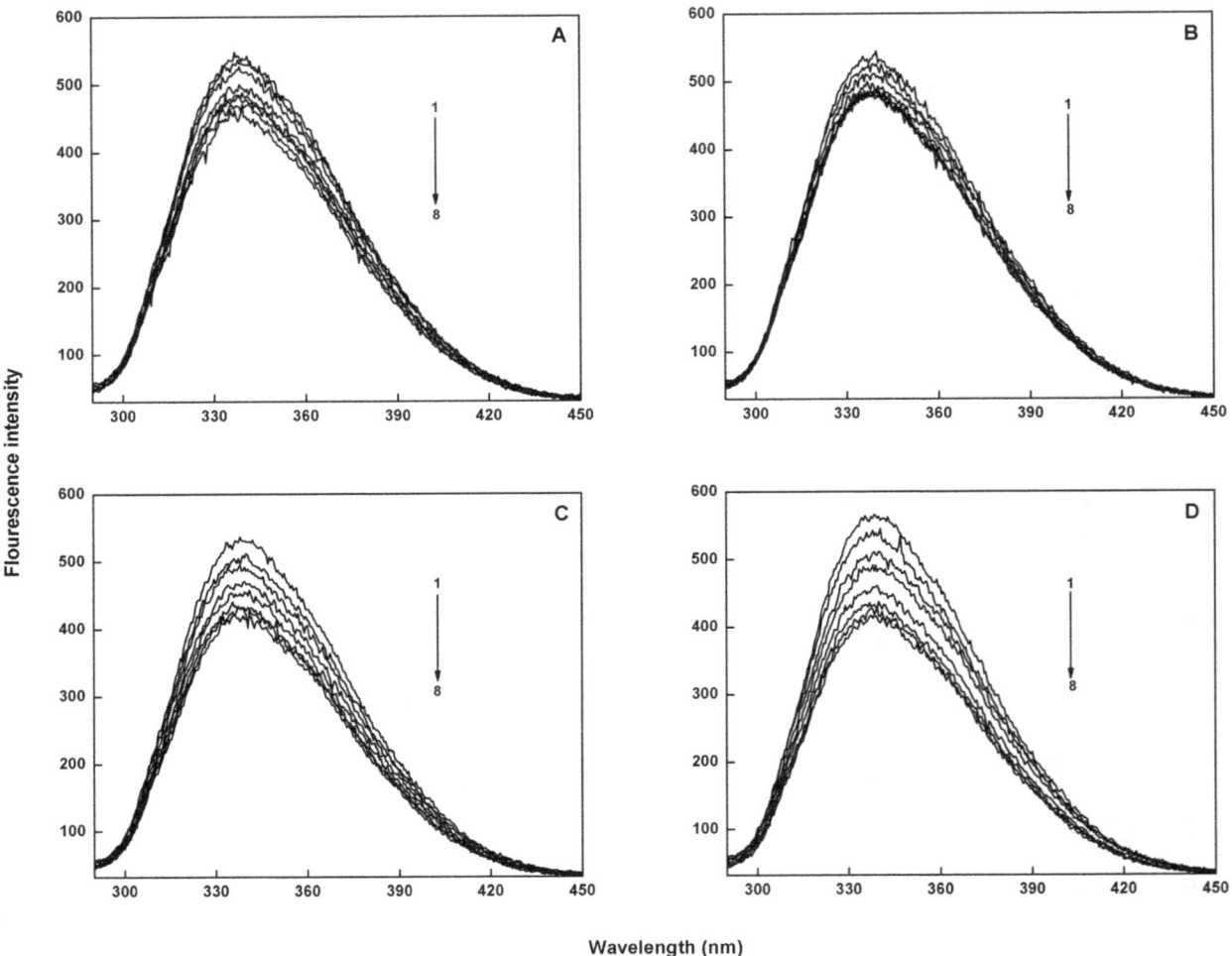

Figure 2. The fluorescence emission spectra of native catalase (5×10^{-7} M) and the quenching effect of catechins on its fluorescence intensity at 25°C ($\lambda_{ex} = 280$ nm): (A) EC-catalase, (B) EGC-catalase, (C) ECG-catalase, and (D) EGCG-catalase. Curves (1 to 8) denote 2.5×10^{-9} M to 1.7×10^{-8} M of catechins.

Absorbance was measured at a wavelength of 570 nm. Cell viability was expressed as a percentage of the control culture.

Measurement of intracellular ROS

Intracellular ROS were estimated by using the DCFDA fluorescent probe [9]. Intracellular H_2O_2 and other peroxides can oxidize DCFDA to highly fluorescent compound dichloro-fluorescein (DCF). K562 cells were incubated at 37°C in absence and presence of EGCG (12.5 – 100 µM) for 24 h. Cells were again incubated with 10 µM DCFDA at 37°C for an additional 30 min, and then washed twice with PBS. Finally the fluorescence intensity of DCF was measured with an excitation wavelength of 485 nm and emission wavelength of 530 nm.

Statistical analysis

Data were analyzed by Origin Software (Version 8) for Windows. The statistical analysis of the samples was undertaken using Student's *t*-test. All data reported are means ± standard deviations for three independent experiments, unless otherwise noted.

Results

Quenching of catalase fluorescence by catechins

Catalase shows a fluorescence emission peak at 340 nm with the excitation wavelength at 280 nm, mainly due to the presence of tryptophan and tyrosine residues [44]. This emission peak exhibited progressive decrease in intensity on addition of catechins suggesting interaction between catalase and catechins (Figure 2 A–D). The binding constants and the number of binding sites involved were calculated according to the relation: log $(F_0-F)/F$ = log K+n log [Q], where K is the binding constant and n is the number of binding sites, F_0 is the fluorescence intensity of free catalase, F is the consecutive fluorescence on addition of catechins, and [Q] is the concentration of the quencher (here catechins). A plot of log $(F_0- F)/F$ and log [Q] was used to determine the values of K and n from the intercept and the slope respectively (Figure 3 A–D). The binding constants of EC, EGC, ECG and EGCG with catalase are calculated to be 1.08×10^5, 7.9×10^4, 1.66×10^6, and 2.27×10^6 M^{-1}, respectively with single binding site.

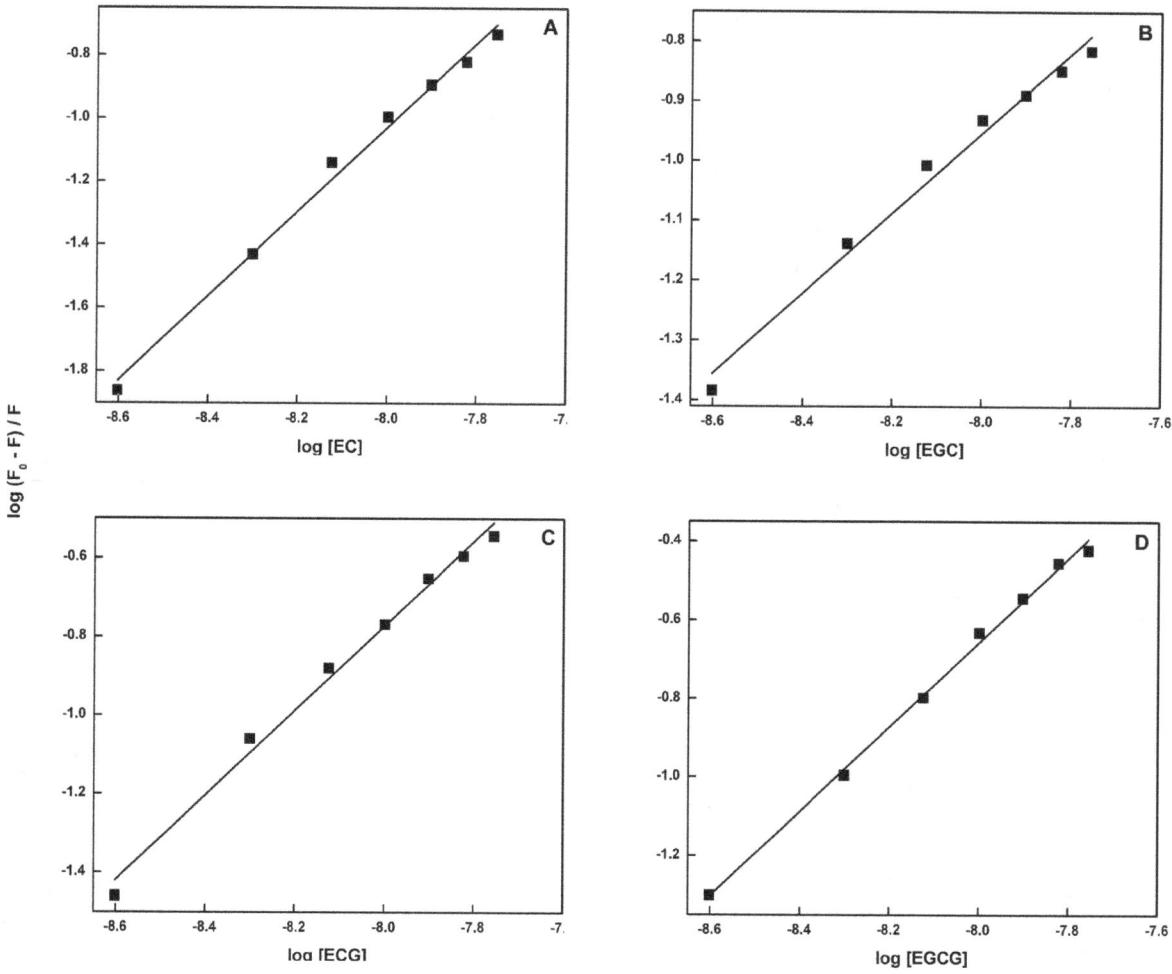

Figure 3. Double logarithmic plot of catalase-catechins system at 25°C (λ_{ex} = 280 nm). (A) EC-catalase, (B) EGC-catalase, (C) ECG-catalase, and (D) EGCG-catalase.

Calorimetric characterization of catalase-catechins interaction

ITC directly measures the heat released during a chemical reaction. ITC can reveal the stoichiometry, enthalpy free energy, and entropy changes that occur over the course of a reaction [45,46]. Data obtained from the titration are presented by the series of peaks corresponding to each injection. Usually, this representation is transformed into the apparent heat change between two injections as a function of the titrant ratio by means of integration of power differential with regard to time. ITC has been used mainly to quantify interactions in various biochemical systems [47–49]. ITC has been used in the present study to determine binding affinity and other thermodynamic parameters of catechins-catalase interaction. Representative isothermal calorimetric heat profiles for the binding of catechins-catalase interaction, at physiological pH and 25°C are shown in Figure 4 (A–D). The data were corrected for the heat of dilution of catechins, which was determined in a separate set of experiments under identical conditions. The titration and analysis were performed using one-site binding model through Origin 7 software provided with the unit. The thermodynamic parameters recorded for the binding of catechins (10 μM each) with catalase (1 μM) are summarized in Table 1. From Figure 4 it is observed that the

titration of catalase with all the four catechins yield negative heat deflections. EC, EGC, ECG, and EGCG bind to catalase with binding constants 2.13×10^5, 1.92×10^5, 6.36×10^5, and 8.19×10^5 M^{-1}, respectively, as determined by the following equation: $\Delta G = -RT \ln K$, where ΔG is the free energy, R and T are the gas constant and temperature, respectively, K is the binding constant.

Catechin induced conformational changes in catalase

CD allows investigation of the conformational changes that occur in a protein upon ligand binding [50]. The CD spectra of catalase at pH 7.4 in the absence and in the presence of EGC, EC, ECG, EGCG, and 3-AT at 25°C are represented in Figure 5 (a – f). The CD spectra of catalase exhibited two negative bands in the far-UV region at 208 and 220 nm, characteristic of an α-helical structure of protein [51]. The negative band at 208 and 220 nm decreased in intensity with the addition of various catechins, which is indicative of the loss of α-helicity upon interaction. The CD spectra of catalase in presence and absence of catechins were observed to be similar in shape, indicating that the structure of catalase is predominantly α-helical [52]. From the CD spectra, the effect of catechins on the secondary structure of catalase is in the following order: 3-AT>EGCG>ECG>EC>EGC.

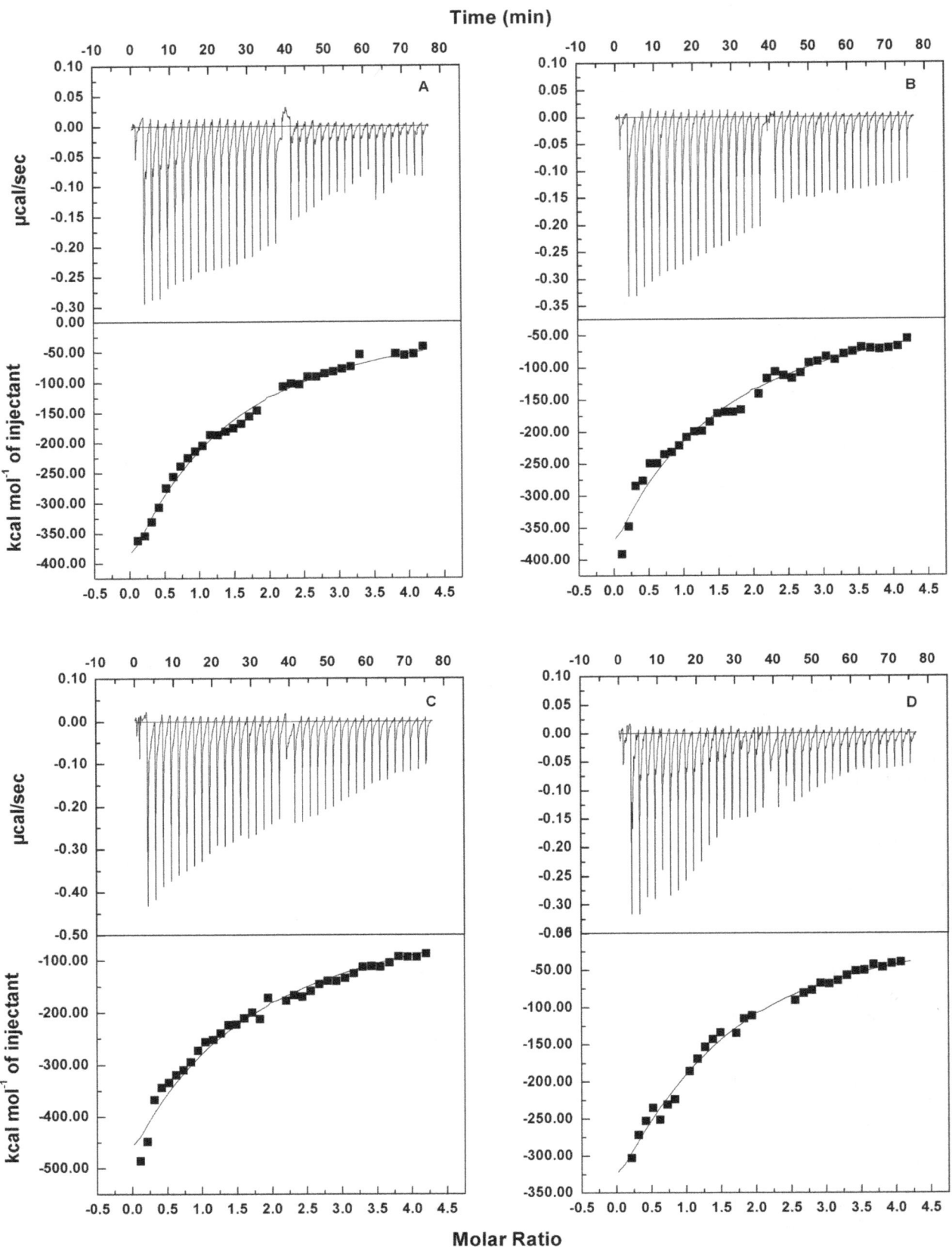

Figure 4. ITC profiles for catalase (1 μM) when titrated with catechins (10 μM each) at 25°C: (A) EC-catalase, (B) EGC-catalase, (C) ECG-catalase, and (D) EGCG-catalase systems.

Table 1. Thermodynamic parameters from ITC experiments for catalase-catechins system at 25°C.

System	ΔG (cal/mol)[a]	ΔH (cal/mol)[b]	ΔS (cal/mol/K)[b]
EGCG – Catalase	-7.95×10^3	$-4.50\times10^5\pm2.25\times10^4$	-1.48×10^3
ECG – Catalase	-7.80×10^3	$-7.56\times10^5\pm3.02\times10^4$	-2.60×10^3
EC – Catalase	-7.17×10^3	$-7.31\times10^9\pm2.15\times10^8$	-2.40×10^7
EGC – Catalase	-7.10×10^3	$-3.20\times10^9\pm1.06\times10^8$	-1.07×10^7

[a]Calculated from the average values of ΔH.
[b]Directly extracted from the experiment.

Inhibition of catalase activity

The spectroscopic approach for determination of catalase activity on H_2O_2 degradation showed continuous breakdown of its substrate, H_2O_2 with time (upto 100 s) at 240 nm. On addition of catechins, a decrease in rate of H_2O_2 degradation i.e. a decrease in pure catalase activity was observed which is represented in Figure 6A. Catechins with galloyl moiety (ECG and EGCG) reduced catalase activity more than its non-gallated counterparts i.e. EGC and EC. However, Inhibition by EGCG is the closest to inhibition by known catalase inhibitor 3-AT. The order of inhibition of catalase activity was found to be 3-AT>EGCG> ECG>EC>EGC. The same experiment with cellular catalase in K562 cells revealed inhibition of cellular catalase activity in the presence of catechins (Figure 6B). The trend of inhibition in this condition was observed to be similar to that of pure catalase. The double reciprocal plot or Lineweaver-Burk plot shows that the inhibition of catalase by EGCG is of uncompetitive type (Figure 7). The y-intercept i.e. $1/V_0$ value for uninhibited and inhibited catalase were found to be 0.0081 and 0.3635 μM^{-1} min, respectively.

MTT assay for cell viability

The concentration (IC_{50}) at which the K562 cell survivability was reduced to 50% was calculated for EGCG and 3-AT by MTT based cell viability assay. The IC_{50} value obtained for EGCG and 3-AT was 54.5 μM and 10.5 μM, respectively.

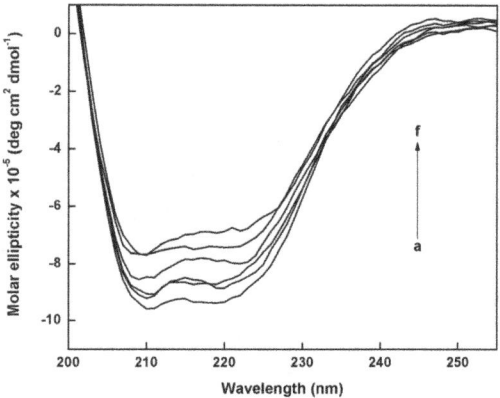

Figure 5. The CD spectra of catalase (1 μM) and its complexes with catechins and 3-AT (3 μM each) at 25°C. The curves (a to f) represent pure catalase, EGC-catalase, EC-catalase, ECG-catalase, EGCG-catalase, and 3-AT-catalase complexes, respectively.

Measurement of intracellular ROS

The basal level of intracellular ROS in K562 cells (255.09±3.69 AU) is increased to 298.01±4.89, 320.83±4.03, 354.07±4.66, and 442.90±7.43 AU in presence of 12.5, 25, 50, and 100 μM concentration of EGCG (Figure 8).

Discussion

Catechins are antioxidants ubiquitously found in green tea and among them EGCG shows highest antioxidant activity that may have therapeutic applications in cancer treatment. EGCG functions as a powerful antioxidant, preventing oxidative damage in healthy cells, but also as an anti-angiogenic and antitumor agent and as a modulator of tumor cell response to chemotherapy. However, neither pro-oxidant nor antioxidant activities have yet been clearly established to occur *in vivo* in humans [53]. Natural polyphenols are reported to be good inhibitors of human dihydrofolate reductase (DHFR) could explain the epidemiological data on their prophylactic effects for certain forms of cancer and open a possibility for the use of natural and synthetic polyphenols in cancer chemotherapy [54].

Fluorescence quenching experiments demonstrated that the ester bond containing tea polyphenols EGCG and ECG are effective inhibitors of DHFR with K_d of 0.9 and 1.8 μM, respectively, while polyphenols lacking the ester bond containing galloyl moiety (e.g., EGC and EC) did not bind to this enzyme [54]. The binding constant of EGCG to catalase was determined fluorimetrically to be 2.27×10^6 M^{-1} (K_d of 0.44 μM). Other catechins were not as effective as EGCG establishing the influence of galloyl moiety for efficient binding. The binding efficiency is attributed to the eight phenolic groups of EGCG which serve as hydrogen bond donors. Involvement of positive charge on the carbon atom of the ester bond with the electronegative groups of amino acid residues like histidine also contributes towards efficient binding. Here it can be reasoned that at this concentration of EGCG it can also bind to other non-specific targets *in vivo* due to the presence of OH groups. To this end, it can be emphasized that binding to catalase inhibits its activity and increases the ROS population which eventually triggers apoptosis. It is known that 3-AT (an efficient inhibitor of catalase) binds with histidine residue (His75) near heme group of catalase forming a non-coplanar adduct (very close to Tyr358). Loss in fluorescence intensity of catalase suggests involvement of EGCG cation with histidine anion (pKa = 6.5) [55] near tyrosine residue (Tyr358) of active site [56]. This identification of the important structural features responsible for inhibition will provide valuable information for designing new inhibitors.

ITC provides valuable information on biomacromolecule-ligand interactions [57,58]. This technique measures the heat released or absorbed (ΔH) when the macromolecule binds to

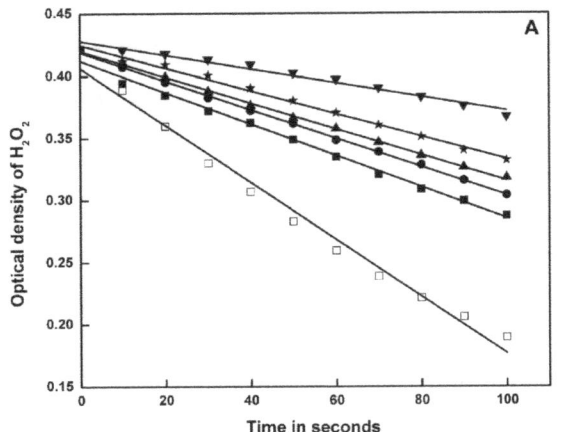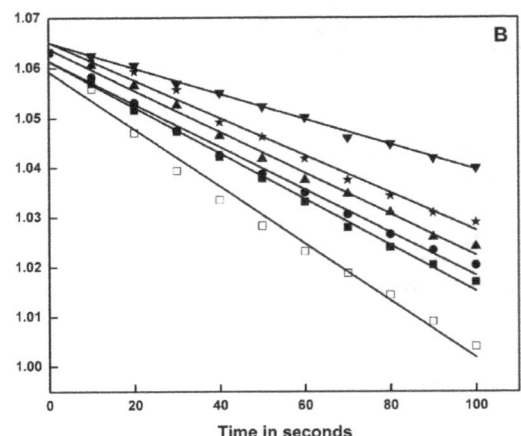

Figure 6. Decay curves of H$_2$O$_2$ by pure catalase (A) and by cellular catalase (B) in absence (□) and presence of EGC (■), EC (●), ECG (▲), EGCG (*), 3-AT (▼) as recorded by spectrophotometer at 240 nm. Initial H$_2$O$_2$ concentration was approximately 0.01 mM.

ligands. This value of ΔH is used to determine other thermodynamic parameters. The thermodynamic parameters measured for the catalase-catechins interactions are summarized in Table 1. From the data, it is evident that the interactions of all four catechins with catalase are spontaneous and exothermic, which is confirmed by the negative values of ΔG. For all the catechins the favorable ΔH and unfavorable $T\Delta S$ values signify dominant forces of interaction to be hydrogen bonding with electrostatic contributions involving carbon cation of galloyl moiety. The catechins devoid of this group show lower affinity to catalase confirming the contribution of the cation towards electrostatic interactions with polar groups of the protein; which on binding are more exposed to interact with the galloyl moiety. The binding constant of EGCG with catalase determined from the thermodynamic parameters as 8.19×10^5 M^{-1} using one-site binding model, is little lower than that obtained from fluorimetric measurements with one binding site. This is because the values observed in fluorescence spectroscopy are usually related to excited state complexes and ITC measures the ground state complexes. The binding constant for catalase-microcystin complex determined fluorimetrically is 6.12×10^4 M^{-1} [40], which is lower than catalase-EGCG complex. Microcystin binding to catalase was reported to influence its physiological functions and conformation [40]. Binding is stronger

with EGCG and more changes are observed in catalase-catechins complexes. Loss of α-helix is identified in the CD spectra by the decrease in negative peak at 222 nm and 208 nm (signature peaks of α-helix) on addition of EGCG as shown in Figure 5. Gain in β-sheet is observed from the increase in negative peak at 218 nm against 222 nm for α-helix. These conformational changes in catalase secondary structure on interaction with catechins influence its active site and physiological functions, rendering it inactive in scavenging ROS.

In the present study, the physiological changes are demonstrated by the inhibition of catalase activity by catechins. EGCG is observed to inhibit more efficiently than ECG and other non-gallated catechins. From the decay curves as shown in Figure 6, it is inferred that the rate of H$_2$O$_2$ degradation is reduced by catechins in cell free (Figure 6A) and cellular state (Figure 6B). From the Lineweaver-Burk plot (Figure 7), it is evident that the maximum inhibition of catalase is achieved by EGCG. The plot also demonstrates uncompetitive type of inhibition suggesting that EGCG binds to catalase at a site other than the active site. As a consequence, the uncompetitive inhibitor (here EGCG) lowers the measured maximum velocity (V$_{max}$) and also decreases apparent K$_m$ value [59]. This inhibition in catalase activity results in

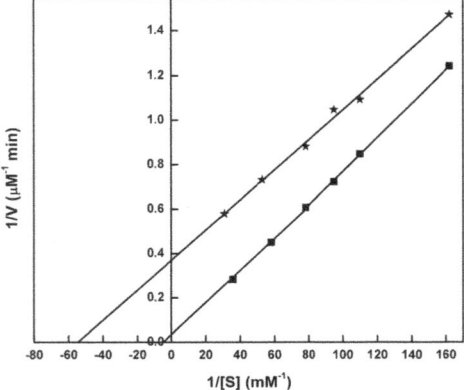

Figure 7. Double reciprocal plot or Lineweaver-Burk plot of catalase – H$_2$O$_2$ (■) and catalase+EGCG – H$_2$O$_2$ (*) reactions.

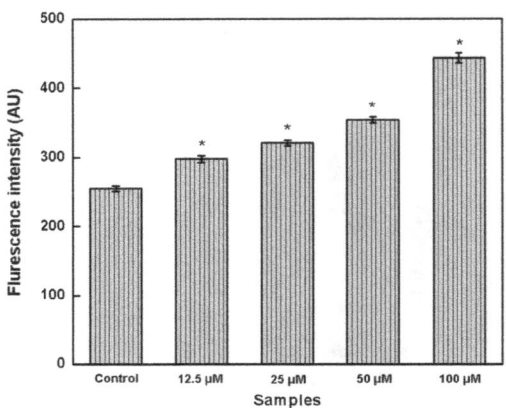

Figure 8. Histograms representing the intracellular ROS content in absence and presence of various concentrations (12.5–100 μM) of EGCG. (*$p < 0.05$ as compared to control group).

accumulation of ROS as illustrated in Figure 8. Higher levels of ROS create oxidative stress which is detrimental to cellular integrity.

The IC_{50} value for suppression of cell viability in K562 cells on EGCG treatment is determined to be 54.5 μM (or 27 μg/ml). Inhibition of human cancer cell growth by tea polyphenols has been observed in H1299, H661, HT-29, H441 and breast cancer cell lines at IC_{50} values between 20–75 μg/ml [39]. Higher EGCG concentration is required in the cell to elicit cellular response than its K_d value to catalase. Here it can be emphasized that suppression of cancer cell viability is not envisaged only by catalase inhibition but other pathways contribute to it. EGCG is capable of generating H_2O_2 in solution by auto-oxidation which contributes to oxidative stress. Superoxide is also reported to be generated in EGCG containing solutions [60]. In such conditions, inhibition of catalase by EGCG further increase the oxidative stress triggering apoptosis and eventually leading to cell death. These phenomena can be therapeutically used for cancer prevention. These revelations suggest that a balance between all the major activities of EGCG like antioxidants, pro-oxidants and enzyme inhibitors leads to cellular suppression, which is again cell specific. Inhibition of cellular enzymes by catechins has found major pharmacological applications as anticancer drugs [61]. Addition of methyl groups to ECG enhanced binding to DHFR and is a new approach to find novel inhibitors from natural resources [54]. A concentration of 27 μg/ml is high for EGCG to be bioavailable by tea consumption. Here EGCG is then treated as drug that can be delivered in various forms; one of which could be nano-formulations.

Conclusion

The efficacy of EGCG to bind with catalase and render its activity in cancer cells definitely represents a fascinating tool in the field of oncology. The present work sheds light on the critical structural features of galloylated catechins and identifies probable mechanism by which they inhibit catalase. These findings provide guidance for designing efficient catalase inhibitors. However, a cancer cell death event is an endpoint of several pathways and further studies in this context are in progress to completely understand the role of catechins in anticancer therapy. This study also demonstrates that natural polyphenols could be used as 'guide compounds' for development of new anticancer drugs.

Acknowledgments

The authors are thankful to Dr. Abhijit Saha at University Grant Commission-Department of Atomic Energy (UGC-DAE) Center for Scientific Research for providing us the ITC facility and his constant cooperation throughout the work. Thanks are also extended to Ms. Somrita Mondal at UGC-DAE Center for her technical support and cooperation during ITC experiments. We are grateful to Central Scientific Service Department of Indian Association for the Cultivation of Science (IACS), Kolkata for providing CD facility.

Author Contributions

Conceived and designed the experiments: CS. Performed the experiments: SP. Analyzed the data: SP CS. Contributed reagents/materials/analysis tools: CS SKD. Contributed to the writing of the manuscript: SP CS.

References

1. Trevisanato SI (2000) Tea and health. Nutr Rev 58: 1–10.
2. Crespy V, Williamson G (2004) A review of the health effects of green tea catechins in vivo animal models. J Nutr 134: 3431S–3440S.
3. Siddiqui AI, Adhami VM, Saleem M, Mukhtar H (2006) Beneficial effects of tea and its polyphenols against prostate cancer. Mol Nutr Food Res 50: 130–143.
4. Procházková D, Boušová I, Wilhelmová N (2011) Antioxidant and prooxidant properties of flavonoids. Fitoterapia 82: 513–523.
5. Oikawa S, Furukawa A, Asada H, Hirakawa K, Kawanishi S (2003) Catechins induce oxidative damage to cellular and isolated DNA through the generation of reactive oxygen species. Free Radic Res 37: 881–890.
6. Halliwell B, Murcia A, Chirico S, Aruoma OI (1995) Free radicals and antioxidants in food and in vivo: what they do and how they work. Crit Rev Food Sci 35: 7–20.
7. Yamashita N, Murata M, Inoue S, Burkitt MJ, Milne L, et al. (1998) Alpha-tocopherol induces oxidative damage to DNA in the presence of copper (II) ions. Chem Res Toxicol 11: 855–862.
8. Yamashita N, Kawanishi S (2000) Distinct mechanisms of DNA damage in apoptosis induced by quercetin and luteolin. Free Radic Res 33: 623–633.
9. Pal S, Saha C, Dey SK (2013) Studies on black tea (Camellia sinensis) extract as a potential antioxidant and a probable radioprotector. Radiat Environ Biophys 52: 269–278.
10. Ghosh D, Pal S, Saha C, Chakrabarti AK, Datta SC, et al. (2012) Black tea extract: a supplementary antioxidant in radiation induced damage to DNA and normal lymphocytes. J Environ Pathol Toxicol Oncol 31: 1–12.
11. Higdon JV, Frei B (2003) Tea catechins and polyphenols: health effects, metabolism, and antioxidant functions. Crit Rev Food Sci 43: 89–143.
12. Richi B, Kale RK, Tiku AB (2012) Radio-modulatory effects of green tea catechin EGCG on pBR322 plasmid DNA and murine splenocytes against gamma-radiation induced damage. Mutat Res 747: 62–70.
13. Nair CKK, Salvi VP (2008) Protection of DNA from gamma-radiation induced strand breaks by epicatechin. Mutat Res 650: 48–54.
14. Yang GY, Liao J, Kim K, Yurkow EJ, Yang CS (1998) Inhibition of growth and induction of apoptosis in human cancer cell lines by tea polyphenols. Cacinogenesis 19: 611–616.
15. Cao YH, Cao RH (1999) Angiogenesis inhibited by drinking tea. Nature 398: 381.
16. Ahmad N, Feyes DK, Nieminen AL, Agarwal R, Mukhtar H (1997) Green tea constituent epigallocatechin-3-gallate and induction of apoptosis and cell cycle arrest in human carcinoma cells. J Natl Cancer Inst 89: 1881–1886.
17. Lyn-Cook BD, Rogers T, Yan Y, Blann EB, Kadlubar FF, et al. (1990) Chemopreventive effects of tea extracts and various components on human pancreatic and prostate tumor cells in vitro. Nutr Cancer 35: 80–86.
18. Yang CS, Wang X, Lu G, Picinich SC (2009) Cancer prevention by tea: animal studies, molecular mechanisms and human relevance. Nat Rev Cancer 9: 429–439.
19. Ishii T, Ishikawa M, Miyoshi N, Yasunaga M, Akagawa M, et al. (2009) Catechol type polyphenol is a potential modifier of protein sulfhydryls: development and application of a new probe for understanding the dietary polyphenol actions. Chem Res Toxicol 22: 1689–1698.
20. Ishii T, Mori T, Tanaka T, Mizuno D, Yamaji R, et al. (2008) Covalent modification of proteins by green tea polyphenol (−)-epigallocatechin-3-gallate through autoxidation. Free Radic Biol Med 45: 1384–1394.
21. Wang X, Song KS, Guo QX, Tian WX (2003) The galloyl moiety of green tea catechins is the critical structural feature to inhibit fatty-acid synthase. Biochem Pharmacol 66: 2039–2047.
22. Sadik CD, Sies H, Schewe T (2003) Inhibition of 15-lipoxygenases by flavonoids: structure-affinity relations and mode of action. Biochem Pharmacol 65: 773–781.
23. Sazuka M, Itoi T, Suzuki Y, Odani S, Koide T, et al. (1996) Evidence for the interaction between (−)-epigallocatechin gallate and human plasma proteins fibronectin, fibrinogen, and histidine-rich glycoprotein. Biosci Biotechnol Biochem 60(8): 1317–1319.
24. Dubeau S, Samson G, Tajmir-Riahi HA (2010) Dual effect of milk on the antioxidant capacity of green, Darjeeling, and English breakfast teas. Food Chem 122: 539–545.
25. Hasni I, Bourassa P, Hamdani S, Samson G, Carpentier R, et al. (2011) Interaction of milk α- and β-caseins with tea polyphenols. Food Chem 126: 630–639.
26. Kanakis CD, Hasni I, Bourassa P, Tarantilis PA, Polissiou MG (2011) Milk β-lactoglobulin complexes with tea polyphenols. Food Chem 127: 1046–1055.
27. Bourassa P, Cote R, Hutchandani S, Samson G, Tajmir-Riahi HA (2013) The effect of milk alpha-casein on the antioxidant activity of tea polyphenols. J Photochem Photobiol B 128: 43–49.
28. Leone M, Zhai D, Sareth S, Kitada S, Reed JC, et al. (2003) Cancer prevention by tea polyphenols is linked to their direct inhibition of antiapoptotic bcl-2-family proteins. Cancer Res 63: 8118–8121.
29. Ermakova S, Choi BY, Choi HS, Kang BS, Bode AM, et al. (2005) The intermediate filament protein vimentin is a new target for epigallocatechin gallate. J Biol Chem 280: 16882–16890.
30. Shim JH, Su ZY, Chae JI, Kim DJ, Zhu F, et al. (2010) Epigallocatechin gallate suppresses lung cancer cell growth through Ras-GTPase-activating protein SH3 domain-binding protein 1. Cancer Prev Res 3: 670–679.
31. Deisseroth A, Dounce AL (1970) Catalase: Physical and chemical properties, mechanism of catalysis, and physiological role. Physiol Rev 50: 319–375.

32. Percy ME (1984) Catalase: an old enzyme with a new role? Can J Biochem Cell Biol 62: 1006–1014.

33. Sund H, Weber K, Molbert E (1967) Dissociation of beef liver catalase in its subunits, Eur J Biochem 1: 400–410.

34. Stern KG (1936) The constitution of the prosthetic group of catalase. J Biol Chem 112: 661–669.

35. Yang L, Zheng XL, Sun H, Zhong YZ, Wang Q, et al. (2011) Catalase suppression-mediated H_2O_2 accumulation in cancer cells by wogonin effectively blocks tumor necrosis factor-induced NF-κB activation and sensitizes apoptosis. Cancer Sci 102: 870–876.

36. Dong Z, Ma W, Huang C, Yang CS (1997) Inhibition of tumor promoter-induced activator protein 1 activation and cell transformation by tea polyphenols, (−)-epigallocatechin gallate, and theaflavins. Cancer Res 57: 4414–4419.

37. Liang YC, Lin-Shiau SY, Chen CF, Lin JK (1999) Inhibition of cyclin-dependent kinases 2 and 4 activities as well as induction of Cdk inhibitors p21 and p27 during growth arrest of human breast carcinoma cells by (-)-epigallocatechin-3-gallate. J Cell Biochem 75: 1–12.

38. Fang MZ, Wang Y, Ai N, Hou Z, Sun Y, et al. (2003) Tea polyphenol (-)-epigallocatechin-3-gallate inhibits DNA methyltransferase and reactivates methylation-silenced genes in cancer cell lines. Cancer Res 63: 7563–7570.

39. Yang G, Liao J, Kim K, Yurkow EJ, Yang CS (1998) Inhibition of growth and induction of apoptosis in human cancer cell lines by tea polyphenols. Carcinogenesis 19: 611–616.

40. Hu Y, Da L (2014) Insights into the selective binding and toxic mechanism of microcystin to catalase. Spectrochim Acta A Mol Biomol Spectrosc 121: 230–237.

41. Lenehan PF, Gutiérrez PL, Wagner JL, Milak N, Fisher GR, et al. (1995) Resistance to oxidants associated with elevated catalase activity in HL-60 leukemia cells that overexpress multidrug-resistance protein does not contribute to daunorubicin manifested by these cells. Cancer Chemother Pharmacol 35: 377–386.

42. Bechtel W, Bauer G (2009) Catalase protects tumor cells from apoptosis induction by intracellular ROS signaling. Anticancer Res 29: 4541–4557.

43. Liu WB, Zhou J, Qu Y, Li X, Lu CT, et al. (2010) Neuroprotective effect of osthole on MPP⁺-induced cytotoxicity in PC12 cells via inhibition of mitochondrial dysfunction and ROS production. Neurochem Int 57: 206–215.

44. Attar F, Khavari-Nejad S, Keyhani J, Keyhani E (2009) Structural and functional alterations of catalase induced by acriflavine, a compound causing apoptosis and necrosis. Ann New York Acad Sci 1171: 292–299.

45. Jelesarov I, Bosshard HR (1999) Isothermal titration calorimetry and differential scanning calorimetry as complementary tools to investigate the energetics of biomolecular recognition. J Mol Recognit 12: 3–18.

46. Wiseman T, Williston S, Brands JF, Lin LN (1989) Rapid measurements of binding constants and heats of binding using titration calorimeter. Anal Biochem 179: 131–137.

47. Pal S, Saha C, Hossain M, Dey SK, Kumar GS (2012) Influence of galloyl moiety in interaction of epicatechin with bovine serum albumin: a spectroscopic and thermodynamic characterization. Plos One 7(8): e43321 doi:10.1371/journal.pone.0043321.

48. Jha NS, Kishore N (2009) Binding of streptomycin with bovine serum albumin: energetics and conformational aspects. Thermochim Acta 482: 21–29.

49. Hossain M, Khan AY, Kumar GS (2011) Interaction of the anticancer plant alkaloid sanguinarine with bovine serum albumin. Plos One 6(4): e18333 doi:10.1371/journal.pone.0018333.

50. Khan MA, Muzammil S, Musarrat J (2002) Differential binding of tetracyclines with serum albumin and induced structural alterations in drug-bound protein. Int J Biol Macromol 30: 243–249.

51. Liu JQ, Tian JN, He WY, Xie JP, Hu ZD, et al. (2004) Spectrofluorimetric study of the binding of daphnetin to bovine serum albumin. J Pharm Biomed Anal 35: 671–677.

52. Ashoka S, Seetharamappa J, Kandagal PB, Shaikh SMT (2006) Investigation of the interaction between trazodone hydrochloride and bovine serum albumin. J Lumin 121: 179–186.

53. Halliwell B (2008) Are polyphenols antioxidants or pro-oxidants? What do we learn from cell culture and in vivo studies? Arch Biochem Biophys 476: 107–112.

54. Sánchez-del-Campo L, Sáez-Ayala M, Chazarra S, Cabezas-Herrera J, Rodríguez-López JN (2009) Binding of natural and synthetic polyphenols to human dihydrofolate reductase. Int J Mol Sci 10: 5398–5410.

55. Al-Shakhshir R, Regnier F, White JL, Hem SL (1994) Effect of protein adsorption on the surface charge characteristics of aluminium-containing adjuvants. Vaccine 5: 472–474.

56. Putnam CD, Arvai AS, Bourne Y, Tainer JA (2000) Active and inhibited human catalase structures: ligand and NADPH binding and catalytic mechanism. J Mol Biol 296: 295–309.

57. Pierce MM, Raman CS, Nall BT (1999) Isothermal titration calorimetry of protein-protein interactions. Methods 19: 213–221.

58. Gohlke H, Klebe G (2002) Approaches to the description and prediction of the binding affinity of small-molecule ligands to macromolecular receptors. Anqew Chem Int Ed Enql 41: 2644–2676.

59. Nelson DL, Cox MM (2004) Lehninger Principles of Biochemistry. Enzymes, Freeman. 190–211.

60. Elbling L, Weiss RM, Teufelhofer O, Uhl M, Knasmueller S, et al. (2005) Green tea extract and (−)-epigallocatechin-3-gallate, the major tea catechin, exert oxidant but lack antioxidant activities. FASEB J 19: 807–809.

61. Trachootham D, Alexandre J, Huang P (2009) Targeting cancer cells by ROS mediated mechanisms: a radical therapeutic approach? Nat Rev Drug Discov 8: 579–591.

5

Phytochemical and Antiproliferative Activity of Proso Millet

Lizhen Zhang[1], Ruihai Liu[2]*, Wei Niu[3]

1 College of Life Science, Shanxi University, Taiyuan, China, 2 Department of Food Science, Cornell University, Ithaca, New York, United States of America, 3 Shanxi Academy of Agricultural Sciences, Taiyuan, China

Abstract

The phytochemical content, antioxidant activity and antiproliferative properties of three diverse varieties of proso millet are reported. The free phenolic content ranged from 27.48 (Gumi 20) to 151.14 (Mi2504-6) mg gallic acid equiv/100 g DW. The bound phenolic content ranged from 55.95 (Gumi20) to 305.81 (Mi2504-6) mg gallic acid equiv/100 g DW. The percentage contribution of bound phenolic to the total phenolic content of genotype samples analyzed ranged between 62.08% and 67.05%. Ferulic acid and chlorogenic acid are the predominant phenolic acid found in bound fraction. Caffeic acid and *p*-coumaric acid were also detected. Syringic acid was detected only in the free fraction. The antioxidant activity was assessed using the hydrophilic peroxyl radical scavenging capacity (PSC) assay. The PSC antioxidant activity of the free fraction ranged from 57.68 (Mi2504-6) to 147.32 (Gumi20) μmol of vitamin C equiv/100 g DW. The PSC antioxidant activity of the bound fraction ranged from 95.38 (Mizao 52) to 136.48 (Gumi 20) μmol of vitamin C equiv/100 g DW. The cellular antioxidant activity (CAA) of the extract was assessed using the HepG2 model. CAA value ranged from 2.51 to 6.10 μmol equiv quercetin/100 g DW. Antiproliferative activities were also studied in vitro against MDA human breast cancer and HepG2 human liver cancer cells. Results exhibited a differential and possible selective antiproliferative property of the proso millet. These results may be used to direct the consumption of proso millet with improved health properties.

Editor: Gianfranco Pintus, University of Sassari, Italy

Funding: The authors have no support or funding to report.

Competing Interests: The authors have declared that no competing interests exist.

* Email: RL23@cornell.edu

Introduction

Proso millet (*Panicum miliaceum L.*) is an important cereal and a valuable component of the human diet, particularly in developing countries. The crop is salt-, alkali-, cold-, and drought-tolerant and can be cultivated in various types of soil and under poor growing conditions [1]. Its grains are mainly used for food in the decorticated form. Traditionally proso millet quality has been evaluated on the basis of nutritional value, such as starch [2] and crude protein contents [3]. Epidemiological studies show that increased consumption of proso millet and its products are associated with reduced risk of chronic diseases, such as elevated serum cholesterol [4], cardiovascular disease [5], type II diabetes [6], and liver injury [7]. These health benefits have been attributed in part to its unique photochemical profile. However, chemistry and biological activities, including antioxidative and antiproliferative effects of proso millet grains have not received as much attention as phytochemicals in fruits and vegetables. Therefore, the phytochemicals contents of edible proso millet need closer examination due to their potential health benefit in the prevention of chronic diseases.

Chandrasekara and Shahidi [8] reported the phenolics in millet whole grain samples, including one proso millet sample. However, millets belong to a range of different species of family *Gramineae*. Proso millet belongs to *Panicum* genus, which possesses a different phytochemical profile to those other genera in *Gramineae*. Further, proso millet germplasm collections have broad genetic variability and vary in kernel color, size, shape, and other characteristics. In China, over 8, 500 accessions (varieties and landraces) of proso millets are conserved in the National Centre for Crop Germplasm Conservation. Some varieties of proso millet seeds can be harvested from 10 to 20 weeks after planting [9], and have many different colors, such as black, red and white, and so on. As a result, a more complete analysis of the phytochemical contents and antioxidant activity of a range of diverse genetype proso millet samples are needed. Here we choosed three varieties based on their different phenotype characters, mean value of nutrients content and their widely usage in production. Therefore, the objectives of this study were to (1) determine the phytochemical profiles of total phenolics, phenloic acid composition, including both free and bound forms; (2) determine the antioxidant activity and antiproliferation in proso millet milled edible fractions; (3) determine the carotenoid content (xanthophyll, zeaxanthin, β-cryptoxanthin) of three diverse proso millet varieties.

Materials and Methods

Chemicals and Reagents

Methanol (MeOH), hydrochloric acid (HCl), sodium carbonate, sodium sulphate, acetone, phosphate buffered saline (PBS) were purchased from Mallinckrodt Chemicals (Phillipsburg, NJ). Folin-Ciocalteu reagent, quercetin, ascorbic acid, ferulic acid, chlorogenic acid, caffeic acid, *p*-coumaric acid and syringic acid, xanthophyll, zeaxanthin, β-cryptoxanthin, dichlorofluorescein-diacetate (DCFH-DA), were purchased from Sigma (St. Louis,

Table 1. Description of proso millet samples.

Cultival name	% of bran to proso millet (g/100g)	Amylose (g/100g)	Amylopectin (g/100g)	crude protein (g/100g)	Fat (g/100g)	moisture content (%)
Gumi20	21.24±0.28[a]	22.03±0.74[a]	33.83±1.07[a]	10.54±0.54[b]	4.09±0.14[a]	7.96±0.14[c]
Mizao52	20.64±0.20[a]	18.85±0.53[b]	27.75±0.15[c]	11.41±0.24[a]	4.40±0.23[a]	8.35±0.18[b]
Mi2504-6	11.13±0.55[b]	20.11±0.88[b]	29.93±1.19[b]	11.62±0.19[a]	3.64±0.16[b]	9.63±0.13[a]

MO). 2, 2-Azobis-amidinopropane (ABAP) was purchased from Wako Chemicals (Richmond, VA). Gallic acid was purchased from ICN Biomedical Inc. (Costa Mesa, CA). Ethyl acetate, triflouroacetic acid, and ethanol were purchased from Mallinckrodt (Paris, KS). Sodium hydroxide, hexane, acetonitrile, magnesium carbonate, tetrahydrofuran were obtained from Fisher Scientific (Pittsburgh, PA). MDA human breast cancer cell lines and HepG2 liver cancer cell lines are provided by the American Type Culture Collection (ATCC, Rockville, MD). Williams' medium E (WME), α-MEM, Hanks' Blanced Salt Solution (HBSS) were purchased from Gibco Life Technologies, and Fetal bovine serum (FBS) was purchased from Atlanta Biologicals (Lawrenceville, GA).

Grain Samples and Sample Preparation

Proso millet varieties (Table 1) used in this study were provided by Shanxi Agriculture Academy. Seeds of Gumi 20, Mizao 52, Mi2504-6 were harvested from plots grown near Taiyuan, Shanxi in 2011. Gumi 20 has dark brown pigmented testa, Mizao 52 has red pigmented testa, Mi2504-6 has white pigmented testa. The three proso millet samples were dehusked to remove inedible husk and aspirated to remove husk, then milled into fine powder, screened though a 60 mesh screen and thoroughly mixed. Each sample was stored at −40°C and used within 2 weeks of milling.

Extraction of Soluble Free Phytochemical Compounds

Soluble free phenolics of proso millet samples were extracted using the method reported previously [10,11]. Briefly, 2 g of proso millet flour was blended for 5 min in 30 mL of 80% chilled acetone (1:8, w/w) using a Waring blender. The mixture was then centrifuged at 2, 500 g for 10 min. The supernatant was removed and the remaining pellet was again extracted with 30 mL of 80% chilled acetone two times. The supernatants were pooled and evaporated at 45°C to dryness. The final extract was diluted to 10 mL MilliQ water, filtered through a 0.45 µm filter, aliquoted into 1 mL per tube, and stored at −40°C until analysis.

Extraction of Bound Phytochemical Compounds

Bound phytochemicals of proso millet samples were extracted using a modification of the method previously described by Adom and Liu [10]. Briefly, bound phenolics were extracted from the residue from the free extraction. The residue was first digested with 20 mL 2 M sodium hydroxide at room temperature for 1 h with shaking under nitrogen. The mixture was then neutralized with appropriate amount of concentrated hydrochloric acid. Hexanes were used to remove the lipids in the mixture. The remaining mixture was then extracted five times with ethyl acetate. The ethyl acetate fractions were pooled and evaporated at 45°C to dryness. The bound phenolics were reconstituted in 10 mL of MilliQ water, filtered through a 0.45 µm filter, aliquoted in 1 mL per tube and stored at −40°C until analysis.

Determination of the Total Phenolic Content

The total phenolic content of each extracts was determined using the method described by Singleton et al [12] and modified by Okarter et al [13,14]. Briefly, the appropriate dilutions of extracts were oxidized with the Folin-Ciocalteu reagent, and the reaction was neutralized with sodium carbonate. The absorbance of the resulting blue solution was measured at 760 nm in a MRX П Dynex plate reader (Dynex Technologies, Inc., Chanilly, VA) after 90 min of incubation at room temperature. Using gallic acid as a standard, the total phenolic content of samples was expressed as *mg* of gallic acid equiv/100 g of sample. Data were reported as mean ±SD for three replicates.

Determination of Phenolic Acid Composition

The determination of the phenolic composition was conducted using an RP-HPLC method reported previously [14,15]. Briefly, the mobile phase was delivered using a Waters 600E quaternary pump at a flow rate of 0.5 mL/min. Isocratic mobile phase was conducted with 20% acetonitrile in water adjusted to pH 2 with triflouroacetic acid. Separation of phenolic compounds was done using a Supelcosil LC-18-DB column (3 µm, 150 mm×4.6 mm). The total run time was 30 min. Twenty microlitres of sample were made in each run using a Water 717 autosampler. Phenolic compounds were detected using a Waters 2487 dural wavelength absorbance Detector. Each injection was monitored at 280 nm. Identification of each peak was confirmed using the retention time and absorbance spectrum of each pure compound. Percent recoveries were determined by spiking a known amount of pure compound into a sample and performing the same extraction and analytical procedures. The percent recovery for ferulic acid, *p*-coumaric acid, syringic acid, caffeic acid, and cholrogenic acid were higher than 90% (n = 3). Data signals were acquired and processed using Waters Empower software (Waters Corp., Milford, MA).

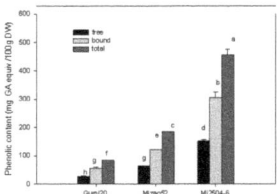

Figure 1. Phenolic contents of proso millet. TPCs were quantified using the Folin-Ciocalteu reagent with gallic acid as the standard. Absorbance was read at 760 nm after 90 min of reaction. Results are expressed as mg gallic acid equivalnts/100 g DW. Analyses were conducted in triplicate, with mean values shown and standard deviation depicted by the vertical bars. Column marked by the same letter are not significantly different from each other ($p < 0.05$).

Table 2. Phenolic acid composition of three diverse varieties of proso millet.

	Free	Bound	Total
Chlorogenic acid			
Gumi 20	6.38 ± 0.38^a (23.54)	20.69 ± 0.24^a (76.46)	27.06 ± 0.28^a
Mizao 52	5.99 ± 0.03^a (23.42)	20.59 ± 5.96^a (76.58)	26.57 ± 5.96^a
Mi2504-6	nd	18.80 ± 1.80^a (100)	18.80 ± 1.80^a
Syringic acid			
Gumi 20	3.05 ± 0.23^a	nd	3.05 ± 0.23^a
Mizao 52	0.74 ± 0.21^b	nd	0.74 ± 0.21^b
Mi2504-6	0.48 ± 0.19^b	nd	0.48 ± 0.19^b
Caffeic acid			
Gumi 20	3.64 ± 0.02^a (48.17)	3.91 ± 0.01^a (51.83)	7.55 ± 0.02^a
Mizao 52	nd	3.98 ± 0.26^a (100)	3.98 ± 0.26^b
Mi2504-6	nd	4.36 ± 0.42^a (100)	4.36 ± 0.42^b
ρ-coumaric acid			
Gumi 20	3.94 ± 0.15^a (47.16)	4.41 ± 0.64^b (52.84)	8.35 ± 0.75^a
Mizao 52	nd	5.18 ± 0.68^{ab} (100)	5.18 ± 0.68^b
Mi2504-6	nd	6.08 ± 0.08^a (100)	6.08 ± 0.08^b
Ferulic acid			
Gumi 20	nd	14.68 ± 1.30^b (100)	14.68 ± 1.30^b
Mizao 52	nd	23.56 ± 0.24^a (100)	23.56 ± 0.24^a
Mi2504-6	nd	24.18 ± 0.10^a (100)	24.18 ± 0.10^a

Values expressed as mg phenolic acid/100 g DW (mean$\pm$SD, n=3). Percent contribution to total phenolic acid content is in parentheses. Values with no letters in common within each column are significantly different ($p<0.05$); nd-not detected.

Determination of the peroxyradical scavenging capacity (PSC)

Hydrophilic peroxyradical scavenging capacity (PSC) assay was developed to determine the total antioxidant capacity of proso millet extracts based on the method described by Adom and Liu [16]. In this assay, the reaction was monitored using the fluorescent dye dichlorofluorescein. Peroxyl radicals generated by ABAP oxidize nonfluorescent dichlorofluorescein (DCFH) to fluorescent dichlorofluorescien (DCF). The degree of inhibition of DCFH oxidation by antioxidants that scavenge peroxyl radicals was used as the basis for calculating the antioxidant activity. Just prior to use in the reaction, 107 μL of 2.48 mM DCFH-DA was hydrolyzed with 893 μL of 1.0 mM KOH for 5 min in the dark to remove the diacetate (DA) moiety and then diluted to a total volume of 8 mL with 75 mM phosphate buffer (pH 7.4). DCFH-DA was stable to oxidation, whereas DCFH was very slowly oxidized at ambient conditions without ABAP. ABAP (200 mM) was prepared fresh in buffer, and each batch was kept at 4°C between runs and discarded after 6 h. The standard or proso millet extracts were appropriately diluted in 75 mM phosphate buffer (pH 7.4) to reach the indicated concentrations. In a run, 100 μL diluted solution of the standard or proso millet extract was transferred into reaction cells in a 96-well plate, and then 100 μL of DCFH was added. The 96-well plate was loaded into the plate holder for the Fluoroskan Ascent fluorescence spectrophotometer (Thermo Labsystems, Franklin, MA), and the solution in each cell was mixed by shaking at 1200 rpm for 20 s. The reaction was then initiated by adding 50 μL of ABAP from the autodispenser of the equipment. The autodispenser was emptied and rinsed with fresh ABAP before each run. Each set of dilutions for a sample and control was analyzed three times in adjacent columns. The reaction was carried out at 37°C, and fluorescence generation was monitored at 485 nm excitation and 538 nm emission with the fluorescence spectrophotometer. The phosphate buffer was used for control reaction. Data were acquired with Ascent software, version 2.6 (Thermo Labsystems, Franklin, MA) running on a PC.

Table 3. Carotenoid (xanthophyll, zeaxanthin, β-cryptoxanthin) content and distribution of proso millet varieties (mean$\pm$SD, n=3).

	xanthophyll	zeaxanthin	β-cryptoxanthin
Gumi20	0.50 ± 13.8^b	1.60 ± 0.4^b	nd
Mizao52	0.49 ± 20.4^b	1.61 ± 2.1^b	nd
M2504-6	1.51 ± 340.3^a	1.68 ± 16.9^a	nd

Values expressed as mg/100 g DW. Values with no letters in common within each column are significantly different ($p<0.05$); nd-not detected.

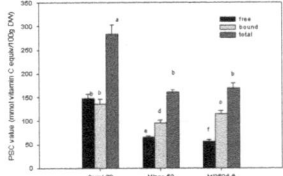

Figure 2. PSC antioxidant activity of proso millet. Peroxyl radical scavenging capacity assay is based on the degree of inhabitation of dichlorofluorescin oxidation by antioxidants that scavenge peroxyl radicals, generated from thermal degradation of 2, 2'azobis (amidino-propane). The median effective concentration (EC$_{50}$) was defined as the dose required to cause a 50% inhibition for each sample extract. Results obtained for sample extract antioxidant activities were expressed as micromoles of vitamin C equivalents/100 g DW. Analyses were conducted in triplicate, with mean values shown and standard deviation depicted by the vertical bars. Column marked by the same letter are not significantly different from each other ($p<0.05$).

Fluorescence values were averaged across columns for each set of dilutions. The areas under the average fluorescence-reaction time kinetic curve (AUC) for both control and samples (up to 36 min) were integrated and used for calculating peroxylradical scavenging capacity (PSC value) according to eq. 1.

$$PSC\ value = 1 - (SA/CA) \qquad (1)$$

Where SA is AUC for the sample or standard dilution and CA is AUC for the control reaction. Compounds or extracts inhibiting the oxidation of DCFH produced smaller SA and higher PSC values. The parameter EC$_{50}$ was defined as the dose required to cause a 50% inhibition (PSC value = 0.5) for each pure compound or sample extract and was used as the basis for comparing different compounds or samples [16]. Results obtained for sample extract antioxidant activities were expressed as μmol of vitamin C equiv/ 100 g of sample ±SD for triplicate analyses.

Cell Culture

MDA human breast cancer cells were grown in α-MEM growth medium supplemented with 10% FBS, 10 mM Hepes, 50 units/ mL penicillin, 50 μg/mL streptomycin, and 100 μg/mL gentamicin. HepG2 liver cancer cells were maintained in Williams' medium E (WME) growth medium with 5% FBS, 10 mM Hepes, 50 units/mL penicillin, 50 μg/mL streptomycin, 100 μg/mL gentamicin, 5 μg/mL insulin and 0.05 μg/mL hydrocortisone. All cells were maintained at 37°C and 5% CO$_2$ in an incubator as described previously [17,18,19]. Cells used in this study were between passages 18 and 32.

Measurement of Cell Cytotoxicity and Inhibition of Proliferation

Cytotoxicity toward MDA and HepG2 cells were measured using the methods as described previously [18,19]. MDA and HepG2 cells in growth media were placed in each well of a 96-well flat-bottom plate at a density of 4.0×10^4 cells/well. After 24 h of incubation at 37°C with 5% CO$_2$, the growth medium was removed, each well washed with 100 μL of PBS, and replaced by media containing different concentrations of sample tested. Control cultures received the extraction solution minus the extracts, and blank wells contained 100 μL of growth medium with no cells. After another 24 h of incubation, cytotoxicity was determined by the methylene blue assay. Cytotoxicity was determined by a 10% reduction of absorbance at 570 nm reading for each concentration compared to the control using an MRX II DYNEX spectrophotometer (DYNEX Technologies, Inc.). A minimum of three replications for each sample was used to determine the cytotoxicity.

Antiproliferative activities of proso millet extracts were measured using the methods described previously [18,19]. MDA cells and HepG2 cells were plated in a 96 well flat-bottom plate at a concentrations of 2.5×10^4 cells/well. After 6 h of MDA cell incubation and 4 h of HepG2 cell incubation, the growth medium was removed and media containing increasing concentrations of proso millet extracts were added to the cells. Control cultures received the extraction solution minus the proso millet extract, and blank wells contained 100 μL of growth medium with no cells. After 72 h of incubation, cell proliferation was determined by the methylene blue assay. Cell proliferation was determined from the absorbance at 570 nm reading for each concentration compared to the control using an MRX II DYNEX spectrophotometer (DYNEX Technologies, Inc.).

Cellular Antioxidant Activity

Extraction of Carotenoids for CAA Samples. Carotenoids were extracted using the method described by Hentschel et al [20] and modified as described previously [21]. The extraction was performed under dim lighting and all sample tubes were wrapped in lightproof paper to protect carotenoids from light-induced degradation. Briefly, 0.6 g samples was mixed with 0.06 g magnesium carbonate and extracted with 3 mL methanol/ tetrahydrofuran (1:1, v/v) solution at 75°C for 5 min in water bath, vortexed again and immediately centrifuged at 2000 g for 5 min. The extraction was repeated three times for complete extraction of carotenoids and the organic solvent phase was collected. The residual was rinsed twice with 2 mL hexane. The hexane and methanol/tetrahydrofuran phases were pooled and vortexed with 1.5 g sodium sulphate. The extracted solvent was evaporated to dryness under a gentle stream of nitrogen. The dry residue was re-dissolved with 1.0 mL methanol/tetrahydrofuran

Table 4. Cellular antioxidant activities of proso millet expressed as EC50 and CAA values (Mean±SD, n=3).

Proso millet	Without PBS wash		With PBS wash	
	EC$_{50}$ (mg/mL)	CAA (μmol of QE/100g)	EC$_{50}$ (mg/mL)	CAA (μmol of QE/100g)
Gumi20	167.57±14.47[a]	5.18±0.29[b]	197.08±2.99[a]	4.38±0.07[a]
Mizao52	187.99±14.50[a]	4.61±0.34[c]	338±14.06[b]	2.51±0.10[c]
M2504-6	142.14±5.11[b]	6.10±0.21[a]	210.80±19.40[a]	3.42±0.61[b]

Values with no letters in common within each column are significantly different (p<0.05).

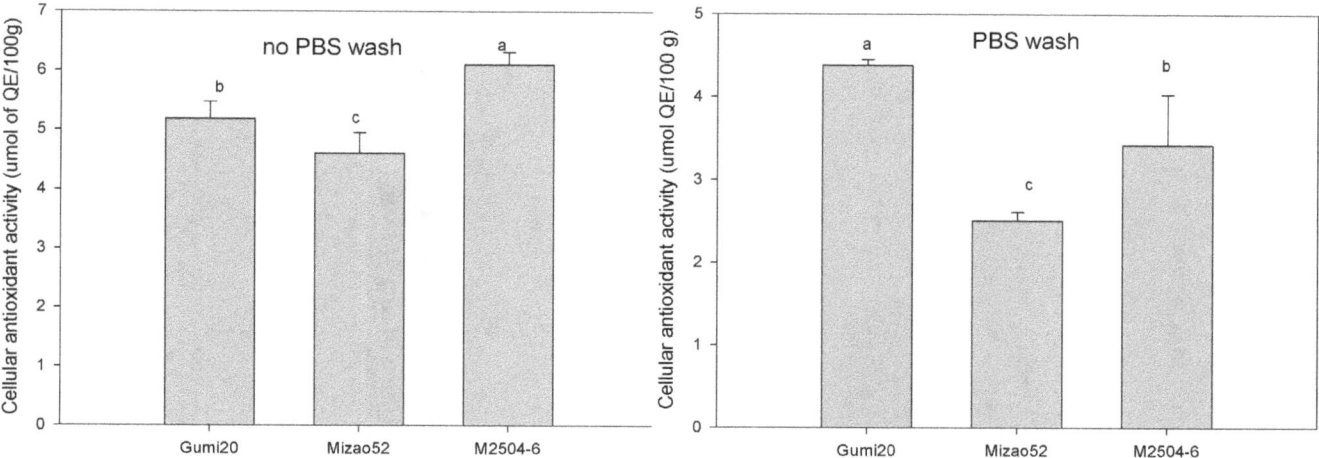

Figure 3. Cellular antioxidant activity of proso millet. Cellular antioxidant activity (CAA) assay is based on the ability of compounds to prevent the formation of DCF by 2,2'-azobis (2-amidinopropane) dihydrochloride (ABAP)-generated peroxyl radicals in human hepatocarcinoma HepG2 cells. Values are based on triplicate tests, with mean values shown and standard deviation depicted by the vertical bars. Column marked by the same letter are not significantly different from each other (p<0.05).

(1:1, v/v), filtered through a 0.45 μm filter, stored under nitrogen at −20°C until CAA analysis within two days.

Quantification of CAA. The CAA of proso millet carotenoids extracts were determined using the protocol described previously [22]. Briefly, HepG2 cells were seeded at a density of 6×10^4/well on a 96-well microplate in 100 μL of complete

medium/well. Twenty-four hours after seeding, the growth medium was removed, and the wells were washed with 100 μL of PBS. Wells were then treated with 100 μL of treatment medium containing solvent control, control extracts, or tested proso millet extracts plus 25 μM DCFH-DA for 1 h. Wells were washed with 100 μL of PBS. Then 600 μM ABAP was applied to the cells in

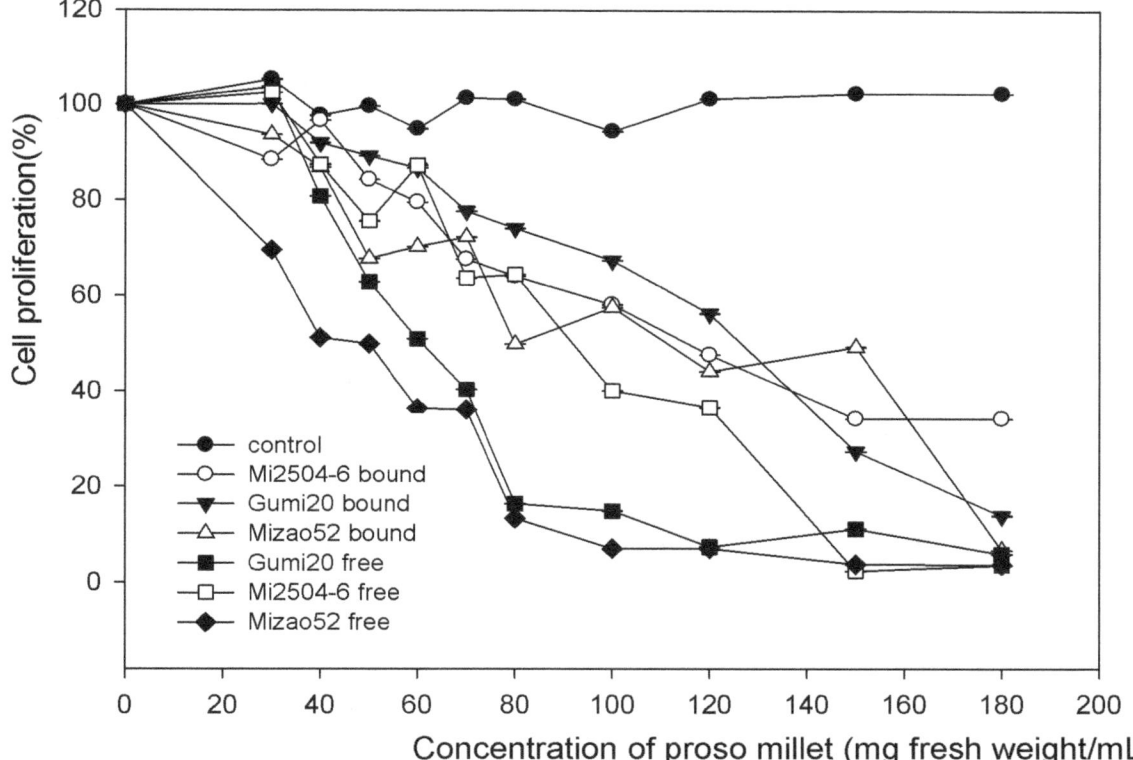

Figure 4. Percentage inhibition of MDA proliferation by proso millet extract. MDA cell (2.5×10^4/mL) were incubated for 6 h to allow sufficient attachment. For the lower level treatment, the initial concentration for samples was 30 mg DW/mL, whereas the high concentration was 180 mg DW/mL. After 72 h of incubation, cell proliferation was determined by the methylene blue assay from the absorbance at 570 nm for each concentration compared to the control. Data were reported as mean ± SD for three replications.

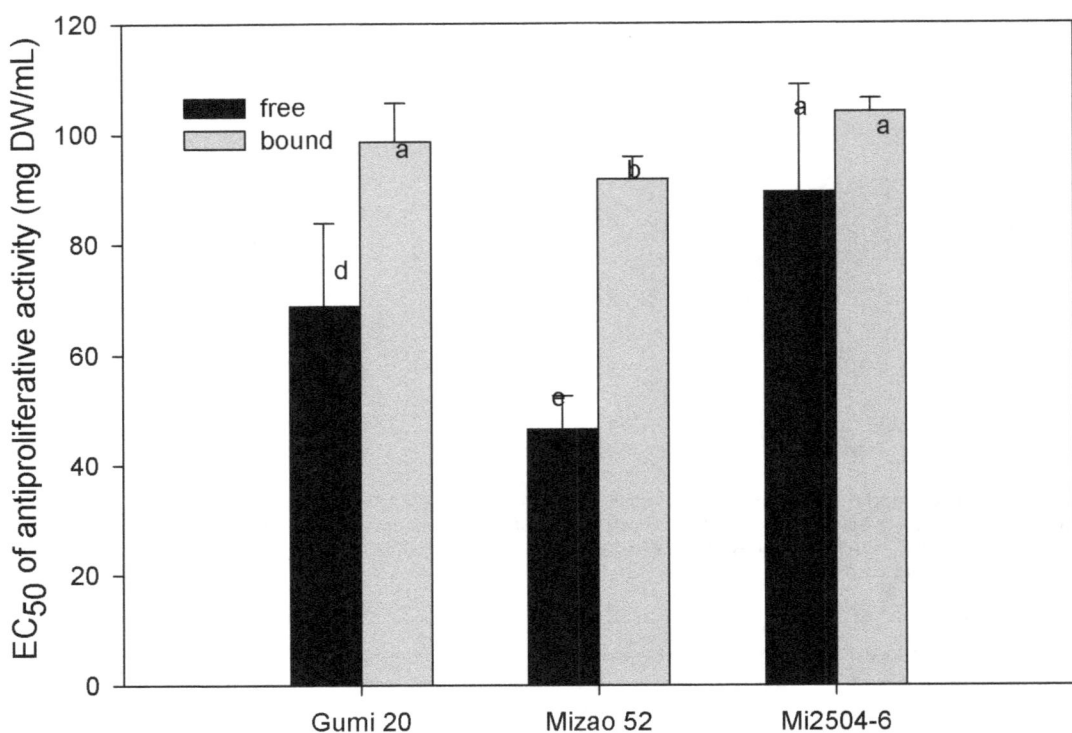

Figure 5. Antiproliferative activities of proso millet against MAD human breast cancer. The antiproliferative activities of proso millets against MAD cells are expressed as the median effective dose (EC$_{50}$). Values are based on triplicate tests, with mean values shown and standard deviation depicted by the vertical bars. Column marked by the same letter are not significantly different from each other ($p < 0.05$).

100 μL of oxidant treatment medium (HBSS with 10 mM Hepes), and the 96-well microplate was placed into a Fluoroskan Ascent FL plate reader at 37°C. Emission at 538 nm was measured after excitation at 485 nm every 5 min for 1 h.

After blank subtraction and subtraction of the initial fluorescence values, the area under the curve for fluorescence versus time was integrated to calculate the CAA value at each concentration of proso millet as described by Wolf and Liu [22].

$$CAA \ unit = 1 - \left(\int SA / \int CA \right)$$

Where $\int SA$ is the integrated area under the sample fluorescence versus time curve and $\int CA$ is the integrated area from the control curve. The median effective dose (EC$_{50}$) was determined for the proso millet extracts from the median effect plot of log (fa/fu) versus log (dose), where fa is the fraction affected (CAA unit) and fu is the fraction unaffected (1-CAA unit) by the treatment. The EC$_{50}$ values were stated as mean± SD for triplicate sets of data obtained from the same experiment. EC$_{50}$ values were converted to CAA values, which are expressed as micromoles of quercetin equiv/100 g sample, using the mean EC$_{50}$ value for quercetin from three replications.

Determination of Carotenoid Content. Carotenoid content was determined using the method described by Hentschel et al [20] and was modified by Liu et al [21]. Briefly, the carotenoid content of each sample was determined using an RP-HPLC procedure employing a 250×4.6 mm YMC C30 column, 3 μm particle size (YMC, Waters Inc., Wilmington, NC). The mobile phase used were methanol/water (95:5, v/v, A) and methyl tert-butyl ether (B). Isocratic elution was performed with 75% solvent

A and 25% solvent B, delivered at 1.0 mL/min using a Water 515 HPLC pump (Water Corp., Milford, MA). A Waters 2487 dual wavelength absorbance detector (Waters Corps, Milford, MA) was used for UV detection of analytes at 450 nm. Data signals were acquired and processed on a PC running the Waters Millennium software, version 3.2 (1999) (Waters Corp, Milford, MA). Percent recoveries for all carotenoids were greater than 90%. The carotenoid content of each sample extract was extrapolated from a pure carotenoid standard curve. All samples were injected via a 20 μL loop and peak heights were used for all calculations. Data were expressed as mg/100 g DW.

Statistical Analysis. Data were expressed as the mean ± standard deviation (SD) of three measurements. One-way analysis of variance (ANOVA) was computed to determine significant differences between the means by SigmaPlot (version 11.0) software. A significant difference was defined at $p < 0.05$.

Results and Discussion

Total Phenolic Content

The free and bound phenolic contents of proso millet and the percentage contribution of each fraction to the total phenolic content of different genotype samples are presented in Fig. 1, expressed as milligrams of gallic acid equiv/100 g DW. The free phenolic content ranged from 27.48 (Gumi 20) to 151.14 (Mi2504-6) mg gallic acid equiv/100 g DW. The percentage contribution of free phenolic to the total phenolic was between 32.93 and 34.72%. The bound phenolic content ranged from 55.95 (Gumi20) to 305.81 (Mi2504-6) mg gallic acid equivalent per 100 g DW. The percentage contribution of bound phenolic to the total ranged between 65.28 and 67.05%. The total phenolic content ranged from 83.44 (Gumi 20) to 456.95 (Mi2504-6) mg gallic acid

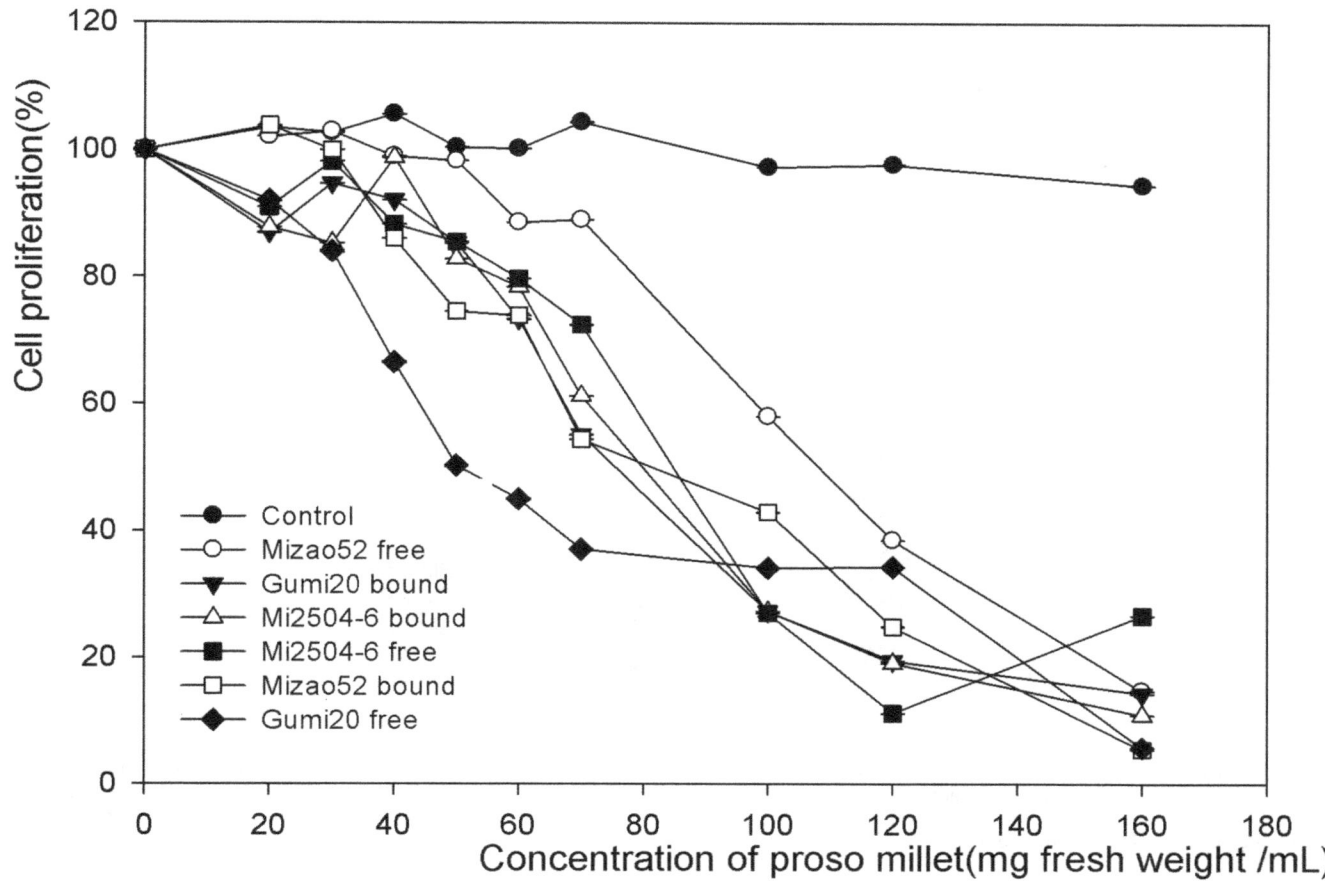

Figure 6. Percentage inhibition of HepG2 proliferation by proso millet extract. HepG2 cells (2.5×10^4/mL) were incubated for 4 h to allow sufficient attachment. For the lower level treatment, the initial concentration for samples was 20 mg DW/mL, whereas the high concentration was 160 mg DW/mL. After 72 h of incubation, cell proliferation was determined by the methylene blue assay from the absorbance at 570 nm for each concentration compared to the control. Data were reported as mean ± SD for three replications.

equivalent per 100 g DW. Those results indicated that bound phenolic content of proso millet was significantly higher than that of free phenolic content.

Similar to other crops [14] and fruits [15], an influence of genetic on the content of phenolics was observed in this study. The free phenolic content, bound phenolic content and total phenolic content were significantly different between three proso millet varieties ($p < 0.05$). Mi2504-6 had the highest phenolic content among the three proso millet varieties, followed by Mizao52. Gumi20 had the lowest phenolic content among the three proso millet varieties. In this study, there was a 5.5-fold difference in free phenolic content, bound phenolic content and total phenolic content between the highest and lowest ranked varieties. Chandrasekara and Shahidi [23] presented the total phenolic of millets with dark brown pigmented testa and pericarp is higher than those with white or yellow testa and pericarp. However, our results showed the phenolic content of different proso millet depends mainly on the varietal differences, not on millet type and color.

Phenolic acid composition

Results for free and bound phenolic acid composition of the tested proso millet varieties are presented in Table 2. Chlorogenic acid was the predominant phenolic acid found in each variety of proso millet tested and was found both in the free and bound

forms. No free chlorogenic acid was detected in Mi2604-6. Free chlorogenic acid content ranged from 5.99 (Mizao52) to 6.38 (Gumi20) mg chlorogenic acid/100 g DW. Bound chlorogenic acid contents ranged from 18.80 (Mi2504-6) to 20.69 (Gumi20) mg chlorogenic acid/100 g DW. Total chlorogenic acid content ranged from 18.80 (Mi2504-6) to 27.06 (Gumi20) mg chlorogenic acid/100 g DW.

Syring acid was found only existed in the free form in the tested proso millet varieties. Free syring acid contents ranged from 0.48 (Mi2504-6) to 3.05 (Gumi20) mg syringic acid/100 g DW.

Caffeic acid was found existed in the bound form in all tested varieties. Free caffeic acid was only found in Gumi20 and the content was 3.64 mg caffeic acid/100 g DW. Bound caffeic acid contents ranged from 3.91 (Gumi20) to 4.36 (Mi2504-6) mg caffeic acid/100 g DW.

ρ-Coumaric acid was found existed in the bound form in all tested varieties. Free ρ-coumaric acid was only found in Gumi20 and the content was 3.94 mg ρ-coumaric acid/100 g DW. Bound ρ-coumaric acid contents ranged from 4.41 (Gumi20) to 6.08 (Mi2504-6) mg ρ-coumaric acid/100 g DW.

Ferulic acid is another the predominant phenolic acid found in all tested edible proso millet varieties and was found existed only in the bound form. The bound ferulic acid contents ranged from 14.68 (Gumi20) to 24.18 (Mi2504-6) mg ferulic acid/100 g DW.

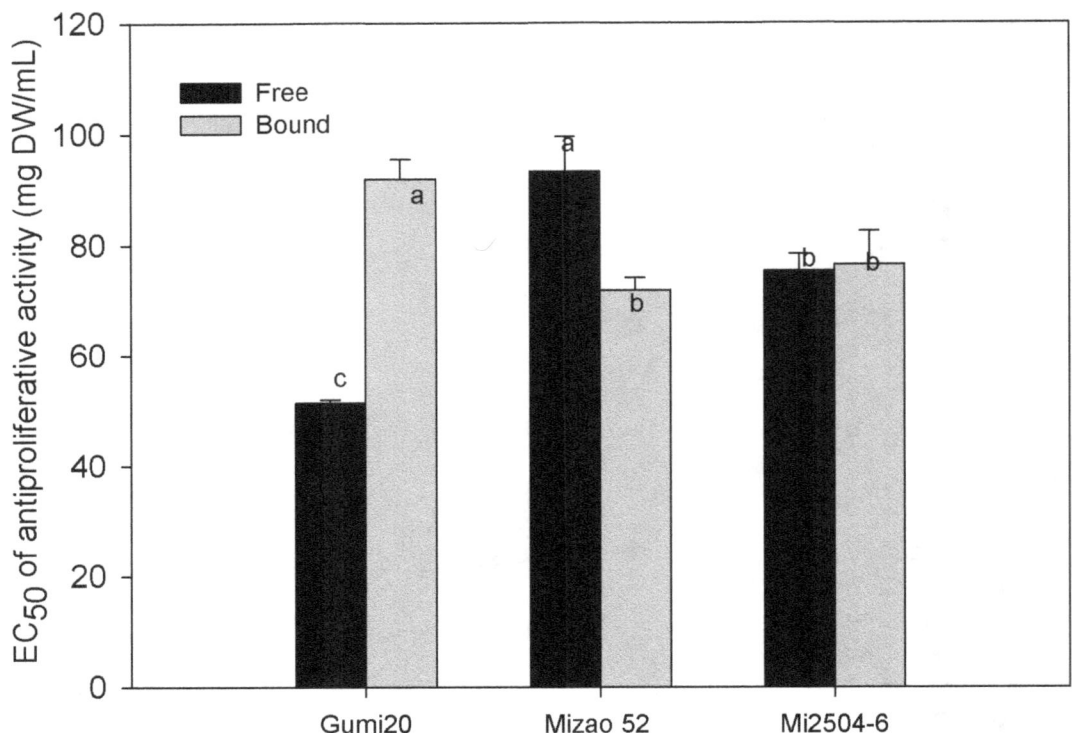

Figure 7. Antiproliferative activities of proso millet against HepG2 human liver cancer cells. The antiproliferative activity of proso millets against HepG2 cells is expressed as the median effective dose (EC$_{50}$), with a lower EC$_{50}$ value signifying a higher antiproliferative acitivity. Values are based on triplicate tests, with mean values shown and standard deviation depicted by the vertical bars. Column marked by the same letter are not significantly different from each other ($p < 0.05$).

Phenolic acids are hydroxylated compounds that are derived from benzoic acid and cinnamic acid. We found the hydroxycinnamic acid derivatives are more prevalent than the hydroxybenzoic acid derivatives in proso millet. The hydroxycinnamic acids found in edible proso millet include chlorogenic acid, ferulic acid, caffeic acid and ρ-coumaric acid. The dexydroxybenzoic acid found in edible proso millet was syring acid, and only found existed in the free form.

RP-HPLC analysis revealed that the content of each hydroxycinnamic acid in the bound fraction was higher than that in the free fraction of three varieties examined in this study. Chandrasekara and Shahidi [9] also reported the bound ferulic acid and ρ-coumaric content of proso millet higher than its soluble counterparts.

Plant-derived phenolic acids received considerable interest because of their potential antioxidant and anticancer properties. McDonough and Rooney [24] reported the ferulic acid, coumaric acid, cinnamic and gentisic acid contents of finger millet, pearl millet, teff millet, fonio millet and foxtail millet. Chandrasekara and Shahidi [23] reported the ferulic acid and ρ-coumaric acid contents of proso millet. Ferulic acid and chlorogenic acid are the predominant phenolic acid found in the bound form. Here, chlorogenic acid, caffeic acid and syringic acid are found in the proso millet for the first time.

Carotenoid content

Results for carotenoid content of the tested proso millet varieties are presented in Table 3. Xanthophyll content ranged from 0.49 (Mizao52) to 1.51 (Mi2504-6) mg/100 g DW. Zeaxanthin ranged from 1.60 (Gumi20) to 1.68 (Mi2504-6) mg/100 g DW. β-

cryptoxanthin was not detected in three variety tested. In this study, xanthophyll and zeaxanthin were significantly different among the three variety tested. Reports on carotenoid in proso millet variety are scanty. Asharani et al [25] reported the total carotenoid content of five different cultivars and the total carotenoid content ranged from 0.518 to 0.249 mg/100 g. However the xanthophyll, zeaxanthin, β-cryptoxanthin content of each sample was not reported. Compared with other grains, carotenoids are rather abound in proso millet. Kean et al [26] reported the carotenoids in sorghum grains. The lutein (xanthophyll) concentrations ranged from 0.149 to 0.301 mg/kg wet weight. The zeaxanthin concentration ranged from 0.126 to 0.362 mg/kg wet weight.

Hulshof et al [27] reported the carotenoids in corn. The lutein (xanthophyll) concentrations ranged from <0.1 mg/100 g to 2.047 mg/100 g. The zeaxanthin concentration ranged from 0.129 to 2.070 mg/100 g. Kimura et al [28] reported the carotenoids in maize. The lutein (xanthophyll) concentrations ranged from 0.148 to 0.360 mg/100 g. The zeaxanthin concentration ranged from 0.401 to 0.565 mg/100 g. Our results indicate that high carotenoid proso millet varieties suitable for production of functional foods for populations at risk of vitamin.

PSC Antioxidant Activity

Results for PSC antioxidant activity of tested proso milllet varieties are presented in Figure 2. The free PSC antioxidant activity ranged from 57.68 (Mi2504-6) to 147.32 (Gumi20) µmol of vitamin C equiv/g. The bound PSC antioxidant activity ranged from 95.38 (Mizao 52) to 136.48 (Gumi 20) µmol of vitamin C equiv/g. The total PSC antioxidant activity of the wheat samples ranged from 161.26 (Mizao52) to 283.82 (Gumi 20) μmol of vitamin C equiv/g.

Asharni et al [25] quantified the total antioxidant activities of the edible flours of proso millet using the phosphomolybdenum reagent and found the antioxidant activity of proso millet varieties ranged from 4.5 to 5.7 mM tocopherol equivalent/g. Chandrasekara and Shahidi [23] evaluated the antioxidant activities of the proso millet on the basis of scavenging capacity of 2,2-diphenyl-1-picrylhydrazyl (DPPH⁻) radicals and reactive oxygen species (ROS) in vitro chemical assays.

Cellular Antioxidant Activity

The cellular antioxidant activities of the carotenoids extracts of proso millet were measured using the CAA assay. Both protocols with and without PBS wash were used to measure the cellular antioxidant activity. The EC_{50} for the extracts are presented in Table 4. The EC_{50} values were converted to CAA values, expressed as micromoles of quercetin equivalent per 100 g of dried proso weight (Fig. 3). When a PBS wash was done between antioxidant and ABAP treatments, the PBS will remove compounds that are loosely associated with the membrane. So for all varieties proso millet tested in our study, the antioxidant quality was lower and the EC_{50} value was higher using the protocol with PBS wash (p<0.05). Significant differences in cellular antioxidant activities of extracts were observed among the three proso millets. In the protocol without PBS wash, Mi2504-6 had the greatest cellular antioxidant activity, with a CAA value of 6.10 μmol quercetin equivalent/100 g DW, followed by Gumi20 and Mizao52, which showed CAA values of 5.18 and 4.16 μmol of quercetin equivalent/100 g DW, respectively. In the protocol with PBS wash, Gumi20 had the greatest cellular antioxidant activity, with a CAA value of 4.38 μmol of quercetin equivalent/100 g DW, followed byMi2504-6 and Mizao52, which had CAA values of 3.42 and 2.51 μmol of quercetin equivalent/100 g DW, respectively.

The cell-based antioxidant assay may be regarded as a more biological relevant method because it accounts for some aspects of uptake, metabolism, and location of antioxidant compounds within cells [22]. The cellular antioxidant activities of varieties were first investigated here. When compared to CAA values of fruits [29], vegetables [30], legumes [31], the carotenoids extracts of proso millet exhibited similar CAA values. These results indicated that proso millet also has strong cellular antioxidant activity.

Both water-soluble and lipid-soluble components of foods are important in combating specific types of radicals and diseases. However, most of the data presented in the literature have been on antioxidant activities of water-soluble food extracts. Here the cellular antioxidant activity of carotenoids extract of proso millet is reported. Proso millets being the primary food in Asian and African countries will provide with a good proportion of carotenoids in the diet. Hence, detailed investigations on carotenoids of proso millets will be very useful for their utilization in health foods.

Antiproliferation Activity

The inhibiting effect of proso millet extracts toward the growth of MDA human breast cells in vitro is presented in Figure 4 and Fig. 5, respectively. The antiproliferative activities of proso millets are expressed as the median effective dose (EC_{50}), with a lower EC_{50} value signifying a higher antiproliferative acitivity. The free extracts of edible proso millet showed relatively higher antiproliferative activities towards MDA cells than bound extracts in a dose-dependent manner. The free extract of Mizao 52 had the highest antiproliferative activity with the lowest EC_{50} of 46.47 mg/mL, followed by the free extract of Gumi20 (68.88 mg/mL), the free extract of Mi2504-6 (89.45 mg/mL). The bound extract of Mizao 52 had the antiproliferative activity with the EC_{50} of 91.78 mg/mL, followed by Gumi 20 (98.65 mg/mL), Mi2504-6 (104.01 mg/mL).

Antiproliferative activity of proso millet extracts on the growth of human HepG2 liver cancer cells in vitro is summarized in Figure 6 and Figure 7. Mizao52, Gumi20 and Mi2504-6 showed relatively potent antiproliferative activities on HepG2 cell growth in a dose-dependent manner. The free extract of Gumi20 had the highest antiproliferative activity with the lowest EC_{50} of 51.37 mg/mL, followed by bound extract of Mizao52 (71.83 mg/mL), free extract of Mi2504-6 (75.34 mg/mL), bound extract of Mi2504-6 (76.45 mg/mL). The free extract of Mizao 52 (93.28) and the bound extract of Gumi20 (91.92) had the lowest antiproliferative activity with the highestest EC_{50}.

In the tests, there was no significant cytotoxicity of both free and bound proso millet extracts against MDA and HepG2 cells up to 120 mg/mL. This suggested that the antiproliferative was not caused by the cytotoxicity.

Traditionally, cereals and its ingredients are accepted as functional foods and nutraceuticals because they provide antioxidants required for human health [32]. In recent years, studies have shown an association between increased consumption of whole-grain cereals and reduced risk of cancers. Several studies have highlighted the contribution of the phenolic acids to their anticancer effect. Other researchers' work also lends support to this hypothesis. Kampa et al [33] reported the antiproliferative activity of caffeic acid, ferulic acid and syringic acid against the T47D human breast cancer cells. Birgit et al [34] reported the antiproliferative activity of ferulic acid and ρ-coumaric acid against the Caco-2 cells. However, further phytochemical and biological investigation is needed to elucidate the active compounds that are responsible for the antiproliferative activity proso millet.

In summary, the present work revealed cellular antioxidant and antiproliferative properties proso millet for the first time. The contents of phenolic acids and antioxidant activity of diverse varieties of proso millet are reported. The bound fraction contributed about 65% of the total phenolic content of the tested proso millet varieties. Proso millet is also rich in bioactive phytochemicals, including ferulic acid, chlorogenic acid, syringic acid, caffeic acid and *p*-coumaric, suggesting its potential benefits to human heath.

Author Contributions

Conceived and designed the experiments: LZ RL. Performed the experiments: LZ RL. Analyzed the data: LZ RL. Contributed reagents/materials/analysis tools: WN. Wrote the paper: LZ RL.

References

1. Wang L, Wang XY, Wen QF (2005) Research and utilization of proso millet germplasm resource in China. J Plant Genet Res 6,471–474 (in Chinese with an English abstract).

2. Yanez GA, Walker CE, Nelson LA (1991) Some chemical and physical propreties of proso millet (Panicum milliaceum) starch. J Cereal Sci 13: 299–305.

3. Kalinova J, Moudry J (2006) Content and quality of protein in proso millet (*Panicum miliaceum* L.) varieties. Plant Foods Human Nutrition 61: 45–49.

4. Nishizawa N, Fudamo Y (1995) The elevation of plasma concentration of high-density lipoprotein chloesterol in mice fed with protein from proso millet (Panicum miliaceum). Biosci Biotech Biochem 59(2), 333–335.

5. Kumari KS, Thayumanavan B (1998) Characterisation of starches of proso, foxtail, barnyard, kodo and little millets. Plant Food Hum Nutr 53: 47–56.

6. Denery-Papini S, Nicolas Y, Popineau Y (1999) Efficiency and limitation of immunochemical assays for the testing of gluten-free foods. J Cereal Sci 30: 121–131.

7. Nishizawa N, Sato D, Ito Y, Nagasawa T, Hatakeyama Y., et al. (2002) Effects of dietary protein of proso millet on liver injury induced by D-galactosamine in rats. Biosci Biotech Biochem 66(1), 92–96.

8. Chandrasekara A, Shahidi F (2010) Content of insoluble bound phenolics in millets and their contribution to antioxidant capacity. J Agric Food Chem 58 (11), 6706–6714.

9. Hu XY, Wang JF, Lu P, Zhang HS (2009) Assessment of genetic diversity in broomcorn millet (Panicum miliaceum L.) using SSR markers. J Genet Genomics 36, 491–500.

10. Adom KK, Liu RH (2002) Antioxidant activity of grains. J Agric Food Chem 50(21), 6182–6187.

11. Adom KK, Sorrells ME, Liu RH (2003) Phytochemical profiles and antioxidant activity of wheat varieties. J Agric Food Chem 51(26), 7825–7834.

12. Singleton VL, Orthofer R, Lamuela-Raventos RM (1999) Analysis of total phenols and other oxidation substrates and antioxidants by means of Folin-Ciocalteu reagent. Oxidants and Antioxidants, Pt A. 299, 152–178.

13. Yang J, Meyers KJ, van der Heide J, Liu RH (2004) Varietal differences in phenolic content and antioxidant and antiproliferative activities of onions. J Agric Food Chem 52, 6787–6793.

14. Okarter N, Liu CS, Sorrells ME, Liu RH (2010) Phytochemical content and antioxidant activity of six diverse varieties of whole wheat. Food Chem 119, 249–257.

15. Yang J, Martinson TE, Liu RH (2009) Phytochemical profiles and antioxidant activities of wine grapes. Food Chem 116, 332–339.

16. Adom KK, Liu RH (2005) Rapid peroxyl radical scavenging capacity (PSC) assay for assessing both hydrophilic and lipophilic antioxidants. J Agric Food Chem 53, 6572–6580.

17. Eberhardt MV, Lee CY, Liu RH (2000) Antioxidant activity of fresh apples. Nature 405, 903–904.

18. Liu M, Li XQ, Weber C, Lee CY, Brown JB, et al. (2002) Antioxidant and antiproliferative activities of raspberries. J Agric Food Chem 50, 2926–2930.

19. Sun J, Liu RH (2008) Apple Phytochemical Extracts Inhibit Proliferation of Estrogen-Dependent and Estrogen-Independent Human Breast Cancer Cells through Cell Cycle Modulation. J Agric Food Chem 56, 11661–11667.

20. Hentschel V, Kranl K, Hollmann J, Lindhauer MG, Bohm V, et al. (2002) Spectrophotometric determination of yellow pigment content and evaluation of carotenoids by high-performance liquid chromatography in durum wheat grain. J Agric Food Chem 50(23), 6663–6668.

21. Liu CS, Glahn RP, Liu RH (2004) Assessment of carotenoid bioavailability of whole foods using a Caco-2 cell culture model coupled with an in vitro digestion. J Agric Food Chem 52(13), 4330–4337.

22. Wolfe KL, Liu RH (2007) Cellular antioxidant activity (CAA) assay for assessing antioxidants, foods, and dietary supplements. J Agric Food Chem. 55, 8896–8907.

23. Chandrsekara A, Shahidi E (2011) Inhibitory activities of soluble and bound millet seed phenolics on free radicals and reactive oxygen species. J Agric Food Chem 59, 428–436.

24. McDonough CM, Rooney LW (1985) Structure and phenol content of six species of millets using fluorescence microscopy and HPLC (Abstract). Cereal Foods World 30: 550.

25. Asharani VT, Jayadeep A, Malleshi NG (2009) Natural antioxidants in edible flours of selected small millets. Inter J Food Properties. 13: 1, 41–50.

26. Kean EG, Bordenave N, Ejeta G, Hamaker BR, Ferruzzi MG (2011) Carotenoid bioaccessibility from whole grain and decorticated yellow endosperm sorghum porridge. J Cereal Sci 54, 450–459.

27. Hulshof PJM, Kosmeijer-Schuil T, West CE, Hollman PCH (2007) Quick screening of maize kernels for provitamin A content. J Food Comp Anal 20, 655–661.

28. Kimura M, Kobori CN, Rodriguez-Amaya DB, Nestel P (2007) Screening and HPLC methods for carotenoids in sweetpotato, cassava and maize for plant breeding trials. Food Chem 100, 1734–1746.

29. Wolfe KL, Kang X, He X, Dong M, Zhang Q, et al. (2008) Cellular antioxidant activity of common fruits. J Agric Food Chem 56, 8418–8426.

30. Song W, Derito CM, Liu MK, He X, Dong M, et al. (2010) Cellular antioxidant activity of common vegetables. J Agric Food Chem 58, 6621–6629.

31. Xu B, Chang SKC (2012) Comparative study on antiproliferation properties and cellular antioxidant activities of commonly consumed food legumes against nine human cancer cell lines. Food Chem 134, 1287–1296.

32. Chaturvedi N, Sharma P, Shukla K, Singh R, Yadav S (2011) Cereals nutraceuticals, health ennoblement and diseases obviation: a comprehensive review. J Applied Pharma Sci 01 (07): 06–12.

33. Kampa M, Alexaki VI, Notas G, Nifli AP, Nistikaki A, et al. (2004) Antiproliferative and apoptotic effects of selective phenolic acids on T47D human breast cancer cells: potential mechanisms of action. Breast Cancer Research 6, r63–r74.

34. Birgit J, Cecilia H, Morten K, Gunilla Ö, Björn Å, et al. (2011) The antiproliferative effect of dietary fiber phenolic compounds ferulic acid and p-coumaric acid on the cell cycle of Caco-2 cells. Nutrition Cancer 63: 4, 611–622.

6

Ozone-Induced Responses in *Croton floribundus* Spreng. (Euphorbiaceae): Metabolic Cross-Talk between Volatile Organic Compounds and Calcium Oxalate Crystal Formation

Poliana Cardoso-Gustavson[1], Vanessa Palermo Bolsoni[2], Debora Pinheiro de Oliveira[2], Maria Tereza Gromboni Guaratini[2], Marcos Pereira Marinho Aidar[3], Mauro Alexandre Marabesi[3], Edenise Segala Alves[4], Silvia Ribeiro de Souza[2]*

1 Programa de Pós-Graduação em Biodiversidade Vegetal e Meio Ambiente, Instituto de Botânica, São Paulo, São Paulo, Brazil, **2** Núcleo de Pesquisa em Ecologia, Instituto de Botânica, São Paulo, São Paulo, Brazil, **3** Núcleo de Pesquisa em Fisiologia e Bioquímica, Instituto de Botânica, São Paulo, São Paulo, Brazil, **4** Núcleo de Pesquisa em Anatomia, Instituto de Botânica, São Paulo, São Paulo, Brazil

Abstract

Here, we proposed that volatile organic compounds (VOC), specifically methyl salicylate (MeSA), mediate the formation of calcium oxalate crystals (COC) in the defence against ozone (O_3) oxidative damage. We performed experiments using *Croton floribundus,* a pioneer tree species that is tolerant to O_3 and widely distributed in the Brazilian forest. This species constitutively produces COC. We exposed plants to a controlled fumigation experiment and assessed biochemical, physiological, and morphological parameters. O_3 induced a significant increase in the concentrations of constitutive oxygenated compounds, MeSA and terpenoids as well as in COC number. Our analysis supported the hypothesis that ozone-induced VOC (mainly MeSA) regulate ROS formation in a way that promotes the opening of calcium channels and the subsequent formation of COC in a fast and stable manner to stop the consequences of the reactive oxygen species in the tissue, indeed immobilising the excess calcium (caused by acute exposition to O_3) that can be dangerous to the plant. To test this hypothesis, we performed an independent experiment spraying MeSA over *C. floribundus* plants and observed an increase in the number of COC, indicating that this compound has a potential to directly induce their formation. Thus, the tolerance of *C. floribundus* to O_3 oxidative stress could be a consequence of a higher capacity for the production of VOC and COC rather than the modulation of antioxidant balance. We also present some insights into constitutive morphological features that may be related to the tolerance that this species exhibits to O_3.

Editor: Martin Heil, Centro de Investigación y de Estudios Avanzados, Mexico

Funding: The authors acknowledge financial support from Fundação de Amparo à Pesquisa do Estado de São Paulo (FAPESP, 12/11663-8). PCG thanks Brazilian CAPES/CNPq (0469-13-0 and 0588/12-1) for funding. DOP and VPB thank Brazilian CNPq for undergraduate scholarships. SRS and ESA thank Brazilian CNPq for their research grant (307238/2013-2 and 36461/2012-1, respectively). The funders had no role in study design, data collection and analysis, decision to publish, or preparation of manuscript.

Competing Interests: The authors have declared that no competing interests exist.

* Email: souzasrd@pq.cnpq.br

Introduction

Tropospheric ozone (O_3) is considered to be the gaseous pollutant that is most damaging to plants due to its strong oxidation capacity [1]. This gas initially enters plants through the stomata, causing biochemical and physiological alterations [2]; based on the concentration, the duration of the exposure, and the responsiveness of the plant species, O_3 can induce disturbances in organelles and tissues, leading to visible morphological symptoms [3–9].

A biochemical balance is required in the cell to counteract the negative effects of O_3. Enzymatic and non-enzymatic antioxidant compounds, including ascorbate peroxidase (APX), superoxide dismutase (SOD), peroxidase (POD), and ascorbic acid (AA), counteract the increase in reactive oxygen species (ROS) promoted by O_3 [2,10–11]. Another possible metabolic response against oxidative stress that is actively modulated by plants is the production of terpenoids, which can constitute the emissions of volatile organic compounds (VOC) [12,13]. Being antioxidant compounds, VOC can remove the ROS formed in the intercellular spaces [11–12,14–22]. Mono- and sesquiterpenes are the most dominant VOC emitted by plants in response to O_3 [1,23–25]; they constitute a large family of plant metabolites, with diverse functions in plant growth, development and stress response [26]. These compounds can be produced by various metabolic routes [27], preferentially in phloem and xylem parenchyma or by secretory cells associated with these tissues [28–30].

Intracellular calcium is also modulated during plant defence responses to O_3 [31], which might lead to the formation of calcium salt crystals. Calcium oxalate crystals (COC) have roles in the defence against herbivores and/or the accumulation of excess calcium –see reviews in [32–34]. It is well known that these constitutive inclusions can be quantitatively induced by biotic stressors such as herbivory [35–37]. Although there is only one report of the induction of crystal formation by O_3 in a gymnosperm species, *Picea abies* [38], there seems to be a significant difference in the quantity of COC produced in *Eugenia uniflora* (Myrtaceae), a tropical species, when polluted and non-polluted environments are considered [39].

There are few studies on the responses of tropical Brazilian tree species to O_3 stress [40]. *Croton floribundus* Spreng. is a pioneer tree species widely distributed in the Brazilian forest and recommended for use in ecological restoration [41]. This species seems to be tolerant to O_3 fumigation, not presenting any structural symptoms of oxidative stress even under 80 ppb of O_3 6 h/day during 53 days [42], although visible symptoms such as hypersensitive-like responses (HR-like), peroxide hydrogen accumulation (H_2O_2) and polyphenol compound accumulation occurred after exposure to 200 ppb of O_3 for 3 h/day for three days [8]. In our ongoing research, O_3 has not been able to change the antioxidant levels in this species, reinforcing the hypothesis that *C. floribundus* is tolerant to oxidative stress caused by this pollutant (at least under less-intense acute exposures). Moreover, in our ongoing research, VOC and COC levels appeared to be O_3-dependent. These preliminary findings suggested that the responsiveness of *C. floribundus* to O_3 might be linked with VOC emission and COC formation. Thus, we raised the hypothesis of which there is a metabolic cross-talk between VOC emission and COC formation that confers defences against oxidative stress.

To test this hypothesis, the aims of this study are (1) to assess ozone oxidative stress by measuring biochemical and physiological responses; (2) to verify whether the variation in the emitted VOC and in the quantity of COC co-occur as a response to this pollutant and if so, (3) to evaluate whether there is cross-talk between the emitted VOC and the COC formation, by the direct application of a compound present in the emitted *bouquet* that is probably involved in the induction of the COC formation.

Materials and Methods

Plant material

Six-month old *C. floribundus* plants were purchased from BIOvida Company (São Paulo, Brazil) and immediately planted in 10 L pots filled with a 3:1 mixture of peat and sand, and watered by capillarity. Plants were kept inside the greenhouse for three weeks and then were transferred from the greenhouse to the fumigation chambers, where they were kept for 2 days before the beginning of fumigation experiment (acclimation period).

Using ozone fumigation to assess leaf responses

The ozone fumigation experiments were performed in closed chambers kept inside a laboratory with temperature and humidity controlled with central air conditioner and under artificial illumination supplied by metallic vapour (400 W) and fluorescent (30 W TL05) lamps [43]. The material was divided into two lots, with half exposed to ozone (FA+O3) and the other half receiving filtered air (FA) only. The chambers were composed of a stainless steel structure covered by a film of Teflon, in the dimensions of 85 cm×94 cm×85 cm (W×D×H). The air was filtered by paper filter to remove gross particulate matter (Whatman 40), followed by silica gel (150 g, Merck), active carbon (250 g, Merck), potassium permanganate (500 g, Purafil Select), and paper filter to remove fine particulate (Whatman QMA). The filtration efficiency was assessed by measuring ozone levels in the air passed through the filtering system. The average ozone levels reached a maximum of 5 ppb after filtration, which indicates an efficiency of filtration of 98.5% [44]. After filtration, the air was enriched with 80 ppb of ozone. Ozone was generated under electrical discharge by the dissociation of oxygen contained in filtered air, using an ozone generator (Ozontechenic). The ozone levels were monitored using an Ecotech 9810B photometric monitor. The ozone monitor was calibrated once before each exposure. During the fumigation experiments, the mean temperature, relative humidity, and photon flux density values were $29\pm1.5°C$, 63 ± 17 and 184.2 $\mu mol/m^2$.s respectively, simulating appropriate conditions for optimal growth of *C. floribundus*. In this experiment, six plants were exposed to filtered ozone-free air (FA) and another six to filtered air plus 80 ppb of ozone (FA+O3) for 4 hours/day per seven consecutive days. Each experiment was performed in triplicate. In order to reduce chamber effects, plants were switched between two chambers in the end of every day of exposure. Thus, both chambers were used for FA and FA+O3 treatment. The position of the plants was also changed to counteract the positional effect [45]. It is important to highlight that there were no differences in the conditions (illumination, humidity, temperature) in which the experiments were carried out.

Table 1. Values (mean and standard deviation) of biochemical and physiological parameters measured in leaves of *Croton floribundus* exposed to ozone (FA+O$_3$) and control plants under filtered air (FA).

Variable	Treatment	
	FA+O$_3$	FA
Biochemical		
AA (mg g^{-1} DW)	7.74±3.39a	7.70±1.85a
POD (10^2 DA min^{-1} DW)	11.13±1.72a	10.55±0.80a
SOD (10^2 U g^{-1} DW)	6.99±1.24a	7.90±0.89a
Physiological		
NPQ	0.657±0.021a	0.077±0.034b
Fv:Fm	0.654±0.017b	0.713±0.004a
ETR	82.7±8.8b	101.9±1.9a

Different letters indicate statistically significant differences among treatments for each parameter analysed (p<0.05).

Table 2. Percentages of volatile organic compounds (means ± S.D) emitted by leaves of *Croton floribundus* exposed to ozone (FA+O₃) and control plants under filtered air (FA), and their linear retention indices in literature (Ri ref) and calculated (Ri cal).

Compounds	FA[a]	FA+O₃[a]	Ri ref	Ri cal[b]
Non-terpenoid Oxygenated				
nonan-2-one	0.67±0.48	0.53±0.52	1091	1091.3
pentadecanoic acid	0.33±0.15	0.50±0.35	1829	1826.3
Nonadecanal	0.33±0.21	0.37±0.24	2105	2105.1
octadecanoic acid	0.34±0.24	0.47±0.17	2200	2200.2
(E)-3-octen-2-one	2.20±1.70	4.07±1.21	1034	1033
Sum	*3.87*	*5.94*		
Aromatic				
(Z)-3-hexenyl benzoate	0.63±0.44	0.48±0.18	1570	1570,27
ethyl benzoate	0.42±0.22	0.62±0.22	1169	1170.68
methyl salicylate	0.30±0.32	0.73±0.13	1191	1191.13
methyl benzyl formate	0.57±0.28	0.57±0.26	1335	1335.56
methyl 3–4 dimethylbenzoate	0.52±0.41	0.34±0.23	1287	1353.73
2,6 dimethylphenol	Nd	0.25±0.20	1510	1504
Sum	*2.44*	*2.99*		
Monoterpene				
Ocimene	0.55±0.48	0.65±0.29	1063	1098
Geraniol	0.39±0.26	0.68±0.22	1276	1235
geranil acetate	0.32±0.12	0.54±0.25	1383	1378
γ terpinene	0.43±0.26	0.50±0.33	1016	1027
Carvone	0.42±0.30	0.45±0.14	1253	1248
Linalool	0.32±0.15	0.26±0.16	1173	1174
α-pinene	0.26±0.20	0.31±0.23	934	--
1,8 cineol	0.70±0.37	1.47±0.18	1027	1030
Myrcene	0.34±0.19	0.40±0.25	988	--
Sum	*3.73*	*5.26*		
Sesquiterpene				
trans-nerolidol	0.53±0.40	0.55±0.18	1566	1564.2
β-gurjunene	0.48±0.37	0.56±0.26	1409	1423
caryophyllene oxide	0.70±0.26	0.84±0.25	1566	1576
aromadendrene	0.48±0.37	0.39±0.26	1636	1638
Copaene	0.55±0.26	0.58±0.29	1369	1372.8
α-caryophyllene	0.23±0.20	Nd	1411	1420
Sum	*2.97*	*2,92*		

Significance of Repeated Measures ANOVA				
Factor[c]	**F**		**P**	
ozone	0.02		0.90	
time	0.54		0.51	
ozone * time	0.94		0.49	

[nd]not detected.
[a]mean percentage and standard deviation of three replicates (including the six young plants after 7 days of exposure).
[b]linear retention index calculated on DB5 capillary column with a homologous series of n-alkanes (C8–C30).
--data insufficient to calculate Ri.
[c]Greenhouse-Geisser values calculated by ANOVA considering all volatiles.

Figure 1. *In situ* **localisation of terpenoids on leaves of** *Croton floribundus.* Control plants under filtered air (A, C) and exposed to ozone for three (D), five (E) or seven (B, F) days. (A, B) Control from histochemical analysis (material without any dye or reagent). (C–F) Positive results for Nadi reagent inside laticifers (C); note a qualitative increase in the terpenoid content inside the parenchyma cells of vascular tissue according to the ozone exposure time (D–F). Arrows indicate laticifers. Bars: 100 µm.

Biochemical and physiological responses to ozone: fluorescence and antioxidant analyses

A Pulse-Amplitude-Modulated Fluorometer (PAM-2100, Heinz Walz GMBH, Effeltrich, Germany) was used to measure leaf steady-state chlorophyll fluorescence parameters after seven days of fumigation and following a 30 min adaptation to the dark. The NPQ was calculated as NPQ = (Fm–Fm')/Fm', where Fm is a maximal fluorescence in the dark and Fm' is a maximal fluorescence in the light. The Fv/Fm values (photosynthetic quantum efficiency) represent averages from 15 measurements taken sequentially on two different fully expand leaves of three plants per experiment. Electron Transport Rate (ETR) is calculated as ETR = yield $\times$ 0.5 $\times$ 0.82 $\times$ PAR, where yield is the quantum yield of the PSII (Fq/Fm), 0.5 represents the light absorbed by the PSII, 0.82 represents the absorbance of the leaf and PAR represents the light intensity used (400 µmol photons/ $m^2.s^2$ using the halogen light source). Values of ETR represent ten measurements in two different fully expand leaves of three plants per treatment. All measurements were taken in the morning, between 10:00 and 11:00 am.

The antioxidant defences were analysed in six individuals per treatment. Collection of leaves and preparation of extracts for analysis of antioxidants always followed the same sequence in time to avoid diurnal variation. The extraction was carried out with a mix of all expanded leaves. Total ascorbic acid was measured in 0.5 g of fresh leaves and homogenised with 12 mL of EDTA-Na$_2$ (0.07%) and oxalic acid (0.5%). The mixture was centrifuged at 40.000 g for 30 min at 2°C. An aliquot of the supernatant was added to 2.5 mL of DCPIP (0.02%), and absorbance was measured with a spectrophotometer at 520 nm. After the addition of 0.05 mL of ascorbic acid (1%), a second absorbance measurement was taken. Both absorbance measurements were used to estimate the total ascorbic acid content following [46].

Superoxide dismutase activity was measured in 0.35 g of fresh leaves homogenised with 12 mL of potassium phosphate buffer (50 mM pH 7.5), EDTA-Na$_2$ 1 mM, NaCl 50 mM and ascorbic acid 1 mM in the presence of 0.4 g of PVPP 2%. This mixture was centrifuged at 22.000 g for 25 min at 4°C. SOD activity was assayed by measuring the SOD inhibition of the NBT photochemical reduction [47]. Each reaction mixture contained 0.5 mL of EDTA-Na$_2$ 0.54 mM, 0.8 mL of potassium phosphate buffer (0.1 M, pH 7.0), 0.5 mL of methionine 0.13 mM, 0.5 mL of NBT 0.44 mM, 0.2 mL of riboflavin 1 mM, and 0.2 mL of leaf extract.

The samples were incubated for 20 min under a fluorescent lamp (80 W). The absorbance of the reaction mixture was measured at 560 nm. A similar mixture lacking the leaf extract was used as a control, and a dark control mixture served as a blank. The enzymatic activity was expressed as the amount of extract needed to inhibit the reduction of NBT by 50%.

Peroxidase activity was measured in 0.3 g of leaves homogenised with 12 mL potassium phosphate buffer (0.1 M, pH 7.0) in the presence of 0.4 g of PVPP 2%. The homogenate was centrifuged at 40.000 g for 30 min at 2°C. Peroxidase activity was also measured in a reaction mixture of plant extracts using 0.1 M potassium phosphate buffer (pH 5.5) and phenylenediamine (1%) to which an aliquot of H$_2$O$_2$ (0.3%) was added. Unspecific POD activity was measured with a spectrophotometer following the increase in absorbance (DA) at 485 nm due to the formation of an H$_2$O$_2$-POD complex at two different times in the linear reaction curve [48].

Volatile Collection

Volatiles from the headspace of six whole plants from each treatment (FA and FA+O3) were collected into steel tubes containing 150 mg charcoal adsorbent (Supelco, PA, USA) at an airflow rate of approximately 200 ml/min for 60 min. To avoid the VOC from soil, the vessels were covered with aluminium foil. During the collection pure air was continuously supplied into the chambers. One tube was fixed into each chamber (FA and FA+ O3) with a line Teflon tubes and connected in the vacuum pumps for volatiles sampling. All collections were performed in the morning after approximately 60 min of the light being switched on and before starting the ozone fumigation. The ozone scrubbers (filters coated with a saturated solution of potassium iodide) were fixed before the adsorbent to avoid any degradation of samples by residual ozone. The collections were made every day after the first day of exposure, comprising a total of six samples per treatment (n = 6). Samples were replicated twice. Blank samples from an empty chamber were also collected twice.

CG-MS analyses

VOC were analysed by gas chromatography-mass spectrometry (GC-MS Hewlett-Packard GC 6890, MSD 5973, Wilmington, DE, USA). Trapped compounds were desorbed chemically with 200 µL of hexane-methylchloride (4:1). The volume was reduced to 50 µL before injection. The separation was performed on a DB-

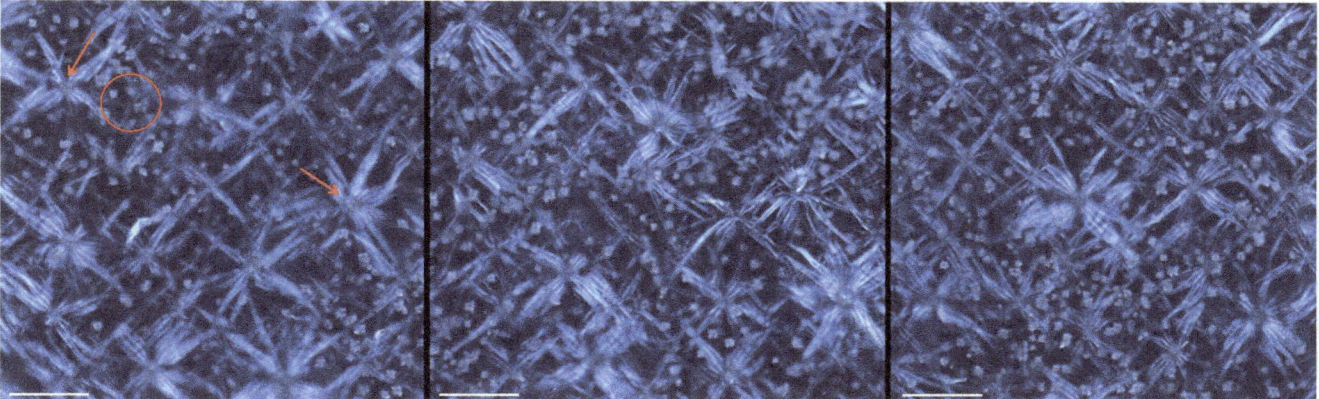

Figure 2. Areas in the adaxial surface of *Croton floribundus* observed under light microscope between crossed polarisers. Leaves before the exposition to O$_3$ (A), and after five (A) and seven (C) days of exposition. Note the abundance of crystals (circle) and non-secretory stellar trichomes (arrows). Bars: 300 µm.

Table 3. Effects of ozone, methyl salicylate and time (days) on the quantity of calcium oxalate crystals from *Croton floribundus* leaves.

Ozone treatment		
	Time (days) of exposure	
	5 (means ± SE)	**7 (means ± SE)**
FA	565±72	501±177
FA+O3	575±90	787±150
Significance of Repeated Measures ANOVA		
Factor[a]	**F**	**P**
ozone	31.42	0.001
time	1.77	0.192
ozone * time	27.12	0.001
MeSA treatment		
	Time (days) of application	
	5 (means ± SE)	**7 (means ± SE)**
Control	72.2±16.2	71.5±19.9
MeSA	105.6±14.7	111.53±21.9
Significance of Repeated Measures ANOVA		
Factor[a]	**F**	**P**
MeSA	165.67	0.001
Time	0.85	3.463
MeSA * time	1.37	0.243

[a]Greenhouse-Geisser values calculated by ANOVA.

5 capillary column (Agilent technologies, USA; 30 m × 0.25 mm ID, 0.25 μm film thickness). Helium was used as a carries gas at a constant flow rate 1.5 ml/min. The inlet temperature was 250°C, splitless injection mode. The oven temperature program was held at 40°C for 1 min and then raised to 210°C at a rate of 5°C/min, and finally to 250°C at a rate of 20°C/min. The mass spectrometer was operated in electron ionization mode at 70 eV, source temperature 230°C and quadrupole temperature 150°C.

Compounds identification and semi-quantification

VOC identification was undertaken by comparing the recorded mass spectra using Wiley library. The peak identification was performed when the similarity of mass spectra was higher than 80%. The linear retention index was used to secure the identification of each molecule. Retention index was calculated by injecting saturated n-alkanes standards solution C7–C30 (Supelco, Belgium) using the definition of Kovats retention [49]. The identification was not confirmed using standards due to limited availability of chemicals. Absolute peak areas were used to calculate the percentage of each compound in the sample. The percentage was performed comparing the sum of peaks areas (hundred percent of compounds, including unidentified compounds) and the individual area of each compound.

Histochemical localisation of terpenoids

For each O_3 exposure time (three, five or seven days), three leaves from the sixth oldest node were used. The medial region of fresh, fully expanded leaf blades was freehand sectioned. Five sections were used for *in situ* localization of terpenoids using Nadi reagent (naphthol+dimethyl-paraphenylenediamine) [50]. Sections were incubated in the dark for 60 min at room temperature in Nadi reagent, prepared immediately prior to staining. After incubation, the sections were rinsed for 2 min in a sodium phosphate buffer (0.1 M, pH 7.2). By oxidation this reagent forms indophenol blue that changes colour with variation in pH and makes it possible to distinguish between essential oils (blue) and resin acids (intense red), in which a purple color is observed when both compounds constitute the secretion [50,51]. Five other sections remained untreated – not submitted to any dye or reagents – for observation of the colour and structure of the cells *in vivo*. Observations and digital images were acquired with an Olympus BX53 compound microscope equipped with an Olympus Q-Color 5 digital camera and Image Pro Express 6.3 software.

Identification and quantification of calcium oxalate crystals

The same leaves used for *in situ* localisation of terpenoids were used for COC quantification purposes. The remainder of each leaf was fixed in FAA_{50} (formalin-acetic acid-50% ethanol, 1:1:18) for

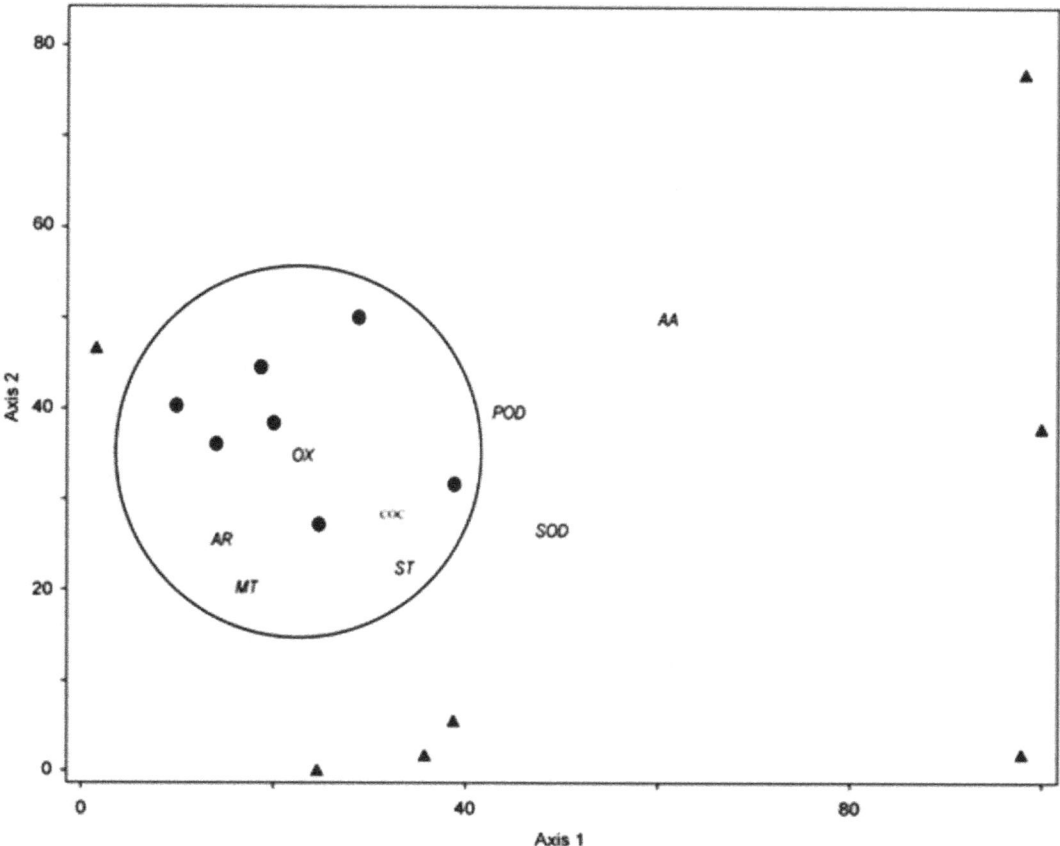

Figure 3. The PCA biplot diagram showing the volatile organic compounds (VOC) and antioxidant defences superimposed on the score of treatments indicated by black circles (ozone exposed, FA +O₃) and black triangles (filtered air, FA). The PC1 axis separates the treatments based on the O_3-induced VOC and calcium oxalate crystals (COC). On the PC2 axis, the treatments were characterised by their antioxidant defences. Abbreviations: VOC as aromatic (AR), monoterpene (MT), oxygenated (OX), sesquiterpene (ST); Ascorbic acid (AA); Peroxidases (POD); Superoxide dismutase (SOD).

24 h and then stored in 70% ethanol [52]. The samples were hydrated and diaphanised using 10% sodium hydroxide and 20% hypochlorite solution (modified from) [52], and the medial region (approximately 3 cm in length) was sectioned and mounted in glycerine. Thirty squares of 5 mm² were observed between crossed polarisers and photographed with an Olympus BX53 microscope. All the crystals present on images were counted using Image J (National Institutes of Health, USA).

For anatomical purposes, we used all the methods above on fully expanded leaves assuming that ozone symptoms initially appear with greater severity in older leaves toward the base of the plant [53] and that total terpenoids significantly increase with plant ontogeny [54].

Is methyl salicylate responsible for the induction of COC?

To verify if methyl salicylate (MeSA) was directly involved in the induction of COC, we performed an independent experiment using young plants acquired from the same company, planted, watered, and two days acclimated in fumigation chambers as aforementioned.

Twelve plants were used for this experiment, separated in two groups of six plants and maintained separately inside two chambers, the same ones used for fumigation experiments, but emitting filtered air only. Plants were kept under the same

conditions as described for filtered air plants from fumigation experiments during seven days.

A solution of 5 mM of MeSA (Sigma-Aldrich, Saint Louis, USA) was sprayed to each plant of the treated group every morning (1 ml/day), while distilled water was applied to each plant of the control group. After five and seven days of application the median region of the expanded leaves from two treated and two control plants were removed and then proceeded the protocols of COC counting as aforementioned. Thirty squares of 1.225 mm² were observed between crossed polarisers, photographed and counted as described above.

Statistical analysis

We used the statistical package SPSS 14.0 for Windows (SPSS; Chicago, IL, USA). Differences between FA and FA+O3 for antioxidant and physiological parameters were determined by the paired t-test with 95% confidence. VOC and COC were analysed using Repeated Measures ANOVA, with time and ozone as factors. Repeated Measures ANOVA was also applied on the independent experiment with MeSA and time. The differences between means were considered to be statistically significant at $P < 0.05$. As the sphericity assumption was violated in all Repeated Measures ANOVA applied on the VOC data, the Greenhouse-Geisser conservative F-test was used [55].

The correlations among treatments, antioxidants, VOC emission and COC formation were investigated by principal component analysis using the data of seventh day of exposure in order to include all the variables in the statistical analyses (PCA, developed by PCord package 6.0), in which seven principal components explained over 90% of the variation.

Results

Which are the effects of ozone on antioxidant defence and fluorescence?

After seven days of exposure, there was a slight increase in total ascorbic acid and peroxidases in plants exposed to FA+O3 treatment, but no significant differences were found when these data were compared to FA treatment (Table 1). Significant decreases in the values of Fv/Fm and ETR (8.3% and 18.9%, respectively) and increase in the values of NPQ were observed in plants submitted to FA+O$_3$ treatment.

What are the emitted volatiles?

All the VOC identified were grouped into their respective chemical classes and summarised on Table 2. The relative quantity of oxygenated non-terpenoids and monoterpene tended to be higher in FA+O3 plants, representing 39 and 27% of total relative quantity of VOC emitted, respectively. MeSA, (E)-3-octen-2-one and 1,8 cineole showed higher levels in FA+O3 plants. No significant effects ($P>0.05$) of time and in the interaction between time and O$_3$ were found.

Where are the terpenes?

Major qualitative results were observed in the midrib of the medial region of the leaf blade. Plants from FA and FA+O3 treatments presented no structural differences in epidermis, ground and vascular tissues (Figures 1A, B). The histochemical tests with Nadi reagent showed a positive reaction inside laticifers and parenchymal cells of xylem and phloem in all plants analysed: there were few cells with terpenoid content in leaves from FA treatment plants (constituent terpenoids, Figure 1C), while a qualitative increase of parenchyma cells involved on terpenoids metabolism was observed with increased exposure time to ozone (induced terpenoids, Figures 1D–F).

Do ozone and MeSA increase the COC abundance?

The abundance of COC and non-secretory stellar trichomes can be visualized in the Figure 2, also representing the areas where the quantification of the crystals was performed. Ozone fumigation resulted in a significant increase in the COC quantity ($P<0.01$) on the fifth and seventh days of exposure (Table 3). No significant effect of time ($P>0.05$) was observed on the COC quantity, whereas the effect of the interaction between O$_3$ and time was significant ($P>0.05$).

A significant increase was observed in the COC quantity on leaves from plants treated with MeSA ($P<0.01$; Table 3): the increase was not time-dependent and there was no interaction effect between time and MeSA.

Correlations among antioxidant defence, volatiles and crystal formation

Principal components analysis was used to study relationships among the antioxidants, volatiles and crystal formation (Figure 3). The first PC was described by the term 'antioxidant defence' and explained 57% of the total variation in the data, characterised by a high score of ascorbic acid rather than SOD and POD. The second PC, designated as 'induced defence', explained 26% of the total variation and was characterised by a lower score in volatile and crystal levels than ascorbic acid. The treatments were separated into specific groups: FA treatment was more related to antioxidant defences, explained by a high level of AA, while FA+O3 treatment was linked to volatile emissions and crystal formation.

Discussion

Our results indicate that O$_3$ induces biochemical, physiological and anatomical responses in a native Brazilian species. When these responses are combined, it is possible to understand how the plant can defend itself and how apparently independent features, such as the emission of VOC and production of COC, may play an important role in these responses.

The antioxidant defences act against O$_3$ to avoid the increase in cellular ROS such that the expression and/or activity levels of enzymes and other compounds change to keep the pro-antioxidant balance [13,56–61]. Changes in the antioxidant levels, such as an increase in the level of ascorbate as a response to O$_3$ fumigation, have been reported in spinach (*Spinacia oleracea*) and beech (*Fagus sylvatica*) [59]. In contrast, when the O$_3$-induced ROS does not result in a reducing environment inside the cell, the antioxidant levels might remain constant. In this case, there can be increases in other metabolic compounds that can act as scavengers of ozone before its strong oxidative action occurs in the cell [1,16,62].

In our study, no significant changes were observed in the levels of antioxidants whereas the inducible defences, terpenoid compounds, tended to increase in O$_3$ exposed leaves. This can be explained by the fact that O$_3$ uptake does not necessarily cause injury in plants due to the production of metabolic compounds, such as volatiles, that remove ROS before they cause serious damage in the cell [63].

In our experiment, although there was a slight increase in the quantity of volatiles emitted, there was no change in volatile profile and *de novo* volatiles were not induced. Monoterpenes, aromatics and non-terpenoid oxygenated compounds, which have been reported to be protective against oxidative damage due to rapid reaction with ROS [15,18–20] increased with FA+O3 treatment. Thus, the increase in these constitutive metabolites in *C. floribundus* might be a defence response against O$_3$ damage, removing ROS through gas-phase chemical reactions in the intercellular spaces of the leaves [2] because isoprene synthesis is stimulated by ozone [11,64]. Although terpenoid compound levels might have inhibited the effects of ozone most likely by restricting the damage caused by ROS, the fumigation of *C. floribundus* with 80 ppb of O$_3$ lead to a significant decrease in Fv:Fm and ETR compared to the FA condition, which might indicate a reduction of photosystem II (PSII) efficiency. A decrease in the ability of PSII to dissipate light energy as non-photochemical quenching (NPQ) can induce the ROS production in chloroplast, signalling the primer mechanism of immune response [65]. In addition, chloroplast-generated ROS are important to up-regulation of defence-relate genes and down-regulation of photosynthesis, such as the regulation of redox state of photosynthetic components, including plastoquinone and glutathione pools [65,66]. Our results suggest no disturbance in enzymes and ascorbic acid, which are compounds of ascorbate-glutathione cycle, indicating that ROS produced by O$_3$ likely not disturbed the redox state of the chloroplast. Our results also indicate that the ROS produced by ozone exposure in *C. floribundus* might have been sufficient to signal the terpenoid pathway but not to cause a strong oxidative stress.

Secretory glands such as laticifers are involved in plant defences against herbivory [67,68]. Laticifers' metabolite contents are biochemical end products (generally cytotoxic); these cells are involved in sequestering toxic compounds or their precursors independent of vascular tissues [69]. It is assumed that laticifers have no functional plasmodesmatal connections with their neighbour cells [70–71], making it impossible to release their contents without physical injury, and they are not modulated in response to oxidative stress. On the other hand, terpenoids are produced by cells of the mesophyll and parenchyma cells from the vascular system, which are also responsible for translocation of these compounds [28–30]. Indeed, FA treatments presented only laticifer cells with terpenoid content (constituent terpenoids), but a progressive recruitment of vascular parenchyma cells involved with terpenoid production and translocation was observed in the course of the time in which the plants were exposed to O_3. Since the terpenoids are volatile, their diffusion from these cells can be confined over and to the leaf (boundary layer) by means of the non-secretory trichomes, which are abundant in both faces of the epidermis. These stellar non-secretory trichomes can act as a container for volatiles analogous to the corona on flowers of *Passiflora* species [72]. Therefore, the concentration of VOC under the leaf might be increased simply by the presence of these non-secretory trichomes. Because VOC can scavenge O_3, this morphological adaptation can also prevent the entrance of this pollutant into the leaf.

Considering that the production of volatiles is systemic [73–74], the O_3-induced VOC in *C. floribundus* might be transported by phloem throughout the plant and act as signalling defences. Methyl salicylate, an oxygenated volatile that increased when *C. floribundus* plants were exposed to O_3, is one of the key messenger molecules synthesised by plants in response to stress [75]. This compound may act as a mobile signal throughout the plant, triggering the systemic acquired resistance by means of its

precursor, salicylic acid, which is able to enhance chemical defences, such as antioxidants [75–76]. It is interesting to note that ROS is mediated by salicylic acid, which plays a key role in the changes in the cytosolic concentration of free calcium ions [77]. Indeed, O_3 stress induces the ROS-mediated opening of calcium channels and increases the intracellular calcium concentration [78], a mechanism involved in COC formation [79]. In our work, *C. floribundus* exposed to O_3 showed slightly increase in MeSA emission and in the number of COC, but there is no evidence of alteration in pro-antioxidant balances. Therefore, we hypothesized that the ozone-induced VOC (such as MeSA) regulate ROS formation in a way that promotes the opening of calcium channels and subsequently, the formation of crystals. To test this hypothesis we performed an independent experiment spraying MeSA on *C. floribundus* leaves and verified that, indeed, MeSA was able to induce an increase on COC number. Thus, we suggest that the increase in the quantity of crystals could be a fast and stable mechanism to stop the consequences of ROS in the tissue, immobilising the calcium excess that can be dangerous to the plant.

Our findings indicate that the tolerance of *C. floribundus* to oxidative stress caused by ozone may be the consequence of a higher capacity to produce volatiles and oxalate crystals rather than the modulation of the antioxidant balance. These metabolic features could be used as biomarkers for ozone tolerance and may also be useful for choosing the functional groups resistant to air pollution and, consequently, for use in ecological restoration plans in urban areas highly impacted by air pollution.

Author Contributions

Conceived and designed the experiments: PCG SRS. Performed the experiments: PCG SRS VPB DPO. Analyzed the data: PCG SRS MTGG MPMA MAM ESA. Contributed reagents/materials/analysis tools: SRS MTGG MPMA ESA. Wrote the paper: PCG SRS. Provided helpful comments on the manuscript: VPB DPO MTGG MPMA MAM ESA.

References

1. Souza SR, Blande J, Holopainen J (2013) Pre-exposure to nitric oxide modulates the effect of ozone on oxidative defenses and volatile emissions in lima bean. Environmental Pollution 179: 111–129.

2. Fares S, McKay M, Holzinger R, Golstein AH (2010) Ozone fluxes in a *Pinus ponderosa* ecosystem are dominated by non-stomatal processes: evidence from long-term continuous measurements. Agricultural and Forest Meteorology 150: 420–431.

3. Günthardt-Goerg MS, Vollenweider P, McQuattie CJ (2003) Differentiation of ozone, heavy metal or biotic stress in leaves and needles. Ekologia Bratislavia 22: 110–113.

4. Vollenweider P, Günthardt-Goerg MS (2006) Diagnosis of abiotic and biotic stress factors using the visible symptoms in the foliage. Environmental Pollution 140: 562–571.

5. Günthardt-Goerg MS, Vollenweider P (2007) Linking stress with macroscopic and microscopic leaf response in trees: New diagnostic perspectives. Environmental Pollution 147: 467–488.

6. Guerrero CC, Günthardt-Goerg MS, Vollenweider P (2013) Foliar Symptoms triggered by ozone stress in irrigated Holm Oaks from the city of Madrid, Spain. PLOS one, DOI 10.1371/journal.pone.0069171.

7. Moura BB, Souza SR, Alves ES (2011) Structural responses of *Ipomoea nil* (L.) Scarlet O'Hara (Convolvulaceae) exposed to ozone. Acta Botanica Brasilica 25: 122–129.

8. Moura BB, Souza SR, Alves ES (2013) Response of Brazilian native trees to acute ozone dose. Environmental Science and Pollution Research 21: 4220–4227.

9. Vollenweider P, Fenn ME, Menard T, Günthardt-Goerg MS, Bytnerowicz A (2013) Structural injury underlying mottling in ponderosa pine needles exposed to ambient ozone concentrations in the San Bernardino Mountains near Los Angeles, California. Trees 27: 895–911.

10. Bagard M, Le Thiec D, Delacote E, Hasenfratz-Sauder MP, Banvoy J, et al. (2008) Ozone-induced changes in photosynthesis and photorespiration of hybrid poplar in relation to the developmental stage of the leaves. Physiologia Plantarum 134: 559–574.

11. Fares S, Park JH, Ormeno E, Gentner DR, McKay M, et al. (2010) Ozone uptake by citrus trees exposed to a range of ozone concentrations. Atmospheric Environment 44: 3404–3412.

12. Loreto F, Pinelli P, Manes F, Kollist H (2004) Impact of ozone on monoterpene emission and evidence for an isoprene-like antioxidant action of monoterpenes emitted by *Quercus ilex* leaves. Tree Physiology 24: 361–367.

13. Vuorinen T, Nerg AM, Holopainen JK (2004) Ozone exposure triggers the emission of herbivore-induced plant volatiles, but does not disturb tritrophic signaling. Environmental Pollution 131: 305–331.

14. Fares S, Loreto F, Kleist E, Wildt J (2008) Stomatal uptake and stomatal deposition of ozone in isoprene and monoterpene emitting plants. Plant Biology 10: 44–54.

15. Logan BA, Monson RK, Potosnak MJ (2000) Biochemistry and physiology of foliar isoprene production. Trends in Plant Science 5: 477–481.

16. Loreto F, Mannozi M, Maris C, Nascetti P, Ferranti F, et al. (2001) Ozone quenching properties of isoprene and its antioxidant role in leaves. Plant Physiology 126: 993–1000.

17. Loreto F, Velikova V (2001) Isoprene produced by leaves protects the photosynthetic apparatus agains ozone damage, quanches ozone products, and reduces lipid peroxidation of cellular membranes. Plant Physiology 127: 1781–1787.

18. Peñuelas J, Llusià J (2002) Linking photorespiration, monoterpenes and thermotolerance in *Quercus*. New Phytologist 155: 227–237.

19. Sharkey TD, Chen X, Yeh S (2001) Isoprene increases thermotolerance of fosmidomycin-fed leaves. Plant Physiology 125: 2001–2006.

20. Sharkey TD, Wiberley AE, Donohue AR (2008) Isoprene emission from plants: why and how. Annals of Botany 101: 5–18.

21. Velikova V, Fares S, Loreto F (2008) Isoprene and nitric oxide reduce damages in leaves exposed to oxidative stress. Plant Cell and Environment 31: 1882–1894.

22. Vickers CE, Gershenzon J, Lerday MT, Loreto F (2009) A unified mechanism of action for volatile isoprenoids in plant abiotic stress. Nature Chemical Biology 5: 283–291.

23. Blande JD, Tiiva P, Oksanen E, Holopainen JK (2007) Emission of herbivore-induced volatile terpenoids from two hybrid aspen (*Populus tremula* x *tremuloides*) clone under ambient and elevated ozone concentration in the field. Global Change Biology 13: 2538–2550.

24. Holopainen JK, Gerhenzon J (2010) Multiple stresses factors and the emission of plant VOC. Trends in Plant Science 15: 1360–1385.

25. Pinto DM, Blande JD, Nykanen R, Dong WX, Nerg AM, et al. (2007) Ozone degrades common herbivore-induced plant volatiles: does this affect herbivore prey location by predators and parasitoids? Journal of Chemical Ecology 33: 683–694.

26. Shah J (2009) Plants under attack: systemic signals in defence. Current Opinion in Plant Biology 12: 459–464.

27. Hampel D, Mosandl A, Wüst M (2005) Induction of de novo volatile terpene biosynthesis via cytosolic and plastidial pathways by methyl jasmonate in foliage of *Vitis vinifera* L. Journal of Agricultural and Food Chemistry 53: 2652–2657.

28. Hudgins JW, Christiansen E, Franceschi VR (2004) Induction of anatomically based defense responses in stems of diverse conifers by methyl jasmonate: a phylogenetic perspective. Tree Physiology 24: 251–264.

29. Köllner TB, Lenk C, Schnee C, Kopke S, Lindemann P, et al. (2013) Localization of sesquiterpene formation and emission in maize leaves after herbivore damage. BMC Plant Biology 13: 15.

30. Martin D, Tholl D, Gershenzon J, Bohlmann J (2002) Methyl jasmonate induces traumatic resin ducts, terpenoids resin biosynthesis and terpenoids accumulation in developing xylem of Norway spruce stems. Plant Physiology 129: 1003–1018.

31. Bowler C, Fluhr R (2000) The role of calcium and activated oxygen as signals for controlling cross-tolerance. Trends in Plant Science 5: 241–246.

32. Franceschi VR, Horner HT (1980) Calcium oxalate crystals in plants. The Botanical Review 46: 361–427.

33. Franceschi VR, Nakata PA (2005) Calcium oxalate in plants: formation and function. Annual Review of Plant Biology 56: 41–71.

34. Nakata PA (2012) Engineering calcium oxalate crystal formation in *Arabidopsis*. Plant Cell Physiology 53: 1275–1282.

35. Molano-Flores B (2001) Herbivory and calcium concentration affect calcium oxalate crystal formation in leaves of *Sida* (Malvaceae). Annals of Botany 88: 387–391.

36. Ruiz N, Ward D, Saltz D (2002) Calcium oxalate crystals in leaves of *Pancratium sickenbergeri*: constitutive or induced defence? Functional Ecology 16: 99–105.

37. Xiang H, Chen J (2004) Interspecific variation of plant traits associated with resistance to herbivory among four species of *Ficus* (Moraceae). Annals of Botany 94: 377–384.

38. Fink S (1991) Unusual patterns in the distribution of calcium oxalate in spruce needles and their possible relationships to the impact of pollutants. New Phytologist 119: 41–51.

39. Alves ES, Tresmondi F, Longui EL (2008) Analise estrutural de folhas de *Eugenia uniflora* L. (Myrtaceae) coletadas em ambiente rural e urbano, SP, Brasil. Acta Botanica Brasilica 22: 241–248.

40. Moraes RM, Bulbovas P, Furlan C M, Domingos M, Meirelles ST, et al. (2006) Physiological responses of saplings of *Caesalpinia echinata* Lam., a Brazilian tree species, under ozone fumigation. Ecotoxicology and Environmental Safety 63: 306–312.

41. Fragoso FP, Varanda EM (2011) Flower-visiting insects of five tree species in a restored area of semideciduous seasonal forest. Neotropical Entomology 40: 431–435.

42. Moura BB (2013) Análises estruturais e ultraestruturais em folhas de espécies nativas sob influência de poluentes aéreos. PhD Thesis, Instituto de Botânica, São Paulo, Brazil.

43. Souza SR, Santana SRM, Rinaldi M, Domingos M (2009) Short-term leaf responses of *Nicotiana tabacum* 'Bel-W3' to ozone under the environmental conditions of Sao Paulo, SE, Brazil. Brazilian Archives of Biology and Technology 52: 251–258.

44. Souza E, Pagliuso JD (2009) Evaluation of the urban air quality in Brazilian Midwest Region. International Journal of Environmental Protection 2: 14–19.

45. Potvin C, Tardif S (1988) Source variability and experiential designs in growth chambers. Functional Ecology 2: 123–130.

46. Keller T, Schwager H (1977) Air pollution and ascorbate. European Journal of Forest Pathology 7: 338–350.

47. Osswald WF, Kraus R, Hipelli S, Bens B, Volpert R, et al. (1992) Comparison of the enzymatic activities of dehydroascorbic acid redutase, glutathione redutase, Catalase, peroxidase and superoxide dismutase of healthy and damaged spruce needles *Picea abies* (L.) Karst. Plant physiology 139: 742–748.

48. Klumpp G, Guderian R, Küppers K (1989) Peroxidase- und Superoxiddismutase-Aktivität sowie Prolingehalte von Fichtennadeln nach Belastung mit O_3, SO_2 und NO_2. European Journal of Forest Pathology 19: 84–97.

49. Lucero M, Estell R, Tellez M, Fredrickson E (2009) A retention index calculator simplifies identification of plant volatile organic compounds. Phytochemical Analysis 20: 378–384.

50. David R, Carde JP (1964) Coloration différentielle dês inclusions lipidique et terpeniques dês pseudophylles du pin maritime au moyen du reactif Nadi. Comptes Rendus de l'Academie des Sciences Paris 258: 1338–1340.

51. Machado SR, Gregorio EA, Guimaraes E (2006) Ovary peltate trichomes of *Zeyheria montana* (Bignoniaceae): developmental ultrastructure and secretion in relation to function. Annals of Botany 97: 357–369.

52. Johansen DA (1940) Plant microtechnique. New York: McGraw-Hill Books.

53. Novak K, Skelly JM, Schaub M, Krauchi N, Hug C, et al. (2003) Ozone air pollution and foliar injury development on native plants of Switzerland. Environmental Pollution 125: 41–52.

54. Goodger JQD, Hesjes AM, Woodrow IE (2013) Contrasting ontogenetic trajectories for phenolic and terpenoids defences in *Eucalyptus froggattii*. Annals of Botany 112: 651–659.

55. Barcikowski RS, Robey RR (1984) Decisions in single group repeated measures analysis: statistical test and three computer packages. The American Statistician 35: 148–150.

56. Hartikainen K, Nerg A-M, Kivimäenpää M, Kontunen-Soppela S, Mäenpää M, et al. (2009) Emissions of volatile organic compounds and leaf structural characteristics of European aspen (*Populus tremula*) grown under elevated ozone and temperature. Tree Physiology 29: 1163–1173.

57. Heiden AC, Hoffman T, Kahl J, Kley D, Klockow D, et al. (1999) Emission of volatile organic compounds from ozone-exposed plants. Ecological Applications 9: 1160–1167.

58. Llusià J, Peñuelas J, Gimeno BS (2002) Seasonal and species-specific response of VOC emissions by Mediterranean woody plants to elevated ozone concentrations. Atmospheric Environment 36: 3931–3938.

59. Luwe M, Heber U (1995) Ozone detoxification in the apoplasm and symplasm of Spinach, broad bean and beech leaves at ambient and elevated concentrations of ozone in air. Planta 197: 448–455.

60. Mehlhorn H, Cottam DA, Lucas PW, Wellburn AR (1987) Indication of ascorbate peroxidase and glutathione reductase activities by interactions of mixtures of air pollutants. Free Radical Research Communications 3: 1–5.

61. Peñuelas J, Llusià J, Gimeno BS (1999) Effects of ozone concentrations on biogenic volatile organic compounds emission in the Mediterranean region. Environmental Pollution 105: 17–23.

62. Loreto F, Schnitzler JP (2010) Abiotic stresses and induced BVOC. Trends in Plant Science 15: 1350–1385.

63. Loreto F, Fares S (2007) Is ozone flux inside leaves only a damage indicator? Clues from volatile isoprenoid studies. Plant Physiology 143: 1096–1100.

64. Fares S, Barta C, Brilli F, Centritto M, Ederli L, et al. (2006) Impact of high ozone on isoprene emission, photoshynthesis and histology of developing *Populus alba* leaves directly or indirectly exposed to the pollutant. Physiologia Plantarum 128: 456–465.

65. Shapiguzov A, Vainonen JP, Wrzaczek M, Kangasjarvi J (2012) ROS-talk – how the apoplast, the chloroplast and the nucleus get the message through. Frontiers in Plant Science 3: 1–9.

66. Pfannschmidt T (2003) Chloroplast redox signals: how photosynthesis controls its own genes. Trends in Plant Science 8: 33–41.

67. Hunter RJ (1994) Reconsidering the functions of latex. Trees 9: 1–5.

68. Optiz S, Kunert G, Gershenzon J (2008) Increased terpenoid accumulation in cotton (*Gossypium hirsutum*) foliage is a general wound response. Journal of Chemical Ecology 34: 508–522.

69. Hagel JM, Yeung EC, Facchini PJ (2008) Got milk? The secret life of laticifers. Trends in Plant Science 13: 631–639.

70. Fat E, Sanier C, Hebant C (1989) The distribution of plasmodesmata in the phloem of *Hevea brasiliensis* in relation to laticifer loading. Protoplasma 149: 155–162.

71. Pickard WF (2008) Laticifers and secretory ducts: two other tube systems in plants. New Phytologist 177: 877–888.

72. Garcia MTA, Galati BG, Hoc OS (2007) Ultraastructure of the corona of the scented and scentless flowers of *Passiflora* spp (Passifloraceae). Flora 202: 302–315.

73. Karban R (2007) Plant behaviour and communication. Ecology Letters 11: 727–729.

74. Pateraki I, Kanellis KA (2010) Stress and developmental responses of terpenoid biosynthetic genes in Cistus creticus subsp. Creticus. Plant Cell Reports 29: 629–641.

75. Baldwin IT, Halitschke R, Paschold A, von Dah CC, Preston C (2006) Volatile signaling in plant-plant interactions: "Talking Trees" in the genomics era. Science 311: 811–813.

76. Blande JD, Korjus M, Holopainen JK (2010) Foliar methyl salicylate emissions indicate prolonged aphid infestation on silver birch and black alder. Tree Physiology 30: 404–416.

77. Kawano T, Boteau F (2013) Crosstalk between intracellular and extracellular salicylic acid signaling events leading to long-distance spread of signals. Plant Cell Reports 32: 1125–1138.

78. McAinsh MR, Evans NH, Montgomery LT, North KA (2002) Calcium signalling in stomatal responses. New Phytologist 153: 441–447.

79. Volk GM, Gossi GM, Franceschi VR (2004) Calcium channels are involved in calcium oxalate crystal formation specialized cells of *Pistia stratiotes* L. Annals of Botany 93: 741–753.

7

Identification of Antioxidant Capacity-Related QTLs in *Brassica oleracea*

Tamara Sotelo*, María Elena Cartea, Pablo Velasco, Pilar Soengas

Group of Genetics, Breeding and Biochemistry of Brassicas, Department of Plant Genetics, Misión Biológica de Galicia, Spanish Council for Scientific Research (MBG-CSIC), Pontevedra, Spain

Abstract

Brassica vegetables possess high levels of antioxidant metabolites associated with beneficial health effects including vitamins, carotenoids, anthocyanins, soluble sugars and phenolics. Until now, no reports have been documented on the genetic basis of the antioxidant activity (AA) in *Brassica*s and the content of metabolites with AA like phenolics, anthocyanins and carotenoids. For this reason, this study aimed to: (1) study the relationship among different electron transfer (ET) methods for measuring AA, (2) study the relationship between these methods and phenolic, carotenoid and anthocyanin content, and (3) find QTLs of AA measured with ET assays and for phenolic, carotenoid and anthocyanin contents in leaves and flower buds in a DH population of *B. oleracea* as an early step in order to identify genes related to these traits. Low correlation coefficients among different methods for measuring AA suggest that it is necessary to employ more than one method at the same time. A total of 19 QTLs were detected for all traits. For AA methods, seven QTLs were found in leaves and six QTLs were found in flower buds. Meanwhile, for the content of metabolites with AA, two QTLs were found in leaves and four QTLs were found in flower buds. AA of the mapping population is related to phenolic compounds but also to carotenoid content. Three genomic regions determined variation for more than one ET method measuring AA. After the syntenic analysis with *A. thaliana*, several candidate genes related to phenylpropanoid biosynthesis are proposed for the QTLs found.

Editor: Sonia Osorio-Algar, University of Malaga-Consejo Superior de Investigaciones Científicas, Spain

Funding: This work was supported by the National Plan for Research and Development (AGL-2009-09922). Tamara Sotelo acknowledges a pre-doctoral research grant (F.P.I.) from the Ministry of Economy and Competitiveness. The funders had no role in study design, data collection and analysis, decision to publish, or preparation of the manuscript.

Competing Interests: The authors have declared that no competing interests exist.

* Email: tsotelo@mbg.csic.es

Introduction

Brassicaceae plants represent one of the major vegetable crops grown worldwide, with *Brassica oleracea* L. (2n = 18) as the main *Brassica* species consumed in Europe and the USA. Cruciferous vegetables, in particular those included in the *Brassica* genus, are an important part of the diet as they provide a multitude of nutrients and bioactive compounds [1]. A high consumption of *Brassica* vegetables reduces the risk of age-related chronic illnesses, degenerative diseases [2] and several types of cancer [3]. Human health benefits associated to *Brassica* consumption could be attributed, in part, to the large amount of constituents having strong antioxidant activity (AA). In fact, AA of *Brassica* vegetable extracts is higher compared to that of other vegetable crops like green pepper, carrot, potato or green bean [4]. Antioxidants have long been recognized to have protective functions against oxidative damage and are associated with a reduced risk of chronic diseases [5]. *Brassica* vegetables possess high levels of antioxidant metabolites associated with beneficial health effects, including vitamins (especially vitamin A, C, E, K and B-6), carotenoids (such as γ- and β-carotene and zeaxanthin), anthocyanins, folate, soluble sugars and phenolic compounds which are known to be the major antioxidants of *Brassica* crops [6–14].

Due to the complexity of food composition, separating each antioxidant compound and studying it individually is costly and inefficient. In addition, there might be synergistic interactions among the antioxidant compounds [15]. There are numerous methods for measuring the total AA of a plant extract *in vitro*. The 2- single electron transfer reaction based assays (ET) measure the reducing capacity of the samples. The ET group includes different methods like the ferric ion reducing antioxidant power assay (FRAP), and the AA measured with the reagents ABTS (2, 2'-azino-bis (3-ethylbenzthiazoline-6-sulphonic acid)) and DPPH (2, 2-diphenyl-1-picrylhydrazyl), among others [15]. Generally speaking, correlations found among these three methods are high in Brassica extracts. Soengas *et al.* [16] found that the correlation between DPPH and FRAP was 0.8 when analyzing several *B. oleracea* crops. Kusznierewicz *et al.* [17] found a correlation of 0.96 between ABTS and DPPH in white cabbage. Zhi *et al.* (2011) [18] found correlations ranging from 0.76 to 0.82 among the three cited methods when analyzing different vegetables, including broccoli. In most studies, several ET methods are often used in order to measure the AA of a sample, but theoretically it could be possible to choose only one because of the high correlations among assays.

Phenolic compounds are known to be the major group with antioxidant capacity in *Brassica* crops [13]. These compounds are able to scavenge reactive oxygen species due to their electron donating properties. The most widespread and diverse group of polyphenols in *Brassica* species are flavonoids and hydroxycinnamic acids. In many *in vitro* studies, phenolic compounds demonstrated higher AA than other antioxidants, such as vitamins and carotenoids [19].

Several studies have demonstrated that highly pigmented cultivars of some vegetables (i.e. cabbage, cauliflower) possess stronger AA than their respective light-colored cultivars [20–22]. This could indicate that pigments '*per se*' have AA. Carotenoids are a diverse group of more than 600 natural pigments that accumulate in the plastids of some vegetables leaves, flowers and fruits [23]. Some carotenoids are essential nutrients for humans, while others have protective effects against several diseases. Anthocyanins are natural pigments responsible for the blue, purple, red and orange colors in the major parts of all higher plants and have attracted much interest due to their impact on the sensorial characteristics of food products, as well as their health-related properties through various biological activities [24,25]. The AA of *Brassica* crops has been mainly related to phenolic compounds and vitamin C. However, carotenoids and anthocyanins could also play an important role.

Comparisons of *in vitro* AA of the main *B. oleracea* crops demonstrated that broccoli, kale and red cabbage show high AA [17,26]. Soengas *et al.* [16] compared the AA of six *Brassica* crops, including broccoli, cabbage, cauliflower, kale, nabicol and tronchuda cabbage, at four different plant stages with DPPH and FRAP assays. They found that kale and broccoli had the highest AA. Nilson *et al.* [27] found that AA of curly kale was at least 10-fold higher than that of cauliflower and white cabbage. At present, there are many studies about AA of *Brassica* crops because of the health related properties of antioxidants. However, as far as we know, there are no repots about genetics and heredity associated with AA in the *Brassica* genus.

QTL analysis is a very important tool in order to study the genetic base of AA. For the last decades, quantitative trait mapping has been the most common approach in order to analyze complex traits and measure the association of genetic markers with phenotypic variation. Identification of QTLs is essential for the understanding of the quantitative genetic control of AA and it is an early step in order to identify and estimate the gene number controlling each trait variation. The high co-linearity between *A. thaliana* and *Brassica* species can be used for identifying candidate genes underlying QTLs that affect AA. To our knowledge, this is the first report on the genetic basis of AA in *Brassica* crops. In other crops, only Jin *et al.* [28] in rice, Dobson *et al.* [29] in raspberry and Hayashi *et al.* [30] in lettuce studied QTLs for total water soluble AA and total phenolic, anthocyanin and carotenoid contents.

For this reason, the aims of our research were 1) to study the relationship among different ET methods for measuring AA, 2) to study the relationship between these methods and phenolic, carotenoid and anthocyanin contents and 3) to find QTLs of AA measured with ET assays and for phenolic, carotenoid and anthocyanin contents in two organs of a DH population of *B. oleracea* as an early step in order to identify genes related to these traits.

Materials and Methods

Chemicals

DPPH (2,20-diphenyl-1-picrylhydrazyl), TPTZ (2,4,6-tripyri-dyl-striazine), Trolox (6-hydroxy-2,5,7,8-tetramethylchroman-2-carboxylic acid), hydrochloric acid, phenolics reagent, ABTS (2, 2'-azino-bis (3-ethylbenzthiazoline-6-sulphonic acid)), potassium persulphate and gallic acid were obtained from Sigma–Aldrich Chemie GmbH (Steinheim, Germany); ferric chloride and

methanol were obtained from Panreacquimica S.A. (Castellar del Vallés, Spain).

Plant material and growing environments

The double haploid (DH) mapping population employed in this study (BolTBDH) was created from an F_1 individual, derived by crossing a DH broccoli line 'Early Big'(P_2) and a DH rapid cycling of Chinese kale line (TO1000DH3,P_1) [31]. Parents and 155 DH lines were grown in autumn 2011 (from September to November) and stored in the greenhouse under controlled conditions: 16 h of daylight and a temperature of $24\pm2°C$; 8 h of darkness having $18\pm2°C$ at night; and a relative humidity of 55% in order to obtain enough seed in the same environmental conditions. Plants were sown in a completely randomized experiment with two replications and four plants per replication. Two sample types were collected and analysed: leaves (one month after sowing) and flower buds (taken sequentially depending on the maturity of each line). Bulks of individual samples were taken from each replication. Samples were frozen *in situ* in liquid N_2, immediately transferred to the laboratory and frozen at $-80°C$. All samples were freeze-dried (BETA 2–8 LD plus, Christ) for 72 h. The dried material was powdered by using an IKA-A10 (IKA-Werke GmbH & Co.KG) mill, and the fine powder was used for methanolic extractions.

Evaluation of AA

Freeze-dried and ground samples (10 mg) were extracted with 1 ml of 80% aqueous methanol in dark maceration for 24 h. After centrifugation (3700 rpm, 5 min), methanolic extracts were employed in order to determine AA (FRAP, DPPH and ABTS) of the mapping population. All AA assays and the content of metabolites with AA were carried out spectrophotometrically by using a microplate spectrophotometer (Spectra MR; Dynex Technologies, Chantilly, VA). Two repetitions were made for each sample and analysis. Standards prepared with different concentrations of Trolox (0, 0.008, 0.016, 0.024, 0.032, 0.04 mM) were measured for FRAP, DPPH and ABTS analyses and AA values were normalized to Trolox equivalents per gram of dry weight.

FRAP assay

The ferric reducing antioxidant activity (FRAP) assay of Benzie and Strain [32] was measured in all samples. Fresh FRAP reagent was prepared by mixing 10 volumes of 300 mM acetate buffer (pH 3.6), one volume of 10 mM TPTZ in 40 mM hydrochloric acid and one volume of 20 mM ferric chloride, and then incubating at 37°C for 5 minutes. For each analysis, 30 µl of methanolic solution of the two organs (leaves and flower buds) were added to 20 µl of distilled water and 250 µl of fresh FRAP solution and mixed thoroughly. The increase in absorbance was recorded at 593 nm after 20 min.

DPPH radical scavenging activity

The antioxidant activity by the DPPH method was determined by monitoring the disappearance of the radical DPPH spectro-photometrically, according to Brand-Williams *et al.* [33]. The working DPPH reagent was prepared by dissolving DPPH in methanol to a final concentration of 75 µM. Fifty microliters of extract for leaves and 35 µl for flower buds were added to 250 µl of freshly prepared DPPH reagent and mixed thoroughly. Readings were taken at 517 nm after 30 min of incubation in the dark at room temperature.

ABTS+ radical scavenging activity

The method of decolorization of free radical ABTS+ employed was a modified version of that used by Samarth et al. [34] and initially reported by Re et al. [35]. ABTS+ was generated by oxidation of ABTS 7 mM with potassium persulphate 2.45 mM in water, at room temperature for 16 h. For each analysis, the ABTS+ solution was freshly diluted with water in order to obtain an initial absorbance around 0.8 at 734 nm. An aliquot of 20 μl methanolic extract for leaves and 30 μl for flower buds were added to 250 μl of ABTS+ solution. Absorbances were measured at 734 nm after 30 min of incubation in the dark at room temperature.

Quantification of phenolic content

The total phenolic content of the extracts was determined according to the phenolic colorimetric method described by Dewanto et al. [36]. The same methanolic extracts employed for AA assays were employed in order to determine phenolic content. Extracts were oxidized with 50 μl of 0.5 M Folin reagent. After 5 min, 200 μl of a 20% Na_2CO_3 solution were added in order to neutralize the reaction. Absorbances were measured at 760 nm after 2 h of incubation in the dark at room temperature. Standards prepared with different concentrations of gallic acid (0, 0.008, 0.016, 0.024, 0.032 and 0.04 mM) were also measured. Results were expressed in terms of micromoles of gallic acid equivalents per gram of dry weight.

Quantification of carotenoid content

Carotenoid content was determined according to Sims & Gamon [37] with minor modifications. Lyophilized samples (10 mg) were ground in 1 ml cold acetone/Tris buffer solution (80:20 vol:vol, pH = 7.8). Samples were mixed overnight in the dark at room temperature; afterwards, the absorbance of samples was measured at 537, 647 and 663 nm. Carotenoid content was computed by following the equations of Sims & Gamon [37] and results were expressed in micromoles per gram of dried weight.

Quantification of anthocyanin content

Anthocyanin content was determined according to Murray et al. [38] with minor modifications. Lyophilized samples (10 mg) were ground in 1 ml of cold methanol/HCL/water (90:1:1, vol:vol:vol). Samples were mixed overnight in the dark at room temperature. The absorbance of samples was measured at 529 and 650 nm and anthocyanin content was determined by using the equation described in Sims & Gamon [37]. Results were expressed in micromoles per gram of dried weight.

Statistical and QTL analysis

A combined analysis of variance across organs and individual analyses of variance for each organ were made for the AA content measured ABTS, DPPH, FRAP assays and for phenolic, carotenoid and anthocyanin contents by using the procedure ANOVA of SAS v 9.2 [39]. Parental differences were analyzed one-tail "t" test by using PROC TTEST of SAS v 9.2 [39]. Simple correlation coefficients were computed with PROC CORR of SAS v 9.2 [39] for each trait.

The genetic map created by Iñiguez-Luy et al. [31] has 279 markers (SSRs and RFLPs) distributed along nine linkage groups (C1–C9) with a total distance of 891.4 cM and a marker density of 3.2 cM/marker. Quantitative trait locus mapping was carried out through a composite interval mapping method [40] by using PLABQTL [41]. Individual analyses were carried out for each trait and organ (leaves and flower buds). A likelihood odds (LOD)

threshold was chosen for each trait in order to declare the putative QTL significant by following a permutation test, with N = 1000, and a critical alpha value of 25%. The confidence intervals were set to 95%. The analysis and cofactor election were carried out by following PLABQTL's recommendations, using an 'F-to-enter' and an 'F-to-delete' value of 7.

The proportion of phenotypic variance explained for a specific trait was determined by the adjusted coefficient of determination of regression (R^2) fitting a model which includes all detected QTLs [42]. Fivefold cross-validation of QTLs was performed by following the procedures described by Utz et al. [43]. The whole data set was randomly split into k = 5 data subsets. Four of these subsets were combined to form the estimation set (ES). The remaining subset formed the test set (TS), in which predictions derived from ES were tested for their validity by correlating predicted and observed data. We used 1,000 replicate CV/G runs. Estimates of medians and percentiles and the frequency of QTL detection in ES and TS were calculated over all replicated CV/G runs. The frequency of QTL detection gives us an estimation of the precision of QTL localization. The PLABQTL [41] software package was used for all calculations. Iñiguez-Luy et al. (2009) identified collinear genomic blocks between the BolTBDH mapping population and A. thaliana by using a synteny analysis. This information was employed in order to locate candidate genes which may directly account for QTLs in B. oleracea. By following this approach, we searched in the database TAIR (the Arabidopsis information resource http://www.arabidopsis.org) genes related to phenylpropanoid biosynthetic process metabolism (phenolic compounds and anthocyanins are synthetized following this pathway) and genes involved in the carotenoid biosynthetic process by including the words 'phenylpropanoid' and 'carotenoid' into the field 'description of the gene in TAIR. Twenty one genes related to phenylpropanoids and 24 genes related to carotenoids were found. We tried to locate these genes on the BolTBDH map by means of in silico mapping.

Results

Quantitative variation for methods measuring AA and the content of metabolites with AA

In this study AA in leaves and flower buds was determined by three ET methods: FRAP, DPPH and ABTS. The content of metabolites with AA (phenolics, anthocyanins and carotenoids) was also determined. We used two ET methods (DPPH and ABTS) where the scavenging was followed by monitoring the decrease in absorbance over time, which occurred due to the AA of the sample [44]. For the FRAP assay, the extract shows an increase of absorbance over time dependent on their AA [45]. A transgressive distribution was found for all traits in both organs (Fig. 1). Results obtained from each analysis are considered below.

FRAP, DPPH and ABTS assays

Mean values for the FRAP and DPPH methods in the population were lower than the corresponding values of ABTS assay in both organs (leaves and flower buds). In leaves, we found mean values of 18.36, 14.04 and 24.78 μmol Trolox g−1 DW in FRAP, DPPH and ABTS assays, respectively. In flower buds, we found values of 15.37, 12.51 and 25.16 μmol Trolox g−1 DW in FRAP, DPPH and ABTS assays, respectively (Table 1).

Population mean values between the two organs present highly significant differences for FRAP (F = 75.95, P = 0.0129) and DPPH (F = 65.09, P = 0.0150) methods.

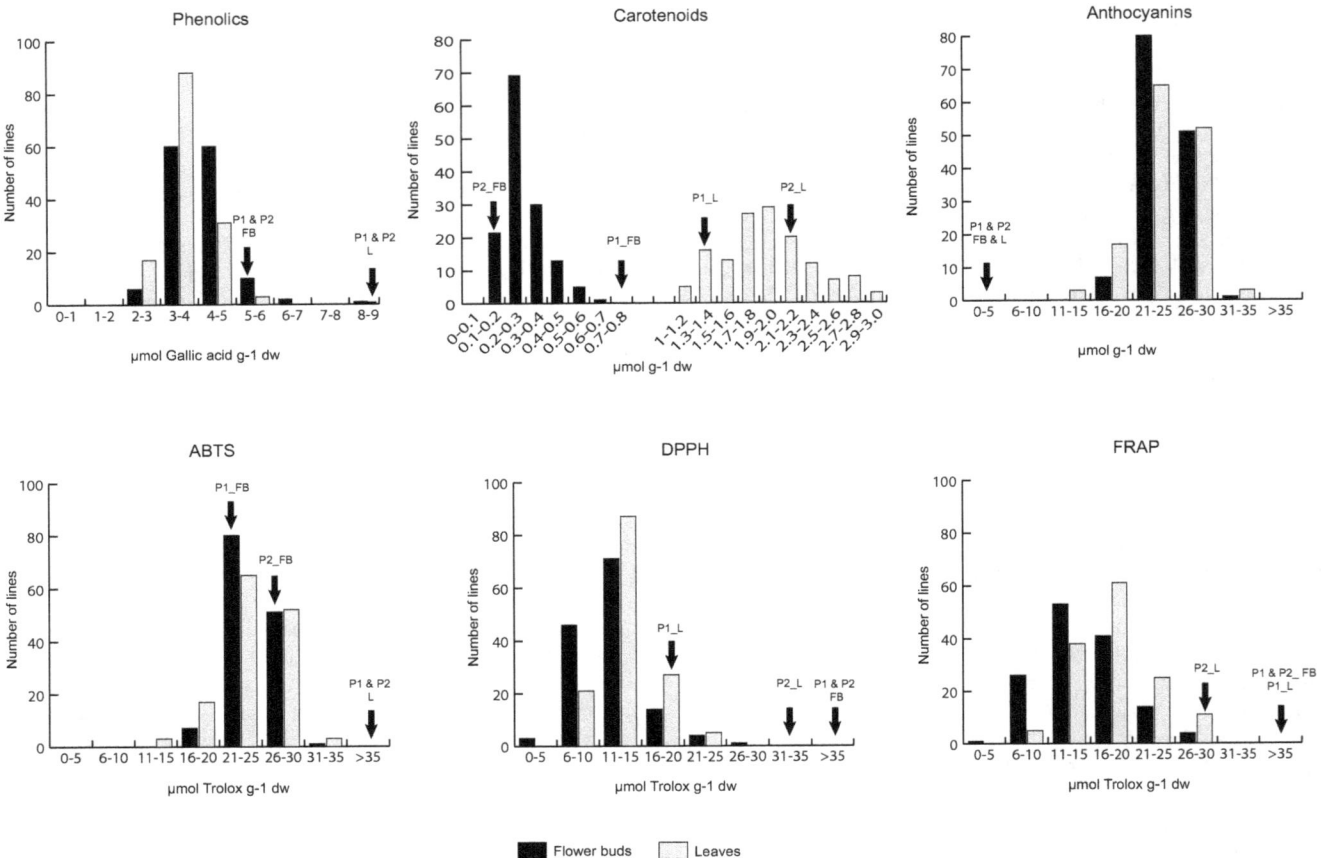

Figure 1. Distribution of the three metabolites with antioxidant activity, carotenoids, anthocyanins and phenolics and the three antioxidant assay methods, ABTS, DPPH and FRAP in the BoITBDH population. Arrows indicate values for the P1 (DH rapid cycling of Chinese kale TO1000DH3) and P2 (DH broccoli line 'Early Big') in the two organs under study, leaves (L) and flower buds (FB).

Metabolites with AA: phenolic, anthocyanin and carotenoid content

Concerning the content of metabolites with AA, we found two different profiles. For the phenolics assay, population showed higher mean values in flower buds than in leaves (4.14 and 3.64 µmol gallic acid g−1 DW, respectively), although differences were not significant. However, both parental lines had higher phenolic content in leaves than in flower buds (Fig. 1).

Leaves of the mapping population had higher anthocyanin and carotenoid content (58.53 µmol g−1 DW and 1.98 µmol g−1 DW, respectively) compared to flower buds (13.2131 µmol g−1 DW and 0.28 µmol g−1 DW, respectively). Mean anthocyanin content of the population represents a strong increase compared to the values found in both parents. As other assays previously described, anthocyanins presented transgressive distributions for both organs (Fig. 1). In the case of carotenoid content, differences between both organs were highly significant (F = 80.44, P = 0.012). Correlation coefficients among methods measuring AA, phenolic

Table 1. Antioxidant activity of parents and population measured in leaves and flower buds with three different antioxidant assay systems and the content of three metabolites with antioxidant activity.

Traits	Leaves			Flower buds		
	P1	P2	Population mean	P1	P2	Population mean
ABTS (µmol Trolox g^{-1} DW)	42.06	44.89	24.78	21.13	30.94	25.16
DPPH (µmol Trolox g^{-1} DW)	20.20	34.18	14.04	50.65	47.84	12.51
FRAP (µmol Trolox g^{-1} DW)	48.17	56.27	18.36	59.40	28.71	15.37
PHENOLICS (µmol Gallic Acid g^{-1} DW)	8.02	8.91	3.64	5.55	5.54	4.14
ANTHOCYANINS (µmol g^{-1} DW)	0.03	0.67	58.53	0.04	0.13	13.31
CAROTENOIDS (µmol g^{-1} DW)	1.48	2.17	1.98	0.84	0.17	0.28

and pigment contents in the BolTBDH population were made. Pairwise correlations between AA measured with three ET assays (FRAP, DPPH and ABTS) were positive and highly significant (P≤0.01) for both leaves and flower buds in the correlation analysis carried out with all lines of the mapping population. However, correlation coefficients were moderately low (Table 2). The highest correlations occurred between DPPH and FRAP assays for both organs. The correlation values were 0.486 in flower buds and 0.526 in leaves. On the other hand, correlation coefficients between the content of phenolic compounds and the three AA methods were positive and significant for both organs (p≤0.01). Significant correlations between the anthocyanin content with DPPH and ABTS were found in leaves. Correlation with DPPH was positive; however, correlation with ABTS was negative (r = −0.339, p≤0.01) (Table 2). Anthocyanin content was significantly and negatively correlated to ABTS assay (Table 2). Carotenoid content showed significant correlation coefficients with the AA measured with ABTS assay (r = 0.140, p≤0.05) in leaves, and significant and positive correlation coefficients with FRAP assay in flower buds (r = 0.305, p≤0.01). Furthermore, correlation between carotenoids and ABTS assay was negative and highly significant in flower buds (r = −0.165, p≤0.01) (Table 2).

QTL mapping for methods measuring AA, phenolic and pigment contents in the BolTBDH population

A total of 19 QTLs were detected for all traits. The number of QTLs by linkage group ranged between one in C9 and five in C3 (Fig. 2). For methods measuring AA, seven significant QTLs were found in leaves. The value of R^2 ranged between 9.8% for FRAP in C3 and 17.4% for DPPH in C4, respectively (Table 3). Three of these QTLs had a frequency of cross-validation higher than 50%. In flower buds, six significant QTLs were found. R^2 value varied between 9.8% for ABTS in C6 and 12.1% for FRAP content in C3, but only two of the QTLs had a frequency of cross-validation higher than 50%.

For the content of metabolites with AA, two significant QTLs for phenolic content were found in leaves. The value of R^2 ranged between 10.3 and 10.4% in C7 and all of them had a frequency of cross-validation higher than 50%. Meanwhile, four significant QTLs were found in flower buds. The value of R^2 ranged between 9.9 and 12.6% for carotenoids in C5 and C9, respectively. Only one of these QTLs presents a frequency of cross-validation higher than 50%. One QTL for anthocyanin content was found on C3 in flower buds, from which a R^2 value of 10.9% and three QTLs for carotenoid content were found on C5, C8 and on C9. R^2 values varied between 9.9 and 12.6% (Table 3).

Based on the position of QTLs and taking into account their confidence interval, three genomic regions determined variability for different traits. The genomic region located on C3, in the interval from marker pW125dE to fito156c & pW133cH (AA-C3), determined variation for the three different methods measuring AA: FRAP in leaves and ABTS and DPPH in flower buds. A second genomic region on C7 from pW225aD to pW104aE (AA-C7) determined variation for the methods measuring AA (ABTS in leaves and FRAP in flower buds) and phenolic content in leaves. Alleles for increasing AA or phenolic content are given by P2 in both genomic regions on C3 and C7. A third genomic region on C5 (AA-C5), from pW209aH to Na10-F06b & fito132a, also determined variation for the methods measuring AA (DPPH in leaves and ABTS in flower buds) and carotenoid content in flower buds. In this case, alleles for increasing AA and carotenoid content are given by P1.

Genes related to phenylpropanoid biosynthesis were located by means of *in silico* mapping in the confidence interval of several QTLs (Table 4). However no gene related to carotenoid biosynthesis could be located.

Discussion

Quantitative variation for methods measuring AA and the content of metabolites with AA

Parents of the DH BolTBDH mapping population showed significant differences for the majority of the methods measuring AA and for the content of metabolites with AA in leaves and flower buds. BolTBDH population was found to be an ideal material in order to study QTLs for the traits under study in *Brassica* genus due to the differences between the two parents of this population. One parent (P2) is a broccoli 'Early Big' line, the *Brassica* crop with one of the highest AA [46], while the other parent (P1) is a DH rapid cycling line (TO1000DH3). Both parents are from different cultivars and as stated before, there is high variability for AA between different *Brassica* crops [16,26,34,47].

The total AA of a sample can be measured by using several methodologies [15–17,26]. The radical scavenging capacity of DH BolTBDH mapping population was measured by using three ET methods: ABTS FRAP and DPPH. The content of metabolites with AA like phenolics, anthocyanin and carotenoid was also measured. Some DH lines exhibited mean values of the traits falling between the values of the two parents, but others exhibited values which were extremely higher or lower than their parents. This phenomenon is referred to as transgressive segregation. Distributions of the methods measuring AA, phenolics and

Table 2. Correlation coefficients for leaves (above the diagonal) and flower buds (below the diagonal) between the three antioxidant assay methods and the content of three metabolites with antioxidant activity (n = 280).

Leaves/Flower buds	ABTS	FRAP	DPPH	PHENOLICS	ANTHOCYANS	CAROTENOIDS
ABTS	–	0.197**	0.267**	0.434**	−0.339**	0.140*
FRAP	0.189**	–	0.526**	0.151*	0.103	0.100
DPPH	0.389**	0.486**	–	0.250**	0.164**	0.051
PHENOLICS	0.633**	0.221**	0.227**	–	−0.110	0.086
ANTHOCYANINS	−0.130*	−0.027	−0.076	−0.100	–	−0.081
CAROTENOIDS	−0.165**	0.305**	0.005	−0.013	0.176**	–

* Significant at p≤0.05, and ** significant at p≤0.01. ABTS: 2, 2'-azino-bis (3-ethylbenzthiazoline-6-sulphonic acid); FRAP: ferric ion reducing antioxidant power assay; DPPH: 2,2-diphenyl-1-picrylhydrazyl.

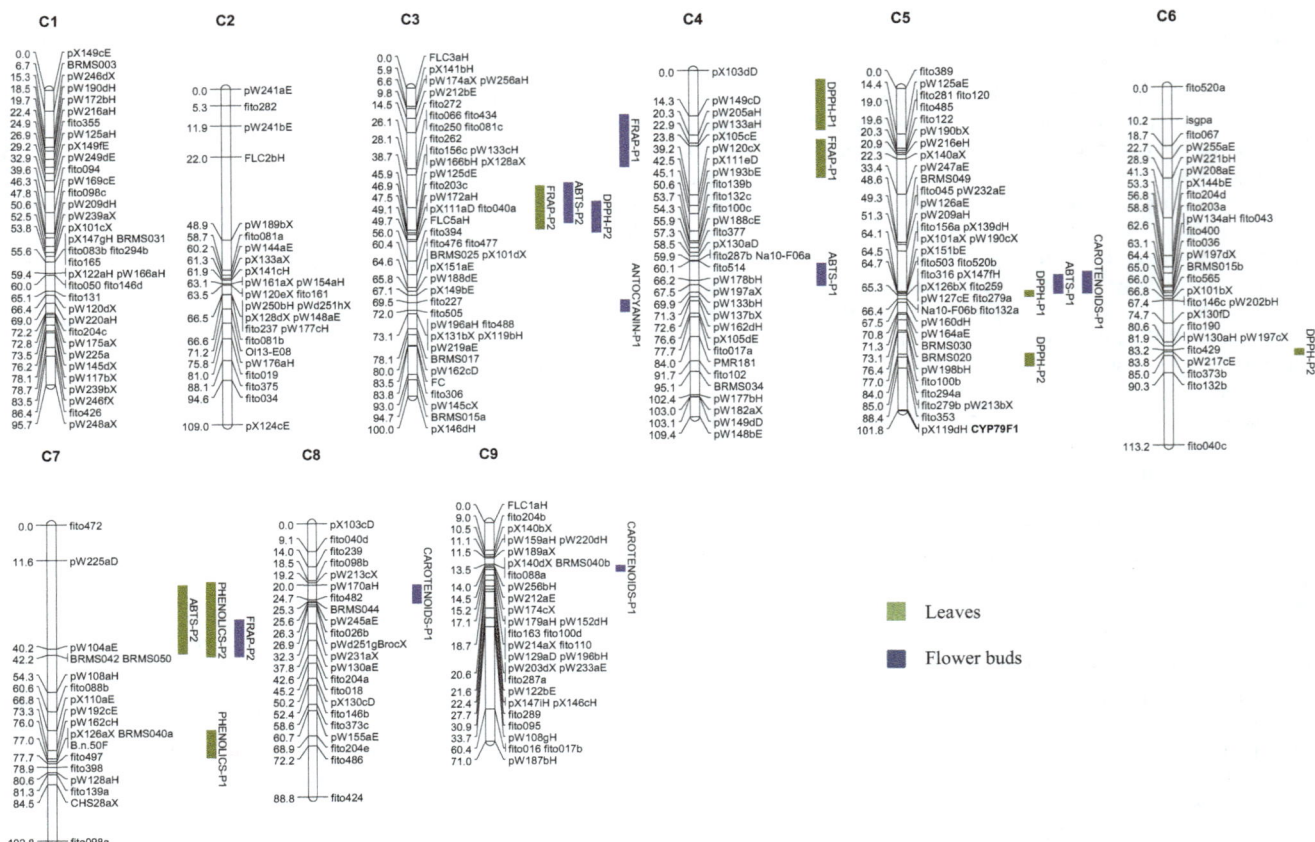

Figure 2. Framework map of DH population showing nineteen metabolic quantitative trait loci (QTL) for individual methods measuring AA. Linkage groups were labeled following the nomenclature of Iñiguez-Luy *et al.* [31]. Bars represent the LOD confidence interval of each QTL. QTLs are in different colors depending on the plant organ: leaves (green) and flower buds (blue). After the name of each QTL P1 indicates allele from, DH rapid cycling of Chinese kale (TO1000DH3) and P2 indicates allele from DH broccoli line 'Early Big'.

pigment content were, in most cases, transgressive. The action of complementary genes may be the primary cause of transgression, although epistasis may also contribute [48]. Further studies could help to explain the transgressive segregation of the traits measured in this study. These studies could use other populations or add more molecular markers to our population.

Total AA varied considerably according to the organ under study. Generally speaking, leaves present higher AA and content of metabolites with AA than flower buds, as it was expected by their photosynthetic complex. This result is in agreement with Soengas *et al.* [16] and Llorach *et al.* [49], who measured the AA of heads and leaves of cauliflower, with the highest values found in leaves. Guo *et al.* [50] found similar values in both organs in broccoli and Soengas *et al.* [16] found that broccoli flower buds have higher AA than leaves. In broccoli and cauliflower, the organs which are consumed are the heads (flower buds) and the leaves surrounding the heads are treated as by-products. Our results show that leaves have more AA and content of metabolites with AA than heads. Therefore, consumption of broccoli by-products, which is one of the parents of the mapping population, could be an interesting option to include in the human diet.

Due to the characteristics of the methods analyzed, AA measured with FRAP and DPPH assays present lower values compared to that of ABTS assay. It is coincident with the results found by Gouveia *et al.* [51] in other species like *Andryala glandulosa*.

Correlation coefficients among methods measuring AA and the content of metabolites with AA

Significant correlation coefficients were found among the three methods measuring AA (FRAP, DPPH and ABTS) in the two organs under study, and ranged between 0.19 and 0.53. These correlations, although significant, were lower than others found in previous studies. Kusznierewicz *et al.* [17] found a correlation of 0.96 between ABTS and DPPH in white cabbage planted in different locations. Soengas *et al.* [16] found a correlation of 0.8 between DPPH and FRAP in extracts of different *Brassica* crops. Zhi *et al.* [18] found correlations ranging from 0.76 to 0.82 between the three cited methods analyzing different vegetables including broccoli. The material studied in our research is much closer genetically than the material studied in previously cited literature, since all the DH lines derive from a single cross. Clearly, correlations among ET methods depend on the material under study and based on our results, we recommend using more than one ET method in order to estimate the AA of a sample as suggested by Kurniereick *et al.* [17] and Gawlik-Dziki [52].

Significant correlations among the three methods measuring AA and the content of metabolites with AA were found in leaves and flower buds. Phenolic content was positively correlated with all the methods measuring AA. The correlation coefficient with ABTS showed the highest value in both organs. Several authors have found significant and high correlations (ranging from 0.7 to 1) between the AA measured with ABTS, DPPH and FRAP assays

Table 3. List of quantitative trait loci (QTL) for antioxidant activity and the content of metabolites with antioxidant activity in two plant organs under study, leaves and flower buds.

	Plant organ	Trait	Linkage group	Peak Position range (cM)	Left marker	Lod theshold	Right marker	LOD score	Frequency	Add	R²%	adj R²%
1	**Leaves**	**ABTS**	7	31 (20–42)	pW225aD	2.89	pW104aE	4.53	829	1.6354	14.2	12.5
2	**Leaves**	**DPPH**	4	12 (3–19)	pX103dD	2.85	pW149cD	5.45	972	−1.794	17.4	27.2
3			5	65 (64–66)	fito316, pX147fH, pX126bX, fito259, pW127cE & fito279a		Na10-F06b & fito132a	4.5	764	−2.112	14.1	
4			5	85 (84–88)	fito294a			3.83	605	2.202	12.2	
5			6	84 (83–85)	pW217cE		fito279b & pW213bX	3.64	249	3.455	11.6	
6	**Leaves**	**PHENOLICS**	7	34 (19–43)	pW225aD	3.02	pW104aE	3.23	278	0.294	10.4	10.5
7			7	73 (67–76)	pX110aE		pW192cE	3.2	213	−0.268	10.3	
8	**Leaves**	**FRAP**	3	41 (32–46)	fito156c, pW133cH, pW166bH & pX128aX	2.86	pW125dE	3.05	299	2.784	9.8	11
9			4	25 (22–34)	pX105cE		pW120cX	4.29	794	−2.518	13.5	
10	**Flower buds**	**ABTS**	3	38 (31–44)	fito262	2.86	fito156c, pW133cH, pW166bH & pX128aX	2.98	260	1.329	9.8	6.6
11			4	64 (61–68)	fito514		pW178bH	3.49	501	−2.441	11.4	
12			5	64 (59–65)	pW209aH		fito156a, pX139dH, pX101aX & pW190cX	3.28	348	−1.141	10.7	
13	**Flower buds**	**DPPH**	3	45 (37–47)	fito156c, pW166bH & pX128aX	2.86	pW125dE	3.11	48	1.402	10.2	1.3
14	**Flower buds**	**FRAP**	3	12 (9–26)	pW212bE	2.83	fito272	3.38	462	−2.325	11	6.8
15			7	40 (31–43)	pW225aD		pW104aE	3.72	631	2.538	12.1	
16	**Flower buds**	**Anthocyanin**	3	72 (69–73)	fito227	3.12	pW196aH, fito488, pX131bX, pX119bH & pW219aE	3.26	361	−4.458	10.9	0.2
17	**Flower buds**	**Carotenoid**	5	64 (58–65)	pW209aH	2.93	fito156a, pX139dH, pX101aX & pW190cX	2.94	226	−0.044	9.9	21.2
18			8	22 (20–26)	pW170aH		fito482	3.27	308	−0.047	10.9	
19			9	15 (14–16)	pW212aE		pW174cX	3.8	583	−0.049	12.6	

Additive effect was calculated as $(P_2-P_1)/2$; R²% coefficient of determination of each QTL. Adj R²%: determination coefficient of each trait.

Table 4. List of phenylpropanoid biosynthesis candidate genes residing within the QTL confidence intervals according to organ and measurement method.

Plant organ	Trait	Markers in the confidence interval	Position in Brassica oleracea (cM)*	Brassica oleracea linkage group	Linkage group and position (bp) in Arabidopsis thaliana*	Genes related to phenylpropanoid biosynthesis located in the interval of Arabidopsis thaliana*	Candidate genes
Leaves	FRAP	fito156c	38.72	3	1(3530200-3530221)		
		pW133cH	38.72	3	2 (15610858-15610982)		
		pX128aX	38.72	3	5 (6804683-6804766)		
		pW125dE	45.85	3	5(2219504-2219693)	AT5G48930	HCT
Flower buds	ABTS	fito250	36.12	3	5(15688217-26579698)	AT5G48930	HCT
Flower buds	DPPH	fito156c	38.72	3	1(3530200-3530221)		
		pW133cH	38.72	3	2 (15610858-15610982)		
		pX128aX	38.72	3	5 (6804683-6804766)		
		pW125dE	45.85	3	5(2219504-2219693)	AT5G48930	HCT
Flower buds	FRAP	pW212bE	9.81	3	3(6427399-6427450)		
		fito066	26.12	3	4(6017387-6017408)	AT4G00040	CHS and SS
Leaves	FRAP	pX105cE	23.78	4	2(16117201-18117509)	AT2G40890	CYP98A3
Leaves	DPPH	pW217cE	83.82	6	1(14257280-18257453)	AT4G30210	
Leaves	ABTS	fito472	0	7	4(18268924-18269031)		
		pW104aE	40.25	7	1(126837519-26837557)	AT1G51680	4CL
Leaves	phenolics	fito472	0	7	4(18268924-18269031)		
		pW104aE	40.25	7	1(126837519-26837557)	AT1G51680	4CL
Flower buds	FRAP	fito472	0	7	4(18268924-18269031)	AT1G51680	4CL
		pW104aE	40.25	7	1(126837519-26837557)	AT1G51680	4CL

* Candidate gene found by means of in silico mapping in the Arabidopsis thaliana TAIR database. CHS and SS: Chalcone and stilbene synthase family protein; 4-CL: 4-coumarate: Co-A ligase 1, 2 or 3; HCT: hydroxycinnamoyltransferase enzyme; C4H: cinnamate 4-hydroxylase.

and phenolic content measured with the Folin–Ciocalteu method in other *Brassica* crops (cabbages, broccoli and Brussels sprouts) [15,18,26,53,54]. These results confirm the hypothesis that phenolic compounds mainly account for the AA of *Brassica* extracts. In the review made by Podsedek *et al.* [26], it is pointed out that phenolic compounds have higher AA in *in vitro* experiments than vitamins and carotenoids.

Furthermore, positive and significant correlations between carotenoid content and methods measuring AA were found in flower buds (FRAP) and in leaves (ABTS) in this study. These correlations are smaller than those of phenolic compounds with AA. Our results confirm that carotenoids are metabolites which contribute to the AA of Brassica extracts. Krinsky *et al.* [55] described that phenolic and carotenoid content is positively correlated with AA. In the case of anthocyanins, our experiments do not show a clear relationship between their content and methods measuring AA.

QTL mapping for methods measuring AA and the content of metabolites with AA

Methods measuring AA on food extracts are extensively used by the scientific community in order to detect potential benefits for human health. Genetic variation for these traits is interesting from the breeder's points of view, since it could allow increasing the AA of *Brassica* foods by selection. As far as we know, no report of QTLs or genetic basis for methods measuring AA has been done before in any *Brassica* crop. This is also one of the first assays, which studies the genetic base of ET methods measuring AA in any crop. Only three recent pieces of research in rice [28], raspberry [29] and in lettuce [30] studied QTLs for total water AA, total phenolic content, anthocyanin and carotenoid content. Knowledge derived from this study can be utilized in order to search for genes underlying these traits.

Ten out of 19 QTLs determine AA or the content of metabolites with AA in only one of the two organs, thus indicating that the regulation of genes underlying several QTLs is organ-dependent. Seven QTLs determined variation for only one method measuring AA, thus indicating that the genetic basis regulation is partially dependent on the method. Genomic regions AA-C3, AA-C5 and AA-C7 determined variation for more than one ET method measuring AA. These genomic regions could be responsible for the significant correlations found between ET methods in this study.

The genomic region AA-C7 determines variation for methods measuring AA and phenolic compounds and the genomic region AA-C5 determines variation for methods measuring AA and carotenoid content. These finding supports the hypothesis that AA of the mapping population is related to phenolic compounds but also to carotenoid content. No QTLs related to methods measuring AA and anthocyanin content were found. Therefore, anthocyanins would not play a significant role in maintaining the AA of extracts in this population. The content of other compounds different from those under study could be responsible for the remaining QTLs, which control variation for methods measuring AA. The core reactions of phenylpropanoid metabolism involve several steps catalyzed by three key enzymes: phenylalanine ammonia-lyase (PAL), cinnamate 4-hydroxylase (C4H) and 4-Coumarate: CoA ligase (4CL) [56]. In *A. thaliana* there are 4CL

different genes. This enzyme has a pivotal role in the biosynthesis of a plant's secondary compounds at the divergence point from general phenylpropanoid metabolism to several major branch pathways [57,58]. After *in silico* mapping analysis, 4CL-1 gene was located in the genomic region AA-C7 which controls AA measured as ABTS and FRAP and phenolic content. The hydroxycinnamoyltransferase enzyme (HCT) appears to be potentially implicated in the pathway both upstream and downstream of the 3-hydroxylation step and it is another key enzyme in phenylpropanoid biosynthesis. HCT enzyme catalyzes reactions both immediately preceding and following the insertion of the 3-hydroxyl group into the monolignol pathway [59–61] realised by the CYP98A3 (C3′H). HCT gene from *A. thaliana* was located by means of *in silico* mapping in the genomic region AA-C3, which controls AA measured with the three ET methods. C3′H gene was located in the interval of pX105cE to pW120cX on chromosome 4 where a QTL for AA measured with FRAP method was found. More candidate genes related to phenylpropanoid biosynthesis, along all the linkage group, were identified as it is the case of the chalcone and stilbene (CHS and SS) family protein which catalyzed the initial steps for flavonoid biosynthesis, route related with the phenylpropanoid biosynthesis [62]. More work is necessary in order to validate and confirm candidate genes for the QTLs found in this study.

Conclusions

No reports on the genetic basis of AA, and the content of metabolites with AA like phenolic, anthocyanin and carotenoid content have been documented before in *Brassica* crops. Results among methods measuring AA suggest that it is necessary to use more than one ET method in order to estimate AA, due to the fact that these methods present low significant correlations between them. Phenolic compounds and carotenoids are responsible for the AA of *Brassica* extracts.

Three genomic regions determined variation for more than one ET method measuring AA. QTL analysis confirms that AA of the mapping population is related to phenolic compounds but also to carotenoid content. It should be pointed out that the experiments have been carried on in one environment and under controlled conditions of temperature and light. Once the existence of QTLs for the traits under study has been proved, new experiments are going to be carried on in different environments to test the stability of the QTLs and the influence of environmental conditions. Several candidate genes related to phenylpropanoid biosynthesis are proposed for the QTLs found. These QTLs and the possible candidate genes identified through syntenic analysis with *A. thaliana* are the first step to understand the genetic basis of AA in the *Brassica* genus.

Acknowledgments

The authors thank Dr. Federico Iñiguez-Luy for supplying seed of the DH population and Rosaura Abilleira, César González for laboratory help and field work.

Author Contributions

Conceived and designed the experiments: MEC PS. Performed the experiments: TS. Analyzed the data: TS PV. Wrote the paper: TS.

References

1. Liu RH (2004) Potential synergy of phytochemicals in cancer prevention: Mechanism of action. The journal of nutrition 134: 3479S–3485S.

2. Kris Etherton PM (2002) Bioactive compounds in foods: Their role in the prevention of cardiovascular disease and cancer. The American journal of medicine 113: 71S–88S.

3. Wang LI, Giovannucci EL, Hunter D, Neuberg D, Su L (2004) Dietary intake of *Cruciferous* vegetables, Glutathione S-transftrase (GST) polymorphisms and lung cancer risk in a caucasian population. Cancer causes & control 15: 977–985.

4. Cao G, Sofic E, Prior R (1996) Antioxidant capacity of tea and common vegetables. Journal of agricultural and food chemistry 44: 3426–3431.

5. Liu R (2010) Health benefits of whole grain phytochemicals. Food engineering & ingredients 35: 18–22.

6. Okmen B, Sigva H, Gurbuz N, Ulger M, Frary A (2011) Quantitative trait loci (QTL) analysis for antioxidant and agronomically important traits in tomato (*Lycopersicon esculentum*). Turkish journal of agriculture & forestry 35: 501–514.

7. Dekker M, Verkerk R, Jongen WMF (2000) Predictive modelling of health aspects in the food production chain: a case study on glucosinolates in cabbage. Trends in food science & technology 11: 174–181.

8. Vallejo F, Tomas Barberan FA, Garcia Viguera C (2002b) Potential bioactive compounds in health promotion from broccoli cultivars grown in Spain. Journal of the science of food and agriculture 82: 1293–1297.

9. Vallejo F, Gil Izquierdo A, Pérez Vicente A, Garcia Viguera C, (2004a) In vitro gastrointestinal digestion study of broccoli inflorescence phenolic compounds, glucosinolates, and vitamin C. Journal of agricultural and food chemistry 52: 135–138.

10. Heimler D, Vignolini P, Dini MG, Vincieri FF, Romani A (2006) Antiradical activity and polyphenol composition of local *Brassicaceae* edible varieties. Food chemistry 99: 464–469.

11. Zhou K, Yu L (2006) Total phenolic contents and antioxidant properties of commonly consumed vegetables grown in Colorado. Food science & technology 39: 1155–1162.

12. Ou B, Huang D, Hampsch Woodill M, Flanagan J, Deemer E (2002) Analysis of antioxidant activities of common vegetables employing oxygen radical absorbance capacity (ORAC) and ferric reducing antioxidant power (FRAP) assays: a comparative study. Journal of agricultural and food chemistry 50: 3122–3128.

13. Podsedek A (2007) Natural antioxidants and antioxidant capacity of *Brassica* vegetables: A review. Food science & technology 40: 1–11.

14. Powers S, De Ruisseau K, Quindry J, Hamilton K (2004) Dietary antioxidants and exercise. Journal of sports sciences 22: 81–94.

15. Huang D, Ou B, Prior R (2005) The chemistry behind antioxidant capacity assays. Journal of agricultural and food chemistry 53: 1841–1856.

16. Soengas P, Cartea ME, Francisco M, Sotelo T, Velasco P (2012) New insights into antioxidant activity of *Brassica* crops. Food chemistry 134: 725–733.

17. Kusznierewicz B, Bartoszek A, Wolska L, Drzewiecki J, Gorinstein S (2008) Partial characterization of white cabbages (*Brassica oleracea* var *capitata* f alba) from different regions by glucosinolates, bioactive compounds, total antioxidant activities and proteins. Food science & technology 41: 1–9.

18. Zhi-Xiang N, Jen-Wai C, Kuppusamy UR (2011) Customized cooking method improves total antioxidant activity in selected vegetables. International journal of food sciences and nutrition 62: 158–163.

19. Vinson JA, Dabbagh YA, Yang J, Serry MM (1995) Pure polyphenols and beverages with an emphasis on tea as antioxidants (aox) for the prevention of heart-disease in vitro and in vivo studies. Abstracts of papers – American chemical society 210: 242–AGFD.

20. Gajewski M, Szymczak P, Elkner K, Dąbrowska A, Kret A, et al. (2007) Some aspects of nutritive and biological value of carrot cultivars with orange, yellow and purple-coloured roots. Vegetable crops research bulletin 67: 149–161.

21. Li H, Deng Z, Zhu H, Hu C, Liu R, et al. (2012) Highly pigmented vegetables: Anthocyanin compositions and their role in antioxidant activities. Food research international 46: 250–259.

22. Pace B, Cefola M, Renna F, Renna M, Serio F, et al. (2013) Multiple regression models and Computer Vision Systems to predict antioxidant activity and total phenols in pigmented carrots. Journal of food engineering 117: 74–81.

23. Paiva SA, Russell RM (1999) Beta-carotene and other carotenoids as antioxidants. Journal of the American college of nutrition 18: 426–433.

24. He F, Mu L, Yan G-L, Liang N-N, Pan Q-H, et al. (2010) Biosynthesis of anthocyanins and their regulation in colored grapes. Molecules 15: 9057–9091.

25. Yang M, Koo SI, Song WO, Chun OK (2011) Food matrix affecting anthocyanin bioavailability: review. Current medicinal chemistry 18: 291–300.

26. Podsedek A, Sosnowska D, Redzynia M, Anders B (2006) Antioxidant capacity and content of *Brassica oleracea* dietary antioxidants. International journal of food science & technology 41: 49–58.

27. Nilsson J, Olsson K, Engqvist G, Ekvall J, Olsson M, et al. (2006) Variation in the content of glucosinolates, hydroxycinnamic acids, carotenoids, total antioxidant capacity and low-molecular-weight carbohydrates in *Brassica* vegetables. Journal of the science of food and agriculture 86: 528–538.

28. Jin L, Xiao P, Lu Y, Shao Y, Shen Y, et al. (2009) Quantitative trait loci for brown rice color, phenolics, flavonoid contents and antioxidant capacity in rice grain. Cereal chemistry 86: 609–615.

29. Dobson P, Graham J, Stewart D, Brennan R, Hackett C, et al. (2012) Over-seasons analysis of quantitative trait loci affecting phenolic content and antioxidant capacity in raspberry. Journal of agricultural and food chemistry 60: 5360–5366.

30. Hayashi E, You Y, Lewis R, Calderon MC, Wan G, et al. (2012) Mapping QTL, epistasis and genotype x environment interaction of antioxidant activity, chlorophyll content and head formation in domesticated lettuce (*Lactuca sativa*). Theoretical and applied genetics 124: 1487–1502.

31. Iñiguez Luy F, Lukens L, Farnham M, Amasino R, Osborn T (2009) Development of public immortal mapping populations, molecular markers and linkage maps for rapid cycling *Brassica rapa* and *B. oleracea*. Theoretical and applied genetics 120: 31–43.

32. Benzie IF, Strain JJ (1996) The ferric reducing ability of plasma (FRAP) as a measure of "antioxidant power": the FRAP assay. Analytical biochemistry 239: 70–76.

33. Brand-Williams W, Cuvelier ME, Berset C (1995) Use of a free radical method to evaluate antioxidant activity. Food science & technology 28: 25–30.

34. Samarth R, Panwar M, Kumar M, Soni A (2008) Evaluation of antioxidant and radical-scavenging activities of certain radioprotective plant extracts. Food chemistry 106: 868–873.

35. Re R, Pellegrini N, Proteggente A, Pannala A, Yang M, et al. (1999) Antioxidant activity applying an improved ABTS radical cation decolorization assay. Free radical biology and medicine 26: 1231–1237.

36. Dewanto V, Wu X, Adom K, Liu R (2002) Thermal processing enhances the nutritional value of tomatoes by increasing total antioxidant activity. Journal of agricultural and food chemistry 50: 3010–3014.

37. Sims DA, Gamon JA (2002) Relationships between leaf pigment content and spectral reflectance across a wide range of species, leaf structures and developmental stages. Remote sensing of environment 81: 337–354.

38. Murray JR, Hackett WP (1991) Dihydroflavonol reductase activity in relation to differential anthocyanin accumulation in juvenile and mature phase *Hedera helix* L. Plant physiology 97: 343–351.

39. SAS (2008) Institute Inc SAS SAS 9.2 SAS. Enhanced Logging Facilities, SAS Inst.: Cary, North Carolina.

40. Zeng ZB (1994) Precision mapping of quantitative trait loci. Genetics 136: 1457–1468.

41. Utz HF, Melchinger AE (2003) A computer program to map QTL.

42. Papst C, Bohn M, Utz HF, Melchinger AE, Klein D (2004) QTL mapping for European corn borer resistance (*Ostrinia nubilalis* Hb.), agronomic and forage quality traits of testcross progenies in early-maturing European maize (*Zea mays* L.) germplasm. Theoretical and applied genetics 108: 1545–1554.

43. Utz HF, Melchinger AE, Schon CC (2000) Bias and sampling error of the estimated proportion of genotypic variance explained by quantitative trait loci determined from experimental data in maize using cross validation and validation with independent samples. Genetics 154: 1839–1849.

44. Fukumoto LR, Mazza G (2000) Assessing antioxidant and prooxidant activities of phenolic compounds. Journal of agricultural and food chemistry 48: 3597–3604.

45. Nilsson J, Pillai D, Onning G, Persson C, Nilsson A, et al. (2005) Comparison of the 2,2′-azinobis-3-ethylbenzotiazo-line-6-sulfonic acid (ABTS) and ferric reducing anti-oxidant power (FRAP) methods to asses the total antioxidant capacity in extracts of fruit and vegetables. Molecular nutrition & food research 49: 239–246.

46. Szeto Y, Benzie IFF, Tomlinson B (2002) Total antioxidant and ascorbic acid content of fresh fruits and vegetables: implications for dietary planning and food preservation. British journal of nutrition 87: 55–59.

47. Samec D, Piljac Zagarac J, Bogovic M, Habjanic K, Gruz J (2011) Antioxidant potency of white (*Brassica oleracea* L. var. *capitata*) and Chinese (*Brassica rapa* L. var. *pekinensis* (Lour.)) cabbage: The influence of development stage, cultivar choice and seed selection. Scientia horticulturae 128: 78–83.

48. Rieseberg LH, Archer MA, Wayne RK (1999) Transgressive segregation, adaptation and speciation. Heredity 83 (4): 363–372.

49. F Llorach R, Espin JC, Tomas Barberan FA, et al. (2003) Valorization of cauliflower (*Brassica oleracea* L. var. *botrytis*) by-products as a source of antioxidant phenolics. Journal of agricultural and food chemistry 51: 2181–2187.

50. Guo JT, Lee HL, Chiang SH, Lin FI, Chang CY (2001) Antioxidant properties of the extracts from different parts of broccoli in Taiwan. Journal of food and drug analysis 9(2): 96–101.

51. Gouveia S, Goncalves J, Castilho P (2013) Characterization of phenolic compounds and antioxidant activity of ethanolic extracts from flowers of *Andryala glandulosa* ssp varia (Lowe ex DC.) R.Fern., an endemic species of Macaronesia region. Industrial crops and products 42: 573–582.

52. Gawlik-Dziki U (2008) Effect of hydrothermal treatment on the antioxidant properties of broccoli (*Brassica oleracea* var. *botrytis italica*) florets. Food chemistry 109: 393–401.

53. Charanjit K, Kumar K, Anil D, Kapoor HC, Kaur C (2007) Variations in antioxidant activity in broccoli (*Brassica oleracea* L.) cultivars. Journal of food biochemistry 31: 621–638.

54. Mrkic V, Cocci E, Dalla Rosa M, Sacchetti G (2006) Effect of drying conditions on bioactive compounds and antioxidant activity of broccoli (*Brassica oleracea* L.). Journal of the science of food and agriculture 86: 1559–1566.

55. Krinsky NI (2001) Carotenoids as antioxidants. Nutrition 17: 815–817.

56. Hahlbrock K (1989) Physiology and molecular biology of phenylpropanoid metabolism. Annual review of plant physiology and plant molecular biology 40: 347–369.

57. Xu B, Escamilla-Treviño LL, Sathitsuksanoh N, Shen Z, Shen H, et al. (2011) Silencing of 4-coumarate:coenzyme A ligase in switchgrass leads to reduced lignin content and improved fermentable sugar yields for biofuel production. New phytologist 192: 611–625.

58. Pietrowska-Borek M, Stuible H-P, Kombrink E, Guranowski A (2003) 4-Coumarate: coenzyme A ligase has the catalytic capacity to synthesize and reuse various (di)adenosine polyphosphates. Plant physiology 131: 1401–1410.

59. Hoffmann L, Maury S, Martz F, Geoffroy P, Legrand M (2003) Purification, cloning, and properties of an acyltransferase controlling shikimate and quinate ester intermediates in phenylpropanoid metabolism. Journal of biological chemistry 278: 95–103.

60. Hoffmann L, Besseau S, Geoffroy P, Ritzenthaler C, Meyer D, et al. (2004) Silencing of hydroxycinnamoyl-coenzyme A shikimate/quinate hydroxycinnamoyltransferase affects phenylpropanoid biosynthesis. Plant Cell 16: 1446–1465.

61. Shadle G, Chen F, Srinivasa Reddy MS, Jackson L, Nakashima J, et al. (2007) Down-regulation of hydroxycinnamoyl CoA: Shikimate hydroxycinnamoyl transferase in transgenic alfalfa affects lignification, development and forage quality. Phytochemistry 68: 1521–1529.

62. Schroder J (2000) The family of chalcone synthase-related proteins: Functional diversity and evolution. In: J. T Romeo, R Ibrahim, L Varin and V DeLuca, editors. Evolution of Metabolic Pathways. Kidlington: Pergamon-Elsevier Science Ltd. pp. 55–89.

8

Flavonoid Fraction of Bergamot Juice Reduces LPS-Induced Inflammatory Response through SIRT1-Mediated NF-κB Inhibition in THP-1 Monocytes

Roberto Risitano[2], Monica Currò[2], Santa Cirmi[1], Nadia Ferlazzo[1], Pietro Campiglia[3], Daniela Caccamo[2], Riccardo Ientile[2], Michele Navarra[1]*

1 Department of Drug Sciences and Products for Health, University of Messina, Messina, Italy, **2** Department of Biomedical Sciences and Morphofunctional Imaging, University of Messina, Messina, Italy, **3** Department of Pharmaceutical and Biomedical Sciences, University of Salerno, Fisciano, Salerno, Italy

Abstract

Plant polyphenols exert anti-inflammatory activity through both anti-oxidant effects and modulation of pivotal pro-inflammatory genes. Recently, *Citrus bergamia* has been studied as a natural source of bioactive molecules with antioxidant activity, but few studies have focused on molecular mechanisms underlying their potential beneficial effects. Several findings have suggested that polyphenols could influence cellular function by acting as activators of SIRT1, a nuclear histone deacetylase, involved in the inhibition of NF-κB signaling. On the basis of these observations we studied the anti-inflammatory effects produced by the flavonoid fraction of the bergamot juice (BJe) in a model of LPS-stimulated THP-1 cell line, focusing on SIRT1-mediated NF-κB inhibition. We demonstrated that BJe inhibited both gene expression and secretion of LPS-induced pro-inflammatory cytokines (IL-6, IL-1β, TNF-α) by a mechanism involving the inhibition of NF-κB activation. In addition, we showed that BJe treatment reversed the LPS-enhanced acetylation of p65 in THP-1 cells. Interestingly, increasing concentrations of Sirtinol were able to suppress the inhibitory effect of BJe via p65 acetylation, underscoring that NF-κB–mediated inflammatory cytokine production may be directly linked to SIRT1 activity. These results suggest that BJe may be useful for the development of alternative pharmacological strategies aimed at reducing the inflammatory process.

Editor: Giovanni Li Volti, University of Catania, Italy

Funding: Research was supported by a grant from Sicily Region (PO FESR Sicilia 2007/2013, CUP G73F11000050004 to MN, project "MEPRA", n° 133 of Linea d'Intervento 4.1.1.1). The funders had no role in study design, data collection and analysis, decision to publish, or preparation of the manuscript.

Competing Interests: The authors declare that the flavonoid fraction of bergamot juice (BJe) has been provided by the company "Agrumaria Corleone" (Palermo, Italy).

* Email: mnavarra@unime.it

Introduction

Stimulation by pathogen-specific ligands, including bacterial lipopolysaccharide (LPS), leads cells of the innate immune system, such as monocytes and macrophages to produce several pro- and anti-inflammatory cytokines [1]. Accumulating evidences suggest that chronic inflammation represents the main contributing factor to several chronic degenerative pathologies, including cardiovascular diseases, neurological disorders and cancer [2–4].

Nuclear factor-kappa B (NF-κB) has been reported to play a pivotal role in inflammatory response through the induction of inflammation-related cytokines (i.e. IL-6, IL-1β, TNF-α) and enzymes such as cyclooxygenase-2 (COX-2) and inducible nitric oxide synthase (iNOS) [5,6]. The best-studied and most prevalent form of NF-κB exists as a heterodimer composed of p50 and RelA/p65 subunits. In unstimulated cells, NF-κB is sequestered in the cytoplasm through its assembly with its inhibitory proteins, which are members of the IκB family [7,8]. In response to various stimuli, such as cytokines, DNA-damaging agents, bacterial wall or viral proteins, IκB dissociates, via phosphorylation by the IκB kinase (IKK), and the activated transcription factor can translocate into the nucleus where is able to induce a large number of target genes involved in cell growth, apoptosis, cell adhesion and inflammation [9,10].

According to several findings, the reversible acetylation of RelA/p65 subunit can modulate NF-κB signaling depending on the acetylation status of specific lysine residues [11]. In particular, acetylation of lysine 310 is required for full transcriptional activity of RelA and for activation of NF-κB complex [12]. Interestingly, SIRT1, a nuclear NAD$^+$-dependent histone deacetylase (HDAC), may play a role in regulating inflammation by directly deacetylating the RelA/p65 protein at lysine 310. Recent studies show how the catalytic activity of SIRT1 can be modulated both positively and negatively through the binding by some agonists. Resveratrol, a polyphenolic compound found in red wine, has been identified as a potent pharmacological activator of SIRT1 [13]. On the other hand, Sirtinol was the first synthetic inhibitor extensively used in literature [14,15].

Many anti-inflammatory drugs (SAID and NSAID) currently used in therapy target NF-κB and COX-2, but because of their frequent and serious associated side effects, result in high rates of morbidity [16,17]. Therefore, during the last years there has been a growing interest towards the anti-inflammatory properties of some natural drugs and their bioactive compounds, like polyphenols (PP).

PP are organic molecules that constitute a numerous and heterogeneous family of secondary metabolites of plant cells, where they exert a protective action against ultraviolet radiation and oxidative stress [18]. PP are present in many edible plants and thus represent an integral part of human supply [19]. Several studies have shown that high regular intake of some phenolic compounds in the diet plays a preventive action against several human diseases, such as cardiovascular pathologies, atherosclerosis, osteoporosis, allergies, diabetes, neurodegenerative diseases and cancer [20,21]. Although the mechanisms by which these natural compounds exert their benefits are not fully understood, anti-inflammatory effects of PP have been attributed primarily to their antioxidant activity, because they were known to scavenge and prevent the formation of reactive oxygen and nitrogen species (ROS and RNS, respectively) [22,23], which represent important hallmarks of inflammation. Furthermore, during recent years, numerous studies have suggested that PP could influence cellular function by direct interaction with several receptors, modulation of intracellular signaling and transcription of gene involved in different pro-inflammatory pathways [24,25].

Citrus bergamia Risso et Poiteau (Bergamot) is a typical fruit of the Southern Italy, belonging to the family of Rutaceae. Bergamot is cultivated almost exclusively on the Ionian coast of the province of Reggio Calabria (Italy), where there are climatic and environmental conditions particularly suitable for its cultivation. Bergamot fruit is mainly used to extract the essential oil obtained from the peel, much employed in the fragrance industry and which have been experimentally studied to evaluate its potential neuroprotective activity [26]. On the contrary, bergamot juice (BJ), obtained from the endocarp of the fruit, is considered just a secondary and discarded product. Studies performed by Miceli et al. [27] have shown that a chronic administration of BJ is effective to prevent the diet-induced hyperlipidemia in rat, suggesting a relationship between the beneficial effect and its antioxidant properties. Moreover, a clinical research showed that the bergamot-derived polyphenolic fraction, given orally in patients suffering from metabolic syndrome, produces significant reduction of serum cholesterol, triglycerides and glycaemia [28], strengthen the finding obtained in animal model. More recently, we demonstrated that BJ is also able to inhibit important molecular pathways related to cancer-associated aggressive phenotype, thus reducing growth, adhesion and migration in different *in vitro* [29] and *in vivo* models [30]. Finally, studies performed on colorectal cancer cell line verified that the antiproliferative effect of BJ is due to its flavonoid fraction which is able to act by multiple mechanisms depending on the concentration [31].

However, to date, the anti-inflammatory potential of BJ has never been evaluated. Therefore, the present study was designed to assess the modulating effects of flavonoid fraction of BJ (BJe) on the expression of inflammation-related cytokines. Considering the involvement of NF-κB pathway on the production of these pro-inflammatory mediators, we focused on the possibility that BJe may regulate NF-κB activation and examined underlying mechanisms associated with NF-κB/SIRT1 crosstalk in LPS-stimulated THP-1 monocytes, a human leukemia monocytic cell line, that have been widely used as a model to study the inflammatory cell response.

Materials and Methods

Materials

The human leukemia monocytic cell line, THP-1, was purchased from American Type Culture Collections (ATCC) (Rockville, MD, USA). RPMI-1640, L-glutamine, HEPES, sodium pyruvate, glucose, 2-mercaptoethanol, penicillin/streptomycin mixture, 3-(4,5-methylthiazol-2-yl)-2,5-diphenyl-tetrazolium bromide (MTT), dimethylsulfoxide (DMSO), phosphate buffered saline solution (PBS), Sirtinol and other chemicals of analytical grade were from Sigma (Milan, Italy). Lipopolysaccharide (LPS) was from InvivoGen (San Diego, CA, USA). Fetal bovine serum (FBS), as well as TRIzol for RNA extraction were from Invitrogen Life Technologies (Milan, Italy). High-capacity cDNA archive kit, TaqMan Gene Expression Mastermix, TaqMan Gene Expression assays (Assays-on-Demand) for human 18S mRNA (ID: Hs99999901_s1), TNF-α (ID: Hs00174128_m1), IL-6 (ID: Hs00985639_m1), IL-1β, (ID: Hs01555410_m1) were from Applied Biosystems (Applera Corp., Milan, Italy). Assay location (midposition of fluorogenic probe), reference sequences and other relevant information are published online by Applied Biosystems (Foster City, CA).

Instant ELISA kits for the quantitative detection of human IL-1β, IL-6 and TNF-α were from eBioscience (Vienna, Austria). Nuclear Extraction Kit and Electrophoretic Mobility Shift Assay (EMSA) for identifying NF-κB were supplied by Panomics (Santa Clara, CA, USA). Protein G PLUS-Agarose Immunoprecipitation reagent were from Santa Cruz Biotechnology (Dallas, TX, USA) and the antibodies rabbit anti-p65, sheep anti-acetylated Lysine and rabbit anti-sheep IgG (HRP) were from Abcam (Cambridge, UK). ECL Plus detection system were from GE Healthcare Bio-Sciences (Pittsburgh, PA, USA).

Bergamot Juice extract and its chemical analysis

The flavonoid fraction of bergamot juice (BJe) has been provided by the company "Agrumaria Corleone" (Palermo, Italy). The fruits of *Citrus bergamia* were coming from crops located in the province of Reggio Calabria (Italy). The extract was centrifuged at 6000 rpm/min for 15 minutes to remove any impurities and successively transformed into a dry powder by the method of spray drying. Small aliquots of BJe were stored at −20°C. Finally, the drug was defrosted, diluted in culture media, pH adjusted to 7.4 and filtered just prior to use.

Chemical composition of BJe was investigated as previously described [32]. Briefly, BJe was solubilized in methanol to a concentration of 1 mg/mL, ultrasonicated and filtered by a 0.2 μm nylon membrane (Millipore, Milan, Italy). Qualitative and quantitative determination of flavonoids in BJe was performed using a UHPLC coupled online to an LCMS–IT-TOF mass spectrometer (Shimadzu, Kyoto, Japan). Flavonoids were identified on the basis of diode array spectra, MS molecular ions and MS/MS fragmentation patterns. Data obtained were compared with those available in scientific literature. Molecular formula was calculated by the Formula Predictor software (Shimadzu).

Cell culture and treatment

THP-1 cells were maintained in RPMI 1640 supplemented with L-glutamine (2 mM), HEPES (10 mM), sodium pyruvate (1 mM), glucose (2.5 g/l), 2-mercaptoethanol (0,05 mM), 10% heat-inactivated fetal bovine serum (FBS), 1% penicillin/streptomycin, at 37°C in a 5% CO_2/95% air humidified atmosphere. Medium was renewed every 2 days and split performed when cells reached maximum density (1×10^6 cells/ml). In our experimental conditions, THP-1 cells were seeded at a density of 5×10^5 cells/ml into culture plates in RPMI complete medium plus 10% FBS and incubated at 37°C with lipopolysaccharide (LPS; 500 ng/ml) for 3 hs, in the presence or absence of BJe (0.05–0.1–0.5 mg/ml) and/or Sirtinol (1–5–10 μM), which were added to the culture medium 30 min prior to LPS treatment. In all experiments, equal volumes of PBS or DMSO were added to the medium of control cultures (controls were performed using non-stimulated cell). Either concentrations of LPS, BJe and Sirtinol were chosen according to our preliminary optimization studies. After incubation, cells were harvested by centrifugation to assess cellular viability, gene expression, activation of transcription factor NF-κB and acetylation status of p65. Media were collected in order to evaluate cytochine release.

Cell viability assay

To assess either LPS and BJe adverse effects on cell viability, we evaluated the mitochondrial activity of living cells by a MTT quantitative colorimetric assay. After treatment, THP-1 cells were harvested by centrifugation and, after counting, they were incubated in 96-well culture plates at a density of 5×10^4 cells/well with fresh red-phenol free medium containing MTT (0.5 mg/mL) at 37°C for 4 hs. Then, insoluble formazan crystals were dissolved in 100 μL of a 0.04 N HCl/isopropanol solution for 1 h. The optical density in each well was evaluated by spectrophotometrical measurement. Absorbance was determined at 570 nm using a microplate reader (Tecan Italia, Cologno Monzese, Italy). All experiments were performed in eightplicate and repeated three times.

Real-Time PCR

After RNA isolation with TRIzol reagent, RNA (3 μg) was reverse transcribed with High- Capacity cDNA Archive kit according to the manufacturer's instructions. Then, mRNA levels of IL-6, IL-1β, TNF-α were analyzed by real-time PCR using TaqMan gene expression assays according to the manufacturer's instructions. 18S mRNA was used as endogenous controls. Quantitative PCR reactions were set up in triplicate in a 96-well plate and were carried out in 10 μl reactions containing 1× TaqMan Gene Expression Mastermix, 1× TaqMan-specific assay and 20 ng RNA converted into cDNA. qPCR was performed in a 7900HT Fast Real-Time PCR System with the following profile: one cycle at 50°C for 2 min, then 95°C for 10 min, followed by 50 cycles at 95°C for 15 s and 60°C for 1 min. Data were collected and analyzed using SDS 2.3 and RQ manager 1.2 software (Applied Biosystems, Foster City, CA) using the $2^{(-\Delta\Delta Ct)}$ relative quantification method. Values are presented as fold change relative to unstimulated cells.

Evaluation of cytokine secretion by ELISA

In order to detect human IL-6, IL-1β and TNF-α, an enzyme-linked immunosorbent assay was performed in cell-free culture supernatants of THP-1 monocytes, using Instant ELISA Kits. Before detection, supernatants recovered from treated and untreated cells were concentrated 10-fold by freeze-drying. All freeze-dried samples were reconstituted by the addition of distilled water. Briefly, according to the manufacturer's guidelines, 50 μl of standards or samples (supernatants recovered from treated and untreated cells) were incubated in 96-well plates at room temperature for 3 hs with shaking. After washing 5 times with 400 μl of wash buffer, 100 μl of the provided substrate solution were added to each well and the plates were incubated in the dark for 10 min. The enzyme reaction was then stopped by pipetting 100 μl of stop solution into each well and the absorbance was determined at 450 nm using a microplate reader (Tecan, Italy). All experiments were performed in triplicate.

Electrophoretic mobility shift assay

At the end of the treatment, THP-1 cells were harvested by centrifugation. After washing twice with cold PBS, the isolation of nuclear cell proteins was performed using a commercial nuclear extraction kit following the manufacturer's guidelines. Protein concentrations were determined using a Bradford method. The presence of NF-κB DNA binding activity in cellular nuclear extracts of LPS-treated and control cells was evaluated by subsequent electrophoretic mobility shift assay, using Affymetrix EMSA Kits according to the manufacturer's instructions. Briefly, nuclear extracts were incubated with the biotin-labeled NF-κB probe and then the protein/DNA complexes were separated on a non-denaturing 6% polyacrylamide gel. After transferring onto nylon membranes bound complexes were detected via streptavidin-HRP and a chemiluminescent substrate and visualized on Kodak film. The bands were scanned and quantified by densitometric analysis with ImageJ 1.47, an open source software freely downloadable from the US National Institute of Health website (http://imagej.nih.gov/ij/).

Immunoprecipitation and immunoblotting analyses

For each sample, 50 μg of nuclear extract were incubated with rabbit anti-p65 for 1 h at 4°C on a rotator. Negative control was set by incubating nuclear proteins under similar conditions but without the immunoprecipitating antibody. Afterwards, 35 μl of resuspended Protein G PLUS-Agarose beads were added to each tube and the samples were incubated at 4°C overnight on a rocker platform. The agarose beads were extensively washed the next day and pellets were resuspended in 40 μl of 1x Laemmli buffer, boiled for 5 min and resolved by SDS-PAGE. Proteins were then transferred onto nitrocellulose membrane and non-specific binding sites were pre-blocked by incubation with 5% non-fat dry milk in Tris-buffered saline containing 0.15% Tween 20 for 1 h at room temperature. The blot was probed overnight at 4°C with primary antibody anti-acetylated Lysine (from sheep, diluted 1:1000), followed by incubation for 2 hs with horseradish peroxidase-conjugated anti-sheep secondary antibody (diluted 1:3000). Final detection was performed by using ECL chemiluminescence system; then, bands were scanned and quantified by densitometric analysis with ImageJ software.

Statistical analysis

Data obtained from three separate experiments were expressed as mean ± SEM, and analyzed by one-way analysis of variance (ANOVA) and the Student-Newman Keuls test using GraphPad Prism software (San Diego, CA). p values lower than 0.05 were considered significant.

Results

The flavonoid profile of the BJe

A chromatogram of the BJe composition is shown in figure 1. Peaks 1–20 correspond to identified flavonoid present in the extract and their amounts are showed in table 1. The main flavonoids identified (mg/g) were Neohesperidin (105.27), Naringin (101.88), Melitidin (75.89), Neoeriocitrin (56.61) and Hesperetin (55.65). Quantitative values were obtained as the average of quintuplicate analyses and were in accordance with tipical BJ flavonoid profile reported in the literature [27,32].

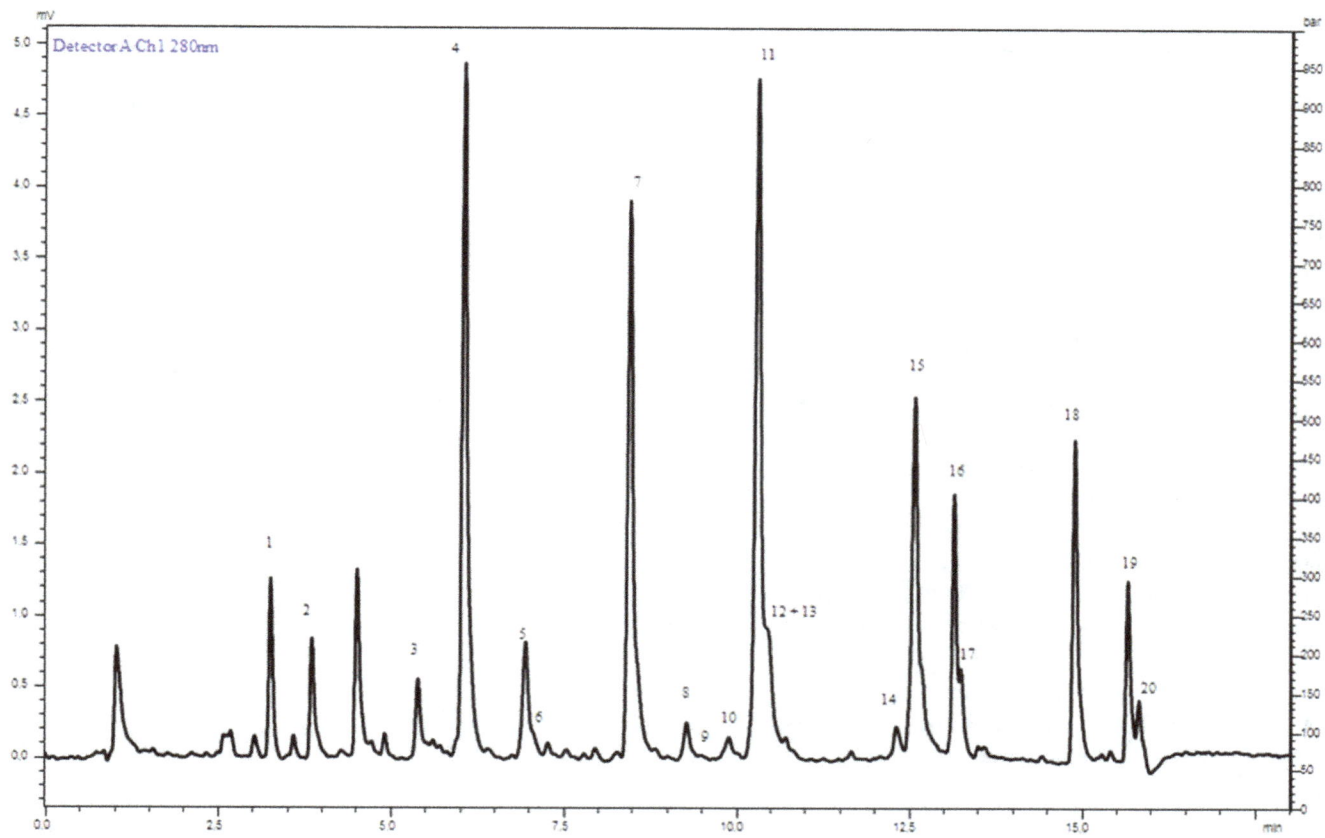

Figure 1. UHPLC chromatogram of BJe. A representative chromatogram of the BJe flavonoid components is shown. The sample was run for five times. For peak identification see Table 1.

Expression of LPS-induced proinflammatory cytokines in presence of BJe

In order to assess BJe toxicity in cell cultures, preliminary experiments were carried out using MTT test. Exposure of THP-1 monocytes to different BJe concentrations (in a range 0.05–0.5 mg/ml) for 3.5 hs didn't show any significant reduction of cell viability when BJe was added in presence or absence of LPS, as shown by MTT data (Fig. 2).

Based on preliminary experiments we used LPS at the concentration of 500 ng/ml that was able to trigger the release of pro-inflammatory factors in THP-1 cells within three hours. Under these conditions, the Real-time PCR analysis showed that LPS stimulation of THP-1 monocytes resulted in a dramatic increase of mRNA of pro-inflammatory cytokines (IL-6, IL-1β, TNF-α), relative to endogenous 18S mRNA levels. In particular, IL-6 mRNA increase was higher than those observed for IL-1β and TNF-α. The IL-6 mRNA levels in LPS-treated cells was almost one hundred-fold higher than those found in untreated cells (Fig. 3A). These effects were reduced in presence of BJe. Specifically, LPS-induced IL-6 up-regulation was strongly reduced in presence of BJe 0.05 mg/ml (by ~70%), with a maximum reduction of ~80% with 0.1 mg/ml of BJe (p<0.001). Even though to a lesser extent, also BJe 0.5 mg/ml was able to reduce IL-6 mRNA levels (~60%; p<0.001). In parallel, the LPS-induced increase in IL-1β gene expression (36-fold higher than controls) was significantly reduced by all BJe concentrations used (from 70 to 80%; p<0.001; Fig. 3B). In a similar way, stimulation of THP-1 monocytes with LPS caused a 5-fold increase of TNF-α mRNA

levels in comparison with untreated cells, while a significant down-regulation of LPS-stimulated TNF-α gene expression was observed in presence of BJe treatment (in the range 0.05–0.5 mg/ml), resulting in mRNA levels similar to those found in control cells (Fig. 3C).

The release of cytokines caused by LPS was reduced by BJe treatment

In order to confirm the LPS-induced up-regulation of pro-inflammatory genes, we assessed the release of the analyzed cytokines by detection in the supernatants collected at the end of the treatment. As shown in figure 4, in comparison to untreated cells, 3 hs of LPS stimulation was able to trigger the secretion of significant amount of IL-1β and especially of TNF-α (4,7 and 37 fold increases, respectively) in THP-1 monocytes, but not of IL-6. These LPS-induced releases were reduced by BJe treatment. In particular, both BJe 0.1 and 0.5 mg/ml decreased significantly the amount of IL-1β secreted by 27 and 25%, respectively (p<0.05 vs LPS-treated cultures; Fig 4B), while all BJe concentrations used affected the release of TNF-α between 17 and 50% (p<0.05 and p<0.001 vs LPS-treated cells; Fig 4C). Interestingly, the most effective concentration of BJe able to diminish the protein levels of both IL-1β and TNF-α was 0.1 mg/ml.

Inhibition of LPS-induced NF-κB activation by BJe

Given the reported pivotal role of NF-κB in inflammatory response induced by various stimuli, we also investigated its role in THP-1 cell response to LPS-induced injury, in presence of the

Table 1. Flavonoid concentration (mg/g) in BJe.

Peak	Compound	Synonyms	Quantity mg/g
1	Apigenin 6,8 di C-glucoside	Vicenin-2	11.98
2	Diosmetin 6,8 di C-glucoside	Lucenin-2 4'-methyl ether	11.34
3	Eriodictyol 7-O-rutinoside	Eriocitrin	8.08
4	Eriodictyol 7-O-neohesperidoside	Neoeriocitrin	56.61
5	5,7-dihydroxy-4' methoxyflavone 7-O-rutinoside	Poncirin	17.01
6	Diosmetin 8-C-glucoside	Orientin 4' methylether	15.06
7	Naringenin 7-O-neohesperidoside	Naringin	101.88
8	Apigenin 7-O-neohesperidoside	Rhoifolin	18.00
9	Hesperetin-7-O-rutinoside	Hesperidin	8.37
10	Quercetin-3-β-glucopyranoside	Isoquercitrin	2.05
11	Hesperetin-7-O-neohesperidoside	Neohesperidin	105.27
12	Diosmetin 7-O-neohesperidoside	Neodiosmin	13.40
13	Apigenin 7-O-neohesperidoside-4'-glucoside	Rhoifolin 4'-glucoside	1.18
14	Naringenin-7-O-rutinoside	Narirutin	6.02
15	Naringenin 7-[2''-α-rhamnosyl-6''-[3''''-hydroxy-3''''-methylglutaryl]-β-glucoside]	Melitidin	75.89
16	Hesperetin 7-[2''-α-rhamnosyl-6''-[3''''-hydroxy-3''''-methylglutaryl]-β-glucoside]	Brutieridin	21.99
17	unknown		10.55
18	5,7-dihydroxy-2-(4-hydroxyphenyl)chroman-4-one	Naringenin	43.22
19	(S)-2,3-dihydro-5,7-dihydroxy-2-(3-hydroxy-4-methoxyphenyl)-4H-1-benzopyran-4-one	Hesperetin	55.65
20	5,7-dihydroxy-2-(3-hydroxy-4-methoxyphenyl)chromen-4-one	Diosmetin	14.01

most effective concentration of BJe (0.1 mg/ml). EMSA analysis of nuclear fractions showed that nuclear translocation and specific DNA binding activity of NF-κB increased in THP-1 cell cultures, after 3 hs of incubation with LPS (Fig. 5). Interestingly, BJe 0.1 mg/ml was able to inhibit the LPS-induced NF-κB activation.

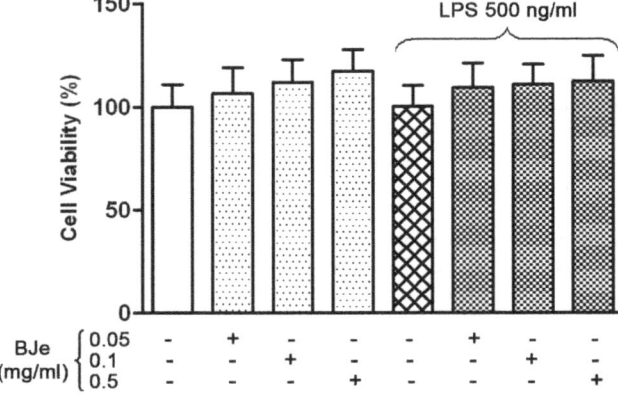

Figure 2. Effect of BJe on THP-1 cell viability in presence or absence of LPS. Different concentrations of BJe (0.01, 0.05, 0.1 and 0.5 mg/ml) were added to the culture medium 30 min before LPS treatment (500 ng/ml for 3 hs) and then cell viability was assessed by the MTT test. Results are expressed as percentages of untreated cultures. Data are means ± SEM from three independent experiments performed in eightplicate.

Involvement of SIRT1 in the inhibition of NF-κB signaling exerted by BJe, via the p65 acetylation

Since it is known that SIRT1 may suppress inflammation by deacetylation of NF-κB subunit, in a subset of experiments we also investigated its regulatory effects on NF-κB activation. As shown in fig. 6, the inhibitory effect of 0.1 mg/ml BJe on the LPS-induced activation of NF-κB was counteracted at highest dose of Sirtinol, the well known inhibitor of SIRT1. In particular, 10 μM of Sirtinol reverted the NF-κB inhibition due to BJe treatment, suggesting a role of SIRT1 in the modulatory effect of BJe on NF-κB signaling.

In order to clarify the involvement of SIRT1's deacetylating activity and the molecular mechanism underlying the BJe-mediated inhibition of NF-κB, an immunoprecipitation assay was carried out to evaluate the acetylation status of RelA/p65 subunit. As shown in Fig.7, BJe 0.1 mg/ml treatment reverted the LPS-enhanced acetylation of p65 in THP-1 monocytes. In addition, increasing concentrations of Sirtinol were able to suppress the inhibitory effect of BJe on the activation of NF-κB, via p65 acetylation, underscoring that NF-κB–mediated inflammatory cytokine production may be directly linked to SIRT1 activity.

Discussion

Several results show that THP-1 monocyte/macrophages are a sensitive *in vitro* model to analyze potential anti-inflammatory activity of different substances, so that it may be a reasonably accurate model to study LPS-dependent inflammatory response [33–35]. Thus, THP-1 cells is considered a suitable and reliable model for screening a variety of compounds prior to a more detailed analysis with human derived cells. According to other results [36,37], the gene expression of pro-inflammatory cytokines

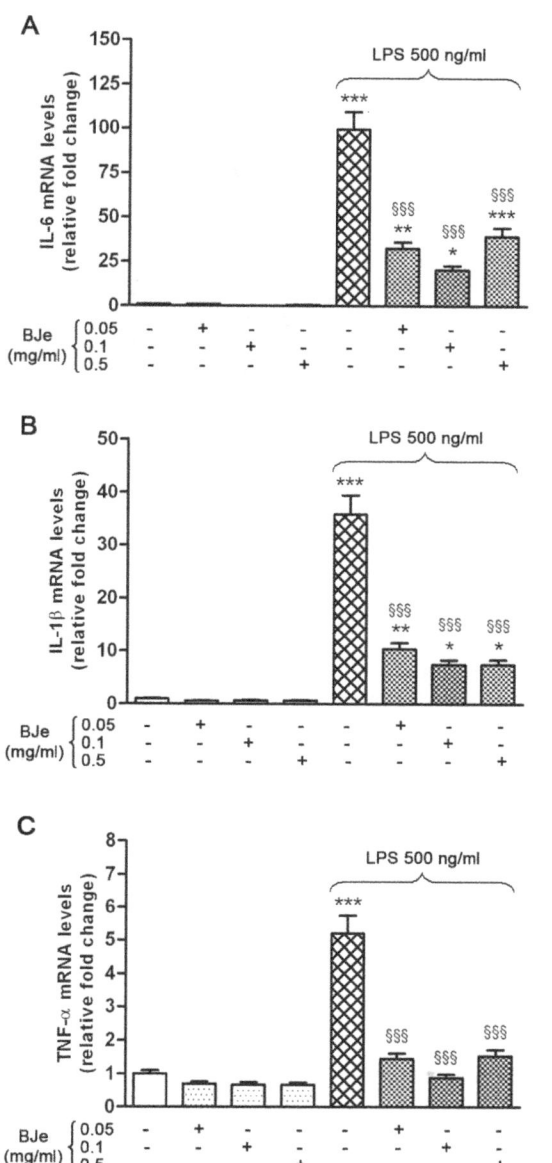

Figure 3. Effects of BJe on cytokine gene expression in THP-1 cells stimulated with LPS. THP-1 cells were treated with different concentrations of BJe (0.05–0.5 mg/ml for 30 min) before exposure to 500 ng/ml of LPS for 3 hs. Results from real-time PCR of IL-6 (A), IL-1β (B) and TNF-α (C) are expressed as a relative fold change compared to untreated cells, after normalization against 18S as endogenous control. Columns and bars represent means ± SEM from triplicate experiments. * p<0.05, ** p<0.01, *** p<0.001, significant values in comparison with control cells; §§§ p<0.001, significant values in comparison with LPS treated cells (ANOVA followed by Student-Newman Keuls multiple comparisons test).

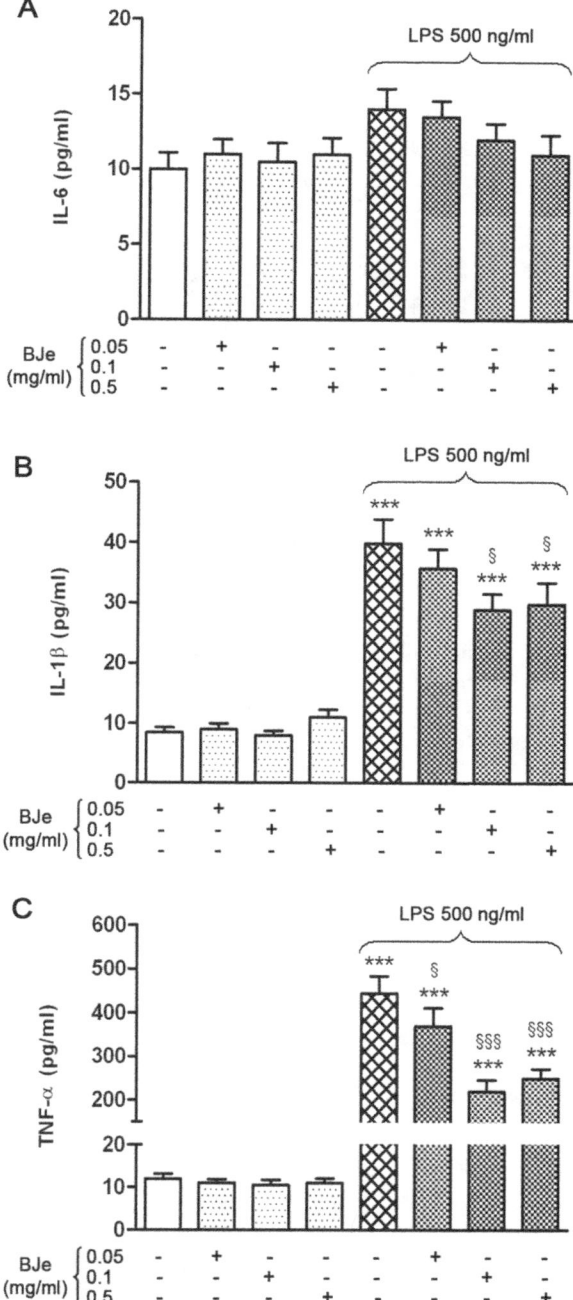

Figure 4. BJe prevents the LPS-stimulated release of IL-1β and TNF-α in THP-1 monocytes. The cells were treated with increasing concentrations of BJe (0.05–0.5 mg/ml for 30 min) prior to add LPS (500 ng/ml; 3 hs). Then, secretion of IL-6 (A), IL-1β (B) and TNF-α (C) in the media was evaluated by ELISA assay. Data are the mean ± SEM of three independent experiments performed in triplicate. *** p<0.001, significant differences vs untreated cultures; § p<0.05 and §§§ p<0.001, significant differences vs LPS treated cells (ANOVA followed by Student-Newman Keuls multiple comparisons test).

such as IL-1β, TNF-α and IL-6 increased within the first hours of the LPS challenge, whereas changes in the secretion pattern of related cytokines did not show similar extent. However, differences between amount of cytokines secreted from THP-1 monocytes and mRNA levels of corresponding genes can be also explained by differences in RNA stability, post-translational modification factors and proteolytic processing events that make the production of individual cytokines different [38]. Under our conditions, the

secreted amount of both TNF-α and IL-1β, but not that of IL-6, were higher than controls, suggesting a general relation between mRNA and protein levels in LPS-stimulated monocyte/macrophages. In this study for the first time we demonstrated that the flavonoid fraction of Bergamot juice is able to reduce significantly

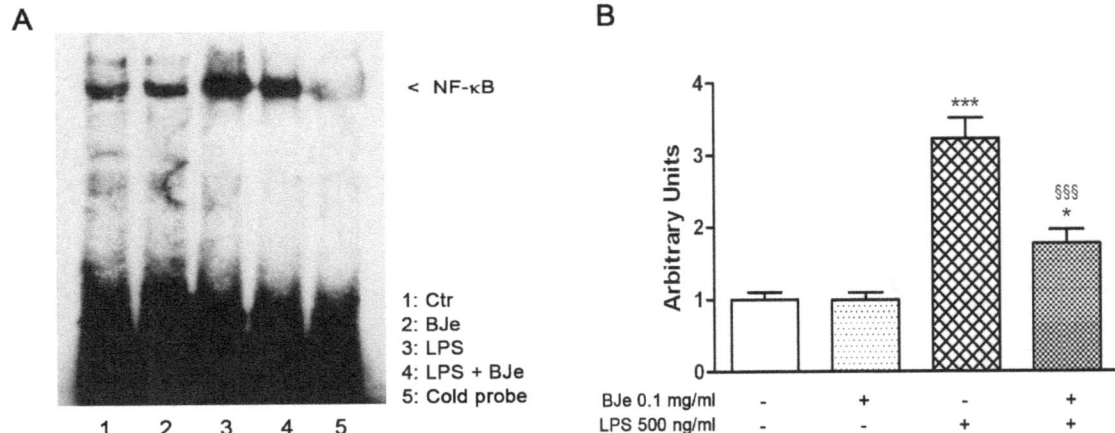

Figure 5. Inibitory effect of BJe on LPS-induced NF-κB activation in THP-1 cells. (A) The cells were exposed to 0.1 mg/ml BJe 30 min before LPS treatment (500 ng/ml for 3 hs), and then NF-κB activation was determined by the electrophoretic mobility shift assay (EMSA). A competition assay was performed using both biotin-labeled and unlabeled specific probe (cold probe, CP). (B) Densitometric analysis of three independent blots (mean ± SEM) is reported. * and *** $p < 0.05$ and $p < 0.001$ *vs* untreated cultures, respectively; §§§, $p < 0.001$ *vs* LPS-treatment (ANOVA followed by Student-Newman Keuls multiple comparisons test).

both transcription profile and protein levels of pro-inflammatory cytokines. Furthermore, we reported a concentration-dependent effect of BJe in a range useful to exclude toxic effects in THP-1 cultures.

As previously reported [6], the induction of most genes involved in the inflammatory response was abolished or attenuated under the inactivation of IκB kinase/nuclear factor-κB (IKK/NF-κB) pathway. NF-κB plays a crucial role in coordinating cellular

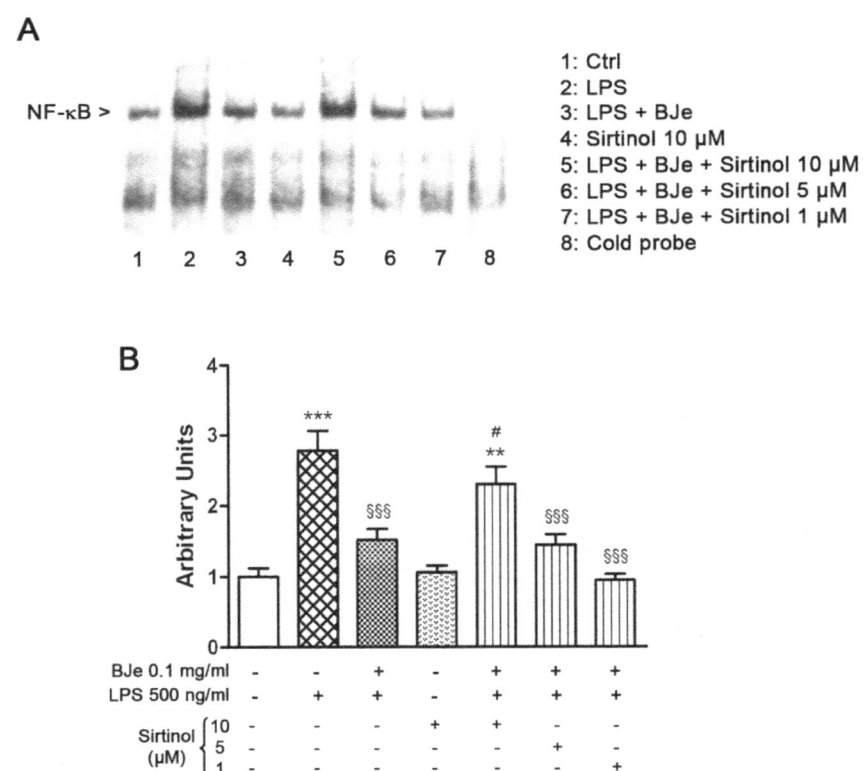

Figure 6. Sirtinol reverts the inhibitory effect of BJe on LPS-induced activation of NF-κB. (A) Exposure of THP-1cells to 0.1 mg/ml BJe reduced LPS-induced NF-κB activation; this effect was reverted by 10 μM Sirtinol, a SIRT1 inhibitor. EMSA analysis was performed using both biotin-labeled and unlabeled specific probe (cold probe). (B) Densitometric analysis of three independent blots (mean ± SEM) is presented. ** $p < 0.01$ and *** $p < 0.001$ *vs* controls; §§§, $p < 0.001$ *vs* LPS treated cells. # $p < 0.05$ *vs* BJe plus LPS treated cells (ANOVA followed by Student-Newman Keuls multiple comparisons test).

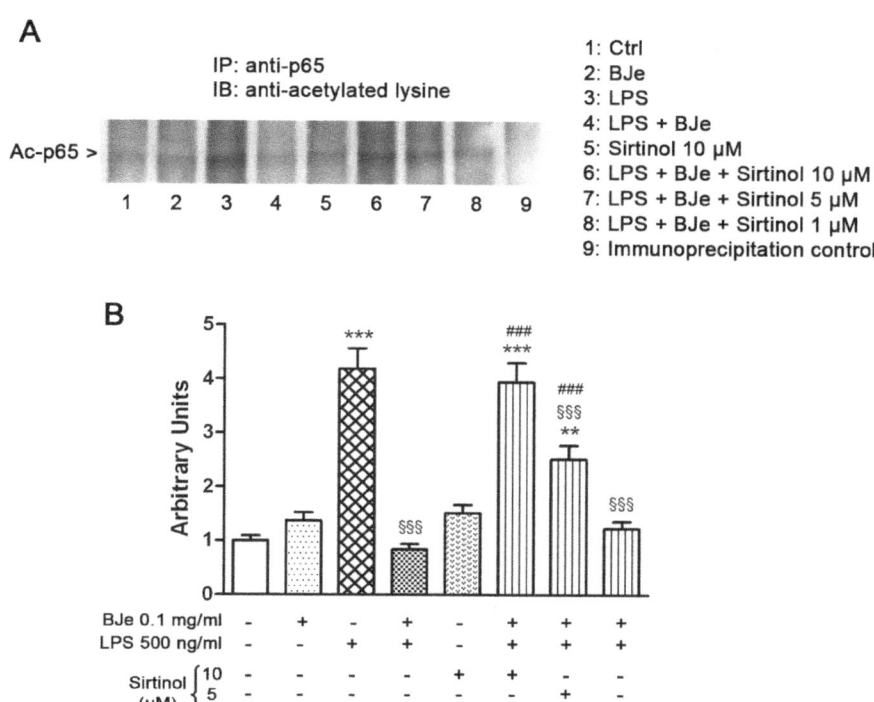

Figure 7. BJe treatment reverts the LPS-enhanced acetylation of p65 in THP-1 cells. (A) After LPS treatment in presence or absence of BJe and/or Sirtinol, THP-1 cells were lysed and proteins were immunoprecipitated using an anti-p65 antibody. Immunoprecipitated proteins were separated by SDS-PAGE and immunoblotted with antibody against acetyl-lysine residues. Immunoprecipitation negative control was set by incubating cell lysates under similar conditions, but without the immunoprecipitating antibody. (B) Densitometric analysis of three independent blots (mean ± SEM) is presented. ** $p<0.01$ and *** $p<0.001$ *vs* control cultures; §§§ $p<0.001$ *vs* LPS treated cells; ### $p<0.001$ *vs* LPS plus BJe treated cells (ANOVA followed by Student-Newman Keuls multiple comparisons test).

response to infections, stress and injury. When the inhibitory protein IκB dissociates, the most common active heterodimer, dealing with RelA/p65 and p50, is able to trigger both innate and adaptive immune responses through the induction of several pro-inflammatory genes [5]. Furthermore, RelA acetylation increases transcriptional activity of NF-κB [39], which in turn regulates genes encoding cytokines and metalloproteases. On the other hand the deacetylation of RelA promotes its effective binding to IκBα leading to IκBα-dependent nuclear export of NF-κB. Indeed, once in the nucleus, the activated transcription factor associates with several histone acetyltransferases (HATs) that catalyze the acetylation of RelA/p65 and lead to "open chromatin" configuration through the acetylation of N-terminal tails of histones.

In this way, the acetylation of specific lysine residues can affects both the DNA-binding ability and transcriptional activity of the protein [12,17].

Therefore, NF-κB is a target for many anti-inflammatory drugs and during the last years several studies have focused on polyphenols, natural compounds that would act as activators of SIRT1, a member of the class III HDAC family that has been implicated in modulating epigenetic gene silencing and cell survival, via acetylation of several both histonic and non-histonic substrates. Indeed several recent studies confirmed that SIRT1-mediated deacetylation of p65/RelA inhibited the NF-κB signaling and the activation of SIRT1 could alleviate a multitude of NF-κB-driven inflammatory and metabolic disorders [40,41]. This implies that SIRT1 activators could exert significant benefits in the treatment of inflammation.

Several findings have shown that polyphenols show anti-inflammatory activity in both *in vitro* and *in vivo* models, by modulation of pro-inflammatory gene expression such as COX-2, iNOS and several pivotal cytokines [42,43]. Due to their properties, flavonoids might be reasonable candidates for the development of new anti-inflammatory drugs, although their mechanism of action remains not fully understood.

On the basis of these results, we investigated the modulatory effect of BJe on a LPS-induced inflammatory response, focusing on SIRT1-mediated NF-κB inhibition. Our results provide evidence for BJe impact on NF-κB pathway, likely via SIRT1 activation. In order to confirm our hypothesis we tested the effects of Sirtinol, a synthetic molecule that inhibits SIRT1 functionality by occupying the site which normally functions as the binding site for the adenine base of NAD+ [15]. Under our conditions, we observed that Sirtinol triggered RelA/p65 binding in the nuclear fraction of LPS-activated cells. Finally, immunoprecipitation assay suggested that Sirtinol and BJe produced opposite effects on SIRT1's deacetylating activity. Considering the Sirtinol-mediated inhibition of SIRT1 and that BJe could act as SIRT1 activators, these results underscore at molecular levels that the BJe-mediated inhibition of NF-κB may be associated to acetylation status of p65/RelA subunit.

Flavonoids are widely recognized as naturally occurring antioxidants and several protective effects against cell injury have been associated to their antioxidant properties. Here we also demonstrated that bioactive molecules present in the BJe are responsible of specific inhibition of intracellular pathway involved in NF-κB-mediated inflammatory response. Noteworthy, our results have been obtained by use of a phytocomplex rather than those of individual components. According other results [27] we believe that the complex mixtures of bioactive molecules present in BJe could be more effective than their individual constituents to induce beneficial effects through both additive and synergistic action.

Our study, for the first time, shows the *in vitro* anti-inflammatory activity of flavonoid fraction from bergamot juice, suggesting the activation of SIRT1 could be a relevant target for novel therapeutic approaches. In addition, our observations (i) demonstrate the inhibitory effects of BJe on LPS-induced increases in mRNA transcripts and protein levels of pro-inflammatory cytokines, (ii) confirm the presence of SIRT1/NF-κB cross-talk in

a model of inflammation such as THP-1 monocyte/macrophages treated with LPS, (iii) give evidence for the involvement of SIRT1 in the production and secretion of cytokines, representing the first data on the interplay existing between SIRT1, NF-κB and BJe effects in LPS-stimulated THP-1 cells.

Author Contributions

Conceived and designed the experiments: MN RI RR. Performed the experiments: RR MC SC NF PC. Analyzed the data: MC DC. Contributed reagents/materials/analysis tools: MN. Wrote the paper: MN RI RR.

References

1. Karima R, Matsumoto S, Higashi H, Matsushima K (1999) The molecular pathogenesis of endotoxic shock and organ failure. Mol Med Today 5(3): 123–32.
2. Shacter E, Weitzman SA (2002) Chronic inflammation and cancer. Oncology (Williston Park) 16(2): 217–26, 229; discussion 230–2.
3. Manabe I (2011) Chronic inflammation links cardiovascular, metabolic and renal diseases. Circ J 75(12): 2739–48.
4. Prasad S, Sung B, Aggarwal BB (2012) Age-associated chronic diseases require age-old medicine: role of chronic inflammation. Prev Med 54 Suppl: S29–37.
5. Tripathi P, Aggarwal A (2006) NF-kB transcription factor: a key player in the generation of immune response. Current Science 90(4): 519–531.
6. Hayden MS, Ghosh S (2008) Shared principles in NF-kappaB signaling. Cell 132(3): 344–62.
7. Baeuerle PA, Baltimore D (1988) I kappa B: a specific inhibitor of the NF-kappa B transcription factor. Science 242(4878): 540–6.
8. Baldwin AS Jr (1996) The NF-kappa B and I kappa B proteins: new discoveries and insights. Annu Rev Immunol 14: 649–83.
9. Karin M, Cao Y, Greten FR, Li ZW (2002) NF-kappaB in cancer: from innocent bystander to major culprit. Nat Rev Cancer 2(4): 301–10.
10. Nakanishi C, Toi M (2005) Nuclear factor-kappa B inhibitors as sensitizers to anticancer drugs. Nature Reviews Cancer 5(4): 297–309.
11. Kiernan R, Bres V, Ng RWM, Coudart MP, El Messaoudi S, et al. (2003) Post-activation turn-off of NF-kappa B-dependent transcription is regulated by acetylation of p65. Journal of Biological Chemistry 278(4): 2758–2766.
12. Chen LF, Mu YJ, Greene WC (2002) Acetylation of RelA at discrete sites regulates distinct nuclear functions of NF-kappa B. Embo Journal 21(23): 6539–6548.
13. Howitz KT, Bitterman KJ, Cohen HY, Lamming DW, Lavu S, et al. (2003) Small molecule activators of sirtuins extend Saccharomyces cerevisiae lifespan. Nature 425(6954): 191–196.
14. Grozinger CM, Chao ED, Blackwell HE, Moazed D, Schreiber SL (2001) Identification of a class of small molecule inhibitors of the sirtuin family of NAD-dependent deacetylases by phenotypic screening. Journal of Biological Chemistry 276(42): 38837–38843.
15. Trapp J, Jochum A, Meier R, Saunders L, Marshall B, et al. (2006) Adenosine mimetics as inhibitors of NAD(+)-dependent histone deacetylases, from kinase to sirtuin inhibition. Journal of Medicinal Chemistry 49(25): 7307–7316.
16. Warner TD, Giuliano F, Vojnovic I, Bukasa A, Mitchell JA, et al. (1999) Nonsteroid drug selectivities for cyclo-oxygenase-1 rather than cyclo-oxygenase-2 are associated with human gastrointestinal toxicity: A full in vitro analysis. Proc Natl Acad Sci U S A 96(13): 7563–7568.
17. Yamamoto Y, Gaynor RB (2001) Therapeutic potential of inhibition of the NF-kappa B pathway in the treatment of inflammation and cancer. Journal of Clinical Investigation 107(2): 135–142.
18. Harborne JB (1990) Role of Secondary Metabolites in Chemical Defense-Mechanisms in Plants. Ciba Foundation Symposia 154: 126–139.
19. Manach C, Scalbert A, Morand C, Remesy C, Jimenez L (2004) Polyphenols: food sources and bioavailability. American Journal of Clinical Nutrition 79(5): 727–747.
20. Bengmark S (2006) Impact of nutrition on ageing and disease. Current Opinion in Clinical Nutrition and Metabolic Care 9(1): 2–7.
21. Manach C, Mazur A, Scalbert A (2005) Polyphenols and prevention of cardiovascular diseases. Current Opinion in Lipidology 16(1): 77–84.
22. Kim DO, Lee KW, Lee HJ, Lee CY (2002) Vitamin C equivalent antioxidant capacity (VCEAC) of phenolic phytochemicals. Journal of Agricultural and Food Chemistry 50(13): 3713–3717.
23. Eberhardt MV, Lee CY, Liu RH (2000) Nutrition - Antioxidant activity of fresh apples. Nature 405(6789): 903–904.
24. Potapovich AI, Lulli D, Fidanza P, Kostyuk VA, De Luca C, et al. (2011) Plant polyphenols differentially modulate inflammatory responses of human keratinocytes by interfering with activation of transcription factors NF kappa B and AhR

and EGFR-ERK pathway. Toxicology and Applied Pharmacology 255(2): 138–149.
25. Kostyuk VA, Potapovich AI, Suhan TO, de Luca C, Korkina LG (2011) Antioxidant and signal modulation properties of plant polyphenols in controlling vascular inflammation. European Journal of Pharmacology 658(2–3): 248–256.
26. Corasaniti MT, Maiuolo J, Maida S, Fratto V, Navarra M, et al. (2007) Cell signaling pathways in the mechanisms of neuroprotection afforded by bergamot essential oil against NMDA-induced cell death in vitro. British Journal of Pharmacology 151(4): 518–529.
27. Miceli N, Mondello MR, Monforte MT, Sdrafkakis V, Dugo P, et al. (2007) Hypolipidemic effects of Citrus bergamia Risso et Poiteau juice in rats fed a hypercholesterolemic diet. Journal of Agricultural and Food Chemistry 55(26): 10671–10677.
28. Mollace V, Sacco I, Janda E, Malara C, Ventrice D, et al. (2011) Hypolipemic and hypoglycaemic activity of bergamot polyphenols: From animal models to human studies. Fitoterapia 82(3): 309–316.
29. Delle Monache S, Sanita P, Trapasso E, Ursino MR, Dugo P, et al. (2013) Mechanisms Underlying the Anti-Tumoral Effects of Citrus bergamia Juice. Plos One 8(4).
30. Navarra M, Ursino MR, Ferlazzo N, Russo M, Schumacher U, et al. (2014) Effect of Citrus bergamia juice on human neuroblastoma cells in vitro and in metastatic xenograft models. Fitoterapia 95: 83–92.
31. Visalli G, Ferlazzo N, Cirmi S, Campiglia P, Gangemi S, et al. (2014) Bergamot Juice Extract Inhibits Proliferation by Inducing Apoptosis in Human Colon Cancer Cells. Anticancer Agents Med Chem 14 (9).
32. Sommella E, Pepe G, Pagano F, Tenore GC, Dugo P, et al. (2013) Ultra high performance liquid chromatography with ion-trap TOF-MS for the fast characterization of flavonoids in Citrus bergamia juice. J Sep Sci 36(20): 3351–5.
33. Sharif O, Bolshakov VN, Raines S, Newham P, Perkins ND (2007) Transcriptional profiling of the LPS induced NF-kappaB response in macrophages. BMC Immunol 8: 1.
34. Sakharwade SC, Sharma PK, Mukhopadhaya A (2013) Vibrio cholerae porin OmpU induces pro-inflammatory responses, but down-regulates LPS-mediated effects in RAW 264.7, THP-1 and human PBMCs. Plos One 8(9): e76583.
35. Kim D, Kim JY (2014) Anti-CD14 antibody reduces LPS responsiveness via TLR4 internalization in human monocytes. Mol Immunol 57(2): 210–5.
36. Chanput W, Mes J, Vreeburg RA, Savelkoul HF, Wichers HJ (2010) Transcription profiles of LPS-stimulated THP-1 monocytes and macrophages: a tool to study inflammation modulating effects of food-derived compounds. Food Funct 1(3): 254–61.
37. Schildberger A, Rossmanith E, Eichhorn T, Strassl K, Weber V (2013) Monocytes, peripheral blood mononuclear cells, and THP-1 cells exhibit different cytokine expression patterns following stimulation with lipopolysaccharide. Mediators Inflamm 2013: 697972.
38. Hegde PS, White IR, Debouck C (2004) Interplay of transcriptomics and proteomics. Drug Discov Today 9(2 Suppl): S53–6.
39. Chen LF, Greene WC (2003) Regulation of distinct biological activities of the NF-kappaB transcription factor complex by acetylation. J Mol Med (Berl) 81(9): 549–57.
40. Yao H, Rahman I (2012) Perspectives on translational and therapeutic aspects of SIRT1 in inflammaging and senescence. Biochem Pharmacol 84(10): 1332–9.
41. Xie J, Zhang X, Zhang L (2013) Negative regulation of inflammation by SIRT1. Pharmacological Research 67(1): 60–7.
42. Middleton E Jr, Kandaswami C, Theoharides TC (2000) The effects of plant flavonoids on mammalian cells: implications for inflammation, heart disease, and cancer. Pharmacol Rev 52(4): 673–751.
43. Gonzalez R, Ballester I, Lopez-Posadas R, Suarez MD, Zarzuelo A, et al. (2011) Effects of flavonoids and other polyphenols on inflammation. Crit Rev Food Sci Nutr 51(4): 331–62.

9

Alteration of Antioxidant Enzymes and Associated Genes Induced by Grape Seed Extracts in the Primary Muscle Cells of Goats *In Vitro*

Tan Yang[1,2], Xiaomin Li[1], Wang Zhu[1], Cheng Chen[1], Zhihong Sun[1]*, Zhiliang Tan[2]*, Jinghe Kang[2]

1 Key Laboratory for Bio-Feed and Molecular Nutrition, Southwest University, Chongqing, P. R. China, **2** Key Laboratory of Agro-Ecological Processes in Subtropical Region, Hunan Research Center of Livestock & Poultry Sciences, South-Central Experimental Station of Animal Nutrition and Feed Science in Ministry of Agriculture, Institute of Subtropical Agriculture, The Chinese Academy of Sciences, Changsha, Hunan, P.R. China

Abstract

This study was conducted to investigate how the activity and expression of certain paramount antioxidant enzymes respond to grape seed extract (GSE) addition in primary muscle cells of goats. Gluteal primary muscle cells (PMCs) isolated from a 3-week old goat were cultivated as an unstressed cell model, or they were exposed to 100 μM H_2O_2 to establish a H_2O_2-stimulated cell model. The activities of catalase (CAT), superoxide dismutases (SOD) and glutathione peroxidases (GPx) in combination with other relevant antioxidant indexes [i.e., reduced glutathione (GSH) and total antioxidant capacity (TAOC)] in response to GSE addition were tested in the unstressed and H_2O_2-stimulated cell models, and the relative mRNA levels of the CAT, GuZu-SOD, and GPx-1 genes were measured by qPCR. In unstressed PMCs, GSE addition at the dose of 10 μg/ml strikingly attenuated the expression levels of CAT and CuZn-SOD as well as the corresponding enzyme activities. By contrast, in cells pretreated with 100 μM H_2O_2, the expression and activity levels of these two antioxidant enzymes were enhanced by GSE addition at 10 μg/ml. GSE addition promoted GPx activity in both unstressed and stressed PMCs, while the expression of the GPx 1 gene displayed partial divergence with GPx activity, which was mitigated by GSE addition at 10 μg/ml in unstressed PMCs. GSH remained comparatively stable except for GSE addition to H_2O_2-stimulated PMCs at 60 μg/ml, in which a dramatic depletion of GSH occurred. Moreover, GSE addition enhanced TAOC in unstressed (but not H_2O_2-stimulated) PMCs. GSE addition exerted a bidirectional modulating effect on the mRNA levels and activities of CAT and SOD in unstressed and stressed PMCs at a moderate dose, and it only exhibited a unidirectional effect on the promotion of GPx activity, reflecting its potential to improve antioxidant protection in ruminants.

Editor: Giovanni Li Volti, University of Catania, Italy

Funding: This study was financially supported by the National Program on Key Basic Research Project of China (2013CB127303 and 2013CB127304), the CAS Presidential Scholarship (Excellence Award), the Natural Science Foundation Project of CQ CSTC (cstc2013jjB0112 and cstc2012jjA80001), the National Science & Technology Pillar Program (2012BAD14B18 and 2014BAD08B11), the Fundamental Research Funds for the Central Universities (XDJK2011C030 and XDJK2011B011), and the National Natural Science Foundation of China (30600436 and 31201826). The funders had no role in study design, data collection and analysis, decision to publish, or preparation of the manuscript.

Competing Interests: The authors have declared that no competing interests exist.

* Email: sunzh2002cn@aliyun.com (ZS); zltan@isa.ac.cn (ZT)

Introduction

Grape seed extract (GSE) has been well-documented for its effect on antioxidation, and it has been extensively used as therapy in substantial diseases of diverse organs and tissues such as asthma, cataracts, cardiovascular diseases, small intestinal mucositis and nephropathy [1–5]. Although the underlying mechanisms relevant to the pathogenesis of these diseases have not yet been completely clarified, reactive oxygen species (ROS) are purportedly an important contributor, and the therapeutic function of GSE is tightly linked to its antioxidant effectiveness [1–5]. Recently, accumulated evidence has demonstrated that replenishing GSE is a feasible way of alleviating the negative effects of ROS within skeletal muscle. It has been found that skeletal muscle cells frequently become the targets of ROS under assorted conditions [6–10]. ROS-mediated oxidative damage and signal interruption both cause cell dysfunction to some extent [11,12]. It has been

observed that the exposure of myotubes to ROS (i.e., H_2O_2 and Doxorubicin) results in muscle protein wasting and fiber atrophy [13–15]. In addition, scores of studies have shown the adverse effects of ROS on skeletal muscle in rats fed high-carbohydrate/high-fat diets [12,16]. However, subsequent testing has indicated that GSE has the ability to alleviate ROS damage by acting as an effective antioxidant [17].

The antioxidant function of GSE is mediated by a vast array of functional components with a phenolic nature such as monomeric flavanols; dimeric, trimeric and polymeric procyanidins; and phenolic acids [4,18]. These phenolic compounds are endowed with the capacity to scavenge singlet oxygen and peroxyl radicals via the reduction of multiple O–H bonds [19]. The effectiveness of phenolic compounds as free radical scavengers was proven to greatly exceed those of vitamins E and C, the major recognized antioxidants in biological systems [20]. However, it is noteworthy that, in the scope of existing GSE research, the beneficial effects of

GSE against oxidative radicals are mainly of a chemical nature. Relatively few studies have addressed whether there is a biological gateway for GSE in relation to the prevention of oxidative challenge.

In fact, phenolic and polyphenolic compounds that act as natural antioxidants are primarily present in and isolated from fruits, some vegetables, tea and herbs [18,21]. Temporarily ignoring the differences between grapeseeds and other botanicals with diverse molecular structures or different amounts of similar polyphenols, it has been found that the protective effect of genistein, an isoflavone mostly found in legumes, against diabetes-induced renal damage is partly dependent on the improvement of glutathione peroxidase (GPx) and superoxide dismutase (SOD) [22]. Our previous study also found that tea catechins can regulate antioxidant enzymes to alleviate H_2O_2-induced damage in goat skeletal muscle cells in vitro [23].

We hypothesized that GSE most likely exerts its antioxidant function by influencing the gene expression levels and production of some or all antioxidative enzymes in various tissues, although chemoprotection is not necessarily exclusively dependent on this. In this study, we aimed to test the putative antioxidative function of GSE from a fresh perspective in ruminants by examining the effect of GSE on antioxidant enzyme activity and evaluating the influence of GSE at the gene level on corresponding enzymes in primary muscle cells of goats.

Materials and Methods

Ethics Statement

This study was carried out in strict accordance with the recommendations in the Animal Care and Use Guidelines of the Institute of Subtropical Agriculture (ISA), Chinese Academy of Sciences. The protocol was approved by the Animal Care Committee on the Ethics of Animal Experiments of ISA. All surgery was performed under sodium pentobarbital anesthesia, and all efforts were made to minimize suffering.

Extracts and Reagents

Commercially available dried and powdered GSE obtained from Tarac Technologies (Nuriootpa, South Australia) contained 5.01% (+)-catechin, 4.78% (−)-epicatechin, 2.35% (−)-epigallo-catechin, 14.1% dimeric proanthocyanidin, 11.60% trimeric proanthocyanidins, 7.69% tetrameric proanthocyanidins and 40.0% polymeric proanthocyanidins. Dulbecco's Modified Eagle Medium (DMEM), fetal bovine serum, penicillin, and streptomycin were purchased from Invitrogen (Rockville, MD). Antioxidant detection kits were ordered from Nanjing Jiancheng Bioengineering Institute (Nanjing, China) and Beyotime Institute of Biotechnology (Shanghai China). Unless otherwise specified, all other chemicals were from Sigma (St. Louis, MO).

Cell Isolation, Culture and Treatment

The primary muscle cells (PMCs) used in our experiments were isolated from the gluteal muscles of one Liu Yang black Goat (a local breed, south of China) according the method of Zhong et al (who modified the existing method of Lynge et al) [23,24]. The experimental goat was carefully chosen from 3-week old weaned healthy goats. The PMCs were incubated in DMEM medium containing 10% FBS and a 1% mixture of penicillin and streptomycin (10 units/mL and 1 mg/mL, respectively) at 37°C under an atmosphere of 5% CO_2/95% air. The growing medium was changed every two days. At 80–90% confluence, the PMCs were washed with phosphate buffered saline (PBS) once and subcultured with 0.25% trypsin–EDTA solution. Without a

specific indication, activated PMCs were used at the age of passage 4–8 and starved to FBS for 24 h before any treatment.

Before the formal treatment, PMCs were exposed to a stepwise series of increasing concentrations of H_2O_2 to induce oxidative stress and oxidative damage. After examining cell activity, 100 μM H_2O_2 was determined as the optimum concentration in the later formal experiment. Likewise, a dose-response study based on a wide range of GSE concentrations from 0.1 to 60 μg/ml medium was performed in the absence and presence of H_2O_2 preincubation. Based on cell viability, GSE levels of 1, 10 and 60 μg/ml were chosen as the ultimate addition levels in this trial, representing mild, moderate and excessive doses, respectively.

To characterize the effects of GSE on PMCs in un-stressed and stressed conditions, two PMC models were separately established as follows: cells were cultured in medium at the GSE levels of 1, 10 and 60 μg/ml for 24 h as the un-stressed cell model; a stressed cell model was established wherein cells were induced by accretion with 100 μM H_2O_2 in the medium for 1 h. It should be noted here, after incubation for 1 h with 100 μM H_2O_2, the cells were washed twice with PBS in order to remove H_2O_2 and to avoid it reacting with GSE directly, followed by the protocol of un-stressed cell model. Untreated cells serving as a control were also run in parallel and subjected to the same medium as the PMC models.

Cell Viability Tests

Cells were seeded in a 96-well plate (1×10^4 cells/well), and after the above-mentioned treatment, they were washed twice with PBS to eliminate interference with the subsequent assay. Cell viability was assessed using the commercial cell counting kit-8 (Dojindo, China) according to the manufacturer's instructions.

Microscopic Analysis

Cells were plated onto 34.8-mm dishes at a density of 2×10^6 cells and cultured to 70–80% confluence at 37°C. Then, they were incubated under different conditions, after which cell morphology was observed using a microscope (CX31, Olympus, Japan).

Total Antioxidant Capacity (TAOC) and Reduced Glutathione (GSH) Measurement

The TAOC in supernatants of cells lysed was measured using the azino-diethyl-benzthiazoline sulfate (ABTS) method. In this assay, incubation of ABTS with H_2O_2 and a peroxidase (metmyoglobin) results in the production of the blue-green radical cation ABTS+. Antioxidants in the sample suppress this color production proportionally to their concentration [25]. The system was standardized using Trolox, a water-soluble vitamin E analogue. The results were expressed as mmol Trolox equivalent/protein concentration of plasma supernatant of lysed cells. Supernatants of lysed cells were analyzed for GSH using the Glutathione Quantification Kit (Jianchen, Co., Nanjing, China) according to the manufacturer's instructions. This kit employs a fundamental reaction in which 5,5′-dithiobis-2-nitrobenzoic acid (DTNB) and GSH react to generate 2-nitro-5-thiobenzoic acid and glutathione disulfide (GSSG). Because 2-nitro-5-thiobenzoic acid is a yellow product, the GSH concentration in a sample can be detected at 405 nm.

Antioxidant Enzyme Activity Assay

Briefly, cells were harvested in a lysate buffer containing 0.1% Triton X-100, then centrifuged to remove the supernatant for the assay [13]. Catalase (CAT) activity in the cell homogenates was determined using a CAT analysis kit purchased from Beyotime Bio-Corporation (Shanghai, China). As per the manufacturer's instructions, samples were placed in a cuvette that received excess hydrogen peroxide for decomposition by CAT in our samples for an exact time between 1 to 5 min, and the remaining hydrogen peroxide coupled with a substrate was treated with peroxidase to generate a red product, N-4-antipyryl-3-chloro-5-sulfonate-p-benzoquinonemonoimine, which absorbs maximally at 520 nm. In this way, CAT activity was determined by measuring the decomposition of hydrogen peroxide spectrophotometrically (Bio-RAD680, Bio-rad Co., USA). The activity of SOD was measured in 96-well microplates using the WST-1 method. WST-1, a 2-(4-iodophenyl)-3-(4-nitrophenyl)-5-(2,4-disulfophenyl)-2H-tetrazolium, monosodium salt, exhibits very low background absorbance and is efficiently reduced by superoxide to a stable water-soluble formazan with high molar absorptivity. As the superoxide production from xanthine oxidase in the SOD kit (from Beyotime Bio-Corporation, Shanghai, China) could be cut down by SOD in cell lysate suspensions, the optical density of formazan was measured at 450 nm. The activity of GPx was estimated using the commercially available GPx assay kit from Beyotime Bio-Corporation (Shanghai, China). The GPx activity was measured indirectly by a coupled reaction with glutathione reductase. Oxidized glutathione produced via the reduction of hydroperoxide by GPx was recycled to its reduced form by the oxidation of triphosphopyridine nucleotide reduced tetrasodium salt to nicotinamide adenine dinucleotide phosphate (NADPH to NADP+), which was accompanied by a decrease in absorbance at 340 nm [26].

RNA Isolation and Real-Time RT-PCR

A total of 1 ml of ice-cold TRIzol reagent (TaKaRa, Japan) was added to cell monolayers. Extraction was carried out according to the manufacturer's instructions. The integrity of the isolated RNA was checked using agarose gel electrophoresis (1%). The RNA concentration was calculated from the absorbance at 260 nm. Protein contamination was further assessed by spectrophotometric determination of the absorbance 260 nm to 280 nm ratio, and only samples with a ratio between 1.8 and 2.1 were used in further experiments. To remove traces of chromosomal DNA, 1000 ng of total RNA was treated with RNase-free DNase I (TaKaRa, Japan) for 5 min at 37°C. Quantitative real-time PCR (qRT-PCR) was performed using the SYBR Premix Ex TaqTM Kit (TaKaRa, Japan) on the Applied Biosystems Prism 7900 HT sequence detection system (Applied Biosystems, Foster, CA). The cDNA was diluted fourfold before equal amounts were added to duplicate qRT-PCR reactions. The tested genes and their sequences designed according to Zhong et al. [23] are listed in Table 1. The thermal profile for all reactions was 30 sec at 95°C, then 40 cycles of denaturation at 95°C for 5 s, annealing at 60°C for 30 s and extension at 72°C for 30 s. At the end of each cycle, the fluorescence monitoring was for 10 s. Each reaction was completed with a melting curve analysis to ensure the specificity of the reaction. Gene expression was quantified using real-time qPCR analyzer software by Eppendorf. The relative expression levels of mRNA species were determined using the comparative Ct method. β-actin was selected as a reference gene. The expression ratio (R) of treatment to control cells was calculated using the following equations: $R = 2^{-\Delta\Delta Ct}$; and $\Delta\Delta Ct = (Ct_{target\ gene} - Ct_{\beta\text{-actin}})$ treatment $- (Ct_{target\ gene} - Ct_{\beta\text{-actin}})$ control. The reactions for each sample were performed in quadruplicate.

Statistical Analyses

Data are presented as the mean ± SEM for all determinations. Significant differences were analyzed by 1-way ANOVA of SAS or Mann Whitney tests as appropriate. A P value <0.05 was considered significant.

Results

A sharp dose-dependent decrease of PMC viability is revealed in Figure 1A. When the concentration of H_2O_2 surpassed 25 μM, this decline continued until the H_2O_2 concentration reached 150 μM. Subsequently, the cell viability remained comparatively stable but at a low level, even with H_2O_2 up to 1000 μM. The cell viability curve indicated that the PMCs were sensitive to H_2O_2 in the range from 25 to 150 μM. We thereby chose 100 μM H_2O_2 as the optimal concentration for inducing oxidative stress in PMCs in the subsequent formal trial, in part because that is in the sensitive range, and a 50% disruption of cells compared to the control was observed. Figure 1B depicts PMC viability after incubation with diverse concentrations of GSE for 24 h. Cell viability was unaffected until the GSE concentration reached 10 μg/ml. GSE at 15 μg/ml resulted in a marked increase of cell viability, but the increasing trend did not continue with higher GSE concentrations. Subsequently, cell viability fell into a recession and reached its minimum at the level of 60 μg/ml.

Figure 2A shows photographs of PMCs (control) and PMCs subjected to the insult of 100 μM H_2O_2 for 1 h and then incubated with 0, 1, 10 or 60 μg GSE/ml for 24 h. Compared with the control, PMCs treated only with H_2O_2 had obvious swelling, showed a dramatic increase in floating cells (non-adhering cells), and the aggregation of floating cells even appeared. For H_2O_2-pretreated PMCs supplemented with GSE at 1 or 10 ug/ml, the adhering cells' morphology remained virtually unchanged, although there was still a significant increase of floating cells compared with the control. As the GSE addition level reached 60 ug/ml for H_2O_2-pretreated PMCs, the amount of cells was visually scarce, even compared with PMCs treated with H_2O_2 alone. Figure 2B reflects the viability of PMCs that initially suffered from the insult of 100 μM H_2O_2 for 1 h then were incubated with various GSE concentrations for 24 h. The viability of PMCs subjected to H_2O_2 challenge was less (P<0.05) than that of the control. However, subsequent GSE supplementation of 0.1–15 ug/ml successfully mitigated the inhibitory effect of H_2O_2 on cell viability. When GSE addition surpassed 15 μg/ml, cell viability markedly declined, and it dropped to its lowest value at the GSE level of 60 ug/ml.

GSE addition at 1 and 60 μg/ml led to a significant increase of TAOC in un-stressed PMCs compared with the control (Figure 3). However, TAOC was not affected by GSE addition to cells pretreated with H_2O_2. GSE addition did not affect the GSH content in un-stressed PMCs with an average value of 49.2±0.9 nmol/mg protein Treatment with H_2O_2 led to a significant depletion of GSH (40.6±3.2 nmol/mg protein). Following GSE addition at levels of 1 and 10 μg/ml, GSH content did not recover, with respective values of 41.2±2.8 and 42.0±3.4 nmol/mg protein, and it even dramatically decreased to 32.2±1.9 nmol/mg protein as the GSE dose reached 60 μg/ml.

Figure 4 shows the effects of GSE addition on the activity and relative mRNA level of CAT in PMCs under un-stressed and stressed conditions. With un-stressed PMCs, GSE addition of 10 μg/ml decreased (P<0.05) CAT activity in comparison with the control and mild-dose groups (Figure 4A). CAT activity was reduced by approximately 45% at the GSE level of 10 μg/ml compared to the control. However, in stressed cells, GSE addition at 10 μg/ml conversely increased (P<0.05) the CAT activity compared to that without GSE addition. Concerning the significant drop of cell viability at the level of 60 μg GSE/ml (Figure. 2), the death of a great number of cells altered the culture environment of the surviving cells. Because it would be no longer correct to compare the enzyme activities and mRNA levels, these

Table 1. Sequences of primers (forward, for; reverse, rev), sizes of primers, and sizes of real-time quantitative PCR products.

Target gene	Primer (5'- 3')	Primer size (bp)	Product size (bp)
CAT, for	TGGGACCCAACTATCTCCAG	20	207
CAT, rev	AAGTGGGTCCTGTGTTCCAG	20	
CuZn-SOD, for	TGCAGGCCCTCACTTTAATC	20	216
CuZn-SOD, rev	CTGCCCAAGTCATCTGGTTT	20	
GPx 1, for	ACATTGAAACCCTGCTGTCC	20	178
GPx 1, rev	TCATGAGGAGCTGTGGTCTG	20	
β-actin, for	CCAACCGTGAGAAGATGACC	20	201
β-actin, rev	CGCTCCGTGAGAATCTTCAT	20	

CAT = catalase; CuZn-SOD = CuZn superoxide dismutase; GPx = glutathione peroxidase.

data at the dose of 60 µg GSE/ml were thereby not shown in the relevant figures. In PCR analysis, the mRNA expression level of the CAT gene was unaffected by GSE addition at 1 µg/ml, but it was down-regulated at 10 µg/ml in un-stressed PMCs, and thus corresponded well to the alterations of CAT activity. Following H_2O_2 treatment, the mRNA level of the CAT gene was less ($P < 0.05$) than that of the control. Subsequent addition of GSE increased the CAT mRNA level only marginally.

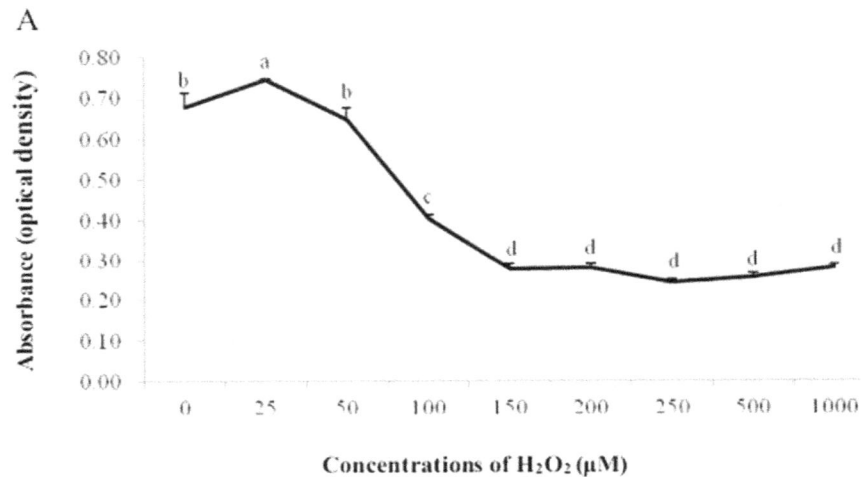

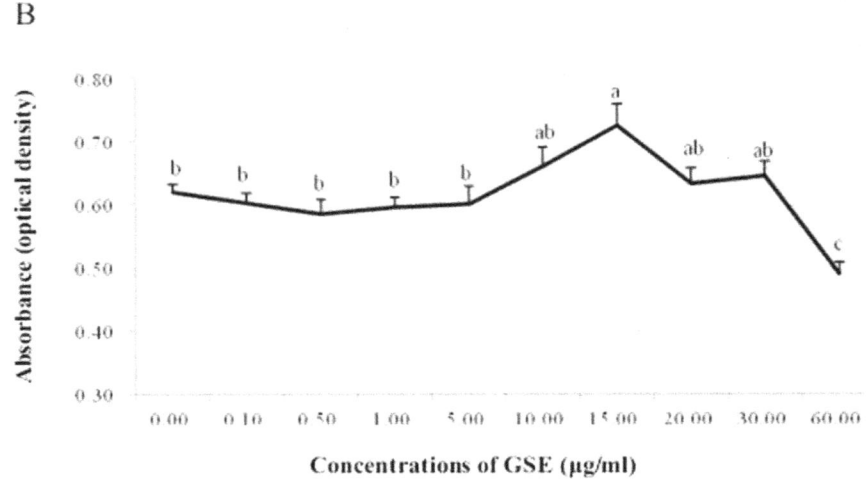

Figure 1. Alteration of primary muscle cells (PMCs) viability after incubation with various concentrations of H_2O_2 for 1 h (A), or GSE for 24 h (B). The results are presented as the means ± SEM (n = 12). Mean values with different letters differ ($P < 0.05$).

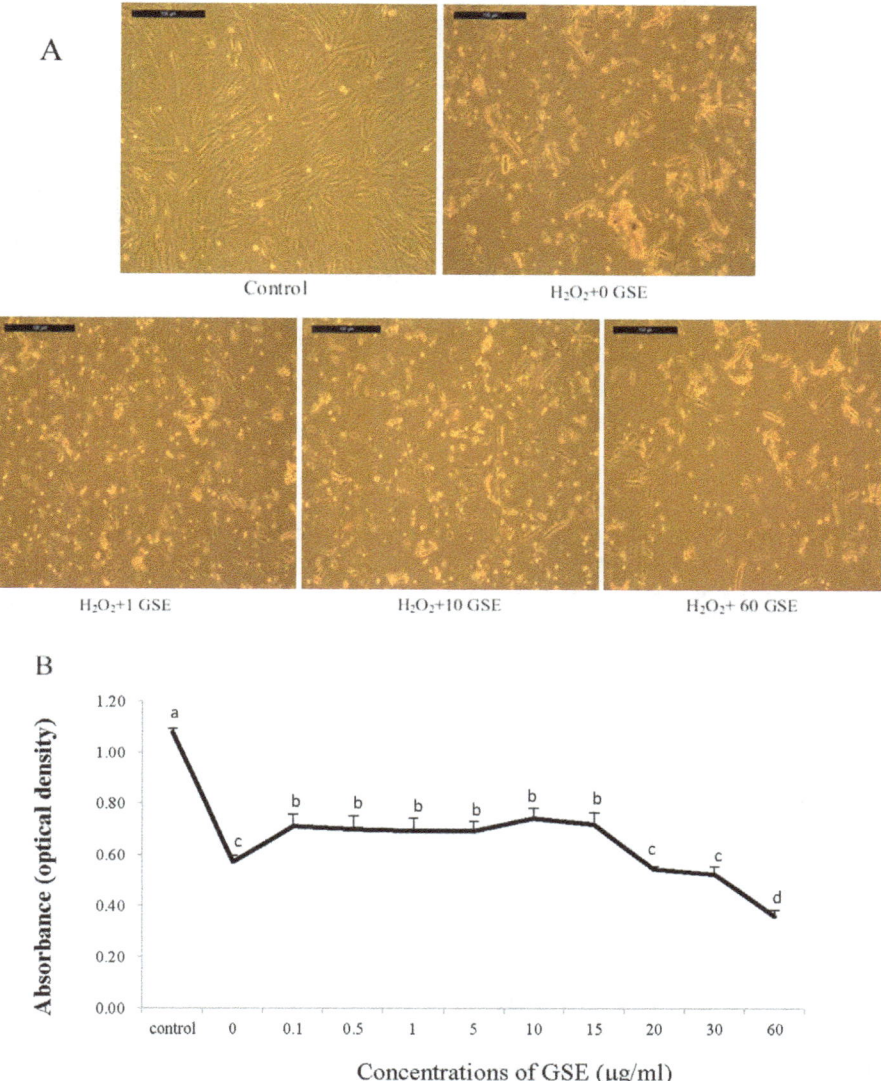

Figure 2. Representative photographs (A) and viability (B) of primary muscle cells (PMCs) pretreated with 100 µM H₂O₂ for 1 h followed by incubation with various concentrations of GSE for 24 h. The values are represented as the means ± SEM (n = 12). Mean values with different letters differ (P<0.05).

Figure 5 illustrates effects of GSE addition on another antioxidant enzyme, SOD, in terms of enzyme activity and CuZn-SOD gene expression. Adding GSE at 10 µg/ml weakened (P<0.05) the SOD activity and the CuZn-SOD mRNA level relative to the control. With respect to the control, the SOD activity and CuZn-SOD mRNA level were inhibited (P<0.05) by 100 µM H₂O₂ treatment alone; however, these values rebounded following the addition of GSE at the dose of 1 or 10 ug/ml.

GSE supplementation at the dose of 1 or 10 µg/ml, regardless of whether the PMCs were un-stressed or stressed, promoted (P< 0.05) GPx activity compared with the control (Figure 6). In regard to the mRNA level of GPx 1, there was a numerical increase in the GPx 1 mRNA level of un-stressed PMCs at the GSE dose of 1 µg/ml, whereas a moderate dose of GSE (10 µg/ml) caused a remarkable reduction (P<0.05). Exposure to H₂O₂ for only 1 h numerically reduced the GPx 1 mRNA expression of PMCs when compared to the control. GSE addition marginally increased GPx

1 mRNA levels of PMCs in contrast with those of H₂O₂-exposed cells.

Discussion

Excess production of ROS has been widely recognized to pose huge threats to the health of humans and animals [22]. It triggers oxidative damage and dysfunction in various cell types, including skeletal muscle cells, and it is strongly linked to the pathogenesis of a broad spectrum of diseases [9,10]. GSE has been extensively reported to effectively scavenge ROS and protect against ROS-associated damage. Acute skeletal muscle damage in rats results in skeletal muscle fiber disruption, oxidative stress and inflammation, while the administration of a grape seed-derived proanthocyani-dolic oligomer before and after the injury facilitates faster effective regeneration of the injured skeletal muscle [27]. Similarly, in the current experiment, direct exposure of PMCs to H₂O₂ led to a

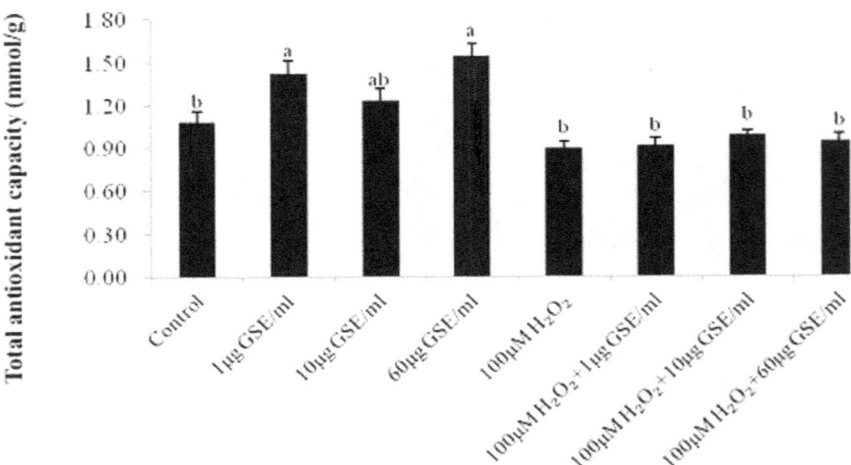

Figure 3. Alteration of total antioxidation capability (TAOC) after incubation of primary muscle cells (PMCs) with various concentrations of GSE for 24 h with or without H₂O₂ pretreatment. Bars represent the mean ± SEM (n = 4). Bars not sharing a common letter differ ($P<0.05$).

serious impairment as assessed by cell viability and morphological observations. However, with the addition of GSE at the proper dose, the viability of H₂O₂-pretreated PMCs clearly recovered, indicating that GSE confers antioxidant protection to goat PMCs.

TAOC is a valuable biomarker now widely used in analyses of serum, feedstuffs and biological tissues [28]. It is generally believed that cells and tissues of the body become more prone to develop dysfunction when antioxidant defenses are weakened. GSE, which

serves as an important natural antioxidant, has been proven to improve TOAC in several experiments. Cyclosporine A is an immunosuppressive drug that has been proposed to expedite the generation of ROS in cardiac muscle cells based on significant increases in oxidative stress index and malondialdehyde. Feeding rats Cyclosporine A together with GSE not only mitigates the levels of oxidative stress and malondialdehyde, it also elevates cardiac TOAC [29]. In the present study GSE addition dramatically raised the TAOC of un-stressed PMCs but not of stressed PMCs. Hence, the maintenance of adequate antioxidant

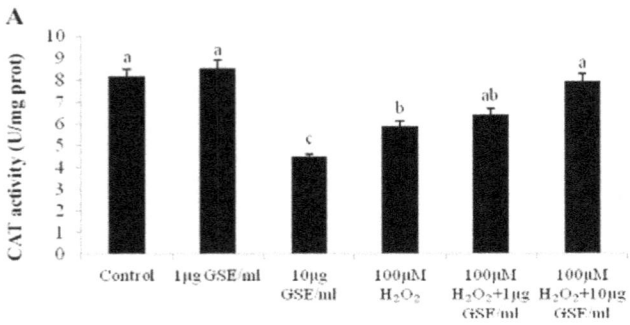

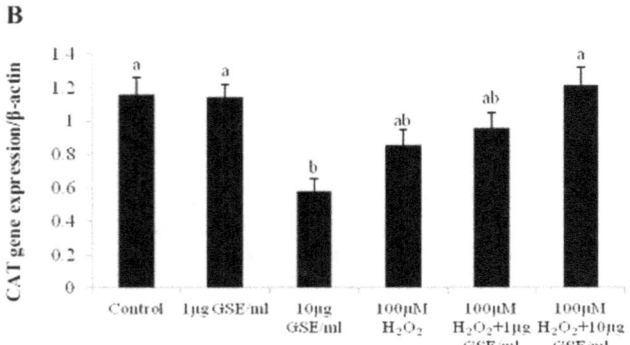

Figure 4. Effects of GSE addition on catalase (CAT) activity (A) and the relative CAT mRNA levels (B) of primary muscle cells (PMCs) pretreated with or without H₂O₂. Bars represent the mean ± SEM (n = 5) (A). Bars represent the mean ± SEM (n = 4) (B). Bars not sharing a common letter differ ($P<0.05$).

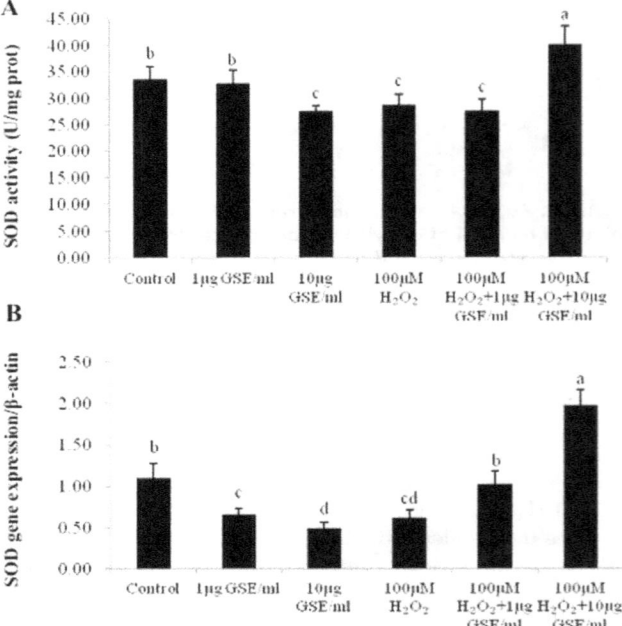

Figure 5. Effects of GSE addition on SOD activity (A) and the relative CuZn-SOD mRNA levels (B) of primary muscle cells (PMCs) pretreated with or without H₂O₂. Bars represent the mean ± SEM (n = 5) (A). Bars represent the mean ± SEM (n = 4) (B). Bars not sharing a common letter differ ($P<0.05$).

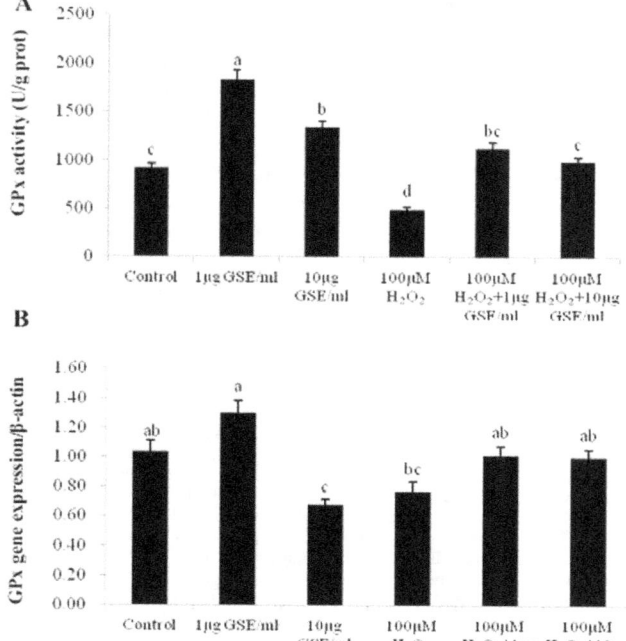

Figure 6. Effects of GSE addition on GPx activity (A) and the relative GPx 1 mRNA levels (B) of primary muscle cells (PMCs) pretreated with or without H₂O₂. Bars represent the mean ± SEM (n = 5) (A). Bars represent the mean ± SEM (n = 4) (B). Bars not sharing a common letter differ (P<0.05).

levels is essential to preventing or merely managing a great number of cell dysfunctions and disease conditions [28].

Previously, the antioxidant properties of GSE were basically attributed to GSE's intrinsic reducing capabilities [30,31]. However, the present study provides some novel evidence that GSE has the ability to modulate the activity and gene expression levels of relevant antioxidant enzymes. We observed severe declines in CAT and SOD activity in un-stressed PMCs incubated with a moderate dose of GSE, and corresponding changes in the relative mRNA levels of CAT and CuZn-SOD genes occurred. Activation of ERK1/2 has been shown to negatively correlate with CAT activity [32]. Additionally, it has been reported that GSE treatment of human colon carcinoma HT29 cells results in strong dose- and time-dependent phosphorylation of ERK1/2 [33]. Presumably, the ERK1/2 signaling pathway provides an effective approach for GSE to attenuate CAT expression and activity in unstressed PMCs. However, striking increases in CAT and SOD activity have been observed in H9C2 (rat heart cell line) cells incubated with catechin or proanthocyanidin B4, two important components of GSE. The inconsistency in comparison with our results is putatively due to the different cells used and some discrepancies in the GSE agents [34].

Interestingly, in H₂O₂-pretreated PMCs, the activities of both CAT and SOD plus the mRNA level of the CuZn-SOD gene were all positively influenced by a moderate dose of GSE. Previous research found that diabetic rats show increased oxidative stress in the body and down-regulated CAT protein expression in the aorta, but CAT protein expression was up-regulated after feeding with proanthocyanidin extracted from grapeseeds [35]. In addition, although decreased CAT and SOD activities have been observed in ischemically injured rat myocardium (which induces

excessive ROS generation), the administration of proanthocyanidin derived from grapeseeds effectively increases the activities of CAT and SOD [36]. Peroxisomal proliferator-activated receptors (PPARs) have been proven to suppress ROS generation through transcriptional up-regulation of a set of antioxidant enzymes such as CAT and CuZn-SOD [37–40]. Moreover, there is evidence demonstrating that some natural polyphenols, such as resveratrol, apigenin, carvacrol and humulon, can activate PPARs-dependent signaling [41]. This finding implies that PPARs are an important transcription factor in the GSE-regulatory signaling pathway leading to antioxidant enzymes.

Unlike the regulatory effects of GSE on the activities of both CAT and SOD, which are likely condition-dependent, GSE boosted GPx activity whether PMCs were pretreated with H₂O₂. In a previous study, the administration of proanthocyanidin extracted from grapeseeds also increased the GPx activity of rat heart under myocardial ischemic injury, but if the heart is not ischemic, increased GPx activity is not seen [36]. GPx has been proven to be regulated by diverse polyphenols via multiple approaches. For instance, β-naphthoflavone can bind and activate the aryl hydrocarbon receptor and subsequently induce the transcription of Nrf2, which is involved in the regulation of various antioxidant enzymes (i.e., GPx) [42–44]. In addition, flavonoids (i.e., cocoa flavonoids) can up-regulate GPx activity via the ERK1/2 signaling pathway [45]. This evidence suggests that GPx is regulated and controlled by GSE similarly to other polyphenols.

As to the reason why there was a partial divergence between GPx 1 mRNA expression and GPx activity, it was not surprising because the GPx superfamily, ignoring the members free from selenium, consists of at least four types of selenium-containing peroxidases that are encoded by separate genes [26]. GPx activity was tested as a whole for all GPx isoforms, while the mRNA of the GPx 1 gene was the only mRNA quantified in our trial. Thus, we could not rule out the possibility that other GPx genes might be upregulated by GSE and ultimately contribute to the increase in GPx enzymatic activity.

The increase of GPx activity in H₂O₂-pretreated PMCs incubated with GSE is believed to accelerate the depletion of intracellular GSH, as GSH provides a redox substrate for GPx as it implements H₂O₂ detoxification. However, GSH content in our study remained relatively stable. One possible explanation for this paradox is that accelerated consumption of GSH is accompanied by accelerated GSH generation. The depletion of cellular GSH could be opposed either by de novo synthesis or by reducing the GSSG formed (the oxidized form of GSH), and γ-glutamylcysteine synthetase (γ-GCS) and glutathione reductase (GRed), respectively, act as the key rate-limiting enzymes in these two process [46]. Several lines of evidence have shown that polyphenols (such as quercetin, epicatechin and epicatechin gallate) have the ability to increase GRed and γ-GCS activity [45,47,48], implying that GSE addition could make up for the increasing loss of GSH via the same mechanism in H₂O₂-pretreated PMCs of goats.

In summary, diverse concentrations of GSE strengthened TAOC to various extents in unstressed PMCs, and cell viability was enhanced by the addition of GSE within the dose range of 0.1 to 15 ug/ml after PMCs underwent H₂O₂ injury. In addition, GSE differentially regulated the mRNA expression levels and activities of CAT and SOD. The gene expression levels and activities of CAT and SOD were markedly low in unstressed PMCs with GSE addition at 10 ug/ml, while in H₂O₂-pretreated PMCs, GSE addition at the same dose induced a prominent increase in the expression levels and activities of CAT and SOD. Moreover, GSE addition enhanced GPx activity in both unstressed and stressed PMCs. These results suggest that GSE possesses the potential to improve antioxidant protection for ruminants by modulating antioxidant enzymes.

Acknowledgments

The authors appreciate Mrs. Zhaoliang Wu, Jing Yang and Min Xiong for their contributions to this study.

Author Contributions

Conceived and designed the experiments: ZS ZT. Performed the experiments: TY. Analyzed the data: XL TY WZ. Contributed reagents/materials/analysis tools: JK CC. Wrote the paper: TY ZS ZT.

References

1. Mellen PB, Daniel KR, Brosnihan KB, Hansen KJ, Herrington DM (2010) Effect of muscadine grape seed supplementation on vascular function in subjects with or at risk for cardiovascular disease: a randomized crossover trial. J Am Coll Nutr 29: 469–75.
2. Zhou DY, Du Q, Li RR, Huang M, Zhang Q, et al. (2011) Grape seed proanthocyanidin extract attenuates airway inflammation and hyperresponsiveness in a murine model of asthma by down regulating inducible nitric oxide synthase. Planta Med 77: 1575–1581.
3. Satyam SM, Bairy LK, Pirasanthan R, Vaishnav RL (2014) Grape seed extract and zinc containing nutritional food supplement prevents onset and progression of age-related cataract in Wistar rats. J Nutr Health Aging 18: 524–530.
4. Cheah KY, Howarth GS, Yazbeck R, Wright TH, Whitford EJ, et al. (2009) Grape seed extract protects IEC-6 cells from chemotherapy-induced cytotoxicity and improves parameters of small intestinal mucositis in rats with experimentally-induced mucositis. Cancer Biol Ther 8: 382–390.
5. Ulusoy S, Ozkan G, Yucesan FB, Ersöz Ş, Orem A, et al. (2012) Anti-apoptotic and anti-oxidant effects of grape seed proanthocyanidin extract in preventing cyclosporine A-induced nephropathy. Nephrology (Carlton), 17: 372–379.
6. Aoi W, Sakuma K (2011) Oxidative stress and skeletal muscle dysfunction with aging. Curr Aging Sci 4: 101–109.
7. Tonon J, Cecchini AL, Brunnquell CR, Bernardes SS, Cecchini R, et al. (2013) Lung injury-dependent oxidative status and chymotrypsin-like activity of skeletal muscles in hamsters with experimental emphysema. BMC Musculoskelet Disord 14: 39.
8. Silva LA, Silveira PC, Ronsani MM, Souza PS, Scheffer D, et al (2011) Taurine supplementation decreases oxidative stress in skeletal muscle after eccentric exercise. Cell Biochem Funct 29: 43–49.
9. Yano CL, Marcondes MC (2005) Cadmium chloride-induced oxidative stress in skeletal muscle cells in vitro. Free Radic Biol Med 39: 1378–1384.
10. Khan N, Mupparaju SP, Mintzopoulos D, Kesarwani M, Righi V, et al. (2008) Burn trauma in skeletal muscle results in oxidative stress as assessed by in vivo electron paramagnetic resonance. Mol Med Rep 1: 813–819.
11. Ohta Y, Kinugawa S, Matsushima S, Ono T, Sobirin MA, et al. (2011) Oxidative stress impairs insulin signal in skeletal muscle and causes insulin resistance in postinfarct heart failure. Am J Physiol Heart Circ Physiol 300: 1637–1644.
12. Yokota T, Kinugawa S, Hirabayashi K, Matsushima S, Inoue N, et al. (2009) Oxidative stress in skeletal muscle impairs mitochondrial respiration and limits exercise capacity in type 2 diabetic mice. Am J Physiol Heart Circ Physiol 297: 1069–1077.
13. McClung JM, Judge AR, Talbert EE, Powers SK (2009) Calpain-1 is required for hydrogen peroxide-induced myotube atrophy. Am J Physiol Cell Physiol 296: 363–371.
14. McClung JM, Judge AR, Powers SK, Yan Z (2010) p38 MAPK links oxidative stress to autophagy-related gene expression in cachectic muscle wasting. Am J Physiol Cell Physiol 298: 542–549.
15. Gilliam LA, Moylan JS, Patterson EW, Smith JD, Wilson AS et al. (2012) Doxorubicin acts via mitochondrial ROS to stimulate catabolism in C2C12 myotubes. Am J Physiol Cell Physiol 302: 195–202.
16. Bonnard C, Durand A, Peyrol S, Chanseaume E, Chauvin MA, et al. (2008) Mitochondrial dysfunction results from oxidative stress in the skeletal muscle of diet-induced insulin-resistant mice. J Clin Invest 118: 789–800.
17. Ding Y, Dai X, Jiang Y, Zhang Z, Bao L, et al. (2013) Grape seed proanthocyanidin extracts alleviate oxidative stress and ER stress in skeletal muscle of low-dose streptozotocin- and high-carbohydrate/high-fat diet-induced diabetic rats. Mol Nutr Food Res 57: 365–369.
18. Mandić AI, Djilas SM, Čanadanović-Brunet JM, Ćetković GS, Vulić JJ, et al. (2009) Antioxidant activity of white grape seed extracts on DPPH radicals. Apteff 40: 1–220.
19. Duthie G, Morrice P (2012) Antioxidant capacity of flavonoids in hepatic microsomes is not reflected by antioxidant effects in vivo. Oxid Med Cell Longev 2012: 165–127.
20. El-Ashmawy IM, El-Nahas AF, Salama OM (2006) Grape seed extract prevents gentamicin-induced nephrotoxicity and genotoxicity in bone marrow cells of mice. Basic Clin Pharmacol Toxicol 99: 230–236.
21. Hanhineva K, Törrönen R, Bondia-Pons I, Pekkinen J, Kolehmainen M, et al. (2010) Impact of dietary polyphenols on carbohydrate metabolism. Int J Mol Sci 11: 1365–1402.
22. Kim MJ, Lim Y (2013) Protective effect of short-term genistein supplementation on the early stage in diabetes-induced renal damage. Mediat Inflamm 2013: 14.
23. Zhong RZ, Zhou DW, Tan CY, Tan ZL, Han XF, et al. (2011) Effect of tea catechins on regulation of antioxidant enzyme expression in H_2O_2-induced skeletal muscle cells of goat in vitro. J Agric Food Chem 59: 11338–11343.
24. Lynge J, Juel C, Hellsten Y (2001) Extracellular formation and uptake of adenosine during skeletal muscle contraction in the rat: Role of adenosine transporters. J Physiol 537: 597–605.
25. Brenes A, Viveros A, Goñi I, Centeno C, Saura-Calixto F, et al. (2010) Effect of grape seed extract on growth performance, protein and polyphenol digestibilities, and antioxidant activity in chickens. Span J Agric Res 8: 326–333.
26. Baud O, Greene AE, Li JR, Wang Hong, Volpe JJ, et al. (2004) Glutathione peroxidase–catalase cooperativity is required for resistance to hydrogen peroxide by mature rat oligodendrocytes. J Neurosci 24: 1531–1540.
27. Myburgh KH, Kruger MJ, Smith C (2012) Accelerated skeletal muscle recovery after in vivo polyphenol administration. J Nutr Biochem 23: 1072–1079.
28. Kusano C, Ferrari B (2008) Total Antioxidant Capacity: a biomarker in biomedical and nutritional studies. J Cell Mol Biol 7: 1–15.
29. Ozkan G, Ulusoy S, Alkanat M, Orem A, Akcan B, et al. (2012) Antiapoptotic and antioxidant effects of GSPE in preventing cyclosporine A-induced cardiotoxicity. Ren Fail 34: 460–466.
30. Yang HB, Xu ZF, Liu W, Wei YG, Deng Y, et al. (2012) Effect of grape seed proanthocyanidin extracts on methylmercury-induced neurotoxicity in rats. Biol Trace Elem Res 147: 156–164.
31. Shao ZH, Becker LB, Vanden Hoek TL, Schumacker PT, Li CQ, et al. (2003) Grape seed proanthocyanidin extract attenuates oxidant injury in cardiomyocytes. Pharmacol Res 47: 463–469.
32. Yang WW, Zhang J, Wang HY, Gao PJ, Singh M, et al. (2011) Angiotensin II downregulates catalase expression and activity in vascular adventitial fibroblasts through an AT1R/ERK1/2-dependent pathway. Mol Cell Biol 358: 21–29.
33. Kaur M, Tyagi A, Singh RP, Sclafani RA, Agarwal R, et al. (2011) Grape seed extract up-regulates p21 (Cip1) through redox-mediated activation of ERK1/2 and post-transcriptional regulation leading to cell cycle arrest in colon carcinoma HT29 cells. Mol Carcinog 50: 553–562.
34. Du Y, Guo H, Lou H (2007) Grape seed polyphenols protect cardiac cells from apoptosis via induction of endogenous antioxidant enzymes. J Agric Food Chem 55: 1695–1701.
35. Li XL, Li BY, Gao HQ, Cheng M, Xu L, et al. (2010) Proteomics approach to study the mechanism of action of grape seed proanthocyanidin extracts on arterial remodeling in diabetic rats. Int J Mol Med 25: 237–248.
36. Guler A, Sahin MA, Yucel O, Yokusoglu M, Gamsizkan M, et al. (2011) Proanthocyanidin prevents myocardial ischemic injury in adult rats. Med Sci Monit Nov 17: 326–331.
37. Girnun GD, Domann FE, Moore SA, Robbins MEC (2002) Identification of a functional peroxisome proliferator-activated receptor response element in the rat catalase promoter. Mol Endocrinol 16: 2793–2801.
38. Chen T, Jin XP, Crawford BH, Cheng H, Saafir TB, et al. (2012) Cardioprotection from oxidative stress in the newborn heart by activation of PPARγ is mediated by catalase. Free Radic Biol Med 53: 208–215.
39. Gong P, Xu HB, Zhang JN, Wang ZY (2012) PPAR expression and its association with SOD and NF-κB in rats with obstructive jaundice. Biomed Res 23: 551–560.
40. Khoo NKH, Hebbar S, Zhao WL, Moore SA, Domann FE, et al. (2013) Differential activation of catalase expression and activity by PPAR agonists: Implications for astrocyte protection in anti-glioma therapy. Redox Biol 1: 70–79.
41. Hotta M, Nakata R, Katsukawa M, Hori K, Takahashi S, et al. (2010) Carvacrol, a component of thyme oil, activates PPARα and γ and suppresses COX-2 expression. J Lipid Res 51: 132–139.
42. Nannelli A, Rossignolo F, Tolando R, Rossato P, Longo V, et al. (2009) Effect of β-naphthoflavone on AhR-regulated genes (CYP1A1, 1A2, 1B1, 2S1, Nrf2, and GST) and antioxidant enzymes in various brain regions of pig. Toxicology 265: 69–79.
43. Zhu H, Itoh K, Yamamoto M, Zweier JL, Li YB (2005) Role of Nrf2 signaling in regulation of antioxidants and phase 2 enzymes in cardiac fibroblasts: Protection against reactive oxygen and nitrogen species-induced cell injury. FEBS Lett 579: 3029–3036.
44. Wang Z, Yang H, Ramesh A, Roberts LJ, Zhou L, et al. (2009) Overexpression of Cu/Zn-superoxide dismutase and/or catalase accelerates benzo(α) pyrene detoxification by upregulation of the aryl hydrocarbon receptor in mouse endothelial cells. Free Radic Biol Med 47: 1221–1229.
45. Martín MÁ, Serrano ABG, Ramos S, Pulido MI, Bravo L, et al. (2010) Cocoa flavonoids up-regulate antioxidant enzyme activity via the ERK1/2 pathway to

protect against oxidative stress-induced apoptosis in HepG2 cells. J Nutr Biochem 21: 196–205.

46. Masella R, Benedetto RD, Varì R, Filesi C, Giovannini C (2005) Novel mechanisms of natural antioxidant compounds in biological systems: involvement of glutathione and glutathione-related enzymes. J Nutr Biochem 16: 577–586.

47. Molina MF, Sanchez-Reus I, Iglesias I, Benedi J (2003) Quercetin, a flavonoid antioxidant, prevents and protects against ethanol-induced oxidative stress in mouse liver. Biol Pharm Bull 26: 1398–402.

48. Chen L, Yang XQ, Jiao H, Zhao B (2004) Effect of tea catechins on the change of glutathione levels caused by Pb^{2+} in PC12 cells. Chem Res Toxicol 17: 922–928.

10

Imine Resveratrol Analogues: Molecular Design, Nrf2 Activation and SAR Analysis

Chang Li[1], Xiaofei Xu[1], Xiu Jun Wang[2]*, Yuanjiang Pan[1]*

1 Department of Chemistry, Zhejiang University, Hangzhou, P. R. China, **2** Department of Pharmacology, School of Medicine, Zhejiang University, Hangzhou, P. R. China

Abstract

Resveratrol is a natural phenol with protective effects against cancer and inflammation-related diseases. Its mechanism of action involves the activation of nuclear factor E2 p45-related factor 2 (Nrf2), which plays a key role in regulation of genes driven by antioxidant response element (ARE). Inspired by the effect of resveratrol, here we synthesized a series of imine resveratrol analogs (IRAs), evaluated their abilities to activate Nrf2 by using cell based ARE-reporter assay. After the first-round screening, preliminary and quantitative structure-activity relationship (SAR) was analyzed, and the structural features determining Nrf2 activation ability were proposed. Two novel IRAs were designed and subsequently synthesized, namely 2-methoxyl-3,6-dihydroxyl-IRA and 2,3,6-trihydroxyl-IRA. They were proved to be the most potent Nrf2 activators among the IRAs.

Editor: Yoshiaki Tsuji, North Carolina State University, United States of America

Funding: This work was financially supported from Joint Sino-German Research Project (GZ689), Ministry of Commerce, PRC (Research and Development of Natural Herbal Medicine Phase II, a Technology Aid Project to Republic of Indonesia), Science and Technology Department of Zhejiang Province, China (2010C33156), the National Natural Science Foundation of China (81172230) and the Zhejiang Natural Science Foundation (LZ12H16001). The funders had no role in study design, data collection and analysis, decision to publish, or preparation of the manuscript.

Competing Interests: The authors have declared that no competing interests exist.

* Email: panyuanjiang@zju.edu.cn (YP); xjwang@zju.edu.cn (XJW)

Introduction

Chemoprevention of cancer, which is a strategy using natural or synthetic small molecules to modulate the metabolism, can reverse, suppress, prevent or delay the process of carcinogenesis [1,2]. The preventive effect could be achieved either by disposition of endogenous and exogenous carcinogens directly, or through upregulation of phase II/III enzymes which can deactivate toxic reactive chemical species. The nuclear factor (erythroid-derived 2)-like 2 (Nrf2), which has emerged as a key regulator of the cancer-preventive genetic program, can regulate defensive enzymes through antioxidant response elements (AREs) [3,4]. These enzymes include phase II/III enzymes (e.g. NQO1, NAD(P)H: quinone oxidoreductase 1) and stress-response proteins (e.g. HO-1, heme oxygenase 1).

Under basal condition, Nrf2 is sequestered in cytoplasm by its major repressor, Kelch-like ECH-associated protein 1 (Keap1) which mediates ubiquitin-mediated degradation of Nrf2. In contrast, electrophile/ROS (reactive oxygen species) can covalently modify Keap1 and further induce release of Nrf2 in stressed cells. Such modifications, which include oxidation of key cysteine residues in Keap1, can lead to the disruption of Keap1–Nrf2 complex. Nrf2 hereby migrates to the nucleus, binds to AREs in the promoters of its target genes, and increases their transcription [5–7]. This process can further contribute to the protection of cells against oxidative stress.

Developing small molecules with Keap1 modifying properties has been considered as a valid strategy to achieve chemoprevention of cancer [8]. Both natural (e.g. sulforaphane [9], EGCG [10] and curcumin [11]) and synthetic small molecules (e. g. butylated hydroxyanisole (BHA) [12], *tert*-butylhydroquinone (tBHQ) [13]

and olitipraz [14]) have been developed as Nrf2 activators in recent years. Among the synthetic approaches, Kumar *et. al.* [15] reported a series of trifluoromethyl-bearing chalcone derivatives based on cell and mouse model tests. Lee *et. al.* [16] developed hydroxyl substituted sulfuretin analogs as potent Nrf2 inducers. Very recently, Vrba *et. al.* [17] discovered that a synthetic flavonoid, 7-O-galloyltaxifolin greatly enhanced the expression of HO-1. Other potent Nrf2-ARE inducers developed included 3-phenyl-1-alpha-pyrones [18], sulforaphane analogs] [19] and a cyano enone (TBE-31) [20].

On the other hand, resveratrol (3,4,5-trihydroxy-trans-stilbene) is abundantly distributed in several dietary sources, such as grapes (as well as wine), berries and peanuts. It has been shown to elicit a broad range of effects that interfere signaling pathways involved in cell proliferation and/or cell death *in vitro* and *in vivo* [21]. Resveratrol has also been shown to be able to upregulate Phase II enzymes [22–24]. The induction likely involves the Keap1-Nrf2-ARE pathway. The activation of Nrf2 by resveratrol is thought to confer protection against phase I enzyme-activated carcinogens and associated carcinogenicity via the transactivation of antioxidant and phase II detoxifying enzymes. However, *in vivo* data indicated that resveratrol has extremely low bioavailability and rapid clearance from the circulation [25]. Therefore, in order to explore the potential of resveratrol as chemopreventive agent, it is a beneficial alternative approach to develop new resveratrol analogs that mimic its effects but with improved bioavailability and higher potency in activating Nrf2.

In prior studies, we synthesized an array of synthetic polyphenols, imine resveratrol analogs (IRAs), by replacing the C = C bond of resveratrol with C = N bond [26]. The simple replacement make it much more easier to generate resveratrol analogs with different substitutions, which can provide broader possibility to develop new analogs possessing higher bioavailability and Nrf2 inducing activity without losing the fundamental properties of resveratrol. Evaluation of the antioxidant activity with chemical model systems revealed that these IRAs are effective DPPH (2,2-diphenyl-1-picrylhydrazyl) radical scavengers and selective singlet oxygen quenchers, but ineffective to react with hydroxyl radicals and superoxide anions. This indicates that, like resveratrol [27] and its oligomers (Pan et. al., unpublished data), IRAs retained the features as direct antioxidants. In this study, we employed a cell-based ARE-reporter assay to examine the ability of the IRAs in activating Nrf2 pathway. Below, we present the results of *in vitro* screening, SAR analysis and molecular design of IRAs as Nrf2 inducers.

Experimental Procedures

Materials and instruments

All chemicals were purchased at the highest commercial quality and used without further purification. Reactions were magnetically stirred and monitored by TLC. NMR spectra were recorded on a Bruker 500 MHz instrument. Mass spectroscopic data were obtained using Waters GCT Premier oa-TOF mass spectrometer. MCF7 cells were obtained from ATCC (Shanghai, China) and cultured as described previously [28]. All media supplements for cell culture were purchased from Invitrogen (Shanghai, China).

Synthesis of IRAs

The process was carried out as described previously [26]. Briefly, differently substituted anilines and benzaldehydes in equal moles were added into small amount of water, forming a suspension. After stirred for 3 hours, the mixture was extracted with ethyl acetate twice. The organic layer was collected and evaporated in vacuum. The residue was then recrystallized in ethyl acetate or methanol twice. (**Figure 1**)

ARE-reporter assay

A stable ARE-reporter cell line was developed after ARE-luciferase reporter plasmid pGL3-10xARE, which contains ten copies of the ARE (5′-GTGACAAAGCA-3′), was stably transfected into MCF7 cells. The cells were seeded in 96-well plates at a density of 1.2×10^4 cells per well. After 24 h, culture medium was replaced with fresh DMEM supplemented with penicillin-streptomycin containing controls or IRAs (dissolved in DMSO to give a final 0.1% (v/v) of vehicle). After 24 h incubation, luciferase activities were measured. Three separate experiments were carried out in each case. The value for cells treated with vehicle DMSO (0.1% v/v) was set at 1. The values of tested compounds were presented as folds to vehicle DMSO control. Statistical analysis was done by unpaired Student's t tests, P<0.05 was considered statistically significant.

Theoretical Calculation of QSAR Descriptors

24 different descriptors were calculated for each individual compound employing ChemBioDraw Ultra 12.0 or Gaussian 03W software package [29]. The calculated descriptors and corresponding software package were listed in **Table S1 in File S1** and the detailed results of theoretical calculation are listed in **Table S2 in File S1**.

QSAR Statistical Analysis with SPSS

The regression analysis of obtained activities and calculated descriptors was carried out using IBM SPSS Statistics 19 software package. The Pearson correlation test was firstly carried out between activity and each individual descriptor, and those with correlation coefficients higher than 0.300 were selected as meaningful descriptors. Linear fitting was then carried out.

Results and Discussion

Initial Screening of IRAs and Preliminary SAR Analysis

The IRAs were prepared as described in previous work [26], with some modifications to Tanaka's method [30] (**Figure 1**). After recrystallization in methanol or ethyl acetate, the isolated yields ranged from 80% to 95%. Synthetic details and characterization data (^{1}H NMR, ^{13}C NMR and HRMS) are available in **Section I of File S1**.

The ARE-dependent firefly luciferase reporter cell assay, which has been used to assess transcriptional activation of Nrf2 previously [28], was employed to evaluate the ability of IRAs to activate Nrf2. The response of ARE-luciferase activity to 34 IRAs (32 first initial entries and 2 designed ones based on preliminary SAR analysis) were expressed as folds to DMSO control and presented in **Table 1**. In order to mimic the structure of resveratrol at the highest level, only hydroxyl, methoxyl and methyl groups were intensely considered.

The preliminary Structure Activity Relationship (SAR) analysis was focused on the substituent effect on Ring B at the first step. Comparing **1** (without substitution), **5** ($R^6 = OH$) and **12** ($R^8 = OH$), it could be inferred that hydroxyl substitution on Ring B could enhance the Nrf2 induction activity at the concentration of 15 µM IRA (this concentration was used as standard index hereinafter) from 1.06 (**1**) to 1.41 (**12**) and 3.03 (**5**). Another inspiring comparison of compounds were **4, 7, 18, 21, 25** and **32**, which share a *meta*-OH substitution on Ring A but distinguish with each other on Ring B. Methyl substitution on Ring B in either position has no significant or even negative effect on the ARE-luciferase activity (comparing **18, 21** and **25** with **4**), and bromo-substitution on Ring B showed negative effect (**32**). Meanwhile, an *ortho*-OH group in Ring B led to 3-fold increase of luciferase activity markedly. Other groups of data sets, as **2, 9, 10, 14, 20, 23** and **24**, as well as **6, 11** and **13**, could also support the

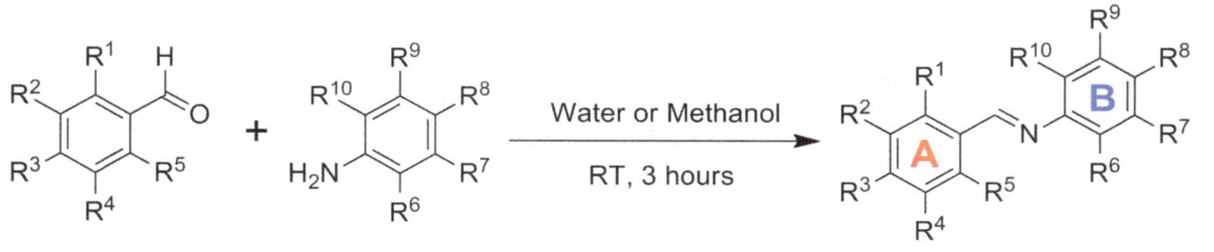

Figure 1. Synthesis of Imine Resveratrol Analogs (IRAs).

Table 1. Synthesized IRAs and their effects on ARE-luciferase activity.

IRAs	Substituents on Ring A				Substituents on Ring B				Relative Fold to Control		
	R^1	R^2	R^3	R^4	R^6	R^7	R^8	R^9	at 7.5 µM	at 15 µM	at 30 µM
1	H	H	H	H	H	H	H	H	0.99±0.17	1.06±0.06	1.08±0.04
2	H	H	OH	H	H	H	H	H	0.78±0.12	0.83±0.05	0.88±0.08
3	H	OMe	H	OMe	H	H	H	H	0.97±0.10	0.92±0.31	1.06±0.25
4	H	OH	H	H	H	H	H	H	1.18±0.27	1.37±0.24	3.67±0.59
5	H	H	H	H	OH	H	H	H	1.83±0.24	3.03±0.13	3.91±0.18
6	OH	H	H	H	OH	H	H	H	2.36±0.32	3.56±0.71	4.71±0.92
7	H	OH	H	H	OH	H	H	H	2.58±0.16	4.16±0.44	8.90±1.20
8	H	OMe	H	H	OH	H	H	H	3.28±0.07	6.17±0.88	10.32±0.40
9	H	H	OH	H	OH	H	H	H	5.69±2.22	12.18±0.68	N.D.[a]
10	H	H	OH	H	H	OH	H	H	0.64±0.03	0.70±0.02	0.69±0.24
11	OH	H	H	H	H	OH	H	H	0.77±0.05	0.78±0.12	0.94±0.12
12	H	H	H	H	H	H	OH	H	1.11±0.13	1.41±0.27	2.78±0.45
13	OH	H	H	H	H	H	OH	H	1.25±0.17	1.79±0.26	4.62±0.46
14	H	H	OH	H	H	H	OH	H	0.96±0.20	1.19±0.07	2.50±0.48
15	H	OMe	H	OMe	H	H	OH	H	1.49±0.16	1.70±0.15	4.58±0.45
16	H	OH	OH	H	H	H	OH	H	0.98±0.16	1.42±0.14	3.14±0.29
17	H	OH	H	OH	H	H	H	H	1.13±0.29	1.14±0.44	1.28±0.23
18	H	OH	H	H	Me	H	H	H	0.71±0.15	0.97±0.09	1.2±0.13
19	H	OMe	H	H	Me	H	H	H	0.99±0.10	1.06±0.03	1.08±0.09
20	H	H	H	H	Me	H	H	H	1.07±0.02	0.92±0.03	1.05±0.10
21	H	OH	H	H	H	Me	H	H	0.97±0.05	0.97±0.05	1.31±0.12
22	H	OMe	H	H	H	Me	H	H	1.07±0.09	1.01±0.05	1.09±0.11
23	H	H	H	H	H	H	Me	H	0.91±0.15	1.01±0.07	1.06±0.05
24	H	H	OH	H	H	H	Me	H	1.34±0.18	1.10±0.09	1.31±0.22
25	H	OH	H	H	H	H	Me	H	1.02±0.15	1.02±0.08	1.25±0.14
26	H	OMe	H	H	H	H	Me	H	1.13±0.10	1.03±0.05	1.20±0.03
27	H	H	Cl	H	H	H	Me	H	0.98±0.08	0.86±0.06	0.87±0.10
28	H	H	OMe	H	H	H	OMe	H	1.12±0.17	1.01±0.08	1.02±0.03
29	H	OMe	OMe	OMe	H	H	OMe	H	1.17±0.09	1.08±0.05	1.20±0.04
30	H	OMe	OMe	OMe	H	OH	OMe	H	1.07±0.08	1.04±0.04	1.03±0.07
31	H	OH	OMe	H	H	OMe	OMe	OMe	1.16±0.13	1.05±0.05	1.15±0.08
32	H	OH	H	H	Br	H	H	H	0.57±0.12	0.57±0.09	0.65±0.15
33	H	OMe	OH	H	OH	H	H	H	4.84±0.50	9.97±3.75	18.61±1.11
34	H	OH	OH	H	OH	H	H	H	6.93±1.55	12.33±4.11	14.28±0.97

Table 1. Cont.

IRAs	Substituents on Ring A				Substituents on Ring B				Relative Fold to Control		
	R^1	R^2	R^3	R^4	R^6	R^7	R^8	R^9	at 7.5 µM	at 15 µM	at 30 µM
resveratrol	-	-	-	-	-	-	-	-	2.54±0.22	3.05±0.62	5.10±2.11

ARE reporter cells were exposed to 7.5 µM, 15 µM or 30 µM IRA for 24 h. The value for cells treated with vehicle DMSO (0.1% v/v) was set at 1. Results are from three separate experiments.
aDue to cytotoxicity of **9**.

same conclusion, that an *ortho*-OH group on Ring B in crucial for effective Nrf2 induction activity.

As *ortho*-hydroxyl group ($R^6 = OH$) is a key substructure in optimization, further investigation was then concentrated on the substitution effect on Ring A. As shown in **Figure 2**, seven 6-OH IRAs with different substituents on Ring A were picked out. In comparison of IRA **5–9**, it could be apparently concluded that 3-OH (**9**) substitution is the most dominant activity increasing factor, followed by 2-OMe (**8**), 2-OH (**7**) and 1-OH (**6**) substitution. However, this conclusion could not be supported by data sets other than 6-OH substituted IRAs, **12–17** for instance. This phenomenon strengthened the conclusion that 6-OH substituent group plays the vital role in Nrf2 activation, while substitutions on Ring A contribute as auxiliary cofactors. The preliminary SAR results generated were summarized in **Figure 3**.

QSAR Analysis

We then investigated the quantitative structure activity relationship (QSAR) of IRAs, as such an approach may reveal the mechanism of the interaction between IRA and Keap1, and possibly provide a statistic guide to design more effective IRAs. 24 different descriptors were calculated using either Gaussian 03W software package or ChemBioOffice Ultra 12.0 (presented in **Section II, Supporting Information**). After Pearson correlation test to all 24 descriptors (detailed results presented in **Table S3 in File S1**), it was found that the most correlated ones were Vertical Detachment Energy (**IP**), Energy of the Lowest Unoccupied Molecular Orbital (**E$_{LUMO}$**), Energy of the Highest Occupied Molecular Orbital (**E$_{HOMO}$**), Electrophilic Frontier Electronic Density (**Fr^E**) and Calculated logP (**clogP**). At last, multiple linear regression (MLR) analysis was employed to search for suitable factors for aforementioned molecular descriptors. **Equation 1** was generated as follows (**LogA$_{15}$** represents the Common logarithm of the activity of IRA at 15 µM concentration).

$$LogA_{15} = 0.01678 + 0.002760IP + 3.230E_{HOMO} + 13.13E_{LUMO}$$
$$+ 0.08958Fr^E + 0.2239clogP$$

(**Equation 1**, correlation coefficient R = 0.792, F-statics = 9.417)

From the generated equation, it could be inferred that **E$_{HOMO}$** and **E$_{LUMO}$** are correlated with the Nrf2 induction activity to a great degree. As these descriptors are key variables in chemical reactions according to Fukui's Frontier Molecular Theory, this result can partly support the assumption that chemical reactions took place between IRAs and Keap1 protein. Moreover, **Fr^E**, which also played an important role in this equation, is a descriptor describing the frontier electron density in electrophilic reactions. The aforementioned results both suggested a electrophilic substitution mechanism in this Nrf2 induction process.

Molecular Design of Effective IRAs and Activity Evaluation

Inspired by the preliminary and quantitative SAR results shown in **Figure 3**, two novel 6-OH IRAs, **33** and **34**, were designed and synthesized. As shown in **Figure 4A**, they are both bearing 3-OH substitutions, while the only difference was at 2-position, on which **33** has a methoxyl group and **34** has a hydroxyl group. ARE-luciferase assay revealed that **33** and **34** induced ARE-luciferase activity in a dose-dependent manner (the dose-dependency curve is shown in **Figure S1 in File S1**). Exposing the reporter cells to 7.5 µM compound **33** for 24 h, ARE-luciferase activity was elevated by nearly 5 folds, whereas that at 15 µM and 30 µM increased the luciferase activity 10 and 19 folds,

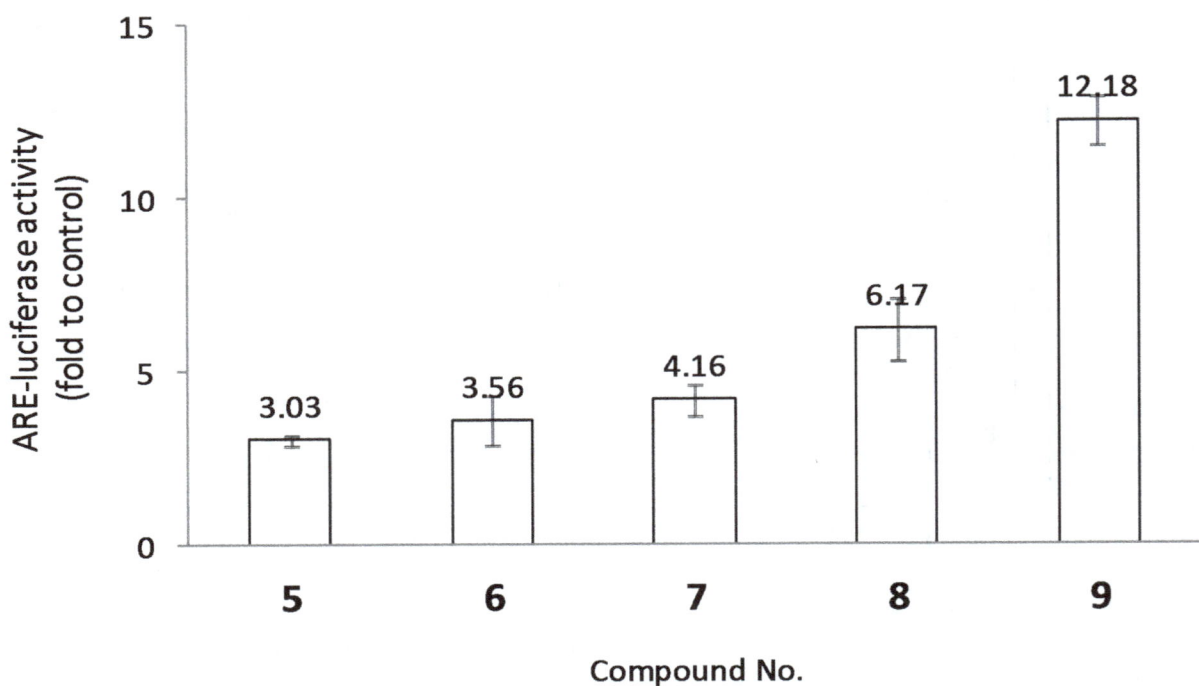

Figure 2. ARE-luciferase activities in response to 6-OH IRAs. ARE reporter cells were exposed to (15 µM) for 24 h. The value for cells treated with vehicle DMSO (0.1% v/v) was set at 1. Results are from three separate experiments.

respectively (**Figure 4B**). Similarly, 7.5 µM of compound **34** induced ARE-luciferase activity by nearly 7 folds, while 15 µM of the compound stimulated more than 12-fold induction of luciferase activity. This is much higher than that of their parent compound resveratrol, whereas under the same culture condition, 30 µM of resveratrol only achieved ~5-fold induction of ARE-luciferase activity (also see **Table 1**). Besides, not like some other Schiff bases, **33** and **34** are stable in a blank ARE luciferase assay

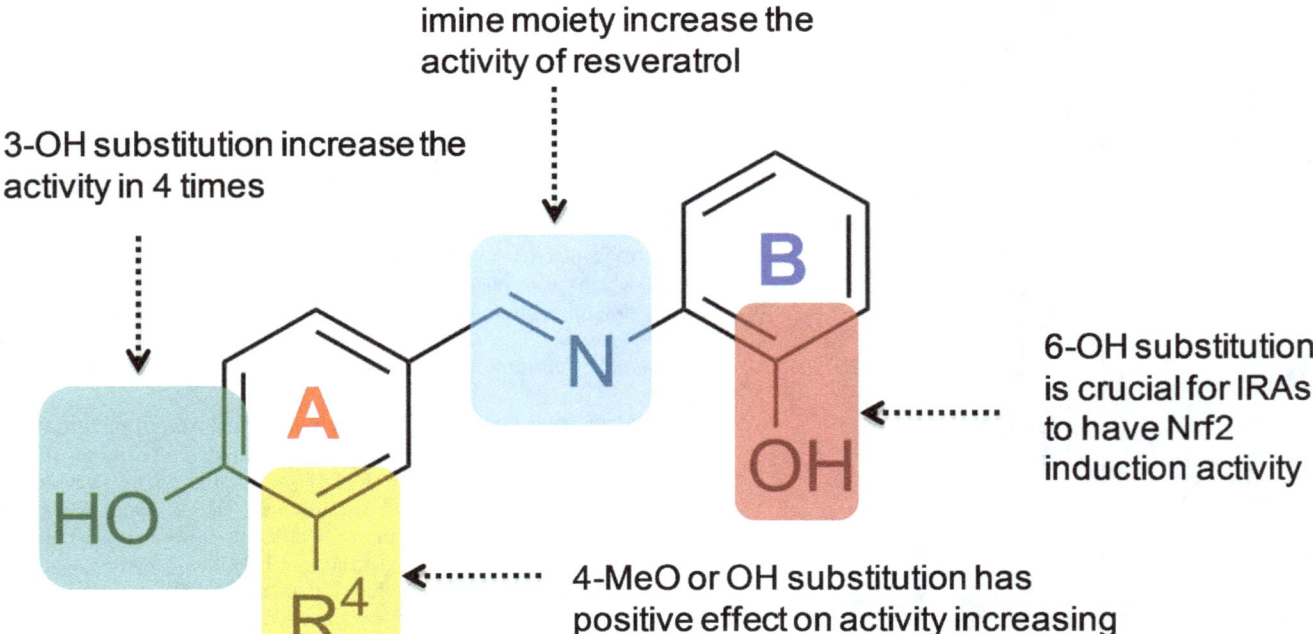

Figure 3. SAR results from initial screening of IRAs.

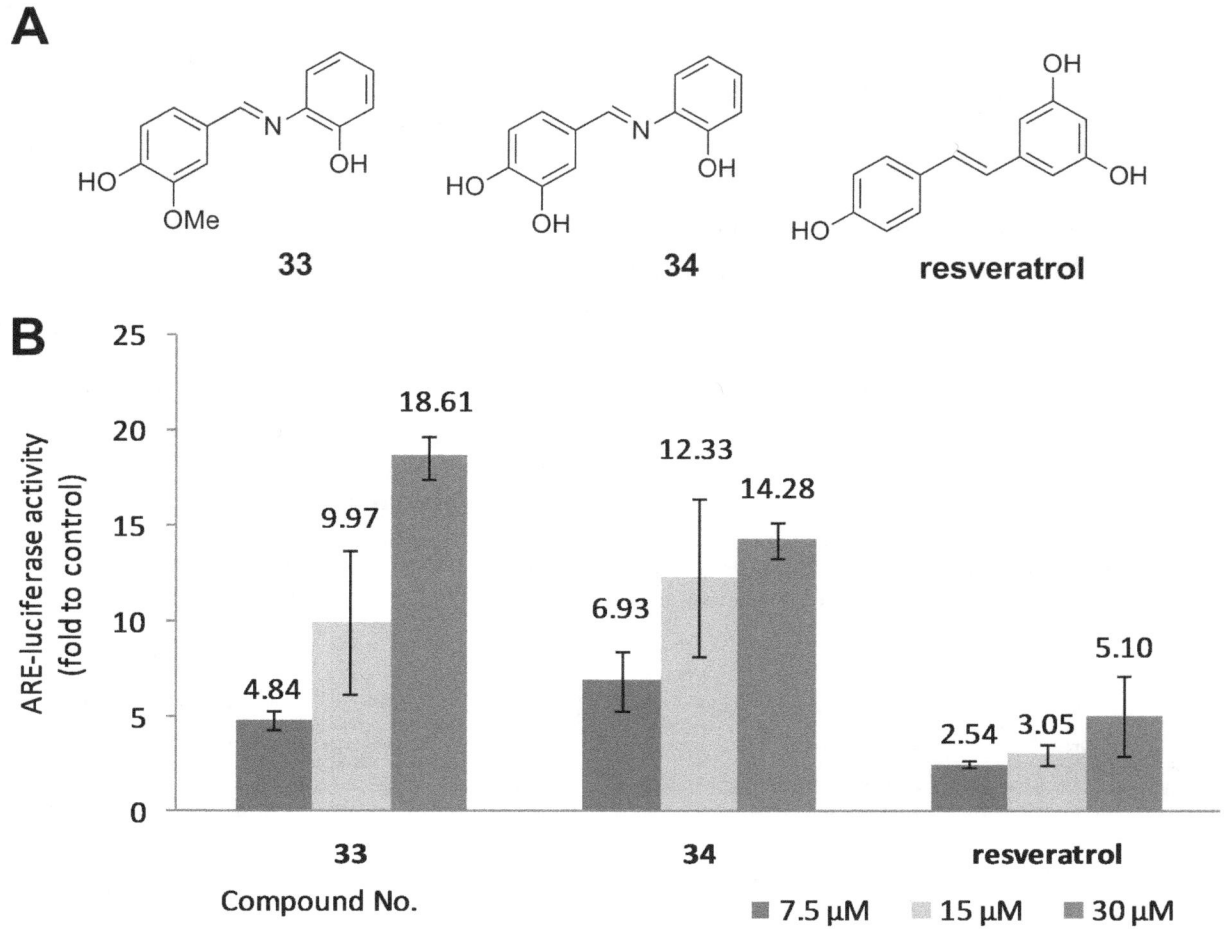

Figure 4. Structures (A) and ARE-luciferase activities (B) of 33, 34 and resveratrol. ARE reporter cells were exposed to **33**, **34** and resveratrol (7.5 μM, 15 μM and 30 μM) for 24 h. The value for cells treated with vehicle DMSO (0.1% v/v) was set at 1. Results are from three separate experiments. Values shown are mean ± SD.

condition. However, their radical scavenging activities are poor, 500 μM of these compounds can only scavenge 4.0% and 9.7% DPPH radical, while resveratrol can achieve complete quench in these concentrations. Therefore, our data demonstrate that **33** and

34 are the most potent Nrf2 activators among the resveratrol analogs.

Most Nrf2 activators, despite the structural diversity, are all electrophiles that have striking propensities to react with

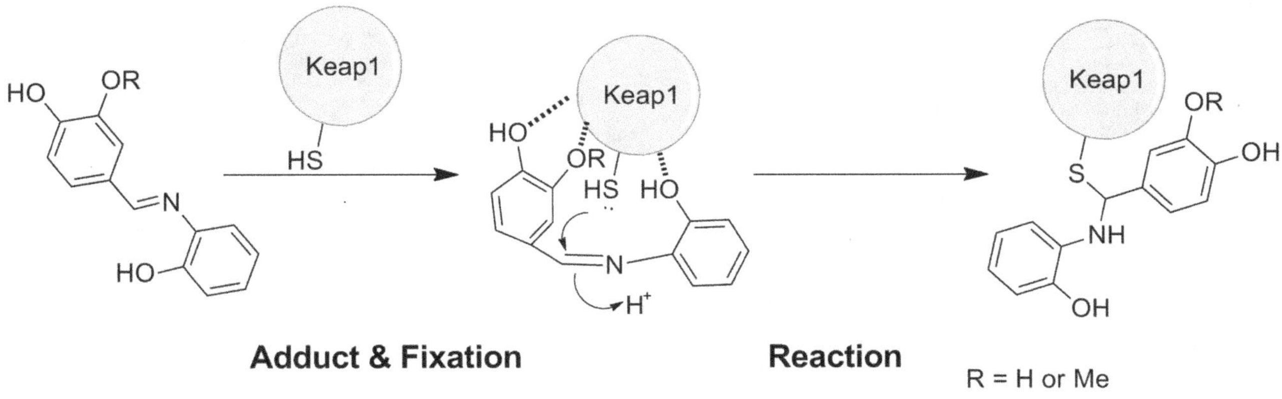

Adduct & Fixation   **Reaction**

R = H or Me

Figure 5. Proposed model for the adduct formation, fixation of IRA, and its reaction with Keap1.

sulfhydryls [31]. Here our theoretical calculation results of Electrophilic Frontier Electronic Density indicate that compound **33** and **34** are reactive electrophiles. As Nrf2 inducers implement their functions by forming covalent bonds with thiol groups of Keap1 [3c], we therefore also classify the IRAs as inhibitors of Keap1-Nrf2 interaction.

Referring to several examples in which imines react as Michael Acceptors in organic chemistry [32,33], a plausible mechanism of IRAs adduct Keap1 protein was proposed and shown in **Figure 5**. 6-OH group played a crucial role in incorporating the small molecule to Keap1 on the surface, the 3-OH group and 2-OH/OMe group also performed auxiliary effect to fix IRA molecule to the protein. Thus, the thiol group in cystein residues in the formed complex were in vicinity of the $C = N$ bond. Thiol group then underwent nucleophilic attack to the carbon in imine moiety and a covalent adduct formed as a result. Once Keap1 was modified, Nrf2 could be released.

Conclusions

In the current study, we have screened the potential of an array of IRAs to be effective Nrf2 activators. Based on SAR analysis, both preliminary and quantitative results were summarized and led us to develop two novel IRAs, 2-methoxyl-3,6-dihydroxyl-IRA and 2,3,6-trihydroxyl-IRA. *In vitro* ARE reporter cell assay indicated that they were both strong Nrf2 activators. A plausible mechanism for their interaction with Keap1 was proposed. Taken together, **33** and **34** are *indirect* antioxidants, which can activate Nrf2-ARE pathway to wake up the defense system in cell. These features warrant further studies in disease models for the development of the IRAs as chemopreventive agents.

Supporting Information

File S1 Figure S1. Compounds **33** and **34** induced ARE-luciferase activities dose-dependently. Table S1. Calculated Descriptors and Corresponding Software Applied. Table S2. Data of Calculated Descriptors. Table S3. Pearson Coefficient Test Results.

Author Contributions

Conceived and designed the experiments: XJW YP. Performed the experiments: CL XX. Analyzed the data: CL. Contributed reagents/materials/analysis tools: CL XX. Wrote the paper: CL XJW YP.

References

1. Lee K, Bode A, Dong Z (2011) Molecular targets of phytochemicals for cancer prevention. Nature Reviews Cancer 11: 211–218.
2. Umar A, Dunn B, Greenwald P (2012) Future directions in cancer prevention. Nature Reviews Cancer 12: 835–848.
3. Itoh K, Chiba T, Takahashi S, Ishii T, Igarashi K, et al. (1997) An Nrf2/small Maf heterodimer mediates the induction of phase II detoxifying enzyme genes through antioxidant response elements. Biochemical and Biophysical Research Communications 236: 313–322.
4. Talalay P, Fahey J. (2001) Phytochemicals from cruciferous plants protect against cancer by modulating carcinogen metabolism. Journal of Nutrition 131: 3027S–3033S.
5. Abiko Y, Kumagai Y (2013) Interaction of Keap1 Modified by 2-tert-Butyl-1,4-benzoquinone with GSH: Evidence for S-Transarylation. Chemical Research in Toxicology.
6. Kobayashi A, Kang M-I, Watai Y, Tong K, Shibata T, et al. (2006) Oxidative and electrophilic stresses activate Nrf2 through inhibition of ubiquitination activity of Keap1. Molecular and Cellular Biology 26: 221–229.
7. Dinkova-Kostova A, Holtzclaw W (2002) Direct evidence that sulfhydryl groups of Keap1 are the sensors regulating induction of phase 2 enzymes that protect against carcinogens and oxidants. Proceedings of the National Academy of Sciences USA 99: 11908–11913.
8. Magesh S, Chen Y, Hu L (2012) Small molecule modulators of Keap1-Nrf2-ARE pathway as potential preventive and therapeutic agents. Medicinal Research Reviews 32: 687–726.
9. Hu C, Eggler A, Mesecar A, van Breemen R (2011) Modification of keap1 cysteine residues by sulforaphane. Chemical Research in Toxicology 24: 515–521.
10. Na H-K, Kim E-H, Jung J-H, Lee H-H, Hyun J-W, et al. (2008) (-)-Epigallocatechin gallate induces Nrf2-mediated antioxidant enzyme expression via activation of PI3K and ERK in human mammary epithelial cells. Archives of Biochemistry and Biophysics 476: 171–177.
11. Farombi E, Shrotriya S, Na H-K, Kim S-H, Surh Y-J (2008) Curcumin attenuates dimethylnitrosamine-induced liver injury in rats through Nrf2-mediated induction of heme oxygenase-1. Food and Chemical Toxicology 46: 1279–1287.
12. Nair S, Xu C, Shen G, Hebbar V, Gopalakrishnan A, et al. (2006) Pharmacogenomics of phenolic antioxidant butylated hydroxyanisole (BHA) in the small intestine and liver of Nrf2 knockout and C57BL/6J mice. Pharmaceutical Research 23: 2621–2637.
13. Cheung K, Yu S, Pan Z, Ma J, Wu T, et al. (2011) tBHQ-induced HO-1 expression is mediated by calcium through regulation of Nrf2 binding to enhancer and polymerase II to promoter region of HO-1. Chemical Research in Toxicology 24: 670–676.
14. Iida K, Itoh K, Kumagai Y, Oyasu R, Hattori K, et al. (2004) Nrf2 is essential for the chemopreventive efficacy of oltipraz against urinary bladder carcinogenesis. Cancer Research 64: 6424–6431.
15. Kumar V, Kumar S, Hassan M, Wu H, Thimmulappa R, et al. (2011) Novel chalcone derivatives as potent Nrf2 activators in mice and human lung epithelial cells. Journal of Medicinal Chemistry 54: 4147–4159.
16. Lee C-Y, Chew E-H, Go M-L (2010) Functionalized aurones as inducers of NAD(P)H: quinone oxidoreductase 1 that activate AhR/XRE and Nrf2/ARE signaling pathways: synthesis, evaluation and SAR. European Journal of Medicinal Chemistry 45: 2957–2971.
17. Vrba J, Gažák R, Kuzma M, Papoušková B, Vacek J, et al. (2013) A novel semisynthetic flavonoid 7-O-galloyltaxifolin upregulates heme oxygenase-1 in RAW264.7 cells via MAPK/Nrf2 pathway. Journal of Medicinal Chemistry 56: 856–866.
18. Xi M-y, Sun Z-y, Sun H-p, Jia J-m, Jiang Z-y, et al. (2013) Synthesis and bioevaluation of a series of α-pyrone derivatives as potent activators of Nrf2/ARE pathway (part I). European Journal of Medicinal Chemistry 66: 364–371.
19. Ahn Y, Hwang Y, Liu H, Wang X. (2010) Electrophilic tuning of the chemoprotective natural product sulforaphane. Proceedings of the National Academy of Sciences USA 107, 9590–9595.
20. Dinkova-Kostova A, Talalay P, Sharkey J, Zhang Y, Holtzclaw W, et al. (2010) An exceptionally potent inducer of cytoprotective enzymes: elucidation of the structural features that determine inducer potency and reactivity with Keap1. The Journal of Biological Chemistry 285: 33747–33755.
21. Baur J, Sinclair D (2006) Therapeutic potential of resveratrol: the in vivo evidence. Nature Reviews Drug discovery 5: 493–506.
22. a) Bishayee A, Barnes K, Bhatia D, Darvesh A, Carroll R (2010) Resveratrol suppresses oxidative stress and inflammatory response in diethylnitrosamine-initiated rat hepatocarcinogenesis. Cancer Prevention Research 3: 753–763.
23. Juan Andrés M, Gilles M, Félix Victor V (2008) Resveratrol protects primary rat hepatocytes against oxidative stress damage. European Journal of Pharmacology 591: 66–72.
24. Hsieh T-c, Lu X, Wang Z, Wu JM (2006) Induction of quinone reductase NQO1 by resveratrol in human K562 cells involves the antioxidant response element ARE and is accompanied by nuclear translocation of transcription factor Nrf2. Medicinal Chemistry (Shāriqah (United Arab Emirates)) 2: 275–285.
25. Walle T, Hsieh F, De Legge M, Oatis J, Walle U (2004) High absorption but very low bioavailability of oral resveratrol in humans. Drug Metabolism and Disposition: the Biological Fate of Chemicals 32: 1377–1382.
26. Lu J, Li C, Chai Y-F, Yang D-Y, Sun C-R (2012) The antioxidant effect of imine resveratrol analogues. Bioorganic &Medicinal Chemistry Letters 22: 5744–5747.
27. Jiang L-Y, He S, Jiang K-Z, Sun C-R, Pan Y-J (2010) Resveratrol and its oligomers from wine grapes are selective (1)O2 quenchers: mechanistic implication by high-performance liquid chromatography-electrospray ionization-tandem mass spectrometry and theoretical calculation. Journal of Agricultural and Food Chemistry 58: 9020–9027.
28. Tang X, Wang H, Fan L, Wu X, Xin A, et al. (2011) Luteolin inhibits Nrf2 leading to negative regulation of the Nrf2/ARE pathway and sensitization of human lung carcinoma A549 cells to therapeutic drugs. Free Radical, Biology and Medicine 50: 1599–1609.

29. Frisch M J, Trucks G W, Schlegel H B, Scuseria G E, Robb M A, et al. (2003) Gaussian 03, Gaussian, Inc., Pittsburgh, PA.

30. Tanaka K, Shiraishi R (2000) Clean and efficient condensation reactions of aldehydes and amines in a water suspension medium. Green Chemistry 2: 272–273.

31. Dinkova-Kostova A T, Fahey J W, Talalay P (2004) Chemical structures of inducers of nicotinamide quinone oxidoreductase 1 (NQO1). Methods in Enzymology 382: 423–448.

32. Takahiko A, Yasuhiro H, Junji I, Kohei F (2008) Vinylogous Mannich-Type Reaction Catalyzed by an Iodine-Substituted Chiral Phosphoric Acid. Advanced Synthesis & Catalysis 350: 399–402.

33. Yamanaka M, Itoh J, Fuchibe K, Akiyama T (2007) Chiral Brønsted acid catalyzed enantioselective Mannich-type reaction. Journal of the American Chemical Society 129: 6756–6764.

11

Phytochemical Profiles and Antioxidant Activities in Six Species of Ramie Leaves

Yongsheng Chen[1], Gaoyan Wang[2], Hong Wang[1], Chaohua Cheng[3], Gonggu Zang[3], Xinbo Guo[1]*, Rui Hai Liu[1,4]*

1 School of Light Industry and Food Sciences, South China University of Technology, Guangzhou, Guangdong, The People's Republic of China, 2 Department of Food Science, Cornell University, Ithaca, New York, United States of America, 3 Institute of Bast Fiber Crops, Chinese Academy of Agricultural Sciences, Changsha, Hunan, The People's Republic of China, 4 Department of Food Science and Institute of Comparative and Environmental Toxicology, Stocking Hall, Cornell University, Ithaca, New York, United States of America

Abstract

Increased consumption of vegetables or plant food has been associated with decreased risk of developing major chronic diseases, such as cancers, diabetes, cardiovascular diseases, and age-related functional decline. Ramie leaves are rich in phenolics and flavonoids, which have been suggested for human health benefits. Phenolic contents, flavonoid contents, phenolic compounds, and anti-cancer properties in six species of ramie leaves were analyzed by Folin-reagent method, sodium borohydride/chloranil-based assay (SBC), HPLC method and antiproliferation, cytoxicity, respectively. Antioxidant activities were measured through peroxyl radical scavenging capacity (PSC) method, oxygen radical absorbance capacity (ORAC) method, and cellular antioxidant activity (CAA). Research indicated that *Boehmeria penduliflora* contained the highest total phenolic content (2313.7±27.28 mg GAE/100 g FW), and flavonoid content (1682.4±27.70 mg CAE/100 g FW). *Boehmeria tricuspis* showed the highest PSC value (9574.8±117.63 µM vit. C equiv./100 g FW), while *Boehmeria penduliflora* indicated the highest ORAC value (330.44±16.88 µmol Trolox equiv./g FW). The antioxidant activities were correlated with phenolic contents and flavonoid contents. *Boehmeria tricuspis* had the highest antiproliferative capacity with the lowest EC$_{50}$ (4.11±0.19 mg/mL). The results for the analyzed ramie for CAA were significantly different from each other ($p<0.05$), *Boehmeria tricuspis* had the highest CAA value (133.63±7.10 µmol QE/100 g). Benzoic acid, 4-coumaric acid, caffeic acid, and ferulic acid were the dominant phenolic ingredients in the ramie leaves according to HPLC analysis. Our research is the first report to study the phytochemical profiles and antioxidant activities in different species of ramie leaves for their health benefit.

Editor: Gianfranco Pintus, University of Sassari, Italy

Funding: The work was supported in part by the Fundamental Research Funds for the Central Universities (2014ZM0072), the Leading Talents Program in Guangdong Province (Rui Hai Liu), the Guangzhou Science and Technology Program (2013J4500036), and the Guangdong Science and Technology Program (2012B050500003). The funders had no role in study design, data collection and analysis, decision to publish, or preparation of the manuscript.

Competing Interests: The authors have declared that no competing interests exist.

* Email: xbg720@gmail.com (XBG); rl23@cornell.edu (RHL)

Introduction

Phytochemicals are a class of bioactive non-nutrient compounds which usually exist in fruits, vegetables, grains, and other based-plant foods. Phytochemicals have been linked to reducing the risk of major chronic diseases [1]. Phytochemicals have potential antioxidant activity and are able to scavenge hydroxyl radicals, capture peroxyl radicals, inhibit hydrogen peroxide, and quench reactive nitrogen species [2–4]. Some phytochemicals have been proved to show higher antioxidant activities than vitamin C and vitamin E [3]. It is common that different plants contain different types of phytochemicals with disparate structures and protective mechanisms, and demonstrate different level of antioxidant activities.

Free radicals and other reactive oxygen species are constantly generated in human cells and organisms as a result of aerobic metabolism, some of which are necessary for life. Their pivotal role is to maintain balance between oxidants and antioxidants to maintain optimal physiological conditions [1]. However, excessive of oxidants may cause oxidative stress and lead to unbalanced physiological conditions. Oxidative stress could cause oxidative damages to large bio-molecules (lipids, proteins, and DNA), eventually resulting in development of many chronic diseases, such as cancer, cardiovascular, and neurodegenerative diseases [5]. Antioxidants are effective substances, which can protect the tissue from oxidative damage by modulating the effects of reactive oxidants [4]. Therefore, natural antioxidants from plant extracts have attracted increasing interests due to consumer concern about the safety of the synthetic antioxidants. Phenolics and flavonoids are biologically active plant compounds, widely spread in fruits, vegetables, teas, and corns, and have also attracted extensive research interests due to their strong antioxidant activities. Phenolics and flavonoids are the secondary metabolites of plants, which is very important in maintaining essential functions in the reproduction and growth of the plants [6], and flavonoids are a kind of phenolics.

Ramie, widely cultivated in China and a member of the *Urticaceae* family *Bochmeria*, is a perennial herbaceous plant commonly grown for its fibers. Ramie has vigorous vegetative growth and can be harvested three times per year in south China areas. Ramie leaves is a abundant by-product and have been consumed in the form of rice-cake or as a tea and are used in traditional medicine for treatment of diarrhea and snake bites [7]. Ramie extracts have high-grade of physiological functions, such as antioxidant, antibacterial and antifungal activities that restrain both bacterial and fungal attacks, and also anticancer effects on lung and liver cancers [8]. Ramie leaves is not only good sources for dietary and medical use; it is also excellent sources for phytochemicals, such as phenolics and flavonoids. They have been reported to show higher antioxidant activity than ordinary diet[9]. Therefore, exploitation for potential value-add usage of those parts would be necessary. In present research, we analyzed and compared the phytochemical profiles (phenolic content, flavonoid content and phenolic compounds) and antioxidant activities of six species of Ramie for evaluating its potential application as health beneficial nutraceuticals.

Materials and Methods

Chemicals

Ascorbic acid (ASA), 2,6-dichloroindophenol sodium salt hydrate, chloranil, catechin hydrate, vanillin, Folin-Ciocalteu reagent, hydrocortisone, penicillin, streptomycin, gentamicin, Gallic acid, 2,2′-Azobis-amidinopropane (ABAP) and dichloro-fluorescin diacetate (DCFH-DA) were purchased from Sigma Chemical Co. (St. Louis, MO, USA). NaOH, KH_2PO_4, KOH, $AlCl_3$, acetic acid, $NaHCO_3$, and K_2HPO_4 were bought from Sangon (Shanghai, China). Chlorogenic acid, Caffeic acid, 4-Coumaric acid, Ferulic acid and benzoic aicd, $NaBH_4$ were bought from Aladdin Co. (Shanghai, China). WME medium, Hank's balanced salt solution (HBSS), epidermal growth factor, heparin, insulin, and other cell culture reagents were purchased from Gibco Biotechnology Co. All reagents used were of analytical grade.

Plant materials

Six species of fresh ramie leaves (Boehmeria nivæa, Boehmeria longispica, Boehmeria clidemioides, Boehmeria macrophylla, Boehmeria tricuspis, Boehmeria penduliflora) were procured from Institute of Bast Fiber Crops (Changsha, China). All samples were individually mixed thoroughly and were stored at −40°C until use.

Extraction of soluble free phytochemicals

Soluble free phytochemicals of ramie leaves were prepared using the modified method as previously reported [10,11]. Briefly, 20 g of samples were first blended in a blender using 150 mL of chilled 80% acetone (1:7.5, w/v) for 3 min. Samples were subsequently homogenized with a IKA homogenizer for another 3 min. Second, the homogenates were centrifuged at 3675 g for 10 min, 4°C. Supernatants were filtered with Whatman no. 2 filter paper, and the filtrate was collected and evaporated at 45°C until 10% of the filtrates had been retained. The filtrates were recovered with water to a final volume of 10 mL and stored at −80°C until further analysis. All the extractions were performed in triplicates.

Extraction of bound phytochemicals

Bound phytochemicals in ramie leaves were extracted using the modified methods as previously reported [10,11]. Briefly, the filter residues from above soluble free extraction were collected, flushed with nitrogen gas, sealed, and hydrolyzed directly with 20 mL of 4 M NaOH at room temperature for 1 hour with shaking. The mixture was acidified with concentrated hydrochloric acid to pH 2, and extracted ten times with ethyl acetate. The ethyl acetate fractions were evaporated at 45°C under vacuum to dryness. Bound phytochemicals were reconstituted in 10 mL water, aliquoted to 1 mL, and stored at −40°C until analysis.

Determination of total phenolic content

Total phenolic content in ramie leaves were determined using the Folin-Ciocalteu colorimetric method with modification [12,13]. In brief, standard curve using gallic acid (0.0–600.0 μg/mL) was made. For each analysis, 100 μL of the standard gallic acid solution or extracts sample were added to 0.4 mL of water in each test tube. Folin-Ciocalteu reagent (0.1 mL) was added to the solution and allowed to react for 6 min to ensure that the Folin-Ciocalteu reagent reacted completely with the oxidizable phenolates in the sample. Second, 1 mL of 7% sodium carbonate solution was added to neutralize the reaction, and 0.8 mL of deionized water was added into the test tubes to adjust the final volume to 2.4 mL. The samples were mixed and allowed to stand for 90 min at room temperature. The absorbance was measured at 760 nm after the color was developed by DU 730 Nucleic Acid/Protein Analyzer (BECKMAN, USA). The absorbance values were calculated based on the standard curve of known gallic acid concentrations and expressed as milligram of gallic acid equivalents (GAE) per 100 g fresh weight (FW). Data were reported as mean ± SD for triplicates.

Determination of water-soluble flavonoid content

The total flavonoid content were determined using the sodium borohydride/chloranil (SBC) method [14]. Briefly, 1 mL of phytochemical extracts of tested samples was added into test tubes (15×150 mm), placed under nitrogen gas to dryness, and reconstituted in 1 mL of terahydrofuran/ethanol (THF/EtOH, 1:1, v/v). Catechin hydrate standard (0.3–10.0 mM) was prepared fresh in 1 mL of THF/EtOH (1:1, v/v). Each test tube with sample or standard was added with 0.5 mL of 50 mM $NaBH_4$ solution and 0.5 mL of 74.6 mM $AlCl_3$ solution. Subsequently, the test tubes were shaken in an orbital shaker at 180 rpm at room temperature for 30 min. An additional 0.5 mL of 50.0 mM $NaBH_4$ solution was added into each test tube with shaking continued for another 30 min at room temperature. Chilled 2.0 mL of 0.8 M acetic acid solution was added into each test tube and kept in the dark for 15 min after being thoroughly mixed. Then, 1 mL 20.0 mM chloranil was added in each tube and heated at 99°C with shaking for 60 min. The reaction solutions were cooled using tap water, and the volume was brought to 4 mL using methanol. In the next step, 1 mL of 16% (w/v) vanillin was added into each tube and mixed. 2 mL of 12 M HCl was subsequently added into each tube and kept in the dark for 15 min after a thorough mix. The reaction solutions were centrifuged at 2500 g for 10 min. The absorbance of the mixture was immediately measured at 490 nm against a prepared blank using a DU 730 Nucleic Acid/Protein Analyzer (BECMAN, USA). Results were calculated by using the standard curve of catechin hydrate concentration. Total flavonoid content was expressed as milligram of catechin equivalents (CAE) per 100 g of fresh weight of ramie. Results were reported as mean ± SD in triplicates.

Hydrophilic peroxyl radical scavenging capacity (PSC) assay

Total antioxidant activity was measured using the hydrophilic peroxyl radical scavenging capacity (Hydro-PSC) assay [2].

Ascorbic acid and phytochemical extracts were diluted in appropriate concentration by using 75 mM phosphate buffer (pH 7.4). Ascorbic acid was prepared fresh and diluted to 6.3, 4.8, 3.2, 2.4, and 1.0 μg/mL. Gallic acid was made fresh and diluted to 5.5, 3.5, 2.7, 1.4, and 0.9 μg/mL. The reaction mix contained 75 mM phosphate buffer at pH 7.4, 40 mM ABAP, 13.26 μM DCFH dye, and the appropriate concentrations of the pure antioxidant compound or sample extracts. The dye was prehydrolyzed with 1 mM KOH to remove the diacetate moiety just prior to use in the reaction, and the reaction was carried out at 37°C, in a total volume of 250 μL using a 96-well plate. Fluorescence generation was monitored (excitation at 485 nm and emission at 535 nm) with a Fluoroskan Ascent fluorescent spectrophotometer (Molecular Devices, USA). Data were acquired with the Ascent Software, version 2.6 (Molecular Devices, USA) running on a PC. The areas under the fluorescence reaction time kinetic curve (AUC) for both control and samples were integrated and used as the basis for calculating peroxyl radical scavenging capacity (PSC) using the equation PSC (value) $= 1 - (SA/CA)$, where SA is AUC for the sample or standard dilution and CA is AUC for the control reaction. Compounds or extracts inhibiting the oxidation of DCFH produced smaller SA and higher PSC values. The parameter EC_{50} was defined as the dose required to cause a 50% inhibition (PSC unit $= 0.5$) for each pure compound or sample extract, and was used as the basis for comparing the antioxidant activities of different compounds or samples. Results obtained for antioxidant activities of sample extracts were presented as micromole of vitamin C equivalents per 100 g of sample $\pm$ SD with triplicates.

Measurement of oxygen radical scavenging capacity (ORAC)

The peroxyl radical scavenging efficacy of selected ramie extracts was determined using the ORAC assay described by Huang et al. [15] and modified in our group [16]. Briefly, 20 μL of blank, Trolox standard, or ramie leave extract dilutions with 75 mM phosphate buffer, pH 7.4 (working buffer), were added to triplicate wells in a blank, clear-bottom, 96-well microplate (Corning Scientific), and incubate at 37°C for 10 min. The outside wells of the plate were not used as there were much more variation from them than the inner wells. In addition, the triplicate samples were distributed throughout the microplate and were not placed side-by-side, to avoid any effects on readings due to location. 200 μL of 0.96 μM fluorescein in working buffer was added to each well and incubated at 37°C for 20 min. Then add 20 μL of freshly prepared 119.4 mM ABAP in working buffer to each well. The microplate was immediately measured using a Fluoroskan Ascent FL plate-reader (Molecular Devices, USA) at 37°C. The decay of fluorescenece at 535 nm was measured with excitation at 485 nm for 35 cycle every 4.5 min. The areas under the fluorescence versus time curve for the samples minus the area under the curve for the blank were calculate and compared to a standard curve of the areas under the curve for 6.25, 12.5, 25, and 50 μM Trolox standards minus the area under the curve for blank. ORAC values were expressed as μmol Trolox equiv./g FW $\pm$ SD for triplicate data from one experiment.

Analyses of phenolic acid compounds by HPLC

The chromatographic analyses were performed according to Malta et al. [17], with slight modifications in our laboratory. Briefly, phenolic acids were determined on a Waters (Waters Corp, Milford, MA) HPLC system, consisting of a Waters 1525 HPLC pump (Waters Corp, Milford, MA), one Waters 2998 dual-wavelength absorbance detector and an intelligent sampler 2707.

The chromatographic data were recorded and processed by Waters software. A Waters C_{18} column (5 μm, 250 mm×4.6 mm) was employed for the separation of ramie extracts. The flow rate of the mobile phase was 1 mL/min. Mobile phase A was water with 0.02% TFA, and phase B was methanol with 0.02% TFA. The gradient conditions were as follows: 0–5 min, 25% B; 5–10 min, 25–30% B; 10–16 min, 30–45% B; 16–18 min, 45% B; 18–40 min, 45–80% B; 40–50 min, 80–25% B. The detection wavelength was 270 nm. Authentic standards of acids included ferulic acid, chlorogenic acid, caffeic acid, 4-coumaric acid and benzoic acids in methanol. The recovery of caffeic acid, chlorogenic acid, 4-coumaric acid, ferulic acid and benzoic acid were 100.1±0.11%, 98.7±0.72%, 101.5±0.02%, 101.2±0.03% and 99.9±0.26% from the phytochemical extracts, respectively. The results were expressed as milligram per 100 g of fresh weight according to the standard curve. Results obtained for sample extracts were expressed as mean $\pm$ SD for triplicates.

Cell culture

Human liver cancer cell line HepG2, purchased from ATCC company (ATCC HB-8065), was grown in growth medium (WME supplemented with 5% FBS, 10 mM Hepes, 2 mM L-glutamine, 5 μg/mL insulin, 0.05 μg/mL hydrocortisone, 50 units/mL penicillin, 50 μg/mL streptomycin, and 100 μg/mL gentmycin), were maintained at 37°C and 5% CO_2.

CAA of ramie leaves extracts

The CAA assay has been previously described [18] and modified by our lab. Briefly, HepG2 cells were seeded at a density of 6×10^4 cells/well on a 96-well microplate in 100 μL of growth medium/well. About twenty-four hours postseeding, the growth medium was removed and the wells were washed with PBS. Microplate were treated for 1 h with 100 μL of medium containing ramie extracts plus 50 μM DCFH-DA. Certain wells were washed with 100 μL of PBS (i.e., PBS wash protocol) and certain wells were not washed (i.e., no PBS wash protocol). PBS wash protocol means that cells are pretreated with ramie extract before the ABAP is added; on the other hand, no PBS wash means that cells are cotreated with ramie extract and ABAP. ABAP (600 μM) in 100 μL of HBSS was added to the cells. The 96-well microplate was placed in a Fluoroskan Ascent fluorescent spectrophotometer (SoftMax systems, Molecular Devices, US) at 37°C. The emission wavelength at 535 nm was measured after an excitation at 485 nm every 5 min for 1 h.

Quantification of CAA

After the subtraction of the blank and the initial fluorescence values, the area under the fluorescence versus time curve was calculated to determine the CAA value at each ramie extract concentration. The following equation was used: CAA (units) $= 1 - (\square SA/\square CA)$ where $\square$ SA is the integrated area under the sample in the fluorescence versus time curve, and $\square$ CA is the integrated area under the control in the flurescence versus time curve. The median effective dose (EC_{50}) of the ramie extracts was calculated from the median effect plot of log (f_a/f_u) versus log (dose), where f_a is the fraction affected by the treatment (CAA unit) and fu is the fraction unaffected (1- CAA unit) by the treatment. The EC_{50} values were expressed as the mean $\pm$ SD using triplicate data sets obtained from the same experiment. EC_{50} values were converted to CAA values, which were expressed as micromoles of quercetin equivalents (QE) per 100 g of ramie, using the mean EC_{50} value for quercetin from three separate experiments.

Cell proliferation inhibiting test

The antiproliferative effects of ramie extracts were assessed in HepG2 cell methylene blue colorimetric method [19]. Briefly, 100 μL of PBS was added to the peripheral wells of the 96-well microplate, and 100 μL of HepG2 cell suspension was seeded at a density of 2.5×10^4/well in the central wells of the 96-well microplate. Incubate the cells for 4 h at 37°C in 5% CO_2 to allow cells to sufficiently attach. Remove the growth medium from the central wells, and 100 μL of fresh medium containing different concentrations of ramie extracts was added. The wells receiving cell suspension without ramie extract served as the control. The plates were incubated for 72 h at 37°C. Following the incubation, the staining was removed, and the 96-well microplates were washed six times in deionized water until the water was clear. Then 100 μL of elution buffer (49% PBS, 50% ethanol, and 1% acetic acid) was added to each well. The 96-well microplates were transferred to a table oscillator for 20 min. Absorbance was measured at 570 nm using a microplate reader. Each sample was measured at least three times. The anti-proliferative effects were assessed by the IC_{50} values, which were expressed as milligrams of ramie extracts per milliliter.

Cytotoxicity test

A cytotoxicity test was performed using the modified methylene blue assay [19,20]. Briefly, HepG2 cells were seeded at a density of 4×10^4 cells/well on a 96-well microplate with 100 μL of growth medium/well. The cells were incubated for 24 h at 37°C and 5% CO_2. After the cells had attached to the wells, the growth medium was removed and the cells were washed with PBS. Then 100 μL of medium with different concentrations of ramie extract was added to each well; wells that received medium without ramie extract served as the control. After 24 h of incubation at 37°C, the medium was removed and the wells were washed with PBS. Add 50 μL of methylene blue solution (98% HBSS, 0.67% glutaraldehyde, and 0.6% methylene blue) to each well of the culture plate. Then incubate plate at 37°C for 60 min. Remove the methylene blue solution and rinse plate in water until the water was clear. Subsequently, add 100 μL of elution buffer (49% PBS, 50% ethanol, and 1% acetic acid) to each well with multiple channel pipette. Place well on a plate rotator at room temperature for 20 min, and read plates using a microplate reader at 570 nm wavelength. Concentrations of ramie extract that decrease the absorbance by > 10% when compared to the control are considered to be cytotoxic.

Statistical analysis

Statistical analyses were performed using Sigmaplot software 11.0 (Sustat Software, Inc., Chicago, IL) and dose-effect analysis was performed using Calcusyn software version 2.0 (Biosoft, Cambridge, U.K.). Results were subjected to ANOVA calculated using SPSS software 21 (SPSS Inc., Chicago, IL, USA) and differences between means were located using Tukey's multiple comparison test. Significance was determined at $p < 0.05$. All data were reported as mean ± SD for triplicates.

Results

Phenolic content in ramie leaves

The total phenolic content of ramie leaves are presented in Figure 1. The free phenolic contents of six tested ramie samples ranged from 291.44 (B. nivea) to 2067.4 (B. ppenduliflora) mg of GAE/100 g of FW. The bound phenolic contents of six tested ramie samples ranged from 60.43 (B. clidemioides) to 246.38 (B. penduliflora) mg of GAE/100 g of FW. The percentage of bound

phenolics to the total ranged from 8.52 (B. macrophylla) to 29.48% (B. nivea). The total phenolic contents of six tested ramie samples ranged from 376.08 (B. longispica) to 2313.7 (B. penduliflora) mg of GAE/100 g of FW. B. penduliflora leaves had the highest total phenolics (2313.7±27.28 mg of GAE/100 g of FW). B. penduliflora leaves had the highest free phenolics (2067.4±22.74 mg of GAE/100 g of FW), up to 89.35% of total phenolics, and had the highest bound phenolics (246.38±4.59 mg of GAE/100 g of FW), accounted for 10.65% of total phenolics.

Flavonoid content in ramie leaves

Total flavonoid content of ramie leaves are presented in Figure 2. The free flavonoid contents of six tested ramie samples ranged from 116.34 (B. nivea) to 1229.8 (B. penduliflora) mg of CAE/100 g of FW. The percentage contribution of free flavonoid to the total ranged from 55.29 (B. nivea) to 82.57% (B. tricuspis). The bound flavonoid contents of six tested ramie samples ranged from 69.09 (B. longispica) to 452.63 (B. penduliflora) mg of CAE/100 g of FW. The percentage contribution of bound flavonoid to the total ranged from 17.43 (B. tricuspis) to 44.71% (B. nivea). The total flavonoid contents of six tested ramie samples ranged from 210.41 (B. nivea) to 1682.4 (B. penduliflora) mg of catechin equiv./100 g of FW. B. penduliflora leaves had the highest total flavonoid (1682.4±27.7 mg of CAE/100 g of FW). B. penduliflora leaves had the highest free flavonoid (1229.81±75.78 mg of CAE/100 g of FW), up to 73.10% of total flavonoid, and had the highest bound phenolics (452.63±57.1 mg of CAE/100 g of FW), accounted for 26.9% of total flavonoids.

Phenolic acid compounds in ramie leaves

The concentrations of represented phenolic acid, chlorogenic acid, caffeic acid, 4-coumaric acid, ferulic acid, and benzoic acid, were analyzed by HPLC and the results are shown in Table 1. Chlorogenic acid was in not detected in all samples of bound form, and was detected in two samples of free form, B. nivea (38.24±2.15 mg/100 g), B. tricuspis (30.19±0.05 mg/100 g).

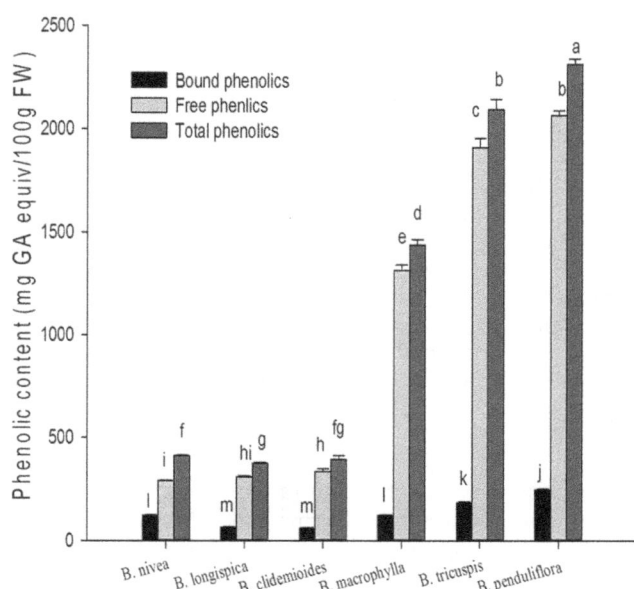

Figure 1. Phenolic contents in six species of ramie leaves (mean ± SD, n = 3). Bars with no letters in common are significantly different ($p < 0.05$).

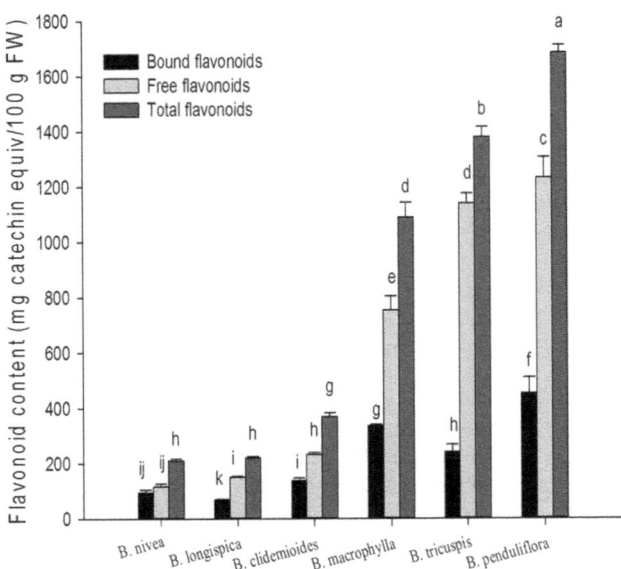

Figure 2. Flavonoid contents in six species of ramie leaves (mean ± SD, n = 3). Bars with no letters in common are significantly different ($p<0.05$).

Caffeic acid was not detected in all samples of free form. *B. tricuspis* had the highest bound caffeic acid (25.79±0.01 mg/100 g), followed by *B.* nivea (18.38±0.02 mg/100 g), *B. macrophylla* (12.04±0.01 mg/100 g), *B. longispica* (4.81±0.01 mg/100 g) and *B. clidemioides*(0). *B. nivea* had the highest bound 4-coumaric acid (9.77±0.12 mg/100 g), followed by *B. tricuspis* (7.33±0.67 mg/100 g), *B. macrophylla* (4.60±0.001 mg/100 g), *B. longispica* (3.74±0.05 mg/100 g), and *B. clidemioides* (2.46±0.1 mg/100 g). Ferulic acid was not detected in all samples of free form. *B. longispica* had the highest bound ferulic acid (21.33±0.21 mg/100 g), followed by *B. penduliflora* (18.76±1.34 mg/100 g), *B. nivea* (13.63±0.17 mg/100 g), *B. macrophylla* (11.22±0.13 mg/100 g), *B. tricuspis* (6.43±0.38 mg/100 g), and *B. clidemioides* (5.28±0.17 mg/100 g). *B. nivea* had the highest bound benzoic acid (28.9±1.29 mg/100 g), followed by *B. tricuspis* (25.88±6.28 mg/100 g), *B. penduliflora* (22.31±1.73 mg/100 g), *B. longispica* (20.76±0.72 mg/100 g), *B. macrophylla* (16.67±0.02 mg/100 g), and *B. clidemioides* (10.98±1.85 mg/100 g). *B. tricuspis* had the highest free benzoic acid (863.22±1.44 mg/100 g), followed by *B. nivea* (151.67±1.20 mg/100 g), *B. clidemioides* (60.95±0.19 mg/100 g), and *B. longispica*, *B. macrophylla* were not detected. *B. tricuspis* had the highest total benzoic acid (889.1±9.66 mg/100 g), followed by *B. nivea* (180.57±1.11 mg/100 g), *B. clidemioides* (71.93±1.62 mg/100 g), *B. penduliflora* (22.31±1.73 mg/100 g), *B. longispica* (20.76±0.72 mg/100 g), and *B. macrophylla* (16.67±0.02 mg/100 g).

In vitro antioxidant activity of ramie leaves

Owing to the complex reactivity of phytochemicals, the antioxidant activities were evaluated using two methods, ORAC and PSC. The PSC values of ramie leaves are presented in Figure 3, expressed as micromoles of vitamin C equivalents per 100 g of fresh weight. The free PSC values ranged from 1216.7 (*B. longispica*) to 7827.9 (*B. tricuspis*) μmol of vitamin C equiv./100 g FW. The bound PSC values ranged from 394.48 (*B. longispica*) to 2139.2 (*B. penduliflora*) μmol of vitamin C equiv./100 g FW. The

total PSC values ranged from 1611.2 (*B. longispica steud*) to 9574.8 (*B. tricuspis*) μmol of vitamin C equiv./100 g FW. *B. tricuspis* had the highest total hydrophilic peroxyl radical scavenging capacity among the six species of ramie, which was 6 times higher than the lowest (*B. longispica*), followed by *B. penduliflora*, *B. macrophylla*, *B. nivea*, *B. clidemioides* and *B. longispica* had the lowest antioxidant activity among the varieties tested.

The ORAC values of ramie leaves are presented in Figure 4, expressed as micromoles of Trolox equivalents per g of fresh weight. The free ORAC values ranged from 62.04 (*B. longispica*) to 289.57 (*B. penduliflora*) μmol of Trolox equiv./g FW. The bound ORAC values ranged from 9.20 (*B. longispica*) to 40.87 (*B. penduliflora*) μmol of Trolox equiv./g FW. The total ORAC values ranged from 59.58 (*B. clidemioides*) to 330.44 (*B. penduliflora*) μmol of Trolox equiv./g FW. *B. penduliflora* had the highest total peroxyl radical scavenging efficacy among the six species of ramie, which was 6 times higher than the lowest (*B. clidemioides*), followed by *B. tricuspis*, *B. macrophylla*, *B. nivea*, *B. longispica*.

In vivo cellular antioxidant activities in ramie leaves

The cellular antioxidant activities of the ramie leaves extracts were measured using the CAA assay. The cellular antioxidant activities were measured using two protocols (PBS wash and no PBS wash). The EC_{50} values ranged from 6.70 (*B. tricuspis*) to 74.54 (*B. nivea*) mg/mL in the PBS wash protocol and from 2.24 (*B. tricuspis*) to 36.08 (*B. clidemioides*)mg/mL in no PBS wash protocol. In both methods, the *B. tricuspis* had the lowest EC_{50} values (6.70±0.36 in the PBS wash protocol and 2.24±0.18 mg/mL in the no PBS wash protocol). The *B. nivea* had the highest EC_{50} values (74.54±16.84 mg/mL) in the PBS wash protocol, followed by *B. longispica* (72.21±9.01 mg/mL), *B. clidemioides* (66.54±9.65 mg/mL), *B. macrophylla* (14.77±1.32 mg/mL), *B. penduliflora* (11.23±0.59 mg/mL), *B. tricuspis* (6.70±0.36 mg/mL). The *B. clidemioides* had the highest EC_{50} values 36.08±5.50 mg/mL in the no PBS wash protocol, followed by *B. longispica* (18.20±2.63 mg/mL), *B. nivea* (11.17±1.53 mg/mL), *B. macrophylla* (6.14±0.99 mg/mL), *B. penduliflora* (3.76±0.29 mg/mL), *B. tricuspis* (2.24±0.18 mg/mL).

We reported the CAA values of ramie leaves firstly. The CAA values for ramie leaves extracts were showed in Figure 5. CAA values ranged from 16.27 (*B. nivea*) to 133.63 (*B. tricuspis*) μmol QE/100 g FW of ramie in the PBS wash protocol. The *B. tricuspis* had the highest CAA value (133.63±7.10 μmol QE/100 g FW of ramie), followed by *B. penduliflora* (79.70± 4.27 μmol QE/100 g FW of ramie), *B. macrophylla* (74.30± 6.55 μmol QE/100 g FW of ramie), *B. clidemioides* (16.63± 2.27 μmol QE/100 g FW of ramie), *B. longispica* (16.45± 1.91 μmol QE/100 g FW of ramie), *B. nivea* (16.27±3.28 μmol QE/100 g FW of ramie). In the no PBS wash protocol, CAA values ranged from 19.88 to 289.60 μmol QE/100 g FW of ramie, *B. tricuspis* had the highest CAA value (289.60±24.35 μmol QE/100 g FW of ramie), followed by *B. penduliflora* (172.59± 13.64 μmol QE/100 g FW of ramie), *B. macrophylla* (117.10± 19.06 μmol QE/100 g FW of ramie), *B. nivea* (64.92±9.40 μmol QE/100 g FW of ramie), *B. longispica* (39.89±5.91 μmol QE/100 g FW of ramie), *B. clidemioides* (19.88±2.79 μmol QE/100 g FW of ramie).

The CAA and EC_{50} values of the different ramie extracts were negatively correlated, the lower the EC_{50} value, the higher the CAA value. Regardless of the protocol, *B. tricuspis* had the lowest EC_{50} value.

Table 1. Phenolic compounds in six species of ramie leaves.

		B. Nivea	B. Longispica	B. Clidemioides	B. Macrophylla	B. Tricuspis	B. Penduliflora
Chlorogenic acid (mg/100 g FW)	bound	ND	ND	ND	ND	ND	ND
	free	38.24±2.15	ND	ND	ND	30.19±0.05	ND
	total	38.24±2.15	ND	ND	ND	30.19±0.05	ND
Caffeic acid (mg/100 g FW)	bound	18.38±0.02	4.81±0.01	ND	12.04±0.01	25.79±0.01	24.78±0.04
	free	ND	ND	ND	ND	ND	ND
	total	18.38±0.02	4.81±0.01	ND	12.04±0.01	25.79±0.01	ND
4-coumaric acid (mg/100 g FW)	bound	9.77±0.12	3.74±0.05	2.46±0.10	4.6±0.001	7.33±0.67	5.86±0.65
	free	ND	10.43±0.92	ND	ND	ND	ND
	total	9.77±0.12	14.17±1.00	2.46±0.10	4.60±0.001	7.33±0.67	5.86±0.65
Ferulic acid (mg/100 g FW)	bound	13.63±0.17	21.33±0.21	5.28±0.17	11.22±0.13	6.43±0.38	18.76±1.34
	free	ND	ND	ND	ND	ND	ND
	total	13.63±0.17	21.33±0.21	5.28±0.17	11.22±0.13	6.43±0.38	18.76±1.34
Benzoic acid (mg/100 g FW)	bound	28.90±1.29	20.76±0.72	10.98±1.85	16.67±0.02	25.88±6.28	22.31±1.73
	free	151.67±1.20	ND	60.95±0.19	ND	863.22±1.44	ND
	total	180.57±1.11	20.76±0.72	71.93±1.62	16.67±0.02	889.10±9.66	22.31±1.73

ND: not detected. FW: fresh weight.

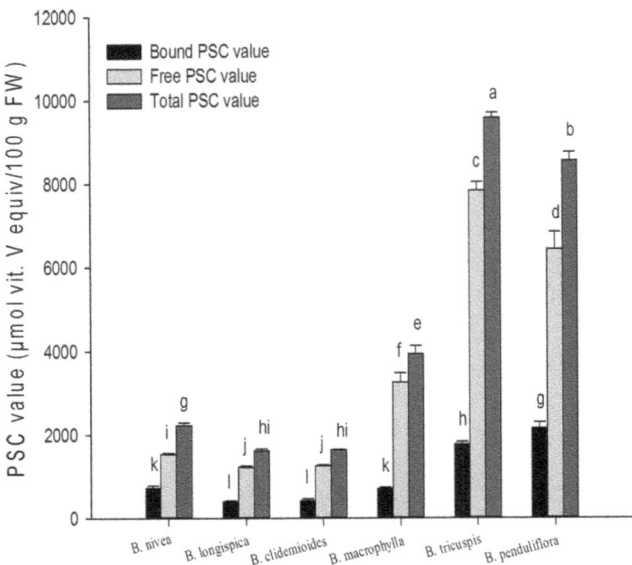

Figure 3. PSC values in six species of ramie leaves (mean ± SD, n = 3). Bars with no letters in common are significantly different ($p<$ 0.05).

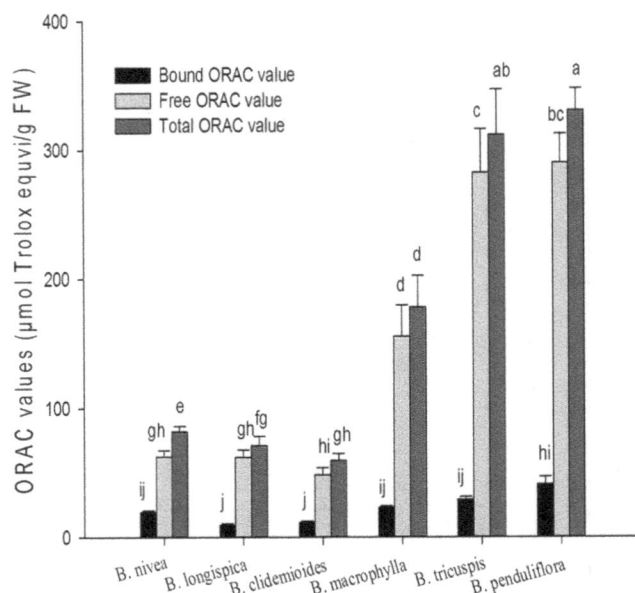

Figure 4. ORAC values in six species of ramie leaves (mean ± SD, n = 3). Bars with no letters in common are significantly different ($p<0.05$).

Antiproliferation activity and cytotoxicity in ramie leaves

The inhibition of HepG2 cell proliferation in vitro by the ramie leaves extracts and the cytotoxic effects are presented in and Table 2. The results were expressed as the median effective dose (IC_{50}), with a lower IC_{50} value indicating a higher antiproliferation activity. The six ramie leaves extracts showed potent inhibit cell proliferation on HepG2 cell growth in a dose-dependent manner, and there were differences among the six ramie extracts. The extracts of *B. tricuspis* had the highest inhibitory effects on cell proliferation, *B. longispica* showed the lowest antiproliferative activity. *B. penduliflora* had cytotoxic effects ($CC_{50} \geq 1.15$ mg/ mL), which showed that the inhibitory effect was due to cytotoxic effect. But others had no cytotoxic effects in the proliferation of concentration, which indicated that the inhibitory effect was attributed to antitumor effects of the extracts.

Discussion

Ramie leaves as food and traditional medicine has a long history. Different varieties of Ramie leaves are not only good sources for dietary and medical use, but also excellent sources of phytochemicals, such as phenolics and flavonoids.

Phenolics compounds may provide health benefits related to decrease the incidence of chronic disease in our diet. This is first report of phenolics in fresh leaves of ramie indicated that phenolics in soluble free forms were significantly higher than bound forms in all selected ramie. Phenolics and flavonoids are present in free and bound (cell wall-associated) forms in plants. We developed a detection method to measure the complete phenolic profiles, this method identified and quantifies the free and bound forms of phenolics [4]. According to the present study, phenolics is one of the main phytochemical in ramie leaves.

Similarly to phenolics, contents of flavonoid in soluble free form were significantly higher than bound flavonoid contents in all selected ramie. Flavonoids, widespread in teas, fruits, vegetables and medicinal plants, are a group of biologically active plant compounds [1,5]. They were also reported to exhibit antioxidant

activity and could inhibit cellular growth [21,22]. Ramie, widely cultivated in China, is a flavonoid rutin rich plant source [23]. There is no published report of flavonoids in ramie leaves. This study was the first report to quantify total flavonoid contents in fresh leaves of ramie using a new method reported recently [13]. The SBC assay, developed by our group, can quantify all types of flavonoids including all subgroups of flavones, flavonones, flavononols, flavonols, flavanols, isoflavonoids and anthocyanins [14]. The occurrence of phenolic acid in food affects stability,

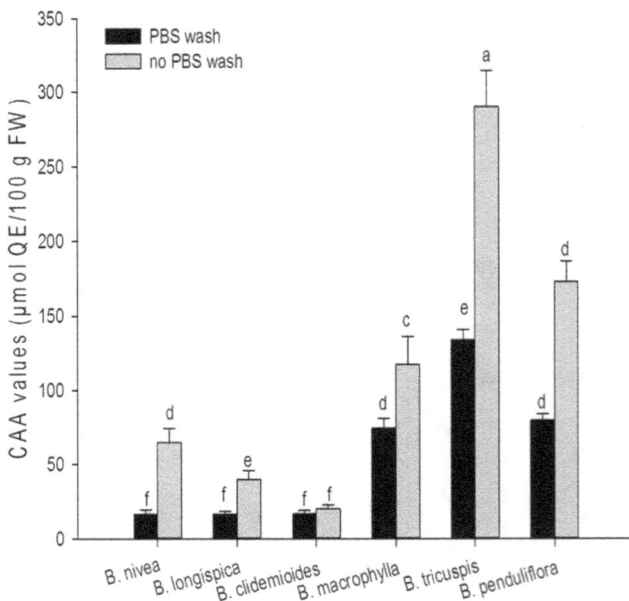

Figure 5. CAA values in six species of ramie leaves (mean ± SD, n = 3). Bars with no letters in common are significantly different ($p<$ 0.05).

Table 2. Antiproliferative Activities (IC_{50}) and Cytotoxicities (CC_{50}) of Ramie Leaves Extracts toward Human HepG2 Cancer Cells.

	IC_{50}(mg/mL)	CC_{50} (mg/mL)
B. longispica	48.06±5.81a	>20
B. macrophylla	6.11±0.03c	>11.9
B. clidemioides	13.85±1.82b	>20
B. tricuspis	4.11±0.19c	>11.3
B. penduliflora	1.46±0.11c	>1.15
B. nivea	ND	>20

Values with no letters in common in each column are significantly different ($p<0.05$). ND: not detected.

color, flavoe, nutritional value, and other qualities. Phenolic acids (*p*-coumaric acid, caffeic acid, ferulic acid and ellagic acid) have been proposed to have beneficial effects on health as antioxidants [24]. The potential of using natural phenolic acid as natural antioxidants has drawn more attention. Ferulic acid, arises from the metabolism of phenylalanine and tyrosine [25], is a ubiquitous plant component also process some biological activities [6]. Ferulic acid is mainly based on the free form and bound in seeds and leaves. Ferulic acid serves an vital antioxidant function in preserving physiological integrity of cells, its addition of foods inhibits lipid peroxidation and subsequent oxidative spoilage [25]. Humans ingest a variety of phenols. Caffeic acid, chlorogenic acid, ferulic acid suppressed the promotion of tumors [26,27]. Caffeic acid has free radical scavenging property and the effect of inhibiting LDL oxidative [28]. To test if the high antioxidant activity of ramie was attributable to the phenolic acid, we analyzed the main phenolic acid. The results reported in this article demonstrated that phenolic acids were showed in bound form, and were in agreement with the reported by S. C. Fry [29].

Various phytochemical components, such as phenolic acids, flavonoids, are contributed to antioxidant capacity in fruits and vegetables [1,5], and have attracted interest because of their potential nutritional and safety. Due to the complex reactivity of phytochemicals, the accurate evaluate of antioxidant activities of phytochemicals to be taken into account. It is suggested that the use of more than one condition of oxidation is required to evaluate antioxidants. In this study, the antioxidant activities were evaluated using two methods, named ORAC and PSC.

There are many kinds of methods detect antioxidant activity. One of these methods is named PSC assay which incorporating dichloroflurescin diacetate (DCFH-DA) as a fluorescent probe to monitor reaction, and is suitable for analyzing both hydroplic and lipophilic antioxidants or food extracts [2]. PSC method is reliable, sensitive, rapid, reproducibility, and precise and can produce acceptable results compare with those obtained with similar published assays [2]. Oxygen radical absorbance capacity (ORAC) assay has been widely accepted as a tool to test the antioxidant activity in the pharmaceutical and food industries. In the typical ORAC assay, the fluorescent loss of probes as phycoerythrin, and then fluorescein is followed kinetics in the absence and presence of antioxidant [30]. The antioxidant activity of *Nivea* extract has been evaluated using DPPH radical scavenging activity [31]. However, it is difficult to compare their results to the antioxidant activity reported here using the PSC/OARC assay. In accordance with previous report that the antioxidant activity of extracts of blueberries were due to phenolic compounds in these berries [32], we reported significant linear relationship between phenolic/flavonoid contents and PSC/ORAC values.

All percent contribution of bound phytochemicals to total were no more than 45%, most percent contribution of bound phytochemicals were about 20%, it was hard to analyze the antioxidant activities in bound phytochemicals, so in cell we used free phytochemicals for analysis.

Cellular antioxidant activity (CAA) assay, quantifying the antioxidant activity of phytochemicals, food extracts, and dietary supplements, perform at physiological pH and temperature, and take into account the bioavailability, uptake, and metabolism of the antioxidant compounds [18]. Cell culture models are cost-effective, relatively fast, and better represent the complexity of biological systems.

The increase in fluorescence from the formation of DCF was inhibited by both quercetin and ramie extracts in a dose-dependent manner. DCFH oxidation was inhibited regardless of whether the cells had been washed with PBS or not between the antioxidant and the ABAP treatments. Compared to the PBS wash protocol, the no PBS wash protocol had significantly lower EC_{50} values and higher antioxidant capacity. This result is attributed to the PBS, which can affect the extracellular antioxidant capacity, thereby reducing the intracellular antioxidant capacity.

The correlations between phenolic content, flavonoid content and antioxidant activity were examined. Significant linear relationship was found between phenolic/flavonoid contents and PSC values of different ramie types ($R^2 = 0.943$, $p<0.05$; $R^2 = 0.771$, $p<0.05$;). A significant linear relationship was found between phenolic/flavonoid content and ORAC values of different ramie types ($R^2 = 0.943$, $p<0.05$; $R^2 = 0.771$, $p<0.05$). A highly significant linear relationship was found between phenolic/flavonoid contents and CAA values (PBS wash) of different ramie types($R^2 = 0.943$, $p<0.01$; $R^2 = 0.943$, $p<0.01$). Significant linear relationship was also found between phenolic/flavonoid contents and CAA values (no PBS wash) of different ramie types ($R^2 = 0.714$, $p<0.05$; $R^2 = 0.714$, $p<0.05$;). The positive correlation shows that antioxidant activities were closely related to phenolics and flavonoids in six species of ramie.

To estimate the effect of ramie leaves on cancer cell growth, we assessed its effect on the proliferation of liver cancer cells. According to our data, ramie leaves extract showed inhibit proliferation activity, the mechanism may be ramie leaves extract enhance anti cancer cytokines activity. In contrast to data, *B. Penduliflora* and B. Tricuspis makino is similar to phytochemical prolifes, however, *B. Penduliflora* showed lowest IC_{50} and CC_{50}, the data suggested that the mechanisms of *B. Penduliflora* inhibit proliferation is associated with apoptosis.

In summary, we found that the primary portion of phytochemicals in the fresh leaves of ramie was showed in the free form, intimating that most phytochemical of fresh leaves of ramie is soluble. There are significant differences in phytochemical content and antioxidant activity among the different varieties of ramie leaves. However, some similarities can be seen within families and genera. Ramie leaves are the potential sources for food and pharmaceutical applications as anticancer agents. However, further research is needed to evaluate the effects of ramie leaf in the prevention of liver cancer.

References

1. Liu RH (2004) Potential synergy of phytochemicals in cancer prevention: Mechanism of action. Journal of Nutrition 134: 3479S–3485S.
2. Adom KK, Liu RH (2005) Rapid peroxyl radical scavenging capacity (PSC) assay for assessing both hydrophilic and lipophilic antioxidants. Journal of Agricultural and Food Chemistry 53: 6572–6580.
3. Eberhardt MV, Lee CY, Liu RH (2000) Nutrition - Antioxidant activity of fresh apples. Nature 405: 903–904.
4. Adom KK, Liu RH (2002) Antioxidant activity of grains. Journal of Agricultural and Food Chemistry 50: 6182–6187.
5. Liu RH (2013) Health-Promoting Components of Fruits and Vegetables in the Diet. Advances in Nutrition 4: 384S–392S.
6. Liu RH (2007) Whole grain phytochemicals and health. Journal of Cereal Science 46: 207–219.
7. Lee H, Joo N (2012) Optimization of Pan Bread Prepared with Ramie Powder and Preservation of Optimized Pan Bread Treated by Gamma Irradiation during Storage. Prev Nutr Food Sci 17: 53–63.
8. IS K (2008) Antibacterial, antioxidant activities and cell viability against cancer of Adenophora remotifloraleaves. University of Wonkwang.
9. Wang F, Zhao S, Li F, Zhang B, Qu Y, et al. (2014) Investigation of Antioxidant Interactions between Radix Astragali and Cimicifuga foetida and Identification of Synergistic Antioxidant Compounds. Plos One 9.
10. Jiratanan T, Liu RH (2004) Antioxidant activity of processed table beets (Beta vulgaris var, conditiva) and green beans (Phaseolus vulgaris L.). Journal of Agricultural and Food Chemistry 52: 2659–2670.
11. Sun J, Chu YF, Wu XZ, Liu RH (2002) Antioxidant and anti proliferative activities of common fruits. Journal of Agricultural and Food Chemistry 50: 7449–7454.
12. Singleton VL, Orthofer R, Lamuela-Raventos RM (1999) Analysis of total phenols and other oxidation substrates and antioxidants by means of Folin-Ciocalteu reagent. In: Packer L, editor. Oxidants and Antioxidants, Pt A. pp. 152–178.
13. Guo X, Li T, Tang K, Liu RH (2012) Effect of Germination on Phytochemical Profiles and Antioxidant Activity of Mung Bean Sprouts (Vigna radiata). Journal of Agricultural and Food Chemistry 60: 11050–11055.
14. He X, Liu D, Liu RH (2008) Sodium Borohydride/Chloranil-Based Assay for Quantifying Total Flavonoids. Journal of Agricultural and Food Chemistry 56: 9337–9344.
15. Huang DJ, Ou BX, Hampsch-Woodill M, Flanagan JA, Prior RL (2002) High-throughput assay of oxygen radical absorbance capacity (ORAC) using a multichannel liquid handling system coupled with a microplate flourescence reader in 96-well format. Journal of Agricultural and Food Chemistry 50: 4437–4444.
16. Zhang MW, Zhang RF, Zhang FX, Liu RH (2010) Phenolic Profiles and Antioxidant Activity of Black Rice Bran of Different Commercially Available Varieties. Journal of Agricultural and Food Chemistry 58: 7580–7587.
17. Malta LG, Ghiraldini FG, Reis R, Oliveira MdV, Silva LB, et al. (2012) In vivo analysis of antigenotoxic and antimutagenic properties of two Brazilian Cerrado fruits and the identification of phenolic phytochemicals. Food Research International 49: 604–611.
18. Wolfe KL, Liu RH (2007) Cellular antioxidant activity (CAA) assay for assessing antioxidants, foods, and dietary supplements. Journal of Agricultural and Food Chemistry 55: 8896–8907.
19. Felice DL, Sun J, Liu RH (2009) A modified methylene blue assay for accurate cell counting. Journal of Functional Foods 1: 109–118.
20. Yoon H, Liu RH (2007) Effect of selected phytochemicals and apple extracts on NF-kappa B activation in human breast cancer MCF-7 cells. Journal of Agricultural and Food Chemistry 55: 3167–3173.
21. He X, Liu RH (2008) Phytochemicals of Apple Peels: Isolation, Structure Elucidation, and Their Antiproliferative and Antioxidant Activities. Journal of Agricultural and Food Chemistry 56: 9905–9910.
22. Yang J, Liu RH (2013) The phenolic profiles and antioxidant activity in different types of tea. International Journal of Food Science and Technology 48: 163–171.
23. Dong JZ, Lu DY, Wang Y (2009) Analysis of Flavonoids from Leaves of Cultivated Lycium barbarum L. Plant Foods for Human Nutrition 64: 199–204.
24. Hakkinen SH, Torronen AR (2000) Content of flavonols and selected phenolic acids in strawberries and Vaccinium species: influence of cultivar, cultivation site and technique. Food Research International 33: 517–524.
25. Graf E (1992) Antioxidant potential of ferulic acid. Free radical biology & medicine 13: 435–448.
26. Wattenberg LW, Coccia JB, Lam LK (1980) Inhibitory effects of phenolic compounds on benzo(a)pyrene-induced neoplasia. Cancer research 40: 2820–2823.
27. Huang MT, Smart RC, Wong CQ, Conney AH (1988) Inhibitory effect of curcumin, chlorogenic acid, caffeic acid, and ferulic acid on tumor promotion in mouse skin by 12-O-tetradecanoylphorbol-13-acetate. Cancer research 48: 5941–5946.
28. Nardini M, D'Aquino M, Tomassi G, Gentili V, Di Felice M, et al. (1995) Inhibition of human low-density lipoprotein oxidation by caffeic acid and other hydroxycinnamic acid derivatives. Free radical biology & medicine 19: 541–552.
29. Fry SC (1979) Phonolic components of the primary cell wall and their possible role in the hormonal regulation of growth. Planta 146: 343–351.
30. Malta LG, Tessaro EP, Eberlin M, Pastore GM, Liu RH (2013) Assessment of antioxidant and antiproliferative activities and the identification of phenolic compounds of exotic Brazilian fruits. Food Research International 53: 417–425.
31. Lee YR, Nho JW, Hwang IG, Kim WJ, Lee YJ, et al. (2009) Chemical Composition and Antioxidant Activity of Ramie Leaf (Boehmeria nivea L.). Food Science and Biotechnology 18: 1096–1099.
32. Liu RH (2008) Blueberry is best antioxidant. Trac-Trends in Analytical Chemistry 27: VIII–IX.

Author Contributions

Conceived and designed the experiments: XBG RHL. Performed the experiments: YSC HW. Analyzed the data: YSC GYW. Contributed reagents/materials/analysis tools: CHC GGZ XBG RHL. Wrote the paper: XBG YSC.

12

Palladium and Platinum Nanoparticles Attenuate Aging-Like Skin Atrophy via Antioxidant Activity in Mice

Shuichi Shibuya[1], Yusuke Ozawa[1], Kenji Watanabe[1], Naotaka Izuo[1], Toshihiko Toda[1], Koutaro Yokote[2], Takahiko Shimizu[1,3]*

1 Department of Advanced Aging Medicine, Chiba University Graduate School of Medicine, Chiba, Japan, **2** Department of Clinical Cell Biology and Medicine, Chiba University Graduate School of Medicine, Chiba, Japan, **3** Molecular Gerontology, Tokyo Metropolitan Institute of Gerontology, Itabashi, Tokyo, Japan

Abstract

Cu-Zn superoxide dismutase (*Sod1*) loss causes a redox imbalance as it leads to excess superoxide generation, which results in the appearance of various aging-related phenotypes, including skin atrophy. Noble metal nanoparticles, such as palladium (Pd) and platinum (Pt) nanoparticles, are considered to function as antioxidants due to their strong catalytic activity. In Japan, a mixture of Pd and Pt nanoparticles called PAPLAL has been used to treat chronic diseases over the past 60 years. In the present study, we investigated the protective effects of PAPLAL against aging-related skin pathologies in mice. Transdermal PAPLAL treatment reversed skin thinning associated with increased lipid peroxidation in $Sod1^{-/-}$ mice. Furthermore, PAPLAL normalized the gene expression levels of *Col1a1*, *Mmp2*, *Has2*, *Tnf-α*, *Il-6*, and *p53* in the skin of the $Sod1^{-/-}$ mice. Pt nanoparticles exhibited marked SOD and catalase activity, while Pd nanoparticles only displayed weak SOD and catalase activity *in vitro*. Although the SOD and catalase activity of the Pt nanoparticles significantly declined after they had been oxidized in air, a mixture of Pd and Pt nanoparticles continued to exhibit SOD and catalase activity after oxidation. Importantly, a mixture of Pd and Pt nanoparticles with a molar ratio of 3 or 4 to 1 continued to exhibit SOD and catalase activity after oxidation, indicating that Pd nanoparticles prevent the oxidative deterioration of Pt nanoparticles. These findings indicate that PAPLAL stably suppresses intrinsic superoxide generation both *in vivo* and *in vitro* via SOD and catalase activity. PAPLAL is a potentially powerful tool for the treatment of aging-related skin diseases caused by oxidative damage.

Editor: Hemachandra Reddy, Oregon Health & Science University, United States of America

Funding: This research was supported by funds from Musashino Pharmaceutical Co. (http://www.musashino-p.co.jp/). The PAPLAL and the Pd and Pt nanoparticles were provided from Musashino Pharmaceutical and Toyo Kosei pharmaceutical (http://toyokosei.co.jp/). The funders had no role in study design, data collection and analysis, decision to publish, or preparation of the manuscript.

Competing Interests: Musashino Pharmaceutical Company made a grant donation to Department of Advanced Aging Medicine, Chiba University Graduate School of Medicine, for the conduct of the animal study and for the purchase of reagents. Musashino Pharmaceutical was not involved with the scientific content of the research projects run by the Department of Advanced Aging Medicine, Chiba University Graduate School of Medicine. The authors were not paid employees and did not receive any kind of fees from Musashino Pharmaceutical Company.

* Email: shimizut@chiba-u.jp

Introduction

Skin aging induced by chronological or intrinsic factors leads to skin atrophy [1]. The amounts of skin collagen components fall in an age-dependent manner in both males and females, resulting in age-related skin thinning in older individuals [2,3]. Evidence suggests that oxidatively modified proteins, DNA, and lipids in the skin and other organs progressively accumulate during aging [4], indicating that reactive oxygen species (ROS) are strongly associated with skin aging. Complex organisms possess multiple antioxidative and repair systems for mitigating oxidative damage. Superoxide dismutase (SOD) plays a central role in antioxidative systems due to its ability to catalyze the conversion of cellular superoxide (O_2^-) to hydrogen peroxide (H_2O_2). H_2O_2 is further degraded to O_2 and H_2O by catalase, glutathione peroxidases and peroxiredoxins. Copper/zinc superoxide dismutase (SOD1) is localized to react with intracellular O_2 in the cytoplasm. Our previous studies have demonstrated that *Sod1*-deficient ($Sod1^{-/-}$)

mice exhibit increased intracellular O_2^- concentrations and various aging-related organ phenotypes, such as age-related macular degeneration [5], fatty deposits in the liver [6,7], skin atrophy [8–11], bone loss and fragility [12,13], progression of Alzheimer's disease [14], infertility [15], dry eye [16,17], and rotator cuff degeneration [18]. These findings suggest that cytoplasmic O_2^--mediated oxidative damage is the primary cause of aging-related changes in various tissues. In particular, *Sod1* insufficiency results in both epidermal and dermal atrophy, which is associated with the downregulation of extracellular matrix related-genes, including *Col1a1* and *Has2* [11]. Therefore, $Sod1^{-/-}$ mice constitute a suitable model for studying skin aging in elderly people.

Noble metal nanoparticles, including those palladium (Pd), platinum (Pt), and gold nanoparticles, display strong catalytic activity, e.g., in hydrogenation, hydration, and oxidation reactions, due to their large surface area and the high proportion of metal atoms located on their surfaces [19–21]. Such noble metal nanoparticle catalysts are considered to function as antioxidants and are potentially useful in material science and engineering as well as in medical science and clinical therapy [22]. A number of studies have reported that Pt nanoparticles exhibit strong antioxidant activity [22–24]. Recently, Elhusseiny and Hassan reported that a complex of Pd and Pt nanoparticles demonstrated highly potent antitumor and antimicrobial activity [25]. In 1915, Dr. Hideyo Noguchi (Rockefeller University) and Dr. Saburo Ishizuka (a dental surgeon at Tokyo Dental College) formulated a plan for creating a solution of Pd and Pt nanoparticles for clinical use. Twenty-one years later, Dr. Ishizuka successfully prepared a Pd and Pt nanoparticle solution called PAPLAL (Toyokose Pharmaceuticals, Japan) [26]. Since then, PAPLAL has been used to treat Japanese patients with burns, frostbite, hives, lung inflammation, gastric ulcers, and rheumatoid arthritis. PAPLAL has been shown to have various beneficial effects on chronic diseases [26]. In addition, PAPLAL has been approved as a treatment for acute gastric inflammation and chronic gastrin catarrh in Japan under the Pharmaceutical Affairs Law, and it has also been patented as an antioxidant with the Japan Patent Office (Patent No. 3411195, 2003). *In vitro* studies have reported that PAPLAL exhibits antioxidant activity against superoxide anions and hydroxyl radicals [27,28]. However, no previous studies have investigated the effects of PAPLAL or other metal nanoparticles on skin aging.

In the present study, we investigated the protective effects of PAPLAL against age-related skin pathologies in model mice. We also analyzed the expression profiles of skin-related genes, including those involved in matrix biosynthesis, inflammation, and aging, in order to clarify the underlying mechanisms of the *in vivo* effects of PAPLAL. In addition, *in vitro* experiments were conducted to evaluate the antioxidant activity of PAPLAL.

Materials and Methods

Nanoparticles

Pd and Pt nanoparticles and PAPLAL were provided by Toyokose Pharmaceutical Co. (Tokyo, Japan) through Musashino Pharmaceutical Co. (Tokyo Japan). PAPLAL is composed of a mixture of 0.3 mg/mL (2.82 mM) of Pd nanoparticles and 0.2 mg/mL (1.03 mM) of Pt nanoparticles.

Mice

$Sod1^{-/-}$ mice were purchased from the Jackson Laboratory (Bar Harbor, ME, USA). Genotyping of the $Sod1^{-/-}$ allele was performed via genomic PCR using genomic DNA isolated from the tail tip, as reported previously [8]. The animals were housed under a 12-hour light/dark cycle and fed *ad libitum*. In addition, they were maintained and studied according to protocols approved by the animal care committee of the Tokyo Metropolitan Institute of Gerontology and Chiba University.

Transdermal administration in $Sod1^{-/-}$ mice

The $Sod1^{+/+}$ and $Sod1^{-/-}$ mice were transdermally administered PAPLAL for four weeks, beginning at four months of age. On the first day of each week, the hair on the backs of the mice was shaved off and then PAPLAL (200 μL/mouse) or MilliQ water (200 μL/mouse) was applied to the exposed skin once a day. MilliQ water was used as a placebo control for PAPLAL.

Histology

To assess the histological morphology of the treated skin, skin specimens were dissected from the back tissue of the mice, fixed overnight in a 20% formalin neutral buffer solution (Wako, Osaka, Japan), embedded in paraffin, and sectioned on a microtome at a thickness of 4 μm according to standard techniques. Hematoxylin and eosin staining was performed as described previously [6,29]. The thickness of the skin tissue was measured using the Leica QWin V3 imaging software (Leica, Germany).

8-isoprostane content

The skin tissue specimens were homogenized with 0.1 M phosphate (pH 7.4) containing 1 mM of ethylenediaminetetraacetic acid (Dojindo Laboratories, Kumamoto, Japan) and 50 μg/mL (w/w) of dibutylhydroxytoluene (Wako). The homogenate was centrifuged at $8,000 \times g$ for 10 minutes at 4°C, and the total supernatant was used for the assay. The 8-isoprostane concentration of the homogenate was measured using an 8-isoprostane enzyme immunoassay kit (Cayman Chemical Company, MI, USA) according to the manufacturer's instructions. The protein concentration of the supernatant was assessed using the DC protein assay kit (BioRad, Hercules, CA, USA), and the 8-isoprostane level was normalized to the protein level.

Outgrowth assay

Mouse back skin samples were sterilized with 70% ethanol and rinsed with phosphate-buffered saline (Takara Bio Inc., Shiga, Japan), and then discs measuring 5 mm in diameter were punched out using a dermal punch (Nipro, Tokyo, Japan). The punched skin discs were placed into a 24-well culture plate (Falcon BD, Franklin Lakes, NJ) and cultured with or without PAPLAL in α-minimum essential medium (α-MEM; Life Technologies Corporation, Carlsbad, CA, USA) containing 20% fetal bovine serum (FBS; Life Technologies Corporation), 100 units/mL of penicillin, and 0.1 mg/mL of streptomycin (Sigma-Aldrich, MO, USA) at 37°C in a humidified incubator under 5% CO_2 and 20% O_2. The number of outgrowing fibroblasts originating from the mouse skin discs was directly counted after 96 h culturing.

Lactate dehydrogenase (LDH) activity

Skin specimens were cultured according to the method described above, and the culture medium was collected after 96 h. The collected medium was centrifuged at $400 \times g$ for 5 min at 4°C, and the total supernatant was used for the subsequent assay. The LDH level was measured using the LDH cytotoxicity assay kit (Cayman Chemical Company) according to the manufacturer's instructions.

Quantitative PCR

Total RNA was extracted from the back skin using Trizol reagent (Life Technologies Corporation) according to the manufacturer's instructions. cDNA was synthesized from 1 μg of total RNA using reverse transcriptase (ReverTra Ace qPCR RT Master MIX, TOYOBO, Osaka, Japan). Real-time PCR was performed on a Mini Opticon™ (Bio-Rad) with SYBR GREEN PCR master mix (Bio-Rad), according to the manufacturer's instructions. All expression data were normalized to the expression level of the housekeeping gene glyceraldehyde-3-phosphate dehydrogenase (*Gapdh*). The primer sequences are listed in Table S1.

Antioxidant activity

The SOD and catalase activity of the Pd and Pt nanoparticles was measured using a WST-based SOD assay kit (Dojindo Laboratories) according to the manufacturer's instructions, whereas the catalase activity of the Pd and Pt nanoparticles was evaluated using an Amplex Red catalase assay kit (Sigma-Aldrich) according to the manufacturer's instructions. SOD derived from bovine erythrocytes (Sigma-Aldrich, lot number 080M76901V) and catalase derived from bovine liver tissue (Sigma-Aldrich, lot number 1232075) were used as positive controls for SOD and catalase activity, respectively. In order to assess the stability of the antioxidant activity of the different nanoparticles, Pd and Pt nanoparticle solutions were incubated at room temperature for four weeks in tubes in which the air had been replaced with N_2. In a further experiment, Pd and Pt nanoparticle solutions were exposed to air by rotating them for 24 h at room temperature. The SOD and catalase activity levels of each solution were then measured as described above.

Statistics

Statistical analyses were performed using the Student's t-test for comparisons between two groups or Tukey's test for comparisons between three groups. Differences were considered significant at p-values of less than 0.05. All data are expressed as mean ± standard deviation (SD) values.

Results

PAPLAL accelerated wound healing in the aged mice

In order to evaluate the protective effects of PAPLAL on the skin, we first investigated the ability of PAPLAL to promote wound healing in aged murine skin (17 months of age). Although the areas of the wounds treated with and without PAPLAL did not differ at two days after wounding, the wounds treated with PAPLAL were significantly smaller than those treated with vehicle at four and six days after wounding (Figure S1A–1B). In addition, PAPLAL treatment was found to have promoted wound healing at 12 days after wounding (Figure S1B). These results demonstrate that the transdermal application of PAPLAL assists in wound healing in aged mice.

PAPLAL attenuated skin atrophy in the Sod1-deficient mice

SOD1, an antioxidant enzyme, plays a pivotal role in cellular antioxidative systems. Therefore, in order to investigate the beneficial effects of PAPLAL on the skin symptoms seen in $Sod1^{-/-}$ mice, PAPLAL was transdermally administered to the skin on the backs of $Sod1^{-/-}$ and wild-type ($Sod1^{+/+}$) mice daily for four weeks, beginning at four months of age. As shown in Figure 1, the skin of the $Sod1^{-/-}$ mice was significantly thinner than that of the $Sod1^{+/+}$ mice, confirming that skin atrophy had occurred in the $Sod1^{-/-}$ mice. The back skin of the $Sod1^{-/-}$ mice that were treated with a high concentration of PAPLAL was significantly thicker (epidermis and dermis: by 46.2% and 19.2%, respectively) than that of the $Sod1^{-/-}$ mice treated with the vehicle. The administration of a low concentration of PAPLAL also improved the skin atrophy observed in the $Sod1^{-/-}$ mice (the thickness of the epidermis and dermis were increased by 42.0% and 21.1%, respectively) compared with that seen in the $Sod1^{-/-}$ mice treated with vehicle. In order to examine the degree of oxidative damage in the skin of the $Sod1^{-/-}$ mice, we measured the concentration of 8-isoprostane as a representative of lipid peroxidation products. The 8-isoprostane content of the $Sod1^{-/-}$ skin was increased by 65.4% compared with that of the $Sod1^{+/+}$ mice, which indicated that lipid peroxidation products had accumulated in it (Figure 1D). Meanwhile, treatment with a high concentration of PAPLAL significantly decreased the 8-isoprostane content of the $Sod1^{-/-}$ mouse skin compared with that of the $Sod1^{-/-}$ mouse skin treated with the vehicle (Figure 1D).

Next, we investigated the protective effects of PAPLAL against skin damage. Organ culture experiments using skin discs demonstrated that the $Sod1^{-/-}$ fibroblasts had a markedly lower outgrowth capacity than the $Sod1^{+/+}$ fibroblasts, which indicated that the migration and proliferation of the $Sod1^{-/-}$ fibroblasts were impaired (Figure 2A). In order to analyze the protective effects of PAPLAL, we added PAPLAL to cultures of $Sod1^{-/-}$ skin discs. Treatment with PAPLAL significantly promoted fibroblast outgrowth from the mutant skin discs (Figure 2A). We also measured the LDH activity, which is a marker of skin damage, in the culture medium. The $Sod1^{-/-}$ skin exhibited significantly increased LDH activity by 5.1-fold compared with the $Sod1^{+/+}$ fibroblasts, which was indicative of skin damage (Figure 2B). Meanwhile, PAPLAL treatment significantly suppressed the LDH activity in the $Sod1^{-/-}$ skin disc culture compared with vehicle treatment (Figure 2B).

In order to investigate the adverse effects of PAPLAL, we transdermally administered PAPLAL into the skin on the backs of $Sod1^{+/+}$ mice. As shown in Figure 3, no significant differences in skin thickness were observed among the $Sod1^{+/+}$ mice treated with or without PAPLAL. In addition, no abnormalities, such as cell infiltration or PAPLAL deposition, were detected in the skin of the $Sod1^{+/+}$ mice, suggesting that PAPLAL does not have any adverse effects on the skin, at least in the short term. In the organ culture analysis, PAPLAL treatment of $Sod1^{+/+}$ skin did not induce any significant change in fibroblast outgrowth capacity or LDH activity detected in skin disc cultures (Figure 2A, 2B). These findings demonstrate that the transdermal application of PAPLAL ameliorates skin atrophy in $Sod1^{-/-}$ mice by suppressing oxidative damage.

PAPLAL normalized the transcriptional profiles of skin-related genes in the Sod1^{−/−} skin

In order to investigate the mechanisms by which PAPLAL treatment counters skin atrophy in $Sod1^{-/-}$ mice, we analyzed the expression patterns of extracellular matrix-related genes in the skin. In the $Sod1^{-/-}$ mice, the skin mRNA expression level of the type I collagen gene ($Col1a1$) was significantly reduced, while that of the matrix metalloproteinase 2 gene ($Mmp2$) were significantly increased, compared with those observed in the $Sod1^{+/+}$ mice, indicating that collagen biosynthesis was reduced in the $Sod1^{-/-}$ mice (Figure 4). Moreover, the mRNA expression of the hyaluronan synthase 2 gene ($Has2$) was significantly downregulated in the skin of the $Sod1^{-/-}$ mice (Figure 4). In contrast, the mRNA expression levels of $Decorin$ and $Ki67$ did not differ between the $Sod1^{+/+}$ and $Sod1^{-/-}$ mice (Figure 4). These results suggest that $Sod1$ deficiency causes skin thinning due to dysregulation of the extracellular matrix. Among the genes that exhibited altered expression levels, PAPLAL treatment significantly normalized the mRNA levels of $Col1a1$, $Mmp2$, and $Has2$, suggesting that PAPLAL treatment increases skin thickness by increasing the concentrations of extracellular matrix components, such as collagen and hyaluronic acid (Figure 4).

The skin of the $Sod1^{-/-}$ mice also exhibited significantly higher expression levels of inflammatory cytokines, including Tnf-α and Il-6, compared with the skin from the $Sod1^{+/+}$ mice (Figure 4). PAPLAL significantly downregulated the mRNA expression levels of Tnf-α and Il-6 in the skin of the $Sod1^{-/-}$ mice, suggesting that a pathological link exists between inflammation and skin thinning in $Sod1^{-/-}$ mice (Figure 4). Furthermore, the expression of the tumor suppresser $p53$ gene, which is known to be associated with DNA damage [30] and skin aging [31], was significantly upregulated in the $Sod1^{-/-}$ mice (Figure 4). In contrast, $Mdm2$ expression tended to be downregulated in the skin of the $Sod1^{-/-}$

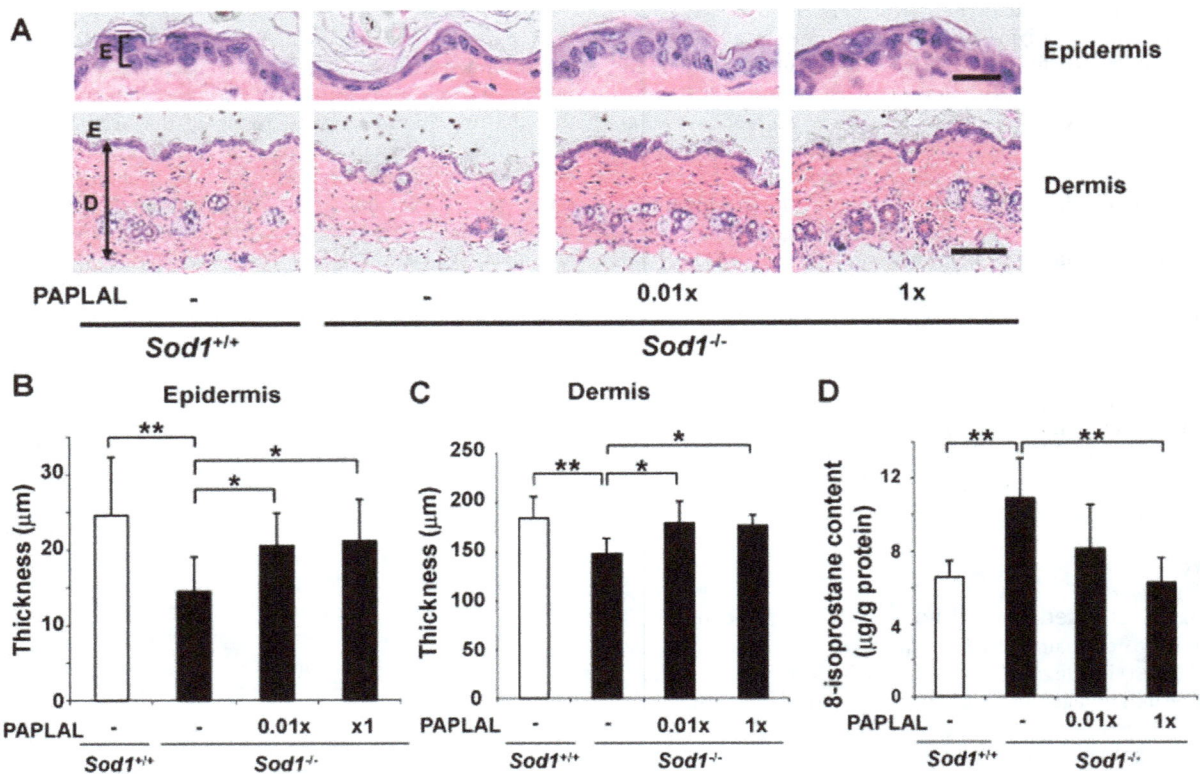

Figure 1. Protective effects of PAPLAL against skin atrophy in the *Sod1*-deficient mice. (A) Hematoxylin and eosin staining of the skin on the backs of the *Sod1*[+/+] and *Sod1*[−/−] mice (five months of age). E, epidermis; D, dermis. The scale bars represent 20 μm (top) or 100 μm (bottom). The thickness of the (B) epidermal and (C) dermal layers of the skin on the backs of *Sod1*[+/+] and *Sod1*[−/−] mice treated with PAPLAL (n = 6–8). (D) 8-isoprostane content of the skin on the backs of *Sod1*[+/+] and *Sod1*[−/−] mice treated with PAPLAL (n = 5–8). 0.01× and 1× PAPLAL indicate 0.01- or 1-fold concentrations of PAPLAL, respectively. Data are shown as the mean ± SD; *$p < 0.05$, **$p < 0.01$.

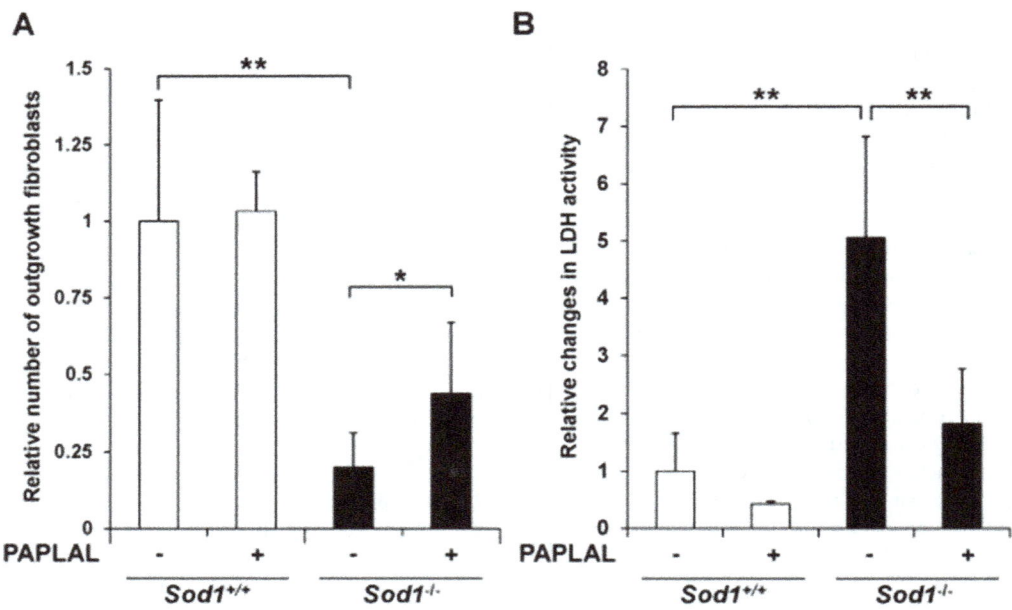

Figure 2. PAPLAL attenuates cellular damage in skin. (A) Relative number of outgrowing fibroblasts in cultured *Sod1*[+/+] and *Sod1*[−/−] skin specimens. (B) LDH activity in the medium used to culture the *Sod1*[+/+] and *Sod1*[−/−] skin specimens. Data are shown as the mean ± SD; *$p < 0.05$, **$p < 0.01$.

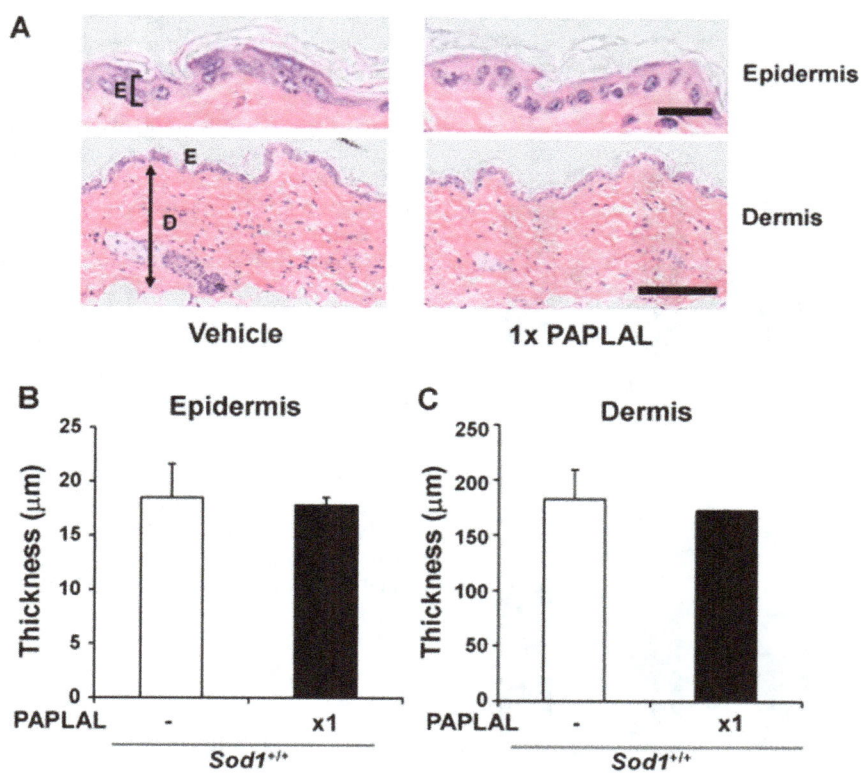

Figure 3. PAPLAL is non-toxic in wild-type mice. (A) Hematoxylin and eosin staining of the skin on the backs of $Sod1^{+/+}$ mice (17–20 weeks of age). E, epidermis; D, dermis. The scale bars represent 20 μm (top) or 100 μm (bottom). The thickness of the (B) epidermal and (C) dermal layers of the skin on the backs of $Sod1^{+/+}$ mice treated with PAPLAL (n = 5).

mice. PAPLAL treatment significantly normalized the mRNA expression level of *p53* of the $Sod1^{-/-}$ mice (Figure 4), suggesting that PAPLAL delays skin aging by inhibiting p53 upregulation in $Sod1^{-/-}$ mice.

Pd nanoparticles prevented the oxidative deterioration of Pt nanoparticles with respect to their SOD and catalase activity

Since Pt nanoparticles display strong antioxidant activity [22–24], we assessed the SOD and catalase activity of PAPLAL and its components *in vitro*. The Pd nanoparticles displayed weak SOD

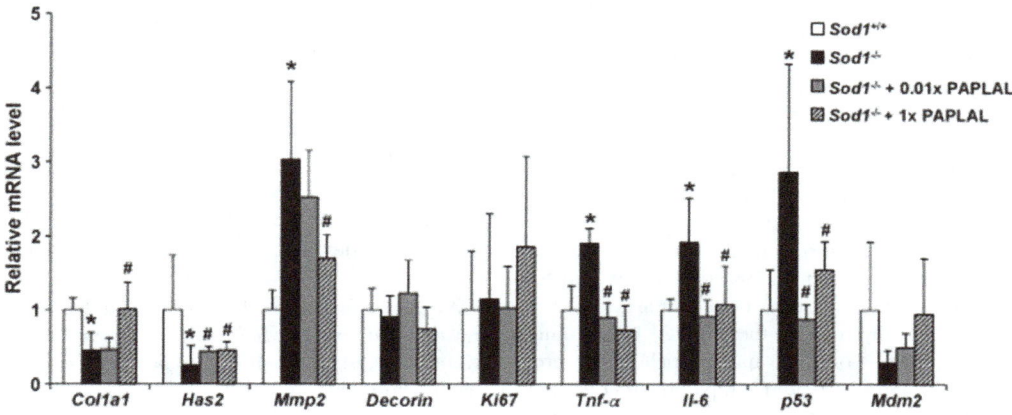

Figure 4. PAPLAL improved the transcriptional profiles of skin-related genes in the skin of $Sod1^{-/-}$ mice. (A) The relative mRNA expression levels of *Col1a1, Has2, Mmp2, Decorin, Ki67, Tnf-α, Il-6, p53,* and *Mdm2*. Each mRNA expression level was determined using qRT-PCR. Data are shown as the mean ± SD; *$p<0.05$ vs. $Sod1^{+/+}$, **$p<0.01$ vs. $Sod1^{+/+}$, #$p<0.05$ vs. $Sod1^{-/-}$.

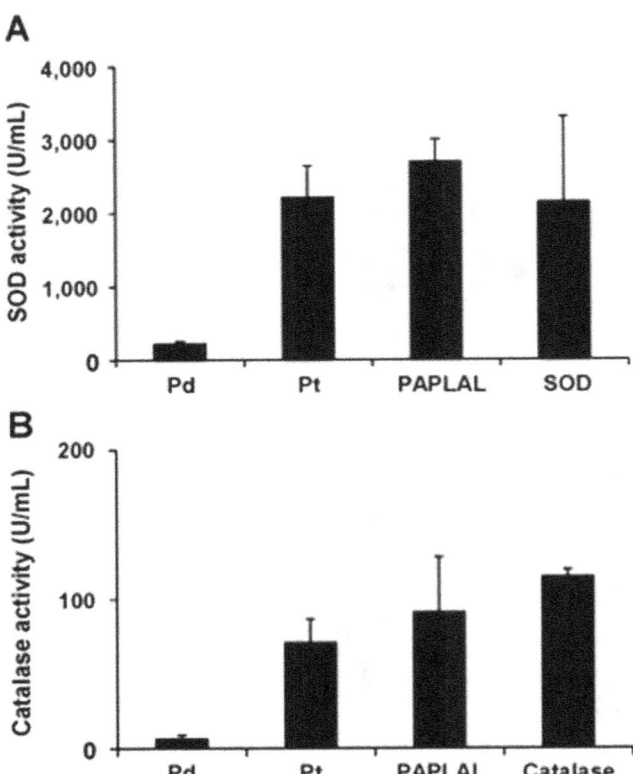

Figure 5. Pt nanoparticles possess SOD and catalase activity. PAPLAL includes 2.82 mM of Pd nanoparticles and 1.03 mM of Pt nanoparticles. (A) The SOD and (B) catalase activity of 2.82 mM Pd nanoparticles, 1.03 mM Pt nanoparticles, and PAPLAL. Five nM of SOD derived from bovine erythrocytes (A) and 0.2 μM of catalase derived from bovine liver tissue (B) were used as positive controls for SOD and catalase, respectively.

and catalase activity, while the Pt nanoparticles exhibited strong SOD and catalase activity (Figure 5A, 5B). PAPLAL, a mixture of Pd and Pt nanoparticles, demonstrated SOD and catalase activity levels that were equivalent to the sum of those exhibited by the Pd and Pt nanoparticles, although no synergistic effects were observed (Figure 5A, 5B).

In order to evaluate the stability of the SOD and catalase activity of the tested nanoparticles, Pd and Pt nanoparticles and PAPLAL were stored at room temperature for four weeks. The SOD and catalase activity levels of the stored Pt nanoparticles were dramatically reduced to 46.5% and 46.2%, respectively, of those exhibited by the freshly prepared Pt nanoparticles (Figure 6A, 6B). In contrast, the SOD and catalase activity levels of stored PAPLAL were only reduced by 22.1% and 8.7%, respectively, compared with those of the freshly prepared PAPAL (Figure 6A, 6B). In order to confirm that the reductions in the SOD and catalase activity of the Pt nanoparticles were due to oxidative deterioration during storage, Pt nanoparticles and PAPAL were exposed to air by rotating them for 24 hours. The SOD and catalase activity levels of the Pt nanoparticles that were exposed to air were significantly decreased compared with those of the freshly prepared nanoparticles (SOD and catalase: 33.0% and 39.7%, respectively; Figure 6C, 6D). In contrast, the PAPLAL exposed to air retained its SOD and catalase activity, indicating that the oxidative deterioration of the Pt nanoparticles within it had been prevented (Figure 6C, 6D).

Furthermore, Pd nanoparticles were mixed with Pt nanoparticles at various concentration ratios and then stored at room temperature for four weeks. When Pd nanoparticles were added to Pt nanoparticles at a molar ratio of 1 or 2 to 1, the SOD and catalase activity levels of the mixture were markedly reduced after their storage at room temperature (Figure 7). However, when Pd nanoparticles were added to Pt nanoparticles at a molar ratio of 3 or 4 to 1, the SOD and catalase activity levels of the mixture were sustained (Figure 7).

We previously reported that $Sod1$ loss significantly enhanced intracellular O_2^- generation in fibroblasts [9,32]. Therefore, we assessed the antioxidant effects of Pd and Pt nanoparticles on $Sod1^{-/-}$ fibroblasts. $Sod1^{-/-}$ fibroblasts were treated with Pd and/or Pt nanoparticles for 16 hours, and then the levels of intracellular O_2^- were assessed. O_2^- generation was significantly decreased (by 23.3%) in the $Sod1^{-/-}$ fibroblasts treated with 10 μM of Pt nanoparticles, but not in those treated with Pd nanoparticles (Figure S2). Notably, treatment with pre-incubated Pt nanoparticles did not decrease intracellular O_2^- generation in the $Sod1^{-/-}$ fibroblasts (Figure S2). These results suggest that Pt nanoparticles possess strong antioxidant effects, such as SOD and catalase activity, and that the Pd in PALAL inhibits the oxidative deterioration of Pt, which enables PALAL to retain strong antioxidant activity over time.

Discussion

PAPLAL attenuates intrinsic skin aging by suppressing oxidative damage

In the present study, we demonstrated that PAPLAL significantly reversed skin thinning by reducing oxidative and cellular damage in $Sod1^{-/-}$ mice (Figures 1 and 2). In addition, PAPLAL and Pt nanoparticles, but not Pd nanoparticles, exhibited SOD and catalase activity (Figure 5). Furthermore, an *in vitro* experiment found that treatment with Pt nanoparticles, but not Pd nanoparticles, significantly reduced O_2^- generation in $Sod1^{-/-}$ fibroblasts (Figure S2). These findings suggest that the antioxidant activity of the Pt nanoparticles in PAPLAL contribute to attenuating age-related skin thinning in $Sod1^{-/-}$ mice. In this context, we previously evaluated the ability of several antioxidants to counteract the *in vivo* age-related changes seen in $Sod1^{-/-}$ mice and found that the administration of ascorbic acid significantly attenuated bone loss and fragility in $Sod1^{-/-}$ mice [12]. Likewise, the transdermal administration of ascorbic acid derivatives was demonstrated to normalize skin thinning in $Sod1^{-/-}$ mice [9,10]. Furthermore, Iuchi *et al.* reported that oral N-acetylcysteine treatment mitigates hemolytic anemia in $Sod1^{-/-}$ mice by suppressing ROS generation in red blood cells [33]. Together, these findings demonstrate that antioxidants, such as PAPLAL, ascorbic acid, and N-acetylcysteine, can improve $Sod1$ loss-induced organ pathologies.

PAPLAL normalizes gene expression, including that related to matrix biosynthesis and inflammation, in the skin

As shown in Figure 4, $Sod1$ loss induced the transcriptional downregulation of $Col1a1$ and $Has2$, as well as the upregulation of $Mmp2$ expression, which were indicative of collagen and hyaluronic acid malformation in the atrophic skin of the $Sod1^{-/-}$ mice. $Sod1$ loss also upregulated the expression of proinflammatory genes, such as Tnf-α and Il-6, in the skin (Figure 4). Tumor necrosis factor (TNF)- α regulates type I collagen expression via the c-Jun N-terminal kinase (JNK) and nuclear factor kappa-light-chain-enhancer of activated B cells (NF-κB) pathways in skin fibroblasts

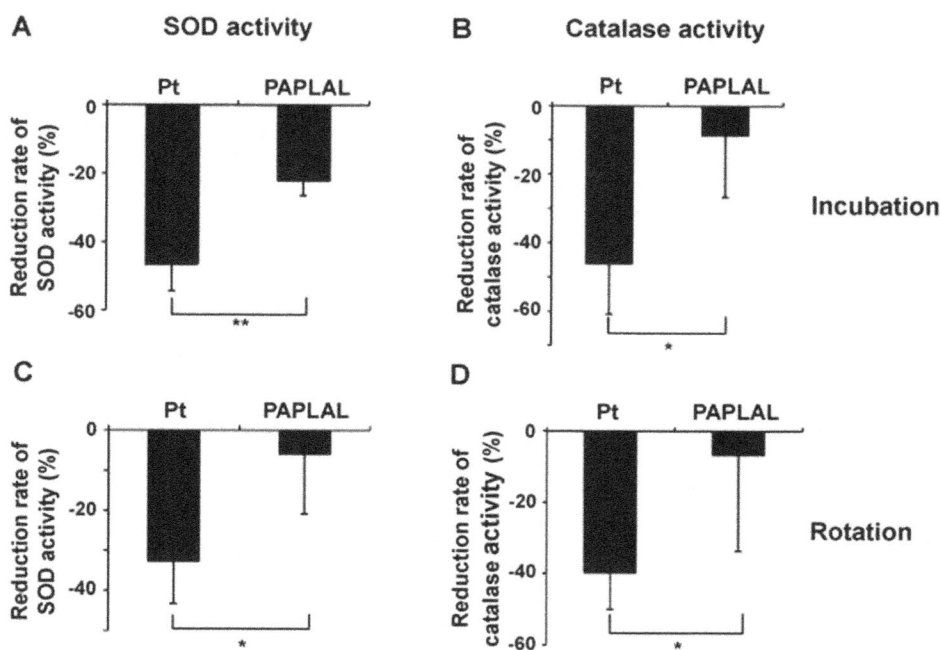

Figure 6. Pd nanoparticles protected the SOD and catalase activity of Pt nanoparticles against oxidative degradation *in vitro*. (A) The SOD and (B) catalase activity of 1.03 mM Pt nanoparticles and PAPLAL that had been stored at room temperature for four weeks. (C) The SOD and (D) catalase activity of Pt nanoparticles and PAPLAL that had been rotated for 24 hours in a tube in order to oxidize the nanoparticles. Data are shown as the mean ± SD; *p<0.05, **p<0.01.

[34]. In addition, Galera *et al.* reported that NF-κB directly suppresses *COL1A1* gene transcription in human dermal fibroblasts [35] and accumulates in the nuclei of aged human fibroblasts in association with the downregulation of the *COL1A1* gene [36]. An IκB kinase (IKK)-β inhibitor has also been shown to suppress interleukin (IL)-1β-induced collagen degradation by inhibiting the activation of NF-κB and upregulation of matrix metalloproteinases [37]. Taken together, the inflammatory response controls collagen homeostasis via transcriptional mechanisms in fibroblasts. In the present study, PAPLAL treatment normalized the gene expression of *Col1a1* and *Tnf-α* in the skin of the $Sod1^{-/-}$ mice (Figure 4). With respect to other organs, Onizawa *et al.* reported that the

intranasal administration of Pt nanoparticles reduced NF-κB activity and inhibited pulmonary inflammation in mice exposed to cigarette smoke [38]. Rehman *et al.* also reported that Pt nanoparticles have anti-inflammatory effects on the lipopolysaccharide-induced inflammatory response by downregulating the expression of IL-1β, TNF-α, and IL-6 in macrophages [39]. Collectively, these findings suggest that PAPLAL and Pt nanoparticles suppress the inflammatory response, resulting in improvements in the anabolic and catabolic regulation of collagen homeostasis.

Pd nanoparticles stabilize the antioxidant activity of Pt nanoparticles by preventing their oxidative deterioration

Noble metal nanoparticles, such as Pd, Pt, and gold, are considered to function as antioxidants by reducing catalysis [19–21]. A number of studies have reported that Pt nanoparticles exhibit a strong ability to scavenge ROS, including O_2^- and H_2O_2 [22–24]. In a lifespan analysis of *C. elegans*, Kim *et al.* reported that Pt nanoparticles extended the lifespans of wild-type N2 and short-lived *mev-1* nematodes, in which intracellular ROS accumulated due to respiratory impairment [40]. The present results demonstrated that Pt nanoparticles, but not Pd nanoparticles, possess SOD and catalase activity (Figure 5), which is consistent with the above results. Okamoto *et al.* reported that Pt nanoparticles were oxidized to PtO by oxygen in the air, which resulted in an time-dependent increase in their ability to degrade ascorbic acid [41]. However, the co-incubation of Pd nanoparticles with Pt nanoparticles effectively prevented PtO formation via Pt nanoparticles oxidation [41]. Since Pd has a lower oxidation/reduction potential than Pt, they proposed that Pd reduces Pt^{2+} to Pt in solution [41]. In order to further investigate the ability of Pd nanoparticles to prevent the oxidative deterioration of Pt, we herein examined the SOD and catalase activity of Pt nanoparticles

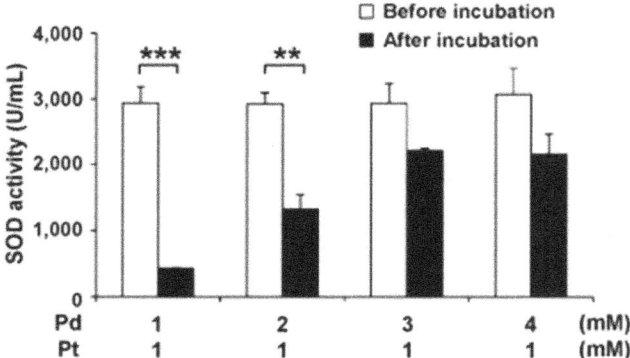

Figure 7. Pd nanoparticles protected the SOD activity of Pt nanoparticles against oxidative degradation at various molar ratios. Pd nanoparticles were added to Pt nanoparticles at various molar ratios, and the SOD activity of each mixture was measured after four weeks of storage at room temperature.

after storage- or rotation-induced oxidization in the presence of Pd nanoparticles (Figure 5). Predictably, the co-storage of Pt nanoparticles with Pd nanoparticles delayed the oxidative inactivation of the SOD and catalase activity of the former nanoparticles, while in the absence of Pd nanoparticles the Pt nanoparticles were inactivated during incubation or rotation (Figure 6). Indeed, Pt nanoparticles that had been oxidized in air failed to suppress intracellular O_2- generation in $Sod1^{-/-}$ cells (Figure S2). In contrast, even after oxidation in air, PAPLAL, a mixture of Pd and Pt nanoparticles, continued to exhibit SOD and catalase activity, and hence, was able to decrease O_2- generation in $Sod1^{-/-}$ fibroblasts (Figure S2).

Notably, when Pd nanoparticles were added to Pt nanoparticles at a molar ratio of 3 or 4 to 1, the SOD and catalase activity levels of the Pt nanoparticles were sustained more effectively than when a molar ratio of 1 or 2 to 1 was employed (Figure 7). In fact, PAPAPL contains Pd and Pt nanoparticles at a molar ratio of 2.74 to 1, suggesting that the excess Pd nanoparticles in PAPLAL effectively protect the Pt nanoparticles from oxidative deterioration. Since long-term storage accelerates the oxidative deterioration of antioxidants in clinical use, the addition of Pd nanoparticles to such antioxidants might efficiently maintain their bioactivity under oxidative conditions.

PAPLAL is a valuable antioxidant for delaying skin aging

In the present study, we directly applied PAPLAL to the skin of $Sod1^{-/-}$ mice. PAPLAL treatment effectively improved the skin thinning seen in the $Sod1^{-/-}$ mice, suggesting that the PAPLAL had been incorporated into epidermal and dermal cells. In a recently electron microscopic study, Okamoto et al. reported that the Pd and Pt nanoparticles in PAPLAL measure 3.59 ± 0.56 and 1.93 ± 0.34 nm, respectively, in diameter [41]. A previous study found that cells that had been treated with Pt nanoparticles exhibited increased Pt concentrations compared with the control cells [24]. In an inductively coupled plasma mass spectrometry-based study of C. elegans, Sakaue et al. reported that treatment with Pt nanoparticles increased the internalization of Pt in nematodes [42]. These results suggest that PAPLAL and/or Pt nanoparticles are transdermally taken in by cells, which inhibits the progression of skin pathologies caused by oxidative stress in mice.

In the present study, PAPLAL treatment did not cause any morphological abnormalities, such as cell infiltration or PAPLAL deposition, or cellular damage in mouse skin (Figures 2, 3). Furthermore, PAPLAL has been shown to be free from adverse effects during its clinical use as a treatment for chronic disease in humans [26]. In agreement with our results, several reports have found that treatment with Pt nanoparticles did not induce any alterations in the biological profiles of wild-type mice [38,43]. On the other hand, Newkirk et al. recently reported that the oral administration of a mixture of Pd, Pt, and rhodium had a synergistic toxic effect on eosinophils in rats [44]. Further research on the dynamic state and safety of PAPLAL in the living body is required.

Vitiligo is an acquired depigmentation disorder characterized by H_2O_2/peroxynitrite-mediated oxidative and nitrative stress in the skin. Salem et al. reported that treatment with PC-KUS, a UVB-activated pseudocatalase, reduces the levels of epidermal H_2O_2 and induces repigmentation in vitiligo patients [45,46]. Since PAPLAL possesses catalase activity, it might also be useful as a treatment for H_2O_2-related skin diseases, including vitiligo.

In conclusion, PAPLAL, which is composed of Pd and Pt nanoparticles at a molar ratio of 2.74 to 1, exhibits potent antioxidant activity (attributed to the effects of Pt nanoparticles) and attenuates aging-related skin pathologies in vivo. The Pd nanoparticles contained in PAPLAL prevent the oxidative deterioration of the Pt nanoparticles by attenuating PtO formation. PAPLAL has been found to have few adverse effects on skin morphology in transdermally treated mice. Consistent with these results, no previous studies have found that PAPLAL induces adverse effects, despite it being used in Japan to treat patients with chronic conditions for over 60 years. Therefore, PAPLAL is considered to be a safe and valuable antioxidant for delaying skin aging in humans.

Supporting Information

Figure S1   PAPLAL improves wound healing in aged mice. (A) Typical pictures of aged C57BL/6 male mice that were treated with or without PAPLAL at six days after wounding. (B) Aged C57BL/6 male mice (17 months of age) were wounded on day 0 and treated with or without PAPLAL for 12 days. Wound size was measured over time. Data are shown as the mean $\pm$ SD; $*p < 0.05$.

Figure S2   PAPLAL suppresses O_2- production in $Sod1$-deficient fibroblasts. $Sod1^{-/-}$ dermal fibroblasts were treated with 10 μM of Pt nanoparticles, 10 μM of Pd nanoparticles, or PAPLAL for 16 hours. Intracellular superoxide generation was detected using a DHE fluorescent probe and calculated as the area of DHE-based fluorescence divided by the number of Hoechst-positive cells. Data are shown as the mean $\pm$ SD. $*p < 0.05$, $**p < 0.01$.

Table S1   qRT-PCR primers

Methods S1   Supporting Methods.

Acknowledgments

We thank Dr. Hirofumi Koyama, Dr. Daichi Morikawa, Dr. Keiji Kobayashi, Masato Koike, Isao Masuda, and Ayami Goda (Chiba University) for their valuable technical assistance.

Author Contributions

Conceived and designed the experiments: SS TS. Performed the experiments: SS YO KW NI TT. Analyzed the data: SS TS. Wrote the paper: SS KY TS.

References

1. Poljsak B, Dahmane RG, Godic A (2012) Intrinsic skin aging: the role of oxidative stress. Acta Dermatovenerol Alp Panonica Adriat 21: 33–36.
2. Shuster S, Black MM, McVitie E (1975) The influence of age and sex on skin thickness, skin collagen and density. Br J Dermatol 93: 639–643.
3. Naylor EC, Watson RE, Sherratt MJ (2011) Molecular aspects of skin ageing. Maturitas 69: 249–256.
4. Finkel T, Holbrook NJ (2000) Oxidants, oxidative stress and the biology of ageing. Nature 408: 239–247.
5. Imamura Y, Noda S, Hashizume K, Shinoda K, Yamaguchi M, et al. (2006) Drusen, choroidal neovascularization, and retinal pigment epithelium dysfunction in SOD1-deficient mice: a model of age-related macular degeneration. Proc Natl Acad Sci U S A 103: 11282–11287.

6. Uchiyama S, Shimizu T, Shirasawa T (2006) CuZn-SOD deficiency causes ApoB degradation and induces hepatic lipid accumulation by impaired lipoprotein secretion in mice. J Biol Chem 281: 31713–31719.

7. Kondo Y, Masutomi H, Noda Y, Ozawa Y, Takahashi K, et al. (2014) Senescence marker protein-30/superoxide dismutase 1 double knockout mice exhibit increased oxidative stress and hepatic steatosis. FEBS Open Bio 4: 522–532.

8. Murakami K, Inagaki J, Saito M, Ikeda Y, Tsuda C, et al. (2009) Skin atrophy in cytoplasmic SOD-deficient mice and its complete recovery using a vitamin C derivative. Biochem Biophys Res Commun 382: 457–461.

9. Shibuya S, Kinoshita K, Shimizu T (2011) Protective effects of vitamin C derivatives on skin atrophy caused by Sod1 deficiency. In: Preedy BS, editor. Handbook of diet, nutrition and the skin. The Netherlands: Wageningen Academic. pp. 351–364.

10. Shibuya S, Nojiri H, Morikawa D, Koyama H, Shimizu T (2014) Protective effects of vitamin C on age-related bone and skin phenotypes caused by intracellular reactive oxygen species. In: Preedy BS, editor. Aging: Oxidative stress and dietary antioxidants. New York: Academic Press. pp. 152–159.

11. Shibuya S, Ozawa Y, Toda T, Watanabe K, Tometsuka C, et al. (2014) Collagen peptide and vitamin C additively attenuate age-related skin atrophy in Sod1-deficient mice. Biosci Biotech Biochem 78: 1212–1220.

12. Nojiri H, Saita Y, Morikawa D, Kobayashi K, Tsuda C, et al. (2011) Cytoplasmic superoxide causes bone fragility owing to low-turnover osteoporosis and impaired collagen cross-linking. J Bone Miner Res 26: 2682–2694.

13. Morikawa D, Nojiri H, Saita Y, Kobayashi K, Watanabe K, et al. (2013) Cytoplasmic reactive oxygen species and SOD1 regulate bone mass during mechanical unloading. J Bone Miner Res 28: 2368–2380.

14. Murakami K, Murata N, Noda Y, Tahara S, Kaneko T, et al. (2011) SOD1 (copper/zinc superoxide dismutase) deficiency drives amyloid beta protein oligomerization and memory loss in mouse model of Alzheimer disease. J Biol Chem 286: 44557–44568.

15. Noda Y, Ota K, Shirasawa T, Shimizu T (2012) Copper/zinc superoxide dismutase insufficiency impairs progesterone secretion and fertility in female mice. Biol Reprod 86: 1–8.

16. Kojima T, Wakamatsu TH, Dogru M, Ogawa Y, Igarashi A, et al. (2012) Age-related dysfunction of the lacrimal gland and oxidative stress: evidence from the Cu,Zn-superoxide dismutase-1 (Sod1) knockout mice. Am J Pathol 180: 1879–1896.

17. Ibrahim OM, Dogru M, Matsumoto Y, Igarashi A, Kojima T, et al. (2014) Oxidative stress induced age dependent meibomian gland dysfunction in Cu, Zn-superoxide dismutase-1 (Sod1) knockout mice. PLoS One 9: e99328.

18. Morikawa D, Itoigawa Y, Nojiri H, Sano H, Itoi E, et al. (2014) Contribution of oxidative stress to the degeneration of rotator cuff entheses. J Shoulder Elbow Surg 23: 628–635.

19. Lewis L, Lewis N (1986) Platinum-Catalyzed Hydrosilylation-Colloid Formation as the Essential Step. J Am Chem Soc 108: 7228–7231.

20. Toshima N, Yonezawa T (1998) Bimetallic nanoparticles-novel mterials for chemical and physical applications. New J Chem 22: 1179–1201.

21. Roucoux A, Schulz J, Patin H (2002) Reduced transition metal colloids: a novel family of reusable catalysts? Chem Rev 102: 3757–3778.

22. Yoshihisa Y, Honda A, Zhao QL, Makino T, Abe R, et al. (2010) Protective effects of platinum nanoparticles against UV-light-induced epidermal inflammation. Exp Dermatol 19: 1000–1006.

23. Kajita M, Hikosaka K, Iitsuka M, Kanayama A, Toshima N, et al. (2007) Platinum nanoparticle is a useful scavenger of superoxide anion and hydrogen peroxide. Free Radic Res 41: 615–626.

24. Yoshihisa Y, Zhao QL, Hassan MA, Wei ZL, Furuichi M, et al. (2011) SOD/catalase mimetic platinum nanoparticles inhibit heat-induced apoptosis in human lymphoma U937 and HH cells. Free Radic Res 45: 326–335.

25. Elhusseiny AF, Hassan HH (2013) Antimicrobial and antitumor activity of platinum and palladium complexes of novel spherical aramides nanoparticles containing flexibilizing linkages: structure-property relationship. Spectrochim Acta A Mol Biomol Spectrosc 103: 232–245.

26. Ishizuka S (1956) In: Ishizuka S, editor. Generation of PAPLAL. Japan: Juseikai. In Japanese.

27. Tajima K, Watabe R, Kanaori K (2005) Antioxidant activity of PAPLAL a colloidal mixture of Pt and Pd metal to superoxide anion radical as studied by quantitative spin trapping ESR measurements. Clinical Phamacology and Therapy 15: 635–642. In Japanese.

28. Tajima K, Sakurai Y, Oishi T, Kanaori K, Morimoto K, et al. (2009) Chemical reactivity of Pd-, and Pt-colloid Involved in PAPLAL to solvated oxygen and hydroxyl radical. Clinical Phamacology and Therapy 19: 397–404. In Japanese.

29. Nojiri H, Shimizu T, Funakoshi M, Yamaguchi O, Zhou H, et al. (2006) Oxidative stress causes heart failure with impaired mitochondrial respiration. J Biol Chem 281: 33789–33801.

30. Lopez-Otin C, Blasco MA, Partridge L, Serrano M, Kroemer G (2013) The hallmarks of aging. Cell 153: 1194–1217.

31. Tyner SD, Venkatachalam S, Choi J, Jones S, Ghebranious N, et al. (2002) p53 mutant mice that display early ageing-associated phenotypes. Nature 415: 45–53.

32. Watanabe K, Shibuya S, Koyama H, Ozawa Y, Toda T, et al. (2013) Sod1 Loss Induces Intrinsic Superoxide Accumulation Leading to p53-Mediated Growth Arrest and Apoptosis. Int J Mol Sci 14: 10998–11010.

33. Iuchi Y, Okada F, Onuma K, Onoda T, Asao H, et al. (2007) Elevated oxidative stress in erythrocytes due to a SOD1 deficiency causes anaemia and triggers autoantibody production. Biochem J 402: 219–227.

34. Verrecchia F, Mauviel A (2004) TGF-beta and TNF-alpha: antagonistic cytokines controlling type I collagen gene expression. Cell Signal 16: 873–880.

35. Beauchef G, Bigot N, Kypriotou M, Renard E, Poree B, et al. (2012) The p65 subunit of NF-kappaB inhibits COL1A1 gene transcription in human dermal and scleroderma fibroblasts through its recruitment on promoter by protein interaction with transcriptional activators (c-Krox, Sp1, and Sp3). J Biol Chem 287: 3462–3478.

36. Bigot N, Beauchef G, Hervieu M, Oddos T, Demoor M, et al. (2012) NF-kappaB accumulation associated with COL1A1 transactivators defects during chronological aging represses type I collagen expression through a -112/-61-bp region of the COL1A1 promoter in human skin fibroblasts. J Invest Dermatol 132: 2360–2367.

37. Kondo Y, Fukuda K, Adachi T, Nishida T (2008) Inhibition by a selective IkappaB kinase-2 inhibitor of interleukin-1-induced collagen degradation by corneal fibroblasts in three-dimensional culture. Invest Ophthalmol Vis Sci 49: 4850–4857.

38. Onizawa S, Aoshiba K, Kajita M, Miyamoto Y, Nagai A (2009) Platinum nanoparticle antioxidants inhibit pulmonary inflammation in mice exposed to cigarette smoke. Pulm Pharmacol Ther 22: 340–349.

39. Rehman MU, Yoshihisa Y, Miyamoto Y, Shimizu T (2012) The anti-inflammatory effects of platinum nanoparticles on the lipopolysaccharide-induced inflammatory response in RAW 264.7 macrophages. Inflamm Res 61: 1177–1185.

40. Kim J, Takahashi M, Shimizu T, Shirasawa T, Kajita M, et al. (2008) Effects of a potent antioxidant, platinum nanoparticle, on the lifespan of Caenorhabditis elegans. Mech Ageing Dev 129: 322–331.

41. Okamoto H, Horii K, Fujisawa A, Yamamoto Y (2012) Oxidative deterioration of platinum nanoparticle and its prevention by palladium. Exp Dermatol 21: 5–7.

42. Sakaue Y, Kim J, Miyamoto Y (2010) Effects of TAT-conjugated platinum nanoparticles on lifespan of mitochondrial electron transport complex I-deficient Caenorhabditis elegans, nuo-1. Int J Nanomedicine 5: 687–695.

43. Takamiya M, Miyamoto Y, Yamashita T, Deguchi K, Ohta Y, et al. (2012) Strong neuroprotection with a novel platinum nanoparticle against ischemic stroke- and tissue plasminogen activator-related brain damages in mice. Neuroscience 221: 47–55.

44. Newkirk CE, Gagnon ZE, Pavel Sizemore IE (2014) Comparative study of hematological responses to platinum group metals, antimony and silver nanoparticles in animal models. J Environ Sci Health 49: 269–280.

45. Salem MM, Shalbaf M, Gibbons NC, Chavan B, Thornton JM, et al. (2009) Enhanced DNA binding capacity on up-regulated epidermal wild-type p53 in vitiligo by H_2O_2-mediated oxidation: a possible repair mechanism for DNA damage. FASEB J 23: 3790–3807.

46. Schallreuter KU, Salem MA, Holtz S, Panske A (2013) Basic evidence for epidermal H_2O_2/ONOO(−)-mediated oxidation/nitration in segmental vitiligo is supported by repigmentation of skin and eyelashes after reduction of epidermal H_2O_2 with topical NB-UVB-activated pseudocatalase PC-KUS. FASEB J 27: 3113–3122.

13

PO$_2$ Cycling Reduces Diaphragm Fatigue by Attenuating ROS Formation

Li Zuo[1,3]*, Philip T. Diaz[6], Michael T. Chien[5], William J. Roberts[1,3], Juliana Kishek[3], Thomas M. Best[4], Peter D. Wagner[2]

1 Radiologic Sciences and Respiratory Therapy Division, School of Health and Rehabilitation Sciences, Davis Heart and Lung Research Institute, The Ohio State University College of Medicine, Columbus, Ohio, United States of America, 2 Department of Medicine, University of California San Diego, La Jolla, California, United States of America, 3 Department of Biological Sciences, Oakland University, Rochester, Michigan, United States of America, 4 Division of Sports Medicine, Department of Family Medicine, Sports Health and Performance Institute, The Ohio State University, Columbus, Ohio, United States of America, 5 Department of Biology, Kalamazoo College, Kalamazoo, Michigan, United States of America, 6 Division of Pulmonary, Allergy, Critical Care & Sleep Medicine, The Ohio State University Wexner Medical Center, Columbus, Ohio, United States of America

Abstract

Prolonged muscle exposure to low PO$_2$ conditions may cause oxidative stress resulting in severe muscular injuries. We hypothesize that PO$_2$ cycling preconditioning, which involves brief cycles of diaphragmatic muscle exposure to a low oxygen level (40 Torr) followed by a high oxygen level (550 Torr), can reduce intracellular reactive oxygen species (ROS) as well as attenuate muscle fatigue in mouse diaphragm under low PO$_2$. Accordingly, dihydrofluorescein (a fluorescent probe) was used to monitor muscular ROS production in real time with confocal microscopy during a lower PO$_2$ condition. In the control group with no PO$_2$ cycling, intracellular ROS formation did not appear during the first 15 min of the low PO$_2$ period. However, after 20 min of low PO$_2$, ROS levels increased significantly by ~30% compared to baseline, and this increase continued until the end of the 30 min low PO$_2$ condition. Conversely, muscles treated with PO$_2$ cycling showed a complete absence of enhanced fluorescence emission throughout the entire low PO$_2$ period. Furthermore, PO$_2$ cycling-treated diaphragm exhibited increased fatigue resistance during prolonged low PO$_2$ period compared to control. Thus, our data suggest that PO$_2$ cycling mitigates diaphragm fatigue during prolonged low PO$_2$. Although the exact mechanism for this protection remains to be elucidated, it is likely that through limiting excessive ROS levels, PO$_2$ cycling initiates ROS-related antioxidant defenses.

Editor: Xiao Su, Chinese Academy of Sciences, China

Funding: Funding provided by Oakland University General Fund G110, Research Excellence Fund of Biomedical Research, and The Ohio State University-Health and Rehabilitation Sciences fund 013000. The funders had no role in study design, data collection and analysis, decision to publish, or preparation of the manuscript.

Competing Interests: The authors have declared that no competing interests exist.

* Email: zuo.4@osu.edu

Introduction

Low oxygen/hypoxic conditions can significantly reduce skeletal muscle contraction [1]. In normal resting muscle, it has been reported that skeletal muscles, such as the diaphragm, produce reactive oxygen species (ROS) including hydrogen peroxide (H$_2$O$_2$) [2]. However, when the diaphragm is repetitively stimulated, these muscle fibers generate excessive ROS leading to oxidative stress with accelerated fatigue development [2]. Moreover, the production of ATP is driven by electron transmission through mitochondrial complex I to complex IV, creating a proton gradient across the inner mitochondrial membrane (IMM) and triggering ATP synthesis [3,4]. Through this mechanism, a small portion of electrons may leak out of the IMM and react with adjacent oxygen molecules to produce superoxide anions, H$_2$O$_2$, and other ROS. Under prolonged low PO$_2$ conditions, the physiological concentration of O$_2$ is altered which results in increased uncoupling between O$_2$ and electron flow, ultimately causing ROS overproduction [5].

A variety of cellular preconditioning pathways associated with muscular protection have been proposed. For instance, ischemic preconditioning (IPC), which consists of ischemic-reperfusion cycles produced by variations in low-high PO$_2$ levels, has been used to prevent cardiac muscle injuries [6]. In addition, IPC initiates intracellular protein kinase pathways, resulting in increased activation of antioxidant enzymes such as catalase [6]. IPC also plays a critical role in protecting the heart against ischemia-reperfusion injuries by opening mitochondrial ATP sensitive potassium channels (mK$_{ATP}$) [7]. The mK$_{ATP}$ channels are regulated by several factors, including adenosine, H$^+$, and/or protein kinase C. Thus, these mediators may partially contribute to the protective response involved in preconditioning therapies [8,9]. Similar to IPC, PO$_2$ cycling preconditioning, which consists of brief periods of lower-higher PO$_2$, significantly protects heart muscle cells subjected to prolonged ischemia by decreasing ROS-induced cell death [7,10]. In addition, human studies have shown that intermittent low oxygen exposure at low altitude significantly increases an aircraft crew's adaptation to low oxygen conditions experienced at high altitude [11]. Since the method of both IPC and PO_2 cycling preconditioning involves brief periods of low and high oxygen levels, it is possible that PO_2 cycling follows a similar molecular pathway as IPC. Furthermore, it has been shown that a protocol consisting of PO_2 cycling provides a protective response against mesenchymal stem cell (MSC) apoptosis through phosphorylation of extracellular regulated kinase (ERK1/2) and protein kinase B (AKT) [12]. Therefore, it is possible that these signaling factors also may be involved in the molecular mechanism of PO_2 cycling in skeletal muscle.

Moreover, lower PO_2 or hypoxic conditions may cause changes in the cytosolic redox equilibrium, resulting in a rise in NADPH. The increase subsequently stimulates inositol triphosphate (IP3) receptor mediated release of Ca^{2+} from the endoplasmic reticulum. This release of Ca^{2+} activates important cell survival signaling pathways, which may potentially contribute to the preconditioning response during lower PO_2 stress [13]. However, the redox mechanism of PO_2 cycling preconditioning particularly in respiratory skeletal muscle has not been fully elucidated. The ultimate importance of the work is to develop treatments for those who may experience respiratory muscle fatigue. It is likely that PO_2 cycling initiates ROS-related protective responses, particularly in a key muscle of respiration such as the diaphragm, which must be active throughout life [14].

In this study, we tested the hypothesis that PO_2 cycling preconditioning decreases intramuscular ROS levels and enhances diaphragm muscle function. Our results demonstrate that PO_2 cycling effectively reduces diaphragm fatigue during a prolonged low PO_2 (40 Torr) condition, which is accompanied by decreased intracellular ROS levels. These findings provide insight into the molecular redox mechanism of PO_2 cycling in diaphragmatic skeletal muscle exposed to a lower PO_2 environment.

Materials and Methods

Animals

Male adult C57BL/6 mice ($\sim$20–30 g, average age of $\sim$3 mo.) were used in accordance with the Ohio State University's and Oakland University's Institutional Laboratory Animal Care and Use Committee (IACUC). We strictly followed the Guide for the Care and Use of Laboratory Animals of the National Institutes of Health and Ethics of Animal Experiments. Mice were anesthetized via intraperitoneal (IP) injection with a combination of ketamine (70 mg/kg) and xylazine (10 mg/kg). The diaphragm was quickly removed from the mouse and muscle strips ($\sim$0.5 cm wide, $\sim$1 cm long, 1–2 strips obtained from each mouse) were dissected from the diaphragm with the corresponding rib attachment and central tendon. After isolation, the muscle strip was kept in Ringer's solution (in mM: 21 $NaHCO_3$, 1.0 $MgCl_2$, 1.2 Na_2HPO_4, 0.9 Na_2SO_4, 2.0 $CaCl_2$, 5.9 KCl, 121 NaCl, and 11.5 glucose), at 37°C.

PO_2 cycling and muscle function measurement

Function experiments were performed in a contraction chamber (model 800 MS; Danish Myo Technology, Denmark), with the central tendon of the muscle strip sutured to a mobile lever, which was used to adjust the muscle length for optimal performance. The opposite end of the strip was secured to a force transducer (detection range 0–1,600 mN). After being mounted, muscle optimal length (L_0) was set as the baseline tension and no adjustments were made throughout the muscle function experiments. All muscles were electrically stimulated (S48 stimulator; Grass Technologies, RI) using square-wave pulses (250-ms train duration, 0.5-ms pulse duration, 70 Hz, 30 V), following previous

skeletal muscle function protocols [15,16]. The A-D board (model ML826; AD instruments, CO) converted the analog signals to digital data, and LabChart 7.3.1 software was used to analyze the function data. The muscle was equilibrated in Ringer's solution for 20–30 min. During the function experiments, the treated muscle strips were switched to a Ringer's solution equilibrated with PO_2 of 40 Torr O_2 (lower) for 2 min, followed by PO_2 of 550 Torr O_2 (higher) for 2 min. This PO_2 cycle was repeated five times, followed by a prolonged 30-min 40 Torr PO_2 period. During this exposure, muscle contractility was evaluated, in order to determine the effect of PO_2 cycling on the muscle function. The chamber solution during lower PO_2 was found to be 40 Torr and during higher PO_2, 550 Torr. In the middle of the 40 Torr PO_2 period (from 15–20 min), the muscle was stimulated for 5 min at 37°C and muscle tension development was recorded. The control group followed an identical protocol as the experimental group except for the one intervention of PO_2 cycling. Following the removal of the attached rib bone and excess tendon, the diaphragm was first air dried which was followed by a 30-min oven drying. The dry mass was then determined using an analytical balance. To reduce random effects due to animal variance, all function data were normalized by dry weight of the muscle strip (mN/mg).

Regarding the H_2O_2 treatment with PO_2 cycling group for low PO_2 contraction measurements, the muscle strips were treated the same as above except that after PO_2 cycling, we added H_2O_2 into the muscle contraction solution. Although H_2O_2 may degrade rapidly, at sufficient levels, it can enter the cell freely and affect the intracellular activity [17]. Specifically, the muscle was loaded with Ringer's solution with adequate H_2O_2 (50 μM) for 15 min prior to the 5-min contractions in low PO_2 conditions. In addition, the time to reach 50% (T_{50}) of the initial tension in contracting diaphragm muscle during a 5-min low PO_2 contraction period was recorded in control, PO_2 cycling, and PO_2 cycling + H_2O_2 groups.

The effect of varying H_2O_2 dosage on muscle tension development was evaluated. The muscle strips were prepared in the same manner as mentioned above. During a high PO_2 (550 Torr) period, each muscle strip was equilibrated with Ringer's solution for 15 min followed by a 15 min incubation with a particular dosage of H_2O_2 (0 μM, 50 μM, 100 μM, 1 mM, to 10 mM, respectively). After incubation, each diaphragm strip was stimulated for 5 min at 37°C and the muscle tension development was recorded.

For the H_2O_2 treatment group with no PO_2 cycling for high PO_2 contraction measurements, the muscle strips were exposed to high PO_2 (550 Torr). Each muscle strip had two independent 5-min contraction periods, which were separated by a 60-min rest period. The muscle was loaded with H_2O_2 (50 μM) for 15 min followed by the first 5-min contractile period. The H_2O_2 was then washed out with fresh Ringer's solution, and the muscle was kept for a 60-min rest period before a second 5-min contraction bout in the absence of H_2O_2. This protocol was also performed in a blocked order to ensure the statistical value.

Confocal studies

To analyze the effects of PO_2 cycling treatment on ROS levels in superfused diaphragm, confocal microscopy was used to measure real-time ROS (H_2O_2) production in both PO_2 cycling-treated and control diaphragm tissue. Specifically, each muscle strip was loaded with a 40 μM solution of dihydrofluorescein diacetate (Hfluor-DA; stock in dimethyl sulfoxide; Sigma-Aldrich) for 30 min. The dye diffused into the intramuscular compartment and was able to chemically react with intracellular ROS (mainly H_2O_2) resulting in enhanced florescence. For statistical purposes, one muscle strip was taken from each mouse. We used five mice for control and five mice for PO$_2$ cycling treatment. A laser scan confocal microscopy system (Nikon confocal microscope D-Eclipse C1 system) recorded fluorescent emission signals from the tissue sample in a glass bottom culture dish (MatTek Corporation, Ashland, MA) in real time. The treated muscle strips were mounted and superfused with Ringer's solution, followed by PO$_2$ cycling treatment, which was a similar method as described above for the muscle function experiments. To ensure an accurate oxygen level, the chamber was sealed except for the tubing inlets containing gas bubbling and superfusate as well as the temperature probe. The superfusate solution was fully saturated with designated gas and preheated to ensure the temperature in the chamber remained at 37°C. The strips were then subjected to a 10-min baseline period (PO$_2$ of 550 Torr) and a subsequent 30-min 40 Torr PO$_2$ period at 37°C. The control muscles followed the same protocol except that there was no PO$_2$ cycling treatment. During the 40-min experimental period (10 min for baseline of 550 Torr PO$_2$ and 30 min 40 Torr PO$_2$ period), we captured an image (512×512 pixels) every 5 min and calculated the mean fluorescence to determine intramuscular ROS levels. To reduce the signal-to-noise ratio, each recorded image was an average from eight sequentially scanned images within 5 s at each time point. The setup parameters for the confocal imaging system were listed as follows: laser, argon; pinhole: medium or large; excitation, 488 nm; emission, 535±25 nm. The baseline autofluorescence was kept at a minimum and did not interfere with the ROS signal in our set-up. To reduce photobleaching or photodamaging, the laser power was set at ~15% without noticeably sacrificing image quality. To reduce imaging saturation due to possible excessive ROS in 40 Torr PO$_2$ conditions, the PMT gain was set as low as possible from the start of each experiment. In order to verify that the increased fluorescence signal was due to ROS, a series of antioxidant experiments were conducted. The glutathione peroxidase mimic, ebselen (30 μM), which is an effective ROS scavenger particularly for skeletal muscle tissue [5], was utilized. The animals were divided into 4 experimental groups, including control, PO$_2$ cycling, ebselen, and PO$_2$ cycling + H$_2$O$_2$. In each experimental group, five muscle samples were directly isolated from five fresh isolated diaphragms. For each experiment, we were able to measure ROS levels from ~8–10 muscle fibers in each image field.

In preliminary studies, we found that the PO$_2$ cycling protocol did not change the fluorescence baseline in the current set-up (data not shown). Each acquired image was analyzed with Adobe Photoshop element 6.0 and the final images were presented in a 300 DPI resolution with LZW compression.

Statistics

By performing the power analysis on the sample, we defined the PO$_2$ cycling effect on the skeletal muscle force development as well as intracellular ROS formation. For instance, we determined the PO$_2$ cycling effect on multiple groups including control, antioxidant (ebselen) treatment and H$_2$O$_2$ application, using the prospective means across these groups. We derived the power when the sample size was ~5 or greater mice per group by calculating the standard deviation. In addition, data were analyzed using a multi-way ANOVA with the animal as a variable, and expressed as means ± SE. The differences between the two treatments were statistically determined by a series of post-ANOVA contrast analyses using JMP software (SAS Institute, Cary, NC). Specifically, the post-ANOVA contrasts involve the comparison among all the groups of subjects and the display of the statistical difference between each pair of data. The treated groups, such as PO$_2$ cycling, antioxidant (ebselen) treatment and

H$_2$O$_2$ application, were used to compare with the control group, revealing any potential significance. $P < 0.05$ was regarded to be significant.

Results

Representative confocal images of the same muscular area in each group are illustrated in Fig. 1. Fluorescence (green color), which represents ROS levels, increased substantially at the end of prolonged 40 Torr PO$_2$ period (30 min, Fig. 1B) compared to baseline in the control group (Fig. 1A). However, after PO$_2$ cycling treatment, fluorescence displayed no significant change at the end of 40 Torr PO$_2$ (Fig. 1D) when compared to baseline (Fig. 1C). The disappearance of increased fluorescence emission in PO$_2$ cycling treatment demonstrated that PO$_2$ cycling effectively suppressed 40 Torr PO$_2$-induced intracellular ROS levels in the diaphragm. In addition, ebselen treated muscle strips showed no fluorescence increase at the end of 40 Torr PO$_2$ exposure (Fig. 1F). Interestingly, exogenous addition of H$_2$O$_2$ (50 μM) mitigated the antioxidant effect of PO$_2$ cycling (Fig. 1H).

Grouped data of mean florescence during 40 Torr PO$_2$ are illustrated in Fig. 2A. At the end of 40 Torr PO$_2$ periods, ROS levels were elevated from baseline in the control group. However, in the PO$_2$ cycling group, ROS levels were kept low compared to control (1.00±0.04 RU $vs.$ 1.48±0.05 RU; n = 5 from five animals; $P < 0.05$). Furthermore, in the control group, intracellular ROS elevation did not appear within the first 15 min of 40 Torr PO$_2$ period. After 20 min, ROS levels were enhanced and these increases lasted until the end of the 30 min period. Fig. 2B displayed the intracellular ROS fluorescence rate (RU/min). In the control group, this rate was close to zero for the first 15 min of 40 Torr PO$_2$, followed by three ROS bursts occurring at 20, 25,

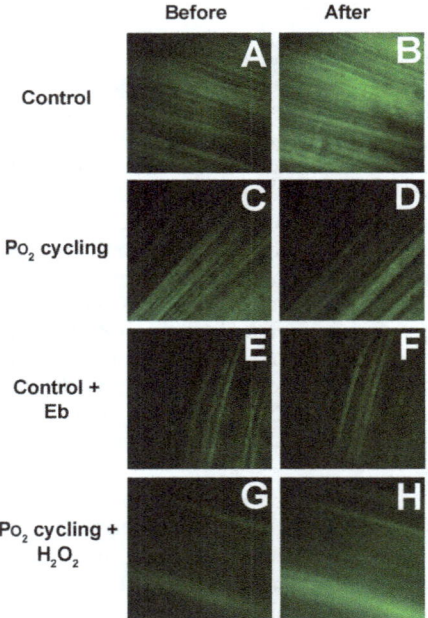

Figure 1. Representative ROS images from Hfluor-loaded diaphragm tissue. A: control muscle before 40 Torr PO$_2$. B: the same area of A at the end of 40 Torr PO$_2$. C: PO$_2$ cycling treated muscle before 40 Torr PO$_2$. D: the same area of C at the end of 40 Torr PO$_2$. E: Ebselen (Eb) treated muscle before 40 Torr PO$_2$. F: the same area of E at the end of 40 Torr PO$_2$. G: PO$_2$ cycling + H$_2$O$_2$ treated muscle before 40 Torr PO$_2$. H: the same area of G at the end of 40 Torr PO$_2$.

and 30 min, represented by positive values shown in Fig. 2B (in RU/min). The first ROS burst was relatively smaller (~50% less) compared to the other two larger bursts. However, in PO$_2$ cycling-treated diaphragm muscles, no ROS burst occurred at 40 Torr PO$_2$. It is important to note that although Fig. 2B looks similar to Fig. 2A, they refer to two separate measurements: fluorescence in Fig. 2A and fluorescence rate in Fig. 2B. In other words, Fig. 2B shows the mathematical slope of the fluorescence increase/decrease, indicating how fast the signal changes while Fig. 2A shows the level or the intensity of fluorescence.

Muscle absolute tension data during 40 Torr PO$_2$ are shown in Fig. 3. In the control group, the tension (in mN/mg muscle dry weight) at 0 min and the tension at each subsequent time point thereafter (1–5 min), was significantly lower than the PO$_2$ cycling group (n = 5 from five animals, $P < 0.05$), suggesting that PO$_2$ cycling ameliorated skeletal muscle resistance to fatigue during the 40 Torr PO$_2$ period (Fig. 3A). The tension decline rate in PO$_2$ cycling treated muscles markedly slowed down in the first 3 min compared to control (n = 5 from five animals for each group, $P < 0.05$, Fig. 3B). However, after 3 min the decline rate of all groups was similar, indicating that PO$_2$ cycling had no effect on the force decline rate for later fatigue development. Furthermore, the time

to 50% of the initial tension (T$_{50}$) from PO$_2$ cycling diaphragm muscle was significantly prolonged when compared to control diaphragm (in seconds, 216.2±33.0 *vs.* 99.5±10.0; n = 5 from five animals, $P < 0.05$, Fig. 4). However, this difference disappeared in the presence of H$_2$O$_2$ in the PO$_2$ cycling + H$_2$O$_2$ group (in seconds, 117.1±7.2 vs. 99.5±10.0; n = 5 from five animals, Fig. 4). Maximal diaphragm force was always measured prior to low PO$_2$ exposure. The corresponding muscle absolute tension values were reported in Table 1.

We observed that ebselen completely quenched the 40 Torr PO$_2$-induced ROS signal in the control (Fig. 1E and F, Fig. 2; n = 4), which demonstrated a similar effect to the PO$_2$ cycling treatment group (Fig. 1C and D, Fig. 2, n = 5). The addition of a small amount of H$_2$O$_2$ (50 μM) entirely abolished the PO$_2$ cycling-induced ROS inhibition effect in the confocal experiments (Fig. 1G and H). Grouped data are shown in Fig. 2A. Following 15 min from the onset of 40 Torr PO$_2$ period, fluorescence was significantly higher in PO$_2$ cycling + H$_2$O$_2$ group than the control, PO$_2$ cycling, and ebselen groups. However, after 30 min of 40 Torr PO$_2$ period, the control group showed higher fluorescence than the PO$_2$ cycling + H$_2$O$_2$ and ebselen groups, respectively (n = 5, $P < 0.05$). In Fig. 2B, at 15 min during 40 Torr PO$_2$ periods, there was a larger fluorescent burst in the PO$_2$ cycling + H$_2$O$_2$ treatment group compared to the other groups. A large fluorescent burst in the control group occurred ~10 min later than in the PO$_2$ cycling + H$_2$O$_2$, while there were no bursts in ebselen treatment groups, respectively (n = 5, $P < 0.05$).

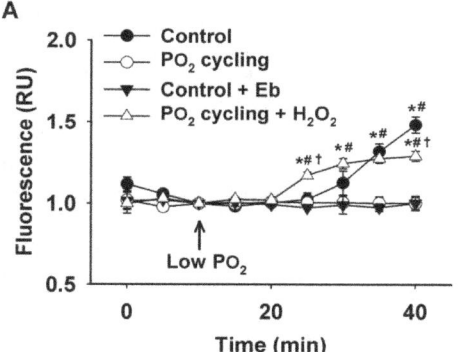

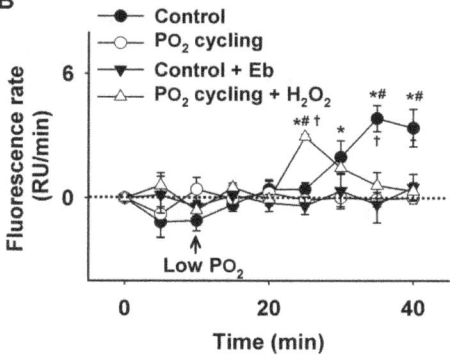

Figure 2. Intracellular ROS fluorescence and fluorescence rate under 40 Torr PO$_2$. A: averaged fluorescence data recorded in a relative unit (RU). Data showed intracellular ROS levels from control, PO$_2$ cycling, ebselen (Eb), and PO$_2$ cycling + H$_2$O$_2$ treated diaphragm muscle (*significantly different from PO$_2$ cycling, $P < 0.05$; #significantly different from Eb treatment, $P < 0.05$; †significantly different from control, $P < 0.05$). B: intracellular ROS burst represented by fluorescence rate. Data was recorded in a relative unit per min (RU/min) from control, PO$_2$ cycling, Eb, and PO$_2$ cycling + H$_2$O$_2$ (50 μM) treated diaphragm muscle under 40 Torr PO$_2$. Fluorescence data was recorded every 5 min (*significantly different from PO$_2$ cycling, $P < 0.05$; #significantly different from Eb treatment, $P < 0.05$; †significantly different from control, $P < 0.05$).

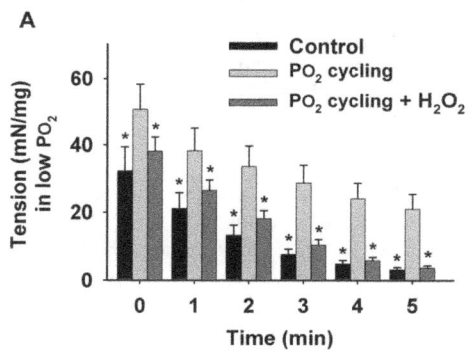

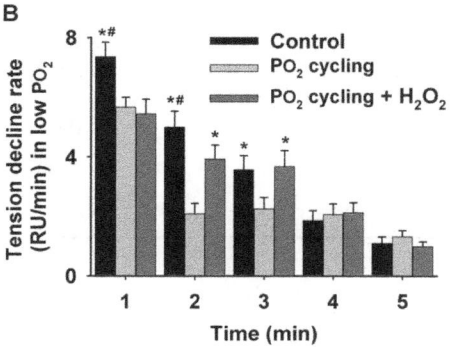

Figure 3. Muscle tension and tension decline rate data under 40 Torr PO$_2$. A: absolute tension (mN/mg) was recorded for 5 min from control, PO$_2$ cycling, and PO$_2$ cycling + H$_2$O$_2$ group (*significantly different from PO$_2$ cycling, $P < 0.05$). B: data showing the tension decline rate (RU/min) from control, PO$_2$ cycling, and PO$_2$ cycling + H$_2$O$_2$ (50 μM) muscles during a 5-min contractile period under 40 Torr PO$_2$ (*significantly different from PO$_2$ cycling, $P < 0.05$; #significantly different from PO$_2$ cycling + H$_2$O$_2$, $P < 0.05$).

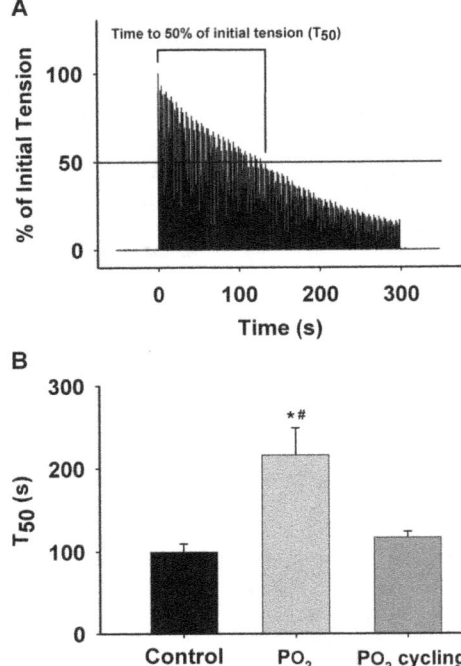

Figure 4. Time to reach 50% (T_{50}) of the initial tension in contracting diaphragm muscle under 40 Torr PO_2. A: a typical chart record illustrating the duration of T_{50} in a contracting diaphragm muscle during a 5-min contraction period. B: summarized T_{50} values from control, PO_2 cycling, and PO_2 cycling + H_2O_2 (50 μM) groups during the contraction (*significantly different from control, $P<0.05$; #significantly different from PO_2 cycling + H_2O_2, $P<0.05$).

Furthermore, under high PO_2 conditions (550 Torr), we investigated the effect of a small amount of H_2O_2 (50 μM) on muscle contraction as shown in Fig. 5 (representative curves) and Fig. 6 (grouped data). The tension development (mN/mg) and the tension decline rate (RU/min) at 1–5 min during the 5-min contraction were recorded in the presence or absence of H_2O_2. Both figures clearly illustrate that H_2O_2 had no marked effect on muscle function at a level of 50 μM (n = 6).

Moreover, H_2O_2 dosage experiments were performed in a range from 0 μM to 10 mM under high PO_2 conditions (550 Torr) as shown in Fig. 7. Muscle tension development was recorded for the maximal contraction during the baseline period and the initial

and end contractions during the 5-min contraction period. There was no significant difference between the control group (0 μM) and the 50 μM group. However, the 100 μM and 1 mM groups (n = 9 for both groups) did show a significant decrease in muscle function during the end contraction when compared to both the control and 50 μM groups ($P<0.05$). The greatest declines in muscle tension were observed in the 10 mM group (n = 8) as both the initial and end contractions showed a marked decrease in muscle tension development in comparison to all other dosage groups ($P<0.05$).

Discussion

The current study provides evidence that the PO_2 cycling preconditioning procedure we used reduces intracellular ROS levels in respiratory skeletal muscle during prolonged low PO_2. The absolute skeletal muscle tension and T_{50} were both greater in the PO_2 cycling group than the control group, but the addition of a small amount of ROS (H_2O_2) reduced these values to control levels. However, this amount of ROS was so marginal that it exerted no significant effect on muscle function during hyperoxia. Collectively, these data indicate that the protection of PO_2 cycling on the diaphragm is related to the reduced levels of intracellular ROS signaling molecules.

Dihydrofluorescein (Hfluor) is a highly sensitive intracellular probe commonly used for fluorescent detection of ROS. Fluorescein (Fluor) formation results when Hfluor reacts with ROS [5]. Previous research has shown that Hfluor is much less sensitive to nitric oxide (NO) compared to its analog dichlorfluorescein (DCFH) and also shows a higher resistance to photobleaching than DCFH [5,18,19]. Since it is superior for detecting ROS (particularly H_2O_2), it was used in our experiments. Our results showed that there was no marked ROS formation in the muscle during the first 15 min of a 40 Torr PO_2 period. In the control group (Fig. 2A), ROS levels were significantly increased after 20 min from the initiation of 40 Torr PO_2, which may suggest that during this timeframe antioxidant defenses were overwhelmed in diaphragm muscle not treated with PO_2 cycling. Moreover, the completely abolished ROS signals in the antioxidant (ebselen) treated control group (Fig. 1 and 2), confirms the existence of ROS, which seems to be quenched by PO_2 cycling treatment in our experiments (Fig. 1 and 2). Interestingly, after PO_2 cycling treatment, extracellular addition of a small amount of ROS (50 μM H_2O_2), which has no effect on normal muscle function (Fig. 5 and 6), completely negated the PO_2 cycling effect. These observations suggest to us that PO_2 cycling may be involved in the initiation of intracellular antioxidant signaling pathways.

Table 1. % of maximal force prior to low PO_2 exposure.

	Control (n = 5)	PO_2 cycling (n = 5)	PO_2 cycling + H_2O_2 (n = 5)
	Force (mN/mg)	Force (mN/mg)	Force (mN/mg)
	23.7	20.8	37.4
	31.1	45.6	41.0
	61.0	59.6	35.2
	29.8	71.5	51.6
	31.1	41.5	42.9
Average ± SE	35.4±6.56	47.8±8.58	41.6±2.84

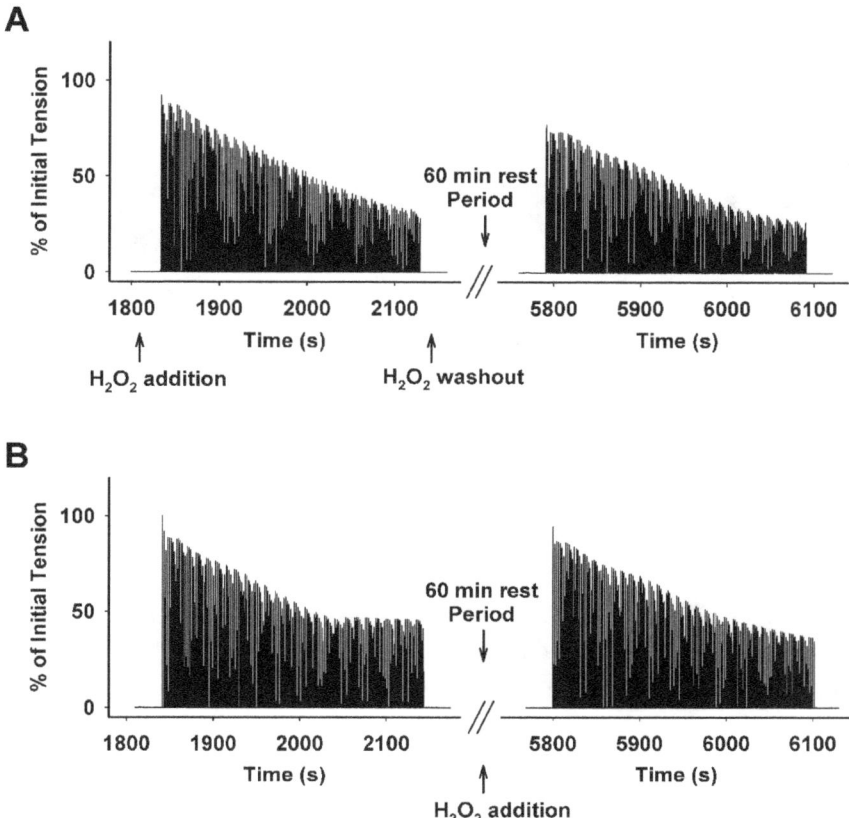

Figure 5. Representative contraction curves showing the effect of H$_2$O$_2$ (50 μM) on the muscle contraction during high PO$_2$ in a blocked order. A: H$_2$O$_2$ was added 15 min prior to the first 5-min contractile period followed by a H$_2$O$_2$ washout and 60 min rest period before the second 5- min contractile period in the absence of H$_2$O$_2$. B: The first 5-min contraction in the absence of H$_2$O$_2$ was followed by the second 5-min contractile period in the presence of H$_2$O$_2$.

It should be noted that the experimental conditions used to detect ROS are different from those used to evaluate skeletal muscle function in our settings. ROS detection was performed in unstimulated muscle, while function was assessed in contracting muscle, for the following reasons: 1) Due to large motion artifact, it is extremely difficult to perform muscle function experiments under confocal microscopy; 2) The muscle function experiment is focused on measurement of maximal force and time to fatigue. However, the confocal experiment is primarily designed to determine intracellular ROS levels in the muscle.

Evidence has shown that PO$_2$ cycling triggers the expression of superoxide dismutase (SOD), an endogenous antioxidant, which may further contribute to reduced ROS levels [7,10,20]. This is highly consistent with our observations of reduced ROS level in PO$_2$ cycling treated skeletal muscle. Similar to heart studies which show that both IPC and PO$_2$ cycling are mediated by ROS [7], intracellular ROS levels were also critical for PO$_2$ cycling efficacy in diaphragm muscle. There is a potential concern that PO$_2$ cycling could have altered mitochondrial function and integrity. However, our muscle function data (Fig. 3 and 4) suggest that fatigue resistance in PO$_2$ cycling treated mouse diaphragm muscle was substantially greater than that of control muscle. Thus, it is likely that mitochondrial activity was not negatively altered by PO$_2$ cycling treatment. Although in the current study design we are unable to determine whether PO$_2$ cycling treatment causes decreased ROS production or increased antioxidant scavenging, it is likely that there is a specific redox mechanism that suppresses

ROS generation during prolonged respiratory muscle exposure to the 40 Torr PO$_2$. The detailed mechanism associated with PO$_2$ cycling protection, however, is still unclear and requires further study. In addition, based on a similar previous study of PO$_2$ levels in myocytes, 40–550 Torr PO$_2$ was an effective setting to initiate intramuscular redox changes in skeletal muscle [21,22].

Interestingly, we noticed that at 15 min during the 40 Torr PO$_2$ period, ROS levels were significantly higher in the PO$_2$ cycling + H$_2$O$_2$ group compared to the other groups; yet, at 30 min, the control group showed a higher ROS level than the PO$_2$ cycling + H$_2$O$_2$ group (Fig. 2A). This suggests that the two treatment plans stimulate a time-dependent intracellular ROS formation mechanism. In addition, we evaluated the ROS generation rate, defined as a ROS burst and represented by the fluorescence rate, as shown in Fig. 2B. H$_2$O$_2$ addition after PO$_2$ cycling treatment induced the first ROS burst at 15 min under the 40 Torr PO$_2$ conditions, which occurred ~10 min earlier than a large ROS burst in the control group. Collectively, these observations suggest a faster diffusion of extracellular ROS (H$_2$O$_2$) into the intramuscular compartment, compared to the intracellular ROS generation in the control muscle. However at a later time (after 25 min during low PO$_2$), the control muscle showed a higher ROS formation rate than the PO$_2$ cycling + H$_2$O$_2$ group. This could be due to leakage of H$_2$O$_2$ into the perfusate. Nevertheless, these data further demonstrate a potential antioxidant-like effect exerted by PO$_2$ cycling, and this effect can be disturbed by a small addition of

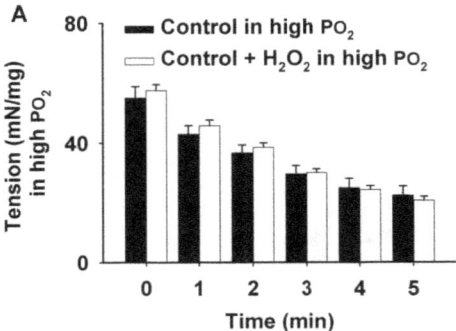

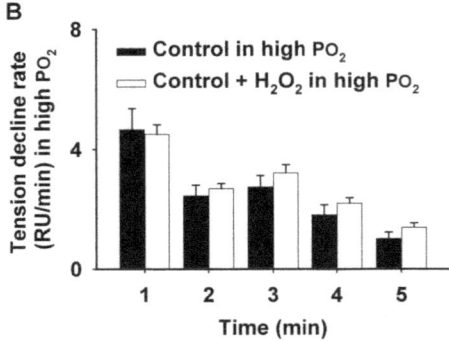

Figure 6. Grouped data showing the effect of H₂O₂ (50 μM) on the muscle contraction during high PO₂ in a blocked order. A: Tension development (mN/mg) at 1–5 min during the 5-min contraction period in the presence vs. absence of H₂O₂. B: Tension decline rate (RU/min) during the 5- min contraction period at 1–5 min in the presence vs. absence of H₂O₂.

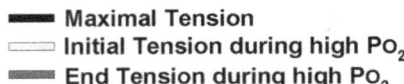

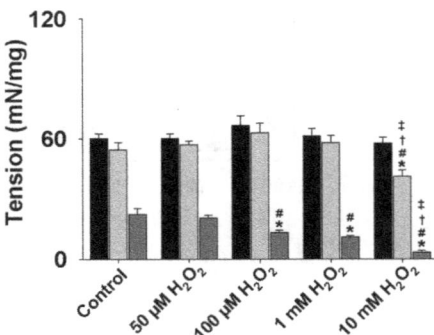

Figure 7. Grouped data showing the effect of varying H₂O₂ dosage on muscle tension development during high PO₂. Data showing the muscle tension development (mN/mg) during the H₂O₂ dosage treatments in mouse diaphragm strips [0 μM (control), 50 μM, 100 μM, 1 mM and 10 mM] in high PO₂ (550 Torr). *Significantly different from control ($P<0.05$). #Significantly different from 50 μM H₂O₂ ($P<0.05$). †Significantly different from 100 μM H₂O₂ ($P<0.05$). ‡Significantly different from 1 mM H₂O₂ ($P<0.05$).

ROS, which was not sufficient to influence muscle contractility (Fig. 5 and 6).

Moreover, it is suggested that PO₂ cycling mitigates fatigue within the diagram during hypoxia. Although the exact mechanism by which this occurs is unclear, it is likely that ROS play a significant role. Accordingly, our data (Fig. 1 and 2) suggest that low levels of ROS negate the benefits of PO₂ cycling and may be involved in other signaling events, including antioxidant cascades. Figures 5 and 6 illustrate the relationship between ROS level and muscular force generation. These results are supported by previous studies in which it was concluded that low levels of H₂O₂ may work more towards signaling pathways since they do not directly impact force generation in the muscle [23,24]. Further research into the effect of both ROS and PO₂ cycling on diaphragm force generation and related muscular mechanisms may lead to potential therapies to mitigate muscle fatigue during hypoxia.

The H₂O₂ dosage experiments (Fig. 7) on the diaphragm function suggest that high levels of ROS (H₂O₂) such as at 1 mM or 10 mM levels markedly reduce muscle function; however, lower levels of ROS, such as 50 μM, do not compromise muscle function as compared to control. In addition, our confocal experiments have clearly shown that this low level of ROS (50 μM) minimized the PO₂ cycling effect confirming that rather than damaging muscle directly, low levels of ROS may be a potential mediator for the signaling events involved in PO₂ cycling preconditioning.

There are a number of intracellular sources of ROS in skeletal muscle, including the mitochondria, xanthine oxidase (XO), peroxisomes, and NADPH oxidase [25]. For example, under respiratory stress, such as ischemia or low PO₂/hypoxia, xanthine

dehydrogenase converts to XO, which is subsequently released into circulation and produces ROS [22,25]. In skeletal muscle myocytes, NADPH oxidase is another likely candidate for ROS formation during hypoxic stress during injuries [23]. Additionally, the mitochondria produce low levels of superoxide anion under normal conditions [26], but in the lack of oxygen, the mitochondria experience excessive oxidant production [5,26,27]. Accordingly in such a condition, the mitochondria function as a source of increased level of superoxide, one of the common precursors to most ROS. This increase in intracellular ROS could potentially overwhelm natural antioxidant defense systems, leading to impaired muscle function [2,28–30]. Our data (Fig. 2B) showed that three ROS oxidative bursts (represented by positive rate values) occurred after 15 min of the 40 Torr period in the control muscle, which may indicate that intracellular antioxidant defenses have a 15 min effective period until they are eventually negated by subsequent ROS formation. This timeframe is within the regular activation time range of mitochondrial antioxidant enzymes [30,31]. Therefore, both XO and mitochondria are possible sources of ROS in this study.

Moreover, our findings have shown considerable evidence regarding the protective effects of PO₂ cycling training on skeletal muscle function (Fig. 3A). During the middle of the 40 Torr PO₂ period, the initial tension of the PO₂ cycling treated muscle strips was higher than that of control, and this trend kept increasing to ~4 fold greater than control from 1- to 5-min suggesting that PO₂ cycling progressively alleviated muscle fatigue. However, after H₂O₂ was loaded into the muscle at a relatively low dosage (Fig. 5 and 6), the protection from PO₂ cycling was diminished. This supports that the ROS signaling molecules may possibly play a negative role in the PO₂ cycling mechanism. PO₂ cycling markedly reduced the tension decline rate at both 1-, 2- and 3-min from the initial contraction compared to control. These rate differences were minimized at 4- and 5-min, respectively. The addition of H₂O₂ also significantly interrupted the PO₂ cycling effect on the tension decline rate at 2- and 3-min in the middle of the 40 Torr PO₂ period (Fig. 3B), confirming that ROS signaling molecules adversely affect PO_2 cycling pathways. A possible correlation among PO_2 cycling, muscle fatigue and ROS implies that PO_2 cycling is able to boost muscle contractility during fatigue, which is consistent with Clanton's and Reid's results [1,32]. Specifically, PO_2 cycling could result in a gradual increase in the production of endogenous antioxidant enzymes, allocating the additional antioxidants to serve as a "reservoir" that can be promptly accessed in response to spontaneous exposure to stressful conditions. This idea is consistent with previous research suggesting that PO_2 cycling significantly increases the expression level of intramuscular antioxidants such as SOD [7].

Previous research has shown that the intramuscular PO_2 during strenuous exercise is ~4 Torr, while ~30 Torr conditions are seen in resting muscle [33]. In the current research, we created a relatively lower PO_2 condition by equilibrating 40 Torr PO_2 and a hyperoxic condition with 550 Torr based on previous studies in skeletal muscle [5,15,21,34]. However, the intracellular PO_2 was difficult to determine under our experimental set-up, particularly in a contracting muscle with marked motion. During exhaustive exercise, skeletal muscle conditions fluctuate between higher PO_2 and lower PO_2, which frequently occur especially during high intensity interval trainings. This cycling in intracellular oxygen exerts an intrinsic preconditioning effect, similar to the PO_2 cycling protocol implemented in our studies. Additionally, it has been shown that increased levels of catalase and SOD are expressed in skeletal muscle during exercise training, resulting in a reduction of ROS level and oxidative stress [35–38]. Similar to exercise training, PO_2 cycling therapy may be an alternative method for increasing muscular endurance. Moreover, it is worth noting that a small amount of ROS (50 μM H_2O_2) has no effect on normal muscle function (Fig. 5 and 6). However, this dosage completely abolished the PO_2 cycling protection in low PO_2 conditions (Fig. 3). The exact mechanism of this response is still unknown, which will be an area of future studies.

Some limitations appear in our study. First, we are unable to determine whether 40 Torr in the solution can cause intramuscular hypoxia. Second, it is possible that our PO_2 cycling protocol, by itself, can directly affect intracellular ROS and muscle fatigue without additional mediators. Third, it is difficult to measure the exact intracellular PO_2 level in a functioning diaphragm. This is mainly because of a marked O_2 diffusion gradient across the multiple layers of diaphragm tissue. Although previous research has shown that in a similar condition to 40 Torr PO_2, intracellular levels of NADH in the diaphragm significantly increase [5], it is not clear whether 40 Torr PO_2 in the superfusate can cause the intracellular compartments hypoxic. Lastly, since O_2 cannot transport across the different layers equally in the whole muscle, it is more likely that a hypoxic condition may occur in the core of the muscle than the peripheral region [39]. In addition, intracellular PO_2 in skeletal muscle is ~10 Torr at rest, but it quickly drops to 3–5 Torr during intense exercise [40]. It is likely that the transition between 550 Torr to 40 Torr triggers mismatches of oxygen supply to the diaphragm, which may be

sufficient to induce a transient ROS formation as described in our earlier research [5]. Our study demonstrated that intracellular ROS is elevated in single myofibers during a similar PO_2 condition [21]. Interestingly, this level of oxygen (3–5 Torr) is regarded as normal for exercising human muscles [33]. Precisely controlling the intramuscular O_2 condition within the whole muscle preparation is highly challenging and therefore should be the focus of future research.

Perspectives and Significance

This study demonstrates that PO_2 cycling mediates beneficial responses through reducing intracellular ROS levels in respiratory muscle. PO_2 cycling is a drug-free treatment that possibly stimulates the diaphragm to activate its own antioxidant defense systems to resist fatigue development. This may be an effective method for enhancing muscular endurance. In addition, the current *in vitro* study provides a redox perspective into mouse respiratory muscle under optimal preconditions.

Supporting Information

Dataset S1   ROS fluorescence (figures 1 and 2).

Dataset S2   Muscle tension and tension decline rate (figure 3).

Dataset S3   Time to reach 50% (T_{50}) of the initial tension (figure 4).

Dataset S4   Representative H_2O_2 (50 μM) contraction curves (figure 5).

Dataset S5   Grouped H_2O_2 (50 μM) data (figure 6).

Dataset S6   Grouped data of varying H_2O_2 dosage (figure 7).

Acknowledgments

We thank Dr. Lan Jiang, Benjamin Pannell, Alexander C. Ziegler, Anthony Re, Shenali Wickramanayake, Tingyang Zhou and Julia Stimpfl for their assistance.

Author Contributions

Conceived and designed the experiments: LZ PTD MTC WJR JK TMB PDW. Performed the experiments: LZ MTC WJR. Analyzed the data: LZ WJR. Contributed reagents/materials/analysis tools: LZ WJR. Wrote the paper: LZ PTD MTC WJR JK TMB PDW. Designed the method of function measurement: LZ. Design of the manuscript writing style: LZ PDW.

References

1. Mohanraj P, Merola AJ, Wright VP, Clanton TL (1998) Antioxidants protect rat diaphragmatic muscle function under hypoxic conditions. J Appl Physiol 84: 1960–1966.
2. Reid MB, Haack KE, Franchek KM, Valberg PA, Kobzik L, et al. (1992) Reactive oxygen in skeletal muscle. I. Intracellular oxidant kinetics and fatigue in vitro. J Appl Physiol 73: 1797–1804.
3. Semenza GL (2011) Hypoxia-inducible factor 1: regulator of mitochondrial metabolism and mediator of ischemic preconditioning. Biochim Biophys Acta 1813: 1263–1268.
4. Chance B, Sies H, Boveris A (1979) Hydroperoxide metabolism in mammalian organs. Physiol Rev 59: 527–605.
5. Zuo L, Clanton TL (2005) Reactive oxygen species formation in the transition to hypoxia in skeletal muscle. Am J Physiol Cell Physiol 289: C207–216.
6. Choi YS, Cho KO, Kim EJ, Sung KW, Kim SY (2007) Ischemic preconditioning in the rat hippocampus increases antioxidant activities but does not affect the level of hydroxyl radicals during subsequent severe ischemia. Exp Mol Med 39: 556–563.
7. Zuo L, Roberts WJ, Tolomello RC, Goins AT (2012) Ischemic and hypoxic preconditioning protect cardiac muscles via intracellular ROS signaling. Front Biol 8: 305–311.
8. Gross GJ, Auchampach JA (1992) Blockade of ATP-sensitive potassium channels prevents myocardial preconditioning in dogs. Circ Res 70: 223–233.

9. Ravingerova T, Lokebo JE, Munch-Ellingsen J, Sundset R, Tande P, et al. (1998) Mechanism of hypoxic preconditioning in guinea pig papillary muscles. Mol Cell Biochem 186: 53–60.

10. Vanden Hoek TL, Becke LB, Shao Z, Li C, Schumacker PT (1998) Reactive oxygen species released from mitochondria during brief hypoxia induced preconditioning in cardiomyocytes. J Biol Chem 273: 18092–18098.

11. Muza SR (2007) Military applications of hypoxic training for high-altitude operations Med Sci Sports Exerc 39: 1625–1631.

12. Wang JA, Chen TL, Jiang J, Shi H, Gui C, et al. (2008) Hypoxic preconditioning attenuates hypoxia/reoxygenation-induced apoptosis in mesenchymal stem cells. Acta Pharmacol Sin 29: 74–82.

13. Bickler PE, Fahlman CS, Gray J, McKleroy W (2009) Inositol 1,4,5-triphosphate receptors and NAD(P)H mediate Ca2+ signaling required for hypoxic preconditioning of hippocampal neurons. Neuroscience 160: 51–60.

14. Zuo L, Hallman AH, Yousif MK, Chien MT (2012) Oxidative stress, respiratory muscle dysfunction, and potential therapeutics in chronic obstructive pulmonary disease. DOI: 101007/s11515-012-1251-x Front Biol 7: 506–513.

15. Zuo L, Nogueira L, Hogan MC (2011) Effect of pulmonary TNF-alpha overexpression on mouse isolated skeletal muscle function. Am J Physiol Regul Integr Comp Physiol 301: R1025–1031.

16. Roberts WJ, Zuo L (2012) Hypoxic preconditioning reduces reoxygenation injuries via PI3K in respiratory muscle. GSTF JBio 2: 42–53.

17. Henriksen EJ (2013) Effects of H2O2 on insulin signaling the glucose transport system in mammalian skeletal muscle. Methods Enzymol 528: 269–278.

18. Zuo L, Clanton TL (2002) Detection of reactive oxygen and nitrogen species in tissues using redox-sensitive fluorescent probes. Methods Enzymol 352: 307–325.

19. Hempel SL, Buettner GR, O'Malley YQ, Wessels DA, Flaherty DM (1999) Dihydrofluorescein diacetate is superior for detecting intracellular oxidants: comparison with 2′,7′-dichlorodihydrofluorescein diacetate, 5(and 6)-carboxy-2′,7′-dichlorodihydrofluorescein diacetate, and dihydrorhodamine 123. Free Radic Biol Med 27: 146–159.

20. Chen CF, Tsai SY, Ma MC, Wu MS (2003) Hypoxic preconditioning enhances renal superoxide dismutase levels in rats. J Physiol 552: 561–569.

21. Zuo L, Shiah A, Roberts WJ, Chien MT, Wagner PD, et al. (2013) Low Po(2) conditions induce reactive oxygen species formation during contractions in single skeletal muscle fibers. Am J Physiol Regul Integr Comp Physiol 304: R1009–1016.

22. Zuo L, Nogueira L, Hogan MC (2011) Reactive oxygen species formation during tetanic contractions in single isolated Xenopus myofibers. J Appl Physiol 111: 898–904.

23. Brotto MA, Nosek TM (1996) Hydrogen peroxide disrupts Ca2+ release from the sarcoplasmic reticulum of rat skeletal muscle fibers. J Appl Physiol (1985) 81: 731–737.

24. Penheiter AR, Bogoger M, Ellison PA, Oswald B, Perkins WJ, et al. (2007) H(2)O(2)-induced kinetic and chemical modifications of smooth muscle myosin: correlation to effects of H(2)O(2) on airway smooth muscle. J Biol Chem 282: 4336–4344.

25. Moylan JS, Reid MB (2007) Oxidative stress, chronic disease, and muscle wasting. Muscle Nerve 35: 411–429.

26. Raedschelders K, Ansley DM, Chen DD (2012) The cellular and molecular origin of reactive oxygen species generation during myocardial ischemia and reperfusion. Pharmacol Ther 133: 230–255.

27. Zhang L, Yu L, Yu CA (1998) Generation of superoxide anion by succinate-cytochrome c reductase from bovine heart mitochondria. J Biol Chem 273: 33972–33976.

28. Andrade FH, Reid MB, Allen DG, Westerblad H (1998) Effect of hydrogen peroxide and dithiothreitol on contractile function of single skeletal muscle fibres from the mouse. J Physiol 509 (Pt 2): 565–575.

29. MacFarlane NG, Miller DJ (1992) Depression of peak force without altering calcium sensitivity by the superoxide anion in chemically skinned cardiac muscle of rat. Circ Res 70: 1217–1224.

30. Jezek P, Hlavata L (2005) Mitochondria in homeostasis of reactive oxygen species in cell, tissues, and organism. Int J Biochem Cell Biol 37: 2478–2503.

31. Zuo L, Christofi FL, Wright VP, Liu CY, Merola AJ, et al. (2000) Intra- and extracellular measurement of reactive oxygen species produced during heat stress in diaphragm muscle. Am J Physiol Cell Physiol 279: C1058–1066.

32. Reid MB (2008) Free radicals and muscle fatigue: Of ROS, canaries, and the IOC. Free Radic Biol Med 44: 169–179.

33. Wagner PD (2012) Muscle intracellular oxygenation during exercise: optimization for oxygen transport, metabolism, and adaptive change. Eur J Appl Physiol 112: 1–8.

34. Zuo L, Hallman AH, Roberts WJ, Wagner PD, Hogan MC (2014) Superoxide release from contracting skeletal muscle in pulmonary TNF-alpha overexpression mice. Am J Physiol Regul Integr Comp Physiol 306: R75–81.

35. Caldarera CM, Guarnieri C, Lazzari F (1973) Catalase and peroxidase activity of cardiac muscle. Boll Soc Ital Biol Sper 49: 72–77.

36. Jenkins RR, Friedland R, Howald H (1984) The relationship of oxygen uptake to superoxide dismutase and catalase activity in human skeletal muscle. Int J Sports Med 5: 11–14.

37. Sen CK (1995) Oxidants and antioxidants in exercise. J Appl Physiol 79: 675–686.

38. Criswell D, Powers S, Dodd S, Lawler J, Edwards W, et al. (1993) High intensity training-induced changes in skeletal muscle antioxidant enzyme activity. Med Sci Sports Exerc 25: 1135–1140.

39. Barclay CJ (2005) Modelling diffusive O(2) supply to isolated preparations of mammalian skeletal and cardiac muscle. J Muscle Res Cell Motil 26: 225–235.

40. Richardson RS, Noyszewski EA, Kendrick KF, Leigh JS, Wagner PD (1995) Myoglobin O2 desaturation during exercise. Evidence of limited O2 transport. J Clin Invest 96: 1916–1926.

14

Improvement of Pro-Oxidant Capacity of Protocatechuic Acid by Esterification

Maria Luiza Zeraik[1,2], Maicon S. Petrônio[3], Dyovani Coelho[4], Luis Octavio Regasini[2], Dulce H. S. Silva[2], Luiz Marcos da Fonseca[3], Sergio A. S. Machado[4], Vanderlan S. Bolzani[2], Valdecir F. Ximenes[1]*

1 Department of Chemistry, Faculty of Sciences, São Paulo State University (UNESP), Bauru, São Paulo, Brazil, 2 Department of Organic Chemistry, Nuclei of Bioassays, Ecophysiology and Biosynthesis of Natural Products (NuBBE), Institute of Chemistry, São Paulo State University (UNESP), Araraquara, São Paulo, Brazil, 3 Department of Clinical Analysis, School of Pharmaceutical Sciences, São Paulo State University (UNESP), Araraquara, São Paulo, Brazil, 4 Institute of Chemistry of São Carlos, São Paulo University (USP), São Carlos, SP, Brazil

Abstract

Pro-oxidant effects of phenolic compounds are usually correlated to the one-electron redox potential of the phenoxyl radicals. Here we demonstrated that, besides their oxidizability, hydrophobicity can also be a decisive factor. We found that esterification of protocatechuic acid (P0) provoked a profound influence in its pro-oxidant capacity. The esters bearing alkyl chains containing two (P2), four (P4) and seven (P7) carbons, but not the acid precursor (P0), were able to exacerbate the oxidation of trolox, α-tocopherol and rifampicin. This effect was also dependent on the catechol moiety, since neither gallic acid nor butyl gallate showed any pro-oxidant effects. A comparison was also made with apocynin, which is well-characterized regarding its pro-oxidant properties. P7 was more efficient than apocynin regarding co-oxidation of trolox. However, P7 was not able to co-oxidize glutathione and NADH, which are targets of the apocynin radical. A correlation was found between pro-oxidant capacity and the stability of the radicals, as suggested by the intensity of the peak current in the differential pulse voltammetry experiments. In conclusion, taking into account that hydroquinone and related moieties are frequently found in biomolecules and quinone-based chemotherapeutics, our demonstration that esters of protocatechuic acid are specific and potent co-catalysts in their oxidations may be very relevant as a pathway to exacerbate redox cycling reactions, which are usually involved in their biological and pharmacological mechanisms of action.

Editor: Dominique Delmas, UMR INSERM U866, France

Funding: The authors gratefully acknowledge financial support from FAPESP (Fundação de Amparo à Pesquisa do Estado de Sao Paulo) grants #2013/08784-0, #010/52327-5, #2011/03017-6 and #2013/07600-3; Conselho Nacional de Desenvolvimento Científico e Tecnológico (CNPq) and Coordenação de Aperfeiçoamento de Pessoal de Nível Superior (CAPES). The funders had no role in study design, data collection and analysis, decision to publish, or preparation of the manuscript.

Competing Interests: The authors have declared that no competing interests exist.

* Email: vfximenes@fc.unesp.br

Introduction

A pro-oxidant effect is a general term used when biological, chemical or physical events are able to initiate or exacerbate, by different mechanisms, a redox imbalance. It can be directly or indirectly related to the generation of reactive oxygen species (ROS) or to the inhibition of endogenous protective antioxidant systems. It is important to note that a pro-oxidant effect is the opposite of an antioxidant effect, but not necessarily related to deleterious and pathological consequences. Indeed, there are numerous and growing evidences that compounds of natural or synthetic origin usually accepted as classical antioxidants can also act as pro-oxidants. In this regard, one of the most well-established substances is the potent antioxidant ascorbic acid, which besides its antiradical capacity is also able to initiate a pro-oxidant reaction by reducing ferric ion Fe(III) to ferrous ion Fe(II), the first step in the generation of the hydroxyl radical (HO•) through the Fenton reaction [1]. Moreover, it is common that their pro-oxidant properties are involved in their beneficial biological effects. This is the case with ascorbic acid and vitamin K_3, which are the active components in Apatone, a drug combination that has been used to kill tumour cells. The mechanism for the pharmacological action of Apatone is linked to the redox cycling of ascorbic acid, which induces a pro-oxidant effect and provokes the death of tumour cells by autoschizis [2].

The pro-oxidant activity of polyphenols is usually related to the generation of ROS, in the form of superoxide anions ($O_2^{•-}$), hydrogen peroxide (H_2O_2) and peroxyl radicals (ROO•), being the last the intermediate species in lipid peroxidation [3–7]. These oxidants are produced by autoxidation reactions, i.e. their direct oxidation by molecular oxygen [8–10]. The typical mechanism of polyphenol autoxidation reactions is demonstrated in the equations below (eq. 1–2). Additionally, these active redox molecules are also able to reduce Fe(III) to Fe(II), hence enabling the production of HO• (eq. 3–4) [11]. Consistent with that, lipoperoxidation, DNA damage, inactivation of antioxidant enzymes, mitochondrial disruption and other ROS-mediated deleterious processes are frequently associated with the antiproliferative effects and induction of apoptosis associated with the use of these molecules [3–6].

$$Ph\text{-}OH + O_2 \rightarrow Ph\text{-}O\bullet + O_2^{\bullet -} + H^+ \qquad (1)$$

$$2O_2^{\bullet -} + 2H^+ \rightarrow H_2O_2 + O_2 \qquad (2)$$

$$Ph\text{-}OH + Fe^{+3} \rightarrow Ph\text{-}O\bullet + Fe^{+2} + H^+ \qquad (3)$$

$$H_2O_2 + Fe^{+2} \rightarrow Fe^{+3} + HO^- + HO^\bullet \qquad (4)$$

Pro-oxidant activity has also been demonstrated for phenolic acids such as gallic, caffeic, ferulic acids and their derivatives. For instance, the cytoxicity of caffeic and gallic acids and their esters against human promyelocytic leukaemia is associated with both pro-oxidant activity and lipophilicity [12]; gallic acid increases ROS levels as well as depleting glutathione in A549 lung cancer cells [13]; caffeic acid phenylethyl ester induces apoptosis of human leukaemic cells by disrupting mitochondrial function via a pathway related to an increase in intracellular ROS [14]; caffeic acid, ferulic acid and their derivatives are effective agents for cleaving DNA in the presence of Cu(II) [15].

Another phytochemical that has been widely studied is protocatechuic acid, a natural phenolic compound found in many edible and medicinal plants. Indeed, besides its antiradical activity, protocatechuic acid has analgesic and anti-inflammatory properties that are comparable with those of standard drugs [16]. Recently, we demonstrated that the esterification of protocatechuic acid increased its efficiency as an inhibitor of the multienzymatic complex NADPH oxidase in neutrophils [17]. The increased hydrophobicity of the ester also altered its antioxidant activity [17,18]. Hence, we hypothesized that esterification could also alter its pro-oxidant potency. For this reason, we synthesized and studied esters of protocatechuic acid. The results confirmed our hypothesis, because greater hydrophobicity resulted in an increased and selective pro-oxidant activity. Besides, this work provides the first study on the specific and potent pro-oxidant activity of a series of protocatechuic acid alkyl esters for hydroquinone and related moieties and target biomolecules.

Materials and Methods

Chemicals

Protocatechuic acid, ethyl protocatechuate, gallic acid, butyl gallate, 3,5-dihydroxybenzoic acid, apocynin, trolox, α-tocopherol, glutathione (GSH), rifampicin, horseradish peroxidase (HRP), reduced nicotinamide adenine dinucleotide (NADH), diethylenetriaminepentaacetic acid (DTPA), 2,2'-Azino-bis(3-ethylbenzthiazoline-6-sulfonic acid) (ABTS), catalase, 5,5'-dithiobis(2-nitrobenzoic acid) (DTNB) and N,N'-dicyclohexylcarbodiimide were purchased from Sigma-Aldrich Chemical Co. (St. Louis, MO, USA). H_2O_2 was prepared by diluting a 30% stock solution and calculating its concentration from its absorbance at 240 nm ($\varepsilon_{240} = 43.6\ M^{-1}\ cm^{-1}$). All reagents used for buffers and mobile phases were of analytical grade. Stock solutions (50 mM) of the tested compounds were prepared in ethyl alcohol. Stock solution (10 mM) of α-tocopherol was prepared in ethyl alcohol. Ultrapure Milli-Q water from Millipore (Belford, MA, USA) was used for the preparation of buffers and solutions.

Preparation of alkyl protocatechuates

The esters of protocatechuic acid were prepared as recently described [17]. A 3 mL solution of N,N'-dicyclohexylcarbodiimide (DCC, 1.0 mmol) in p-dioxane was added to a cooled (5°C) solution of 0.2 mmol protocatechuic acid and 20 mmol of butyl, heptyl, or decyl alcohols in 6 mL p-dioxane. The solution was stirred for 48 hours and the solvent was removed under reduced pressure. The residue was then partitioned 3 times with ethyl acetate and filtered. The filtrate was washed successively with saturated aqueous citric acid solution (3 times), saturated aqueous $NaHCO_3$ (3 times), water (2 times), dried over anhydrous $MgSO_4$, and evaporated under reduced pressure. The crude products were purified over a silica gel column (0.06–0.20 mm, ACROS Organics, USA) and eluted isocratically with $CHCl_3$–CH_3OH (98:2). All compounds were identified using ^1H and ^{13}C NMR spectral data obtained from a Varian DRX-500 spectrometer (11.7 T). Chemical shifts (δ) were expressed in ppm. Coupling constants (J) were expressed in Hz, and splitting patterns are described as follows: s = singlet; br s = broad singlet; d = doublet; t = triplet; m = multiplet; and dd = doublet of doublets.

Butyl protocatechuate. It was obtained as a white solid in 93% yield. ^1H NMR (500 MHz, DMSO-d_6): δ 0.93 (t, J = 6.5 Hz, 3H), 1.30–1.82 (m, 4 H), 4.23 (t, J = 6.5 Hz, 2H), 6.79 (d, J = 8.5 Hz, 1H), 7.29 (dd, J = 2.0, 8.5 Hz, 1H), 7.34 (d, J = 2.0 Hz, 1H). ^{13}C NMR (125 MHz, DMSO-d_6) δ 13.5, δ 18.7, δ 30.3, δ 63.8, δ 165,6, δ 112.5, δ 123.3, δ 151.5, δ 147.4, δ 120.8, δ 115.2.

Heptyl protocatechuate. It was obtained as a white solid in 89% yield. ^1H NMR (500 MHz, DMSO-d_6): δ 0.84 (t, J = 6.5 Hz, 3H), 1.20–1.70 (m, 10 H), 4.15 (t, J = 6.5 Hz, 2H), 6.79 (d, J = 8.5 Hz, 1H), 7.29 (dd, J = 2.0, 8.5 Hz, 1H), 7.34 (d, J = 2.0 Hz, 1H). ^{13}C NMR (125 MHz, DMSO-d_6) δ 13.9, δ 22.0, δ 25.5, δ 28.2, δ 28.3, δ 31.2, δ 64.0, δ 165,7, δ 116.2, δ 121.7, δ 150.4, δ 145.1, δ 120.8, δ 115.3.

Decyl protocatechuate. It was obtained as a yellow solid in 73% yield. ^1H NMR (500 MHz, DMSO-d_6): δ 0.87 (t, J = 6.5 Hz, 3H), 1.25–1.76 (m, 16 H), 4.17 (t, J = 6.5 Hz, 2H), 6.79 (d, J = 8.5 Hz, 1H), 7.29 (dd, J = 2.0, 8.5 Hz, 1H), 7.34 (d, J = 2.0 Hz, 1H). ^{13}C NMR (125 MHz, DMSO-d_6) δ 13.7, δ 22.1, δ 25.3, δ 28.2, δ 28.3, δ 28.2, δ 28.3, δ 29.5, δ 31.3, δ 64.0, δ 165,7, δ 116.2, δ 121.7, δ 150.4, δ 145.1, δ 120.8, δ 115.3.

Electrochemical measurements

Voltammetric studies were performed and oxidation potential, measured as anodic peak potential (Epa), was determined using an Autolab PGSTAT 30 potentiostat/galvanostat (Eco-Chemie, Utrecht, Netherlands). Voltammetric curves were recorded at room temperature using a 3-electrode setup cell. The working electrode was a glassy carbon disk electrode (GC electrode, 3 mm diameter). The counter electrode was a platinum plate and the reference an Ag/AgCl electrode saturated with 3 M KCl. The working electrode surface was carefully polished with 0.5 μm alumina slurries before each experiment and thoroughly rinsed with distilled water. A solution of sodium phosphate buffer 0.2 M (pH = 7) was used as a supporting electrolyte. The solutions were purged with nitrogen for 5 min before recording the voltammograms. The ethanolic solutions (5 mM) of the compounds were diluted in the electrochemical cell to final concentrations of 0.1 mM using the supporting electrolyte solution. The cyclic voltammograms were recorded at a potential scan rate of 5 mV s^{-1}. Differential pulse voltammograms were obtained at modulation time 0.05 s, pulse amplitude 0.05 V, step potential 0.45 mV and scan rate 4.5 mV/s [19].

Oxidation of trolox

A 100 μM trolox solution was incubated with 50 μM H_2O_2 and 0.01 μM HRP in 50 mM phosphate buffer, pH 7.0, with 50 μM DTPA at 25°C in the absence (control) or presence of the test compounds. The rate of trolox oxidation was measured at 272 nm. The blank for absorbance measurements consisted of phosphate buffer and the same concentration of the tested compounds. An HP8452 diode array spectrophotometer (Agilent, SC, CA, USA) was used to measure absorbance in these assays [20].

Oxidation of α-tocopherol

A 100 μM α-tocopherol solution was incubated with 100 μM H_2O_2 and 0.01 μM HRP in 50 mM phosphate buffer, pH 7.0, with 50 μM DTPA at 25°C in the absence (control) or presence of the test compounds. The reactions were stopped after 10 min by the addition of 10 μg/mL catalase and an aliquot of 50 μL was injected into the HPLC. The analyses were carried out on a Luna C18 reversed-phase column (250×4.6 mm, 5 μm) using an HPLC–PDA detection system set at 295 nm (Jasco, Easton, MD, USA). The mobile phase consisted of 95% methanol and 5% water and the flow rate was 1.2 mL/min.

Oxidation of rifampicin

A 100 μM rifampicin solution was incubated with 100 μM H_2O_2 and 0.01 μM HRP in 50 mM phosphate buffer, pH 5.5, at 25°C in the absence (control) or presence of the test compounds. The rate of rifampicin oxidation was measured at 472 nm. The blank for absorbance measurements consisted of phosphate buffer. An HP8452 diode array spectrophotometer (Agilent, SC, CA, USA) was used to measure absorbance in these assays [21].

Oxidation of Glutathione

A 1 mM GSH solution was incubated with 100 μM H_2O_2 and 0.1 μM HRP in 50 mM phosphate buffer, pH 7.0 at 25°C in the absence (control) or presence of the test compounds. The reactions were stopped after 5 min by the addition of 10 μg/mL catalase. The concentration of GSH remaining was measured using the DTNB method, as previously described [22]. Briefly, the supernatant (0.45 mL) was combined with an equal volume of 300 mM Na_2HPO_4 and 0.1 mL of DTNB solution (0.2 mg/mL DTNB in 1% citrate). The absorbance at 412 nm was calculated relative to a blank containing 0.45 mL PBS, 0.45 mL 300 mM Na_2HPO_4, and 0.1 mL DTNB. A standard curve was generated to calculate the concentration of GSH. An HP8452 diode array spectrophotometer (Agilent, SC, CA, USA) was used to measure absorbance in these assays.

Oxidation of NADH

A 100 μM NADH solution was incubated with 10 μM H_2O_2 and 0.01 μM HRP in 50 mM phosphate buffer, pH 7.0, with 50 μM DTPA at 25°C in the absence (control) or presence of the test compounds. The rate of NADH oxidation was measured at 340 nm. The blank for absorbance measurements consisted of phosphate buffer and the same concentration of the tested compounds. An HP8452 diode array spectrophotometer (Agilent, SC, CA, USA) was used to measure absorbance in these assays [20].

Statistical analysis

The assays were conducted in duplicate and on three different days using different solutions. The results are expressed as mean ± SEM. Analysis of variance and significant differences among the means were tested by one-way ANOVA and Student-Newman-Keuls Multiple Comparison Test. Values of $p<0.05$ were regarded as significant.

Results and Discussion

Pro-oxidant activity (trolox as the target)

The molecular structures of protocatechuic acid (**P0**) and its esters bearing alkyl chains containing two (**P2**), four (**P4**), seven (**P7**) and ten (**P10**) carbons are shown in Figure 1. We found that, in addition to the improved antiradical capacity [17,18,23], the esterification of **P0** was also crucial for its action as a pro-oxidant compound. First of all, it must be emphasized that for phenolic and other redox active compounds, the distinction between anti- and pro-oxidant effects might be related to the stability and reactivity of the transient phenoxyl radicals generated during their oxidation. In this regard, an antioxidant effect is obtained when the transient free radical undergoes subsequent reactions leading to stable and non-harmful end products. On the other hand, a pro-oxidant effect will be observed when the transient free radical is

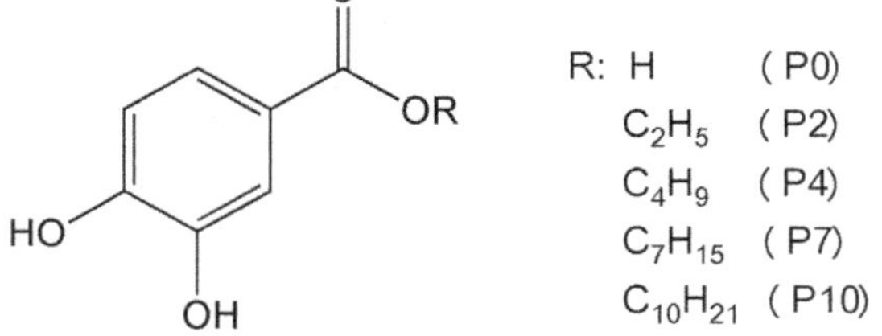

Figure 1. Molecular structures of protocatechuic acid and its alkyl esters.

R: H (P0)
C₂H₅ (P2)
C₄H₉ (P4)
C₇H₁₅ (P7)
C₁₀H₂₁ (P10)

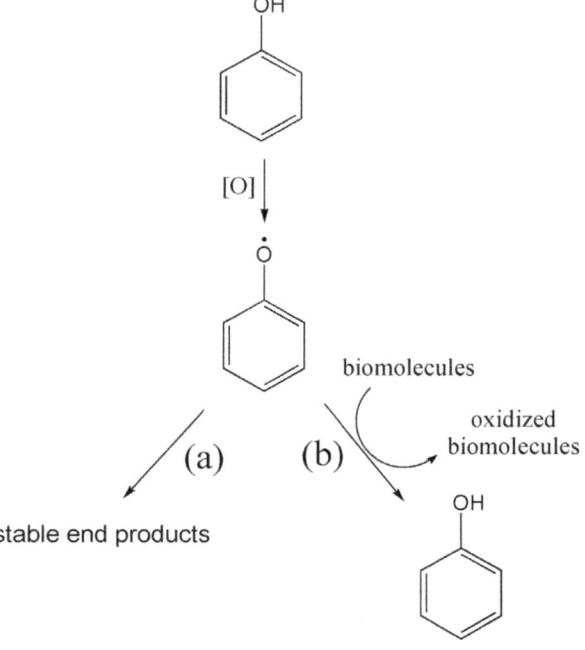

Figure 2. General pathways for phenol derivatives as (a) antioxidant or (b) pro-oxidant.

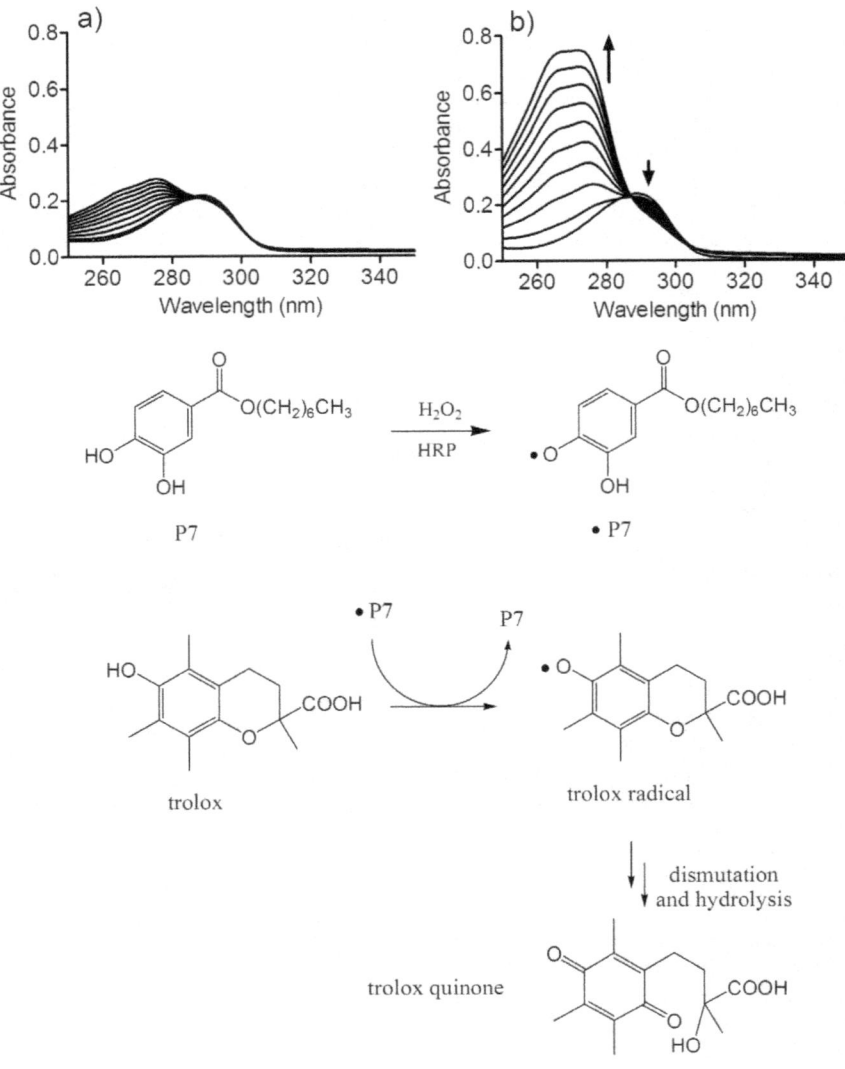

Figure 3. Effect and proposed pathway for the pro-oxidative action of P7 on trolox oxidation. The reaction mixture consisted of 100 μM trolox, 0.01 μM HRP and 50 μM H_2O_2 in the absence (a) or presence (b) of 2 μM P7. Scans were obtained at 5 s intervals.

able to promote downstream oxidations. To reinforce this important concept, which is the major point in this report, the chemical equations in Figure 2 illustrate the action of a generic phenolic compound acting either as an antioxidant (pathway **a**) or a pro-oxidant (pathway **b**). Consistent with pathway (**b**), the pro-oxidant activity of transient free radicals has been demonstrated in many publications; for instance: oxidation of NADH by raloxifene radicals [24], oxidation and depletion of GSH in HL-60 cells by etoposide radicals [25], and the induction of microsomal lipid peroxidation, ascorbate and GSH by phenoxyl radicals of vitamin E and its analogues [26]. On the other hand, most of the phenolic acids, which usually act as antioxidants, follow the pathway (**a**).

Following this concept, our experimental approach was to assess the pro-oxidant activity of transient phenoxyl radicals of proto-catechuic acid and its esters. The phenoxyl radicals were produced by their reactions with H_2O_2 in a reaction catalysed by HRP. This is a well-established procedure for the generation and study of the pro-oxidant capacity of phenoxyl radicals [20–21,24–26]. In this experimental model, it is important that the target biomolecules that are subjected to the pro-oxidative effect either do not react, or

react with low efficiency, with the HRP/H_2O_2 enzymatic system. This is the case for trolox, a water-soluble derivative of vitamin E that was one of the targets in our studies.

The results depicted in Figure 3 show time-dependent changes in the absorbance of trolox when subjected to oxidation by HRP/H_2O_2 in the absence or presence of **P7**. It can be observed that the addition of a catalytic amount of **P7** provoked a significant increase in the rate of oxidation of trolox, which is indicative of its co-catalytic effect in this oxidation. It is noteworthy that the absorption band with a maximum at 272 nm and an isosbestic point at 286 nm is consistent with the formation of trolox quinone, which is the product of dismutation and hydrolysis of the transient trolox radical, as previously demonstrated [27,28]. Figure 3 also shows our proposal for the effect of **P7** on trolox oxidation, consistent with a typical pro-oxidant effect.

Unlike with **P7**, the results depicted in Figure 4[A] and 4[B] show that the addition of **P0** did not alter the oxidation of trolox. Moreover, except for **P10**, the pro-oxidant capacity was depen-dent on the length of the carbon chain, reaching its maximum with **P7**. The reduced efficacy of **P10** in this pro-oxidative model can

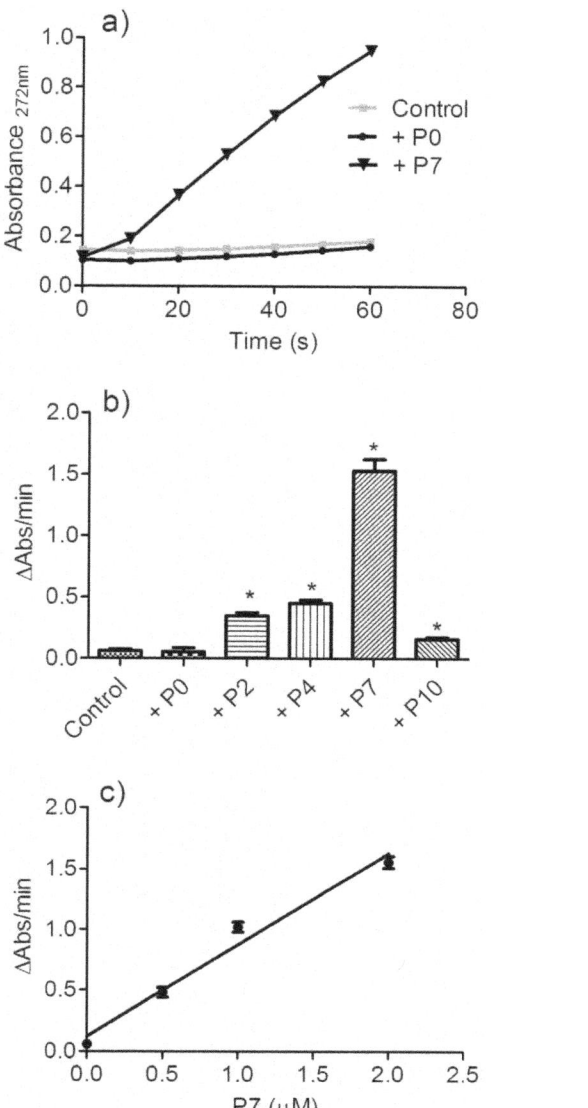

Figure 4. The length of the protocatechuate's carbon chain and the rate of oxidation of trolox. (a) The reaction mixture consisted of 100 μM trolox, 0.01 μM HRP and 50 μM H_2O_2 in the absence (control) or presence (2 μM) of the tested compounds. (b) Rate of oxidation. Data represent the mean and SEM of three experiments. *$p < 0.05$ relative to the control. (c) Effect of the concentration of P7 on the rate of oxidation ($r^2 = 0.9766$).

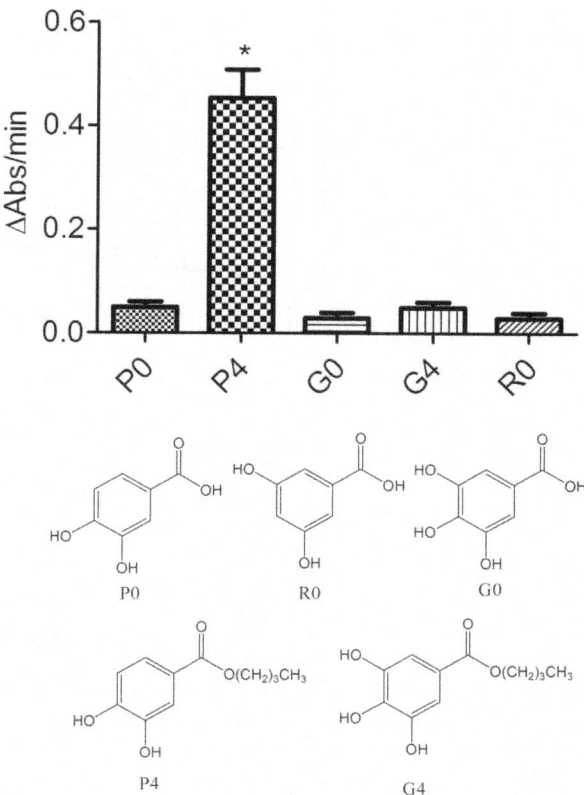

Figure 5. Effect of the number and position of hydroxyl groups and hydrophobicity on the rate of oxidation of trolox. The reaction mixture consisted of 100 μM trolox, 0.01 μM HRP and 50 μM H_2O_2 and 2 μM of the tested compounds. Data represent the mean and SD of three experiments. *$p < 0.05$ relative to the P0.

be explained by its lower aqueous solubility, which decreases its effective concentration in the medium. The pro-oxidant effect of **P7** was also dependent on its concentration (Figure 4[C]). It is noteworthy that using 100 μM trolox and only 0.5 μM **P7**, the rate of oxidation of trolox was increased about eight-fold, which is a clear indication that **P7** was not consumed during the reaction course, hence confirming its co-catalytic effect in the oxidation of trolox.

The next step was to test the importance of the catechol moiety for the pro-oxidant effect. For that, we compared **P0** with gallic acid (**G0**) and the acid derivative of resorcinol (**R0**) (Figure 5). These compounds were chosen because there is a relationship between the number and position of hydroxyl groups in the

benzene ring and the value of the oxidation potential. For instance, a recent study using a modified DNA-based glassy carbon electrode reported the following values: pyrogallol (0.39 V), catechol (0.42 V) and resorcinol (0.9 V) [29], which are related to **G0**, **P0** and **R0**, respectively. Hence, a comparison among **G0**, **P0** and **R0** could indicate the relationship between pro-oxidant activity and oxidation potential. From the results described in Figure 5, we concluded that the oxidation potential was not relevant to the pro-oxidant effect, since neither **G0** nor **R0** was able to act as a co-catalyst in the oxidation of trolox.

From another perspective and considering the importance of the esterification to the pro-oxidant activity of the tested compounds, we also measured and compared the pro-oxidant activities of butyl gallate (**G4**) and the equivalent **P4** (Figure 5). However, it can be seen that only **P4** was able to promote a co-catalytic effect. Hence, we concluded that the presence of the catechol moiety and the increased hydrophobicity are both necessary for the pro-oxidative activity demonstrated here.

Pro-oxidative activity (GSH and NADH as targets)

So far, we have demonstrated that transient radicals generated during the oxidation of alkyl protocatechuates are able to oxidize trolox. However, besides trolox, several biomolecules have been identified as potential targets for pro-oxidative effects of transient radicals [21–32]. Hence, we studied the capacity of **P7** as a pro-oxidant of GSH and NADH and compared its efficacy with that of apocynin, a methoxy-catechol widely used as an inhibitor of the multienzymatic NADPH oxidase complex, and that has also been

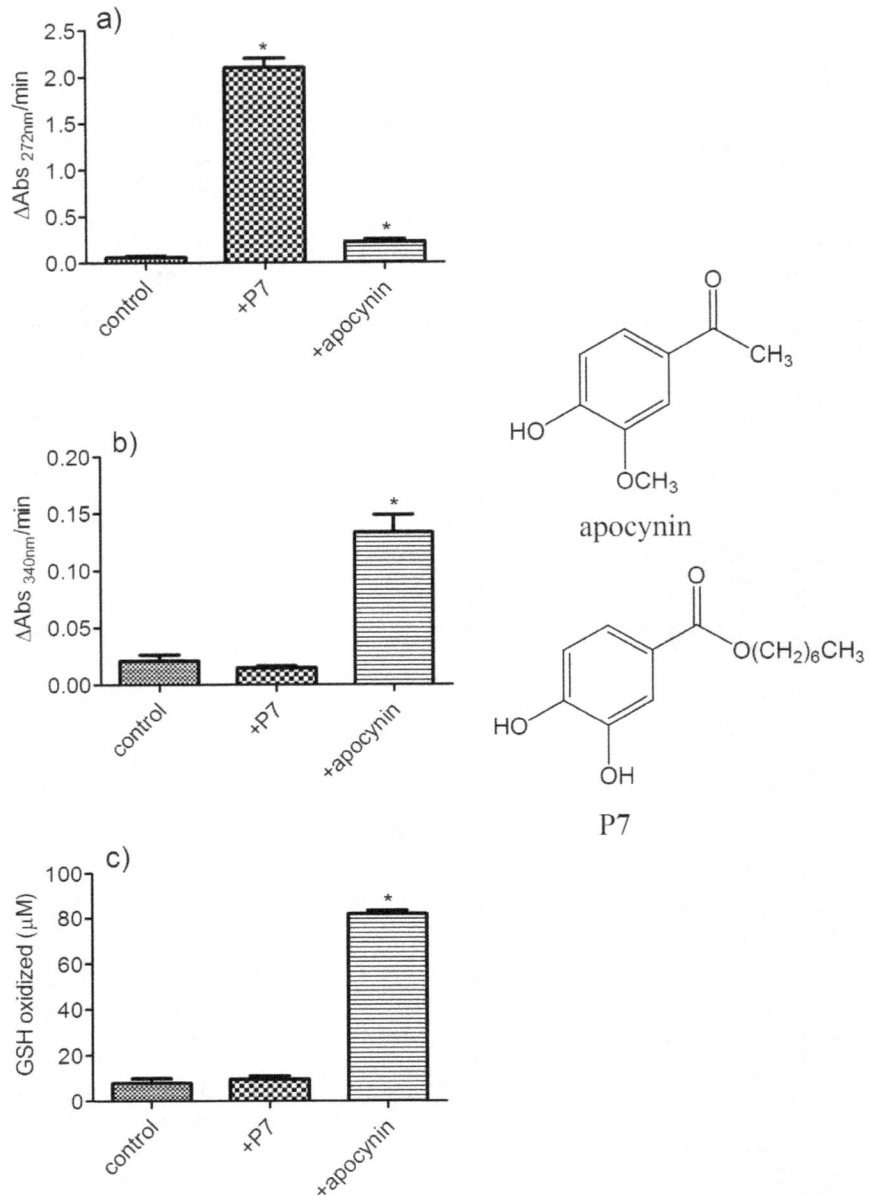

Figure 6. Apocynin versus P7 as pro-oxidants. (a) Oxidation of trolox, (b) oxidation of NADH and (c) oxidation of GSH. When present, apocynin and P7 were added at 5 µM. See material and methods for further experimental details. Data represent the mean and SD of three experiments. *p<0.05 relative to control.

used as an anti-inflammatory [33], neuroprotective [34], and vasorelaxant [35], among many other biological applications. The apocynin radical is able to oxidize trolox, GSH, NADH and sulfhydryl residues in proteins [20]. The results depicted in Figure 6 show that apocynin was significantly more potent than **P7** as a pro-oxidant of NADH and GSH; however, **P7** was more effective regarding the oxidation of Trolox. These results suggested that the pro-oxidative activity of **P7** could be selective for the presence of hydroquinone or related functional group in the target molecules. These are relevant findings, since a specific pro-oxidant action could be useful as a modulator for quinone-base chemo-therapeutics, where redox reactions are responsible by generation of cytotoxic oxidant species [36].

Pro-oxidative activity (rifampicin and α-tocopherol as targets)

Considering the above results, we studied the effect of the test compounds as pro-oxidants using rifampicin as target. The reason for the choice of this antibiotic was the presence of a hydroquinone moiety in its molecular structure, low reactivity with H_2O_2/HRP and the ease of monitoring its oxidation at 472 nm [21]. The results depicted in Figure 7 confirmed our expectation, since the addition of catalytic amount of the esters increased the rate of rifampicin oxidation significantly and, again, **P7** was the most effective compound.

For an additional confirmation of the pro-oxidant effect of the protocatechuates, α-tocopherol, the antioxidant present in cell

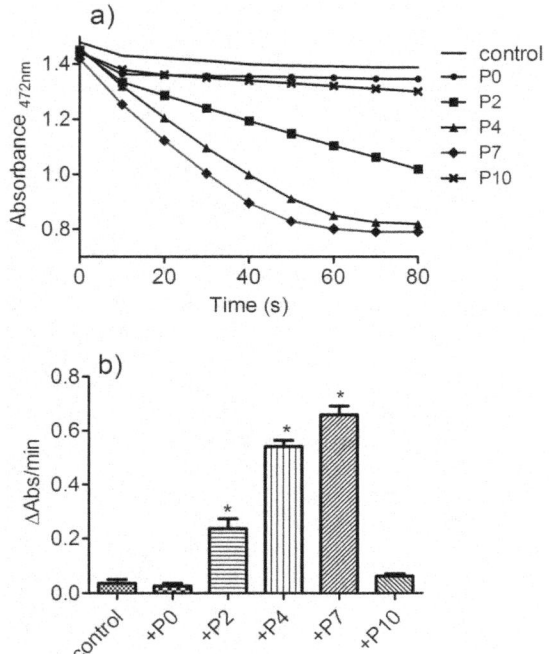

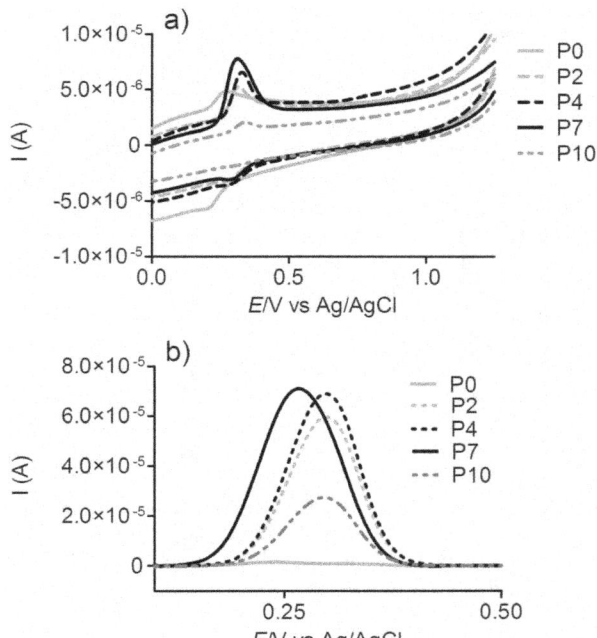

Figure 7. The length of a protocatechuate's carbon chain and the rate of oxidation of rifampicin. (a) The reaction mixture consisted of 100 μM rifampicin, 0.01 μM HRP, and 50 μM H_2O_2 in the absence (control) or presence (2 μM) of the tested compounds. (b) Reaction rate. Data represent the mean and SD of three experiments. *$p < 0.05$ relative to the control.

Figure 9. Cyclic (a) and differential pulse voltammograms (b) of 0.1 mM solutions of protocatechuic acid and its esters at pH 7.0. The cyclic voltammograms were recorded at a potential scan rate of 5 mV s^{-1}. Differential pulse voltammograms were obtained at modulation time 0.05 s, pulse amplitude 0.05 V, step potential 0.45 mV and scan rate 4.5 mV/s.

membranes, was also used as a target. In this case the reaction was performed in the presence or absence of **P7** and after 10 minutes the reaction mixture was analysed by HPLC. Figure 8 shows that α-tocopherol was totally non-reactive with H_2O_2/HRP alone or after the addition of **P0**; however, the addition of **P7** caused its almost complete oxidation.

Pro-oxidant activity and oxidation peak potential

It has been demonstrated that the capacity of phenolic compounds as pro-oxidants is partly correlated with the one-electron redox potential of their phenoxyl radicals [37]. For

instance, naringenin (Epa = 0.600 V) was about five-fold more efficient than rutin (Epa = 0.180 V) as a co-catalyst in the oxidation of NADH by HRP/H_2O_2 [37]. Similarly, apocynin (Epa = 0.760 V [38]) was four-fold more efficient than the structurally related compound vanillic acid (Epa = 0.494 V, [39]) as a co-catalyst in the oxidation of GSH mediated by HRP/H_2O_2 [20]. Hence, it was not surprising to find that apocynin was more effective than **P7** (Epa = 0.266 V) regarding co-oxidation of NADH and GSH (Figure 6). However, the relationship between the oxidation peak potential and pro-oxidant activity did not explain the results obtained for the oxidation of trolox, rifampicin and α-tocopherol, since **P7** was more efficient than apocynin with these compounds. These findings and the large difference between

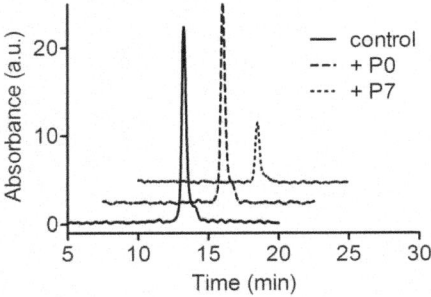

Figure 8. Oxidation of α-tocopherol and the effect of P0 and P7. The reaction mixture consisted of 100 μM α-tocopherol, 0.01 μM HRP, 50 μM H_2O_2 and 5 μM of the tested compounds. The reactions were stopped after 10 min and injected into the HPLC. See materials and methods for further experimental details. The control is a standard of α-tocopherol (100 μM).

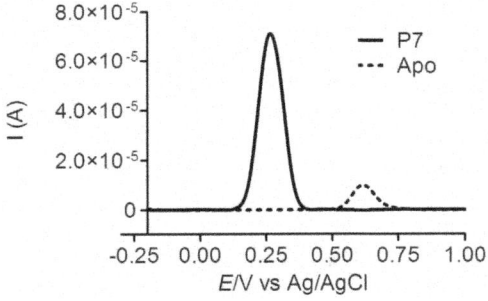

Figure 10. Differential pulse voltammograms of 0.1 mM solutions of heptyl protocatechuate and apocynin at pH 7.0. Differential pulse voltammograms were obtained at modulation time 0.05 s, pulse amplitude 0.05 V, step potential 0.45 mV and scan rate 4.5 mV/s.

P7 and **P0** in oxidizing the hydroquinone related compounds motivated us to conduct a more detailed study of the effect of esterification on the oxidation peak potential of protocatechuic acid.

The test compounds were studied looking for a relationship between their pro-oxidative capacities and their oxidation peak potential. Figure 9 shows the results of electrochemical experiments obtained using a glassy carbon working electrode at pH 7.0. The cyclic and differential pulse voltammetry experiments showed a single and well-defined oxidation peak for the protocatechuic alkyl esters, which ranged from 0.266 to 0.298 V. Similar values were reported by Reis et al. [18], who studied the relationship between antioxidant activity and the electrochemical properties of protocatechuic acid and its methyl, ethyl and propyl esters. Actually, the absence of a significant difference in the oxidation peak potentials for these compounds was not unexpected, since the esterification of a carboxylic group does not provoke a significant alteration in its redox potential. Comparisons can also be made between gallic acid (half-wave potential, $E_{1/2}$, 0.52 V) and its esters, propyl gallate ($E_{1/2}$ 0.51 V) and butyl gallate ($E_{1/2}$ 0.51 V) [40]; caffeic acid (Epa 0.183 V) and ethyl caffeate (Epa 0.175 V); and ferulic acid (Epa 0.335 V) and ethyl ferulate (Epa 0.368 V) [41].

By comparing Epa values, it can be concluded that oxidation peak potential cannot explain the large and significant differences seen in the pro-oxidant effect. However, the same cannot be said for the intensity of the peak current as obtained in the differential pulse voltammetry experiments (Figure 9[B]). Here it can be observed that the peak intensity correlated with the pro-oxidant activity, since **P7** produced the higher current, whereas **P0** had the lowest.

As is well known, in differential pulse voltammetry experiments the peak current is related to the concentration, mass transport, chemical properties of the studied compound and physicochemical properties of the electrode. Hence, considering that the concentration of the tested compounds and the electrode properties were the same, only mass transport or the physicochemical properties of the analyte could be responsible for the differences observed in peak current. However, mass transport, which is proportional to the diffusion coefficient and, most straightforwardly, to the size of the molecules under investigation, seems to be very similar for each molecule, since **P0**, **P2**, **P4** and **P7**, although very different in molecular volumes, present similar peak current values. Hence, we suggest that the higher peak current intensity for **P2**, **P4** and **P7** must be related to the stability (lifetime) of the particular transient phenoxyl radical formed during the oxidation process, which could affect its efficacy as an oxidant species. An additional point must be considered regarding the relationship between the peak current intensity and pro-oxidative activity. How can be observed by comparing the Figure 4 and 9, the difference in the peak current intensity for **P2**, **P4** and **P7** was not proportional to their efficacy as pro-oxidant. This is an indication that additional factors must be involved in the efficacy of the esters compared to

protocatechuic acid. However, there is no doubt that esterification was the most important factor.

It is noteworthy that in a classical peroxidase-catalysed reaction the substrates are oxidized by the active redox forms of the peroxidase, named compound I and II; then, the formed transient radicals diffuse from the enzyme active site and undergo subsequent reactions leading to dimers, trimers and higher oligomers [42]. Hence, taking into account these facts: i) the higher stability of the transient radicals of the esters of protocatechuic acid suggested by the differential pulse voltammetry experiments; and ii) the classical pathway for its formation through HRP-catalysed oxidation, the following proposal could explain our findings. Namely, after its formation in the active site of the enzyme and due to its higher lifetime, the radicals of **P2**, **P4** and **P7** could diffuse and interact with the target molecule, promoting its oxidation. On the other hand, the unstable **P0** radical would follow the usual pathway for free radicals, i.e. dimerization, oligomerization, etc. To reinforce this proposal, we also measured the differential pulse voltammetry of apocynin under the same experimental conditions. Figure 10 shows the results for **P7** and apocynin. Here it can be observed that, despite its higher oxidation peak potential, the intensity of the peak current for apocynin was significantly lower compared to **P7** and, as demonstrated above, less pro-oxidant.

Conclusions

The involvement of transient pro-oxidant free radicals as deleterious species has long been recognized. For instance, the oxidation of LDL mediated by MPO-generated paracetamol radicals [43], the depletion of GSH by AAPH-induced melatonin oxidation [44], the oxidation of NADH by raloxifene radical [24], the oxidation and depletion of GSH in HL-60 cells by etoposide radicals [25], the induction of microsomal lipid peroxidation, ascorbate and GSH by phenoxyl radicals of vitamin E and analogues [26], etc. In this context, the magnitude of the oxidation peak potential is usually accepted as the major factor determining whether the transient free radicals will act as a pro-oxidant or antioxidant substance [45]. Herein, we have demonstrated for the first time that besides the magnitude of the oxidation potential, hydrophobicity can also be an important factor. Our findings that esters of protocatechuic acid are specific and potent co-catalysts in the oxidation of hydroquinone related compounds may be very relevant, since this moiety is wide present in biomolecules (ubiquinones) and pharmaceuticals (quinone-based chemotherapeutics). Hence, its interaction with hydroquinones could exacerbate redox cycling reactions that are involved in their biological and pharmacological mechanisms of action.

Author Contributions

Conceived and designed the experiments: VFX. Performed the experiments: VFX MLZ MSP DC LOR. Analyzed the data: VFX MLZ SASM DHSS VSB DC LMF. Contributed reagents/materials/analysis tools: VFX DHLS SASM. Wrote the paper: VFX MLZ SAS.

References

1. Putchala MC, Ramani P, Sherlin HJ, Premkumar P, Natesan A (2013) Ascorbic acid and its pro-oxidant activity as a therapy for tumours of oral cavity – a systematic review. Arch Oral Biol 58: 563–574.
2. Gilloteaux J, Jamison JM, Neal DR, Loukas M, Doberzstyn T, et al. (2010) Cell damage and death by autoschizis in human bladder (RT4) carcinoma cells resulting from treatment with ascorbate and menadione. Ultrastruct Pathol 34: 140–160.
3. Martins LA, Coelho BP, Behr G, Pettenuzzo LF, Souza IC, et al. (2014) Resveratrol induces pro-oxidant effects and time-dependent resistance to

cytotoxicity in activated hepatic stellate cells. Cell Biochem Biophys 68: 247–257.
4. Aggeli IK, Koustas E, Gaitanaki C, Beis I (2013) Curcumin acts as a pro-oxidant inducing apoptosis via JNKs in the isolated perfused Rana ridibunda heart. J Exp Zool A Ecol Genet Physiol 319: 328–339.
5. Du Q, Hu B, An HM, Shen KP, Xu L, et al. (2013) Synergistic anticancer effects of curcumin and resveratrol in Hepa1-6 hepatocellular carcinoma cells. Oncol Rep 29: 1851–1858.

6. Gandhy SU, Kim K, Larsen L, Rosengren RJ, Safe S (2012) Curcumin and synthetic analogs induce reactive oxygen species and decreases specificity protein (Sp) transcription factors by targeting microRNAs. BMC Cancer 12: 564–576.

7. Porter NA (2013) A perspective on free radical autoxidation: the physical organic chemistry of polyunsaturated fatty acid and sterol peroxidation. J Org Chem 78: 3511–3524.

8. Russell LHJr, Mazzio E, Batista RB, Zhu ZP, Agharahimi M, et al. (2012) Autoxidation of gallic acid induces ROS-dependent death in human prostate cancer LNCaP cells. Anticancer Res 32: 1595–1602.

9. Metzler M, Pfeiffer E, Schulz SI, Dempe JS (2013) Curcumin uptake and metabolism. Biofactors 39: 14–20.

10. Murias M, Jäger W, Handler N, Erker T, Horvath Z, et al. (2005) Antioxidant, prooxidant and cytotoxic activity of hydroxylated resveratrol analogues: structure-activity relationship. Biochem Pharmacol 69: 903–912.

11. Ligeret H, Barthelemy S, Zini R, Tillement JP, Labidalle S, et al. (2004) Effects of curcumin and curcumin derivatives on mitochondrial permeability transition pore. Free Radic Biol Med 36: 919–929.

12. Locatelli C, Rosso R, Santos-Silva MC, de Souza CA, Licínio MA, et al. (2008) Ester derivatives of gallic acid with potential toxicity toward L1210 leukemia cells. Bioorg Med Chem 16: 3791–3799.

13. Park WH, Kim SH (2013) MAPK inhibitors augment gallic acid-induced A549 lung cancer cell death through the enhancement of glutathione depletion. Oncol Rep 30: 513–519.

14. Cavaliere V, Papademetrio DL, Lombardo T, Costantino SN, Blanco GA, et al. (2014) Caffeic acid phenylethyl ester and MG132, two novel nonconventional chemotherapeutic agents, induce apoptosis of human leukemic cells by disrupting mitochondrial function. Target Oncol 9: 25–42.

15. Fan GJ, Jin XL, Qian YP, Wang Q, Yang RT, et al. (2009) Hydroxycinnamic acids as DNA-cleaving agents in the presence of Cu (II) ions: mechanism, structure-activity relationship, and biological implications. Chem Eur J 15: 12889–12899.

16. Lende AB, Kshirsagar AD, Deshpande AD, Muley MM, Patil RR, et al. (2011) Anti-inflammatory and analgesic activity of protocatechuic acid in rats and mice. Inflammopharmacology 19: 255–263.

17. de Faria CM, Nazaré AC, Petrônio MS, Paracatu LC, Zeraik ML, et al. (2012) Protocatechuic acid alkyl esters: hydrophobicity as a determinant factor for inhibition of NADPH oxidase. Curr Med Chem 19: 4885–4893.

18. Reis B, Martins M, Barreto B, Milhazes N, Garrido EM, et al. (2010) Structure-property-activity relationship of phenolic acids and derivatives. Protocatechuic acid alkyl esters. J Agric Food Chem 58: 6986–6993.

19. Zeraik ML, Ximenes VF, Regasini LO, Dutra LA, Silva DH, et al. (2012) 4'-Aminochalcones as novel inhibitors of the chlorinating activity of myeloperoxidase. Curr Med Chem 19: 5405–5413.

20. Castor LR, Locatelli KA, Ximenes VF (2010) Pro-oxidant activity of apocynin radical. Free Radic Biol Med 48: 1636–1643.

21. dos Santo FJ, Ximenes VF, da Fonseca LM, de Faria Oliveira OM, Brunetti IL (2005) Horseradish peroxidase-catalyzed oxidation of rifampicin: reaction rate enhancement by co-oxidation with anti-inflammatory drugs. Biol Pharm Bull 28: 1822–1826.

22. Ximenes VF, Lopes MG, Petrônio MS, Regasini LO, Silva DH, et al. (2010) Inhibitory effect of gallic acid and its esters on 2,2'-azobis(2-amidinopropane)-hydrochloride (AAPH)-induced hemolysis and depletion of intracellular glutathione in erythrocytes. J Agric Food Chem 58: 5355–5362.

23. Saito S, Kawabata J (2006) DPPH (= 2,2-Diphenyl-1-picrylhydrazyl) radical-scavenging reaction of protocatechuic acid (= 3,4-Dihydroxybenzoic Acid): difference in reactivity between acids and their esters. Helv Chim Acta 89: 1395–1407.

24. Martins-Maciel ER, Campos LB, Salgueiro-Pagadigorria CL, Bracht A, Ishii-Iwamoto EL (2013) Raloxifene affects fatty acid oxidation in livers from ovariectomized rats by acting as a pro-oxidant agent. Toxicol Lett 217: 82–89.

25. Kagan VE, Kuzmenko AI, Tyurina YY, Shvedova AA, Matsura T, et al. (2001) Pro-oxidant and antioxidant mechanisms of etoposide in HL-60 cells: role of myeloperoxidase. Cancer Res. 61: 7777–7784.

26. Tafazoli S, Wright JS, O'Brien PJ (2005) Prooxidant and antioxidant activity of vitamin E analogues and troglitazone. Chem Res Toxicol 18: 1567–1574.

27. Núñez-Delicado E, Sánchez-Ferrer A, García-Carmona F (1997) A kinetic study of the one-electron oxidation of Trolox C by the hydroperoxidase activity of lipoxygenase. Biochim Biophys Acta 1335: 127–134.

28. Thomas MJ, Bielski BHJ (1989) Oxidation and reaction of trolox c, a tocopherol analog, in aqueous solution. A pulse-radiolysis study. J Am Chem Soc 111: 3315–3319.

29. Zou L, Xu Y, Luo P, Zhang S, Ye B (2012) Electrochemical detection of dihydromyricetin using a DNA immobilized ethylenediamine/polyglutamic modified electrode. Analyst 137: 414–419.

30. Galati G, Sabzevari O, Wilson JX, O'Brien PJ (2002) Prooxidant activity and cellular effects of the phenoxyl radicals of dietary flavonoids and other polyphenolics. Toxicology 177: 91–104.

31. Hadzi-Tasković Sukalović V, Vuletić M, Vucinić Z, Veljović-Jovanović S (2008) Effectiveness of phenoxyl radicals generated by peroxidase/H2O2-catalyzed oxidation of caffeate, ferulate, and p-coumarate in cooxidation of ascorbate and NADH. J Plant Res 121: 115–123.

32. Ximenes VF, Kanegae MP, Rissato SR, Galhiane MS (2007) The oxidation of apocynin catalyzed by myeloperoxidase: proposal for NADPH oxidase inhibition. Arch Biochem Biophys 457: 134–141.

33. Ghosh A, Kanthasamy A, Joseph J, Anantharam V, Srivastava P, et al. (2012) Anti-inflammatory and neuroprotective effects of an orally active apocynin derivative in pre-clinical models of Parkinson's disease. J Neuroinflammation 9: 241–257.

34. Simonyi A, Serfozo P, Lehmidi TM, Cui J, Gu Z, et al. (2012) The neuroprotective effects of apocynin. Front Biosci 4: 2183–2193.

35. Senejoux F, Girard-Thernier C, Berthelot A, Bévalot F, Demougeot C (2013) New insights into the mechanisms of the vasorelaxant effects of apocynin in rat thoracic aorta. Fundam Clin Pharmacol 27: 262–270.

36. Wondrak GT (2009) Redox-Directed Cancer Therapeutics: Molecular Mechanisms and Opportunities. Antioxid Redox Signal 11: 3013–3069.

37. Chan T, Galati G, O'Brien PJ (1999) Oxygen activation during peroxidase catalysed metabolism of flavones or flavanones. Chem Biol Interact 122: 15–25.

38. Petrônio MS, Zeraik ML, Fonseca LM, Ximenes VF (2013) Apocynin: Chemical and Biophysical Properties of a NADPH Oxidase Inhibitor. Molecules 18: 2821–2839.

39. Galato D, Ckless K, Susin MF, Giacomelli C, Ribeiro-do-Valle RM, et al. (2001) Antioxidant capacity of phenolic and related compounds: correlation among electrochemical, visible spectroscopy methods and structure-antioxidant activity. Redox Rep 6: 243–250.

40. Gunckel S, Santander P, Cordano G, Ferreira J, Munoz S, et al. (1998) Antioxidant activity of gallates: an electrochemical study in aqueous media. Chem Biol Interact 114: 45–59.

41. Gaspar A, Garrido EM, Esteves M, Quezada E, Milhazes N, et al. (2009) New insights into the antioxidant activity of hydroxycinnamic acids: Synthesis and physicochemical characterization of novel halogenated derivatives. Eur J Med Chem 44: 2092–2099.

42. Veitch NC (2004) Horseradish peroxidase: a modern view of a classic enzyme. Phytochemistry 65: 249–259.

43. Kapiotis S, Sengoelge G, Hermann M, Held I, Seelos C, et al. (1997) Paracetamol catalyzes myeloperoxidase-initiated lipid oxidation in LDL. Arterioscler Thromb Vasc Biol 17: 2855–2860.

44. Ximenes VF, Pessoa AS, Padovan CZ, Abrantes DC, Gomes FH, et al. (2009) Oxidation of melatonin by AAPH-derived peroxyl radicals: evidence of a pro-oxidant effect of melatonin. Biochim Biophys Acta 1790: 787–792.

45. Simić A, Manojlović D, Šegan D, Todorović M (2007) Electrochemical behavior and antioxidant and prooxidant activity of natural phenolics. Molecules 12: 2327–2340.

15

Deletion of Thioredoxin Interacting Protein (TXNIP) Augments Hyperoxia-Induced Vaso-Obliteration in a Mouse Model of Oxygen Induced-Retinopathy

Mohammed A. Abdelsaid[1,3,4], Suraporn Matragoon[1,2,4], Adviye Ergul[1,3,4], Azza B. El-Remessy[1,2,4]*

1 Clinical and Experimental Therapeutics, University of Georgia, Augusta, Georgia, United States of America, 2 Culver Vision Discovery Institute, Augusta, Georgia, United States of America, 3 Department of Physiology, Georgia Regents University, Augusta, Georgia, United States of America, 4 Charlie Norwood VA Medical Center, Augusta, Georgia, United States of America

Abstract

We have recently shown that thioredoxin interacting protein (TXNIP) is required for VEGF-mediated VEGFR2 receptor activation and angiogenic signal. Retinas from TXNIP knockout mice (TKO) exhibited higher cellular antioxidant defense compared to wild type (WT). This study aimed to examine the impact of TXNIP deletion on hyperoxia-induced vaso-obliteration in ischemic retinopathy. TKO and WT pups were subjected to oxygen-induced retinopathy model. Retinal central capillary dropout was measured at p12. Retinal redox and nitrative state were assessed by reduced-glutathione (GSH), thioredoxin reductase activity and nitrotyrosine formation. Western blot and QT-PCR were used to assess VEGF, VEGFR-2, Akt, iNOS and eNOS, thioredoxin expression, ASK-1 activation and downstream cleaved caspase-3 and PARP in retinal lysates. Retinas from TKO mice exposed to hyperoxia showed significant increases (1.5-fold) in vaso-obliteration as indicated by central capillary drop out area compared to WT. Retinas from TKO showed minimal nitrotyrosine levels (10% of WT) with no change in eNOS or iNOS mRNA expression. There was no change in levels of VEGF or activation of VEGFR2 and its downstream Akt in retinas from TKO and WT. In comparison to WT, retinas from TKO showed significantly higher level of GSH and thioredoxin reductase activity in normoxia but comparable levels under hyperoxia. Exposure of TKO to hyperoxia significantly decreased the anti-apoptotic thioredoxin protein (~50%) level compared with WT. This effect was associated with a significant increase in activation of the apoptotic ASK-1, PARP and caspase-3 pathway. Our results showed that despite comparable VEGF level and signal in TKO, exposure to hyperoxia significantly decreased Trx expression compared to WT. This effect resulted in liberation and activation of the apoptotic ASK-1 signal. These findings suggest that TXNIP is required for endothelial cell survival and homeostasis especially under stress conditions including hyperoxia.

Editor: Jing Chen, Children's Hospital Boston, United States of America

Funding: Source of research support: American Heart Association predoctoral fellowship award #10PRE3660004 to M.A.A., VDI pilot grant to A.B.E., Career Development Award from Juvenile Diabetes Research Foundation (2-2008-149) to A.B.E. and R01-EY022408 to A.B.E. A.E. is a Research Career Scientist at the Charlie Norwood Veterans Affairs Medical Center in Augusta, Georgia. This work was supported in part by VA Merit Award (BX000347), VA Research Career Scientist Award, and NIH (R01NS083559) to A.E. The contents do not represent the views of any of the funding agencies, the Department of Veterans Affairs or the United States Government. The funders had no role in study design, data collection and analysis, decision to publish, or preparation of the manuscript.

Competing Interests: The authors have declared that no competing interests exist.

* Email: aelremessy@gru.edu

Introduction

Imbalance in cellular redox system has been linked to be several cardiovascular disorders such as ischemic heart disease, inflammation, atherosclerosis and aberrant angiogenesis (reviewed in [1-3]). The thioredoxin (Trx) system, major regulator of antioxidant defense represents an attractive target for oxidative stress-associated disorders. Trx is a multifunctional protein that acts as a protein disulfide reductase and participates in redox-dependent processes [3], including protein folding, regulation of apoptosis and antioxidant protection from oxidative stress. Trx has 2 isoforms, cytosolic/nuclear (Trx-1) and mitochondrial (Trx-2). Overexpression of Trx in transgenic mice increases the resistance to various oxidative stresses and attenuates ischemic damage in the brain [4] and myocardial infarction [5]. Expression and activity of Trx are tightly regulated by the endogenous inhibitor thioredoxin-interacting protein (TXNIP), where TXNIP binds Trx and limit its ability to interact with other proteins [6]. As such, TXNIP regulates Trx-dependent cellular redox state and increases reactive oxygen species and oxidative stress [7].

TXNIP is a stress sensor and its expression can be induced to a various number of exogenous and endogenous stimuli including inflammation, metabolic stress, changes in calcium levels, and also in response to changes in oxygen levels [8-13]. Increased levels of TXNIP bind more thioredoxin limiting its anti-apoptotic effects by releasing the apoptosis signal–regulating kinase 1 (ASK-1) from the inhibitory complex [14,15]. We and others showed that TXNIP plays a pro-apoptotic role as it binds and inhibits Trx releasing free ASK-1 resulting in cell death [8,13,16–19]. Although TXNIP lacks specific pharmacological inhibitor, studies using calcium channel blockers, quercetin or T-resveratrol demonstrated neuro- and vascular protective effects that were associated with TXNIP inhibition [13,17,20–22]. Genetic deletion of TXNIP (TKO) demonstrated significant increases in antioxidant defense compared to wild-type mice [23,24]. Retinas from TKO showed similar vascular density to WT littermates as recently characterized by our group [23]. Interestingly, TXNIP expression is required to achieve homeostasis of redox state in endothelial cells [23,25]. Silencing TXNIP expression impaired VEGF receptor activation and angiogenic response VEGF via redox-dependent and independent pathways [23,25]. Here we examined the impact of TXNIP deletion on hyperoxia-induced vaso-obliteration. Oxygen induced retinopathy model is a well-established model that utilize high oxygen to induce oxidative stress, endothelial cell ischemia and apoptosis in the developing retina [26]. Our initial hypothesis was that TKO mice will be protected against the hyperoxia induced vaso-obliteration. Instead, our results showed that TKO mice are more vulnerable to oxygen induced retinopathy model. The current study investigates the molecular mechanism involved to better understand the complex nature of redox regulation.

Materials and Methods

Animals

Experiments were approved by the Institutional Committee for Animal Use in Research and Education at Charlie Norwood VA medical Center (ACORP # 04-12-044) and conformed to the ARVO Statement for the Use of Animals in Ophthalmic and Vision Research. All experiments were performed using age-matched wild type (WT) mice C57Bl/6 mice (Jackson Laboratory, Bar Harbor, Maine) and TXNIP knockout mice (TKO) that was provided as a kind gift from Dr. AJ Lusis and Dr. ST Hui at the BioSciences Center, San Diego State University, San Diego, CA. TKO mice have a global knockout of the expression of functional TXNIP as characterized previously [24]. TKO mice are similar in weight and activity to WT or heterozygous littermates, with no differences in food consumption or litter sizes.

TKO breeding and genotyping

Littermates of WT and homozygous TKO were used and genotyping was performed as described previously [24]. Briefly, DNA was prepared by incubating ear tissue with proteinase K and digestion buffer for one hour at 95°C. A mixture of primer sequence (5′-TGA-GGT-GGT-CTT-CAA-CGA-CC-3′. 5′GGA-AAG-ACA-ACG-CCA-GAA-GG-3′ and 5′-CCT-TGA-GGA-AGC-TCG-AAG-CC-3′ (IDT San Diego, CA)), buffer and 2 mM $MgCl_2$ and polymerase enzyme (GoTAG Hot start polymerase, Promega, Madison, WI) were added to the DNA template. DNA segments were amplified using the Master plex-RealPlex2 (Eppendorf, Germany) and were detected with 1% agarose gel electrophoresis. Deleted TXNIP allele was detected at 530 bp while wild type was detected at 699 bp.

Oxygen induced retinopathy model

Oxygen induced retinopathy was performed as described previously [27]. Briefly, on postnatal day 7 (p7), mice were placed along with their dam into a custom-built chamber (Biospherix, Redfield, NY) in which the partial pressure of oxygen was maintained at 70% for 5 days. At p12 pups were deeply anesthetized by IP injection of Avertin 240 mg/kg and sacrificed by jugular vein cut. One eye was enucleated and fixed in 2% paraformaldehyde overnight to be flat- mounted for vascular density. For the other eye, retinas were isolated and snap frozen for biochemical assays.

Assessment of retinal vascular density and central capillary dropout areas

We examined the effect of TXNIP deletion on retinal vascular development at post-natal day (p7) in normoxic animals. Retinal capillary dropout areas were analyzed at p12 after hyperoxic exposure. Retinas of both wild type (WT) and TXNIP knockout (TKO) mice were inoculated and fixed in 4% paraformalhyde and flat-mounted. Retinas were labeled with the red fluorescent Alexa Fluor 594 isolectin GS-IB₄ conjugate (Molecular Probes, Life Technology, Grand Island, NY) to quantify retinal vascular density. Retinas were viewed and imaged with fluorescence AxioObserver Zeiss Microscope (Germany). Images were then processed using Image J software to skeletonize and quantify the vascular density as described previously [23,27]. Results were expressed as percentage of the total retinal area.

Oxidized- and reduced-glutathione

Reduced glutathione was measured using the Northwest Life Science kit (Vancouver, WA) as described before [23,27]. Briefly, reduced-GSH was calculated by subtracting the oxidized-GSSG from the total glutathione. For total glutathione, cells were lysed in phosphate buffer (100 mM potassium phosphate and 1 mM EDTA) and were mixed with an equal volume of 10 mM 5, 5′-dithiobis 2-nitrobenzoic acid (DTNB) in the presence of glutathione reductase and NADPH producing a yellow color measured at 412 nm. To detect GSSG, samples were treated with 10 mM 2-vinylpyridine (Sigma) in ethanol to sequester all the reduced GSH then measured using the same protocol as the total glutathione.

Thioredoxin reductase activity (Trx-R)

Thioredoxin reductase activity was performed using a kit (Sigma) as described previously [8,17]. Briefly, retinal samples were homogenized in assay buffer followed by the addition of DTNB with NADPH. Reduction of DTNB produced a strong yellow color that was measured at 412 nm. Thioredoxin reductase activity was measured as the difference between DTNB-reaction measurement of each sample in the presence and absence of a selective thioredoxin reductase inhibitor (provided in the kit) and expressed as unit/µg/min.

Quantitative real time PCR

The One-Step qRT-PCR kit (Invitrogen) was used to amplify 10 ng retinal mRNA and quantification was performed as described previously [23]. PCR primers (listed in Table 1) were purchased from Integrated DNA Technologies Inc. (IDT, Coralville, IA). Quantitative PCR was performed using a Realplex Master cycler (Eppendorf, Germany). Expression of TXNIP, Trx-1, Trx-2, VEGF, eNOS, iNOS was normalized to the 18S level and expressed as relative expression to control.

Western blot analysis

Protein expression in retina lysate was analyzed as described previously [23]. For VEGF, retinal lysates were subjected to heparin beads (Sigma) as described before [27,28]. The beads were pelleted at 5000 × g for 1 min, washed in 400 mM NaCl and 20 mM Tris and loaded onto a 4-20% gradient Trisglycine pre-cast gel (BioRad). The primary antibodies were purchased as follow: VEGF (Rabbit polyclonal, EMD-Millipore), phosphor-VEGFR2, VEGFR2, phospho-Akt, Akt, phospho-ASK-1, ASK-1, cleaved caspase-3 (Rabbit polyclonal, Cell Signaling Tech, Danvers, MA), total Trx (Mouse monoclonal, Santa Cruz, Dallas, TX), and TXNIP (Rabbit polyclonal, Invitrogen, Grand Island, NY), cleaved PARP (BD Bioscience Pharmingen, San Diego, CA).

Table 1. Primer sequence used to quantify mRNA expression levels using PCR analysis.

Gene	Forward primers	Reverse primers
TXNIP	5′AAGCTGTCCTCAGTCAGAGGCAAT3′	5′ATGACTTTCTTGGAGCCAGGGACA3′
VEGF	5′TGAGCCTTGTTCAGAGCGGAGAAA3′	5′TTCGTTTAACTCAAGCTGCCTCGC3′
Trx-1	5′TCAAGCCCTTCTTCCATTCC3′	5′GTCGGCATGCATTTGACTTC 3′
Trx-2	5′ CGCGGCTAGAGAAGATGGTC3′	5′TTGATGGCTAGCACGGTAGG3′
eNOS	5′ GCAGTGAAGATCTCTGCCTCA3′	5′AGAATGGTTGCCTTCACACG3′
iNOS	5′ CACCTTGGAGTTCACCCAGT3′	5′ ACCACTCGTACTTGGGATGC3′
18S	5′CGCGGTTCTATTTTGTTGGT3′	5′AGTCGGCATCGTTTATGGTC3′

Primary antibodies were detected using a horseradish peroxidase-conjugated antibody and enhanced chemiluminescence (GE Healthcare, Piscataway NJ).The films were scanned, and band intensity was quantified using densitometry software (Alpha Innotech Fluorchem, Santa Clara, CA).

Detection of nitrotyrosine

Relative amounts of proteins nitrated on tyrosine were measured by use of slot-blot techniques as described previously [27]. In brief, retinal homogenate was immobilized onto nitrocellulose membrane by use of a slot-blot microfiltration unit (Bio-Rad Laboratories). After being blocked, membranes were incubated with a 1:500 dilution of nitrotyrosine (Calbiochem, San Diego, CA, USA) antibody followed by HRP-conjugated sheep anti-rabbit antibody and enhanced chemiluminescence (GE Healthcare). Relative levels of nitrotyrosine immunoreactivity were determined by densitometry software (Alpa Innotech).

Data analysis

All the results were expressed as mean ± SE and the data were evaluated for normality and appropriate transformations were used when necessary. Vaso-obliteration was evaluated by analysis of variance, and the significance of difference between groups was assessed by the post-hoc test (Fisher's PLSD) and significance was defined as $P < 0.05$. A two-way ANOVA was used to examine the effect of manipulation gene (WT vs. TKO) and oxygen levels (normoxia vs. hyperoxia) and their interaction on mRNA and expression of VEGF, TRX-1, TRX-2, eNOS and iNOS; expression of total TRX, NY, pAkt, pVEGFR2, pASK-1, Caspase-3 and PARP; TrxR activity and reduced GSH levels. Two-way ANOVA followed by Bonferroni post-tests were used to adjust for the multiple comparisons used to assess significant effects. NCSS 2007 was used for all analyses (NCSS, version 07.1.14 LLC, Kaysville, UT). Statistical significance was determined at alpha = 0.05.

Results

Deletion of TXNIP augments hyperoxia-induced retinal vaso-obliteration

Exposure of the developing rodent retina (post-natal day p7 to p12) to high levels of oxygen causes vascular cell death as indicated by central capillary dropout [26]. Increases in oxidative stress and peroxynitrite have been shown to cause vascular cell loss in ischemic retinopathy model [27,29–31]. Retinas from TKO demonstrate comparable vascular density to their WT littermates at normal oxygen level at p7 (Fig. S1) and at p12 as recently characterized by our group [23]. We and others have previously showed that deletion of TXNIP enhanced antioxidant defense and decreased oxidative stress [8,23,24]. Therefore, TKO mice were predicted to show higher vascular protection against hyperoxia compared to wild type (WT) mice. In contrast, deletion of TXNIP aggravated hyperoxia-induced vaso-obliteration as indicated by significant increase (1.6-fold) in central capillary dropout areas compared to WT at p12 (Fig. 1A–C). Of note, hyperoxia drives retinal TXNIP mRNA expression (1.5-fold) in WT mice but not TKO (Fig. 1D).

Deletion of TXNIP decreases nitrative stress under normoxia and hyperoxia

We next examined the levels of nitrotyrosine (NY), the footprint of peroxynitrite which is believed to mediate the detrimental effects of hyperoxia. As shown in Fig. 2A, hyperoxia significantly increased retinal NY formation (3-fold) in WT compared to normoxia. TKO showed minimal level of NY (10%) at normoxia and 20% at hyperoxia compared to WT. The two-way ANOVA showed a significant interaction between TKO and WT in response to high levels of oxygen (Fig. 2A). We also examined the expression of eNOS and iNOS at the mRNA level. Our results showed a significant interaction between TKO and WT in eNOS mRNA levels. TKO retinas showed 1.6-fold increase under normoxic conditions compared to WT-normoxia. Hyperoxia induced significant reduction in retinas from WT and TKO 52% and 51%, respectively compared to retinas from WT-normoxia (Fig. 2B). There was no significant interaction between TKO and WT in iNOS mRNA levels. Hyperoxia caused significant decrease in iNOS mRNA levels in both WT and TKO 54% and 51%, respectively compared to normoxia (Fig. 2C). These results suggest that TKO retinas had less tyrosine nitration at both normoxia and hyperoxia compared to WT.

Deletion of TXNIP does not alter VEGF levels under normoxia and hyperoxia

We next examined the effect of TXNIP deletion on VEGF levels under hyperoxia. Our 2×2 analysis showed no significant interaction between TKO and WT in the VEGF mRNA levels. We detected a significant decrease in the VEGF mRNA in groups exposed to hyperoxic conditions compared to the normoxia (Fig. 3A). On the other hand, we did not detect any significance interaction in the VEGF protein expression between groups (Fig. 3B). Our results show that TXNIP deletion did not alter VEGF protein levels compared to WT under normoxic or hyperoxic conditions.

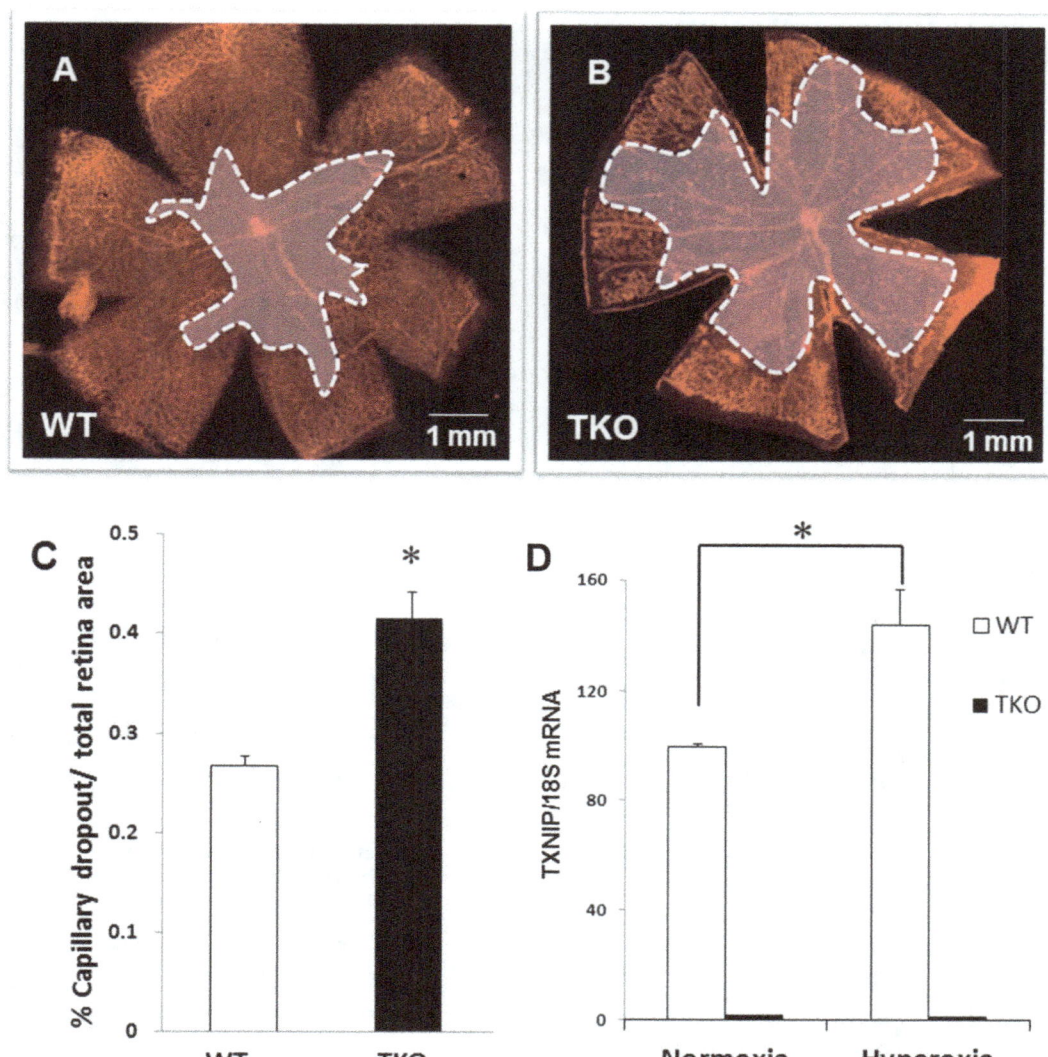

Figure 1. Deletion of TXNIP augments hyperoxia-induced vaso-obliteration compared to WT. Wild type (WT) and TXNIP knockout (TKO) mice were subjected to hyperoxia (75% O2, p7–p12). Retinas were fixed and stained with iso-lectin B4 to quantify oxygen induced vaso-obliteration. **A–C**) Retinas from TKO mice exposed to hyperoxia showed significant increases in vaso-obliteration compared to WT. (*P<0.05 vs WT, n = 12). **D**) Hyperoxia stimulates TXNIP expression mRNA in WT but not in TKO mice. (*P<0.05 vs WT normoxia, n = 4)

Deletion of TXNIP impairs VEGFR2/Akt activation in normoxia but not under hyperoxia

Our recent studies demonstrated that retinas from TKO mice showed similar level of VEGF but less VEGFR2 activation compared to WT under hypoxic condition [23]. Indeed, we detected a significant reduction in VEGFR2 phosphorylation in normoxic TKO retinas compared to normoxic WT (Fig. 4A). This effect was paralleled with significant decreases in the activation of survival protein Akt (Fig. 4B). Similar to our previous findings [27], hyperoxia decreased activation of Akt in WT compared to normoxic controls at p12 (Fig. 4B). Hyperoxia increased activation of VEGFR2 and its downstream target Akt in TKO mice to a comparable level with control WT. Two-way ANOVA did not show any significant interaction between TKO and WT in activation of VEGFR2 or Akt under hyperoxic conditions. Together, these results exclude the possibility that alteration in

VEGF or VEGFR2 activation causes the aggravated vaso-obliteration response of TKO to hyperoxia.

Deletion of TXNIP increases antioxidant defense and thioredoxin reductase activity

We next examined reduced-glutathione levels (GSH) as marker of retinal antioxidant defense. As shown in Fig. 5A, TKO showed a significant interaction when compared to WT in the levels of GSH under normoxic (5-fold) and hyperoxic (2.25-fold) conditions. Similar trend was observed when we measured thioredoxin reductase (Trx-R) activity (Fig. 5B). Our results showed a significant interaction between TKO and WT in Trx-R activity when exposed to hyperoxia. TKO showed a significant increase in TrxR activity both under normoxic (1.6-fold) and hyperoxic conditions (1.25-fold) when compared to WT. These results confirmed that retinas from TKO had a higher antioxidant defense compared to WT.

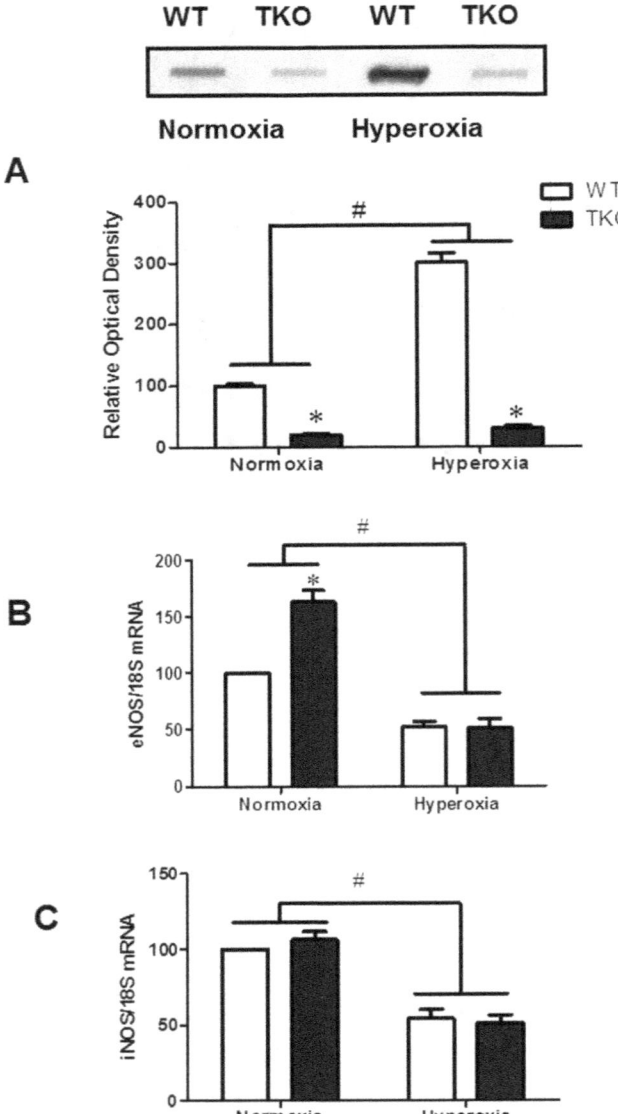

Figure 2. Deletion of TXNIP decreases nitrative stress under normoxia and hyperoxia. A) Retinas of TKO showed significantly less nitrotyrosine levels at normoxia or hyperoxia compared to WT. **B)** TKO showed higher eNOS mRNA level in normoxia. Hyperoxia significantly reduced eNOS mRNA levels. A 2×2 analysis showed a significant interaction between the gene (TKO) and the oxygen level (hyperoxia) in both nitrotyrosine and eNOS levels. C) We did not detect difference between TKO and WT in the expression of iNOS. Hyperoxia caused significant reduction of iNOS compared to normoxia. (#P<0.05 Hyperoxia vs Normoxia, *P<0.05, TKO vs WT, n=4–6).

Deletion of TXNIP decreases Trx mRNA and protein levels under hyperoxia

As the Trx-R activity increases, the availability of free thioredoxin increases. We examined the effect of TXNIP deletion on levels of Trx levels in response to hyperoxia both on the transcriptional and expression levels. Trx has 2 isoforms, cytosolic/nuclear (Trx-1) and mitochondrial (Trx-2). Our results showed a significant interaction in the levels of Trx-1 and Trx-2 mRNA between gene (TKO and WT) under different oxygen level (normoxia vs hyperoxia). At normoxia, retinas from TKO had

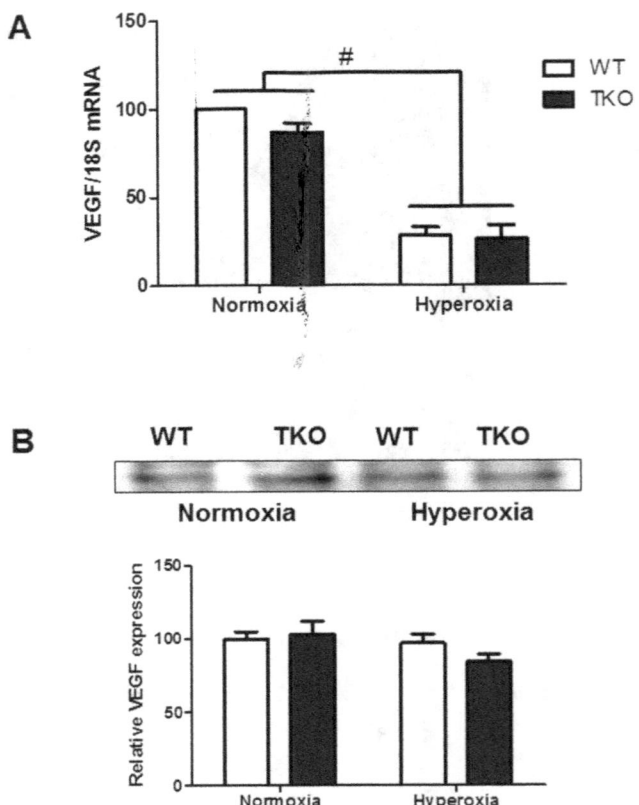

Figure 3. Deletion of TXNIP does not alter VEGF levels under normoxia or hyperoxia. (A) VEGF mRNA levels were detected from various groups using rt-PCR. (B) VEGF protein expression was examined using heparin-bound beads from p12 WT and TKO retinas. There was no change in levels of VEGF mRNA or VEGF expression between WT and TKO under normoxia. Hyperoxia caused significant decrease in VEGF mRNA compared to normoxia. Hyperoxia did not alter VEGF protein levels from normoxia in WT and TKO. (#P<0.05 Hyperoxia vs Normoxia, n=4–6).

significantly higher mRNA expression of Trx-1 (1.5-fold) and Trx-2 (1.4-fold), respectively compared to WT (Fig. 6A–B). Hyperoxia induced mRNA levels 1.5-fold and 1.25 fold in Trx-1 and Trx-2, respectively in WT (Fig. 6A–B). In contrast, hyperoxia significantly decreased mRNA levels for Trx-1 (50%) and Trx-2 (45%) compared to TKO under normoxia (Fig. 6A–B). Next we examined total Trx protein expression and 2×2 analysis showed a significant interaction between TKO and WT in Trx protein expression. While total Trx expression was increased in TKO under normoxia (1.52 fold), hyperoxia caused a significant increase (1.5-Fold) in WT but a 40% reduction in the total Trx expression in TKO mice (Fig. 6C).

TXNIP deletion increases ASK-1 activation and pro-apoptotic signal under hyperoxia

Trx is a negative regulator of the Apoptosis signal-regulating kinase-1 (ASK-1) pro-apoptotic pathway through direct binding to the N-terminal region of ASK-1 [32]. Hyperoxia decreased Trx expression in TKO animal, therefore, we examined the activation of ASK-1 and downstream apoptotic signal. A 2×2 analysis showed a significant interaction between TKO and WT in ASK-1 activation. While no significant difference was detected at normoxia, hyperoxia caused a significant increase in ASK-1 phosphorylation in TKO compared to WT (Fig. 7A). We next

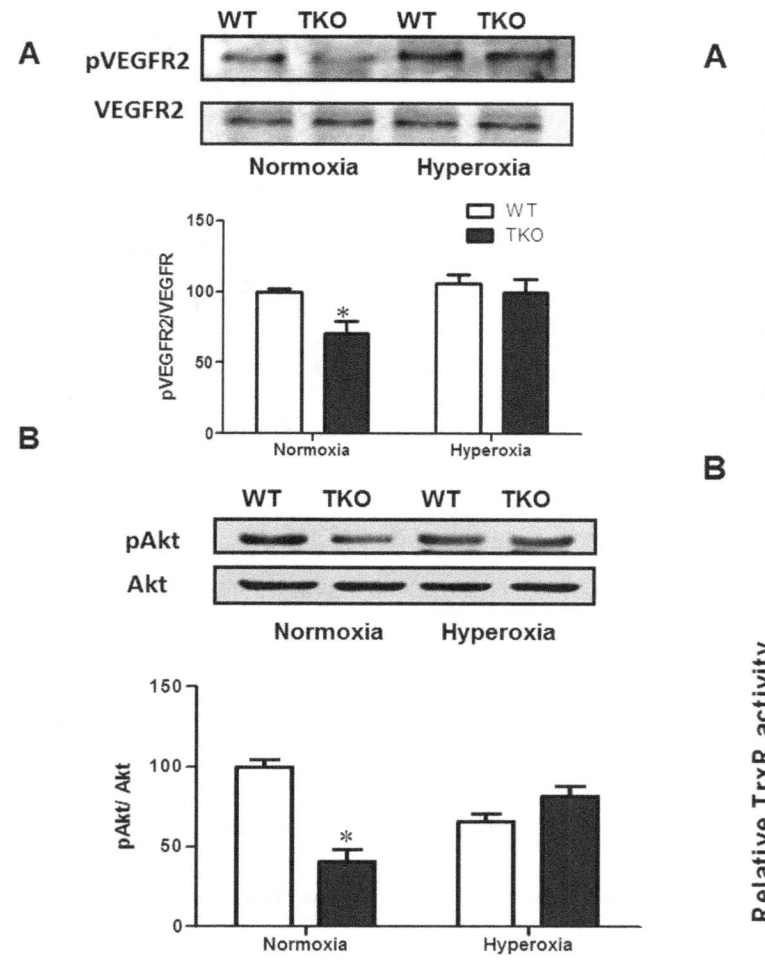

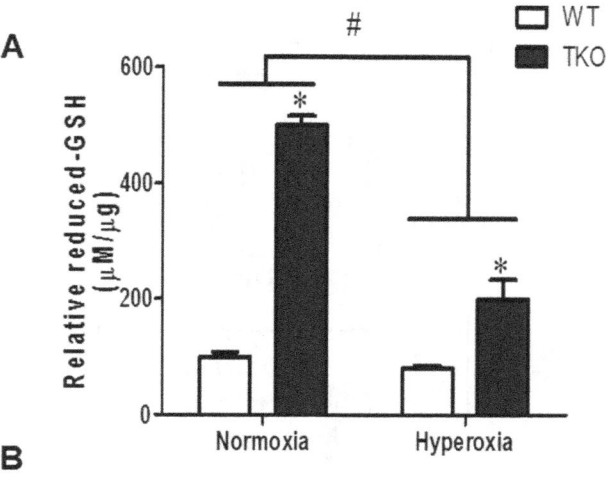

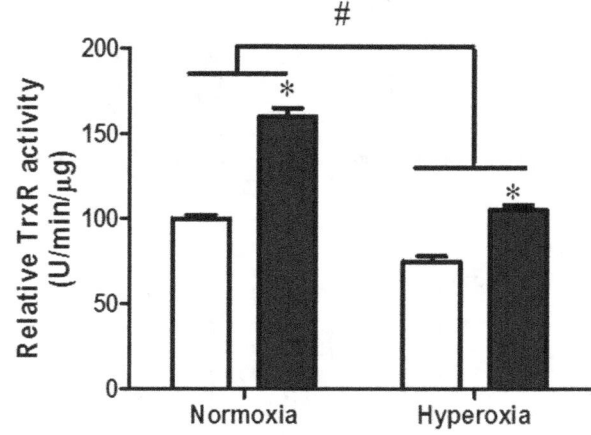

Figure 4. Deletion of TXNIP does not alter VEGFR2/Akt activation under hyperoxia. Wild type (WT) and TXNIP knockout (TKO) mice were subjected to hyperoxia (75% O2, p7–p12). Activation of VEGFR2 (**A**) and Akt (**B**) were examined as downstream signal of VEGF in p12 WT and TKO retinas. TKO showed significant decrease in phosphorylation of VEGFR-2 and Akt compared to WT under normoxic condition. We did not detect significant change in the activation of VEGFR2 and its downstream Akt in retinas from TKO and WT in response to hyperoxia. (*P<0.05, TKO vs WT, n = 4–6).

Figure 5. Deletion of TXNIP increases antioxidant defense level under normoxia and hyperoxia. In comparison to WT, retinas from TKO showed significantly higher level of reduced-GSH (**A**) and thioredoxin reductase activity (**B**) under normoxia. A 2×2 analysis showed a significant interaction between gene (TKO) and oxygen levels (hyperoxia) in both reduced-GSH and thioredoxin reductase activity measurements. (#P<0.05 Hyperoxia vs Normoxia, *P<0.05, TKO vs WT, n = 6–8).

examined the downstream apoptotic signal in TKO mice. Although no significant interaction was detected between TKO and WT, hyperoxia caused increased apoptotic signal in both WT and TKO mice. In WT, hyperoxia caused 2.5-fold and 2.65-fold increase in caspase-3 cleavage and PARP expression respectively. In TKO, hyperoxia caused 3-fold and 3.5-fold increase in caspase-3 cleavage and PARP expression respectively.

Discussion

The main finding of the present study is: Despite increased antioxidant defense and decreased nitrative stress, TKO mice were more vulnerable to hyperoxia and had aggravated retinal vascular cell death (Fig. 1,2,5). These effects were not associated with changes in either retinal VEGF expression or activation of VEGFR-2/Akt (Fig. 3,4). Hyperoxia caused significant decrease in thioredoxin expression that was associated with activation ASK-1 apoptotic signal in TKO mice compared to WT (Fig. 6,7). These

findings suggest that TXNIP expression is required for homeostasis of anti-apoptotic function of thioredoxin-ASK-1 complex in the retina in response to hyperoxia as depicted in Fig. 8.

Our current and previous work [23] demonstrated that retinas from TKO mice show similar level of vascular density at p7 and p12 to their WT-littermates, respectively. At normoxia, retinas from TKO have similar VEGF to WT mice, however higher level of Trx and less VEGFR2 activation suggesting compensatory mechanism of different angiogenic pathways (Trx-mediated pathway versus VEGFR2) that eventually resulted in normal retina vascular development. Nevertheless, exposure of TKO to drastic changes in oxygen level demonstrated different response of TKO in comparison to WT. Exposure of TKO to relative hypoxia further decreased VEGFR2 activation and angiogenic response [23]. Here, exposure of TKO to hyperoxia aggravated vascular cell death despite increasing activation of VEGFR2 to a comparable level of WT, suggesting differential response of TKO and activation of different signaling pathways in response

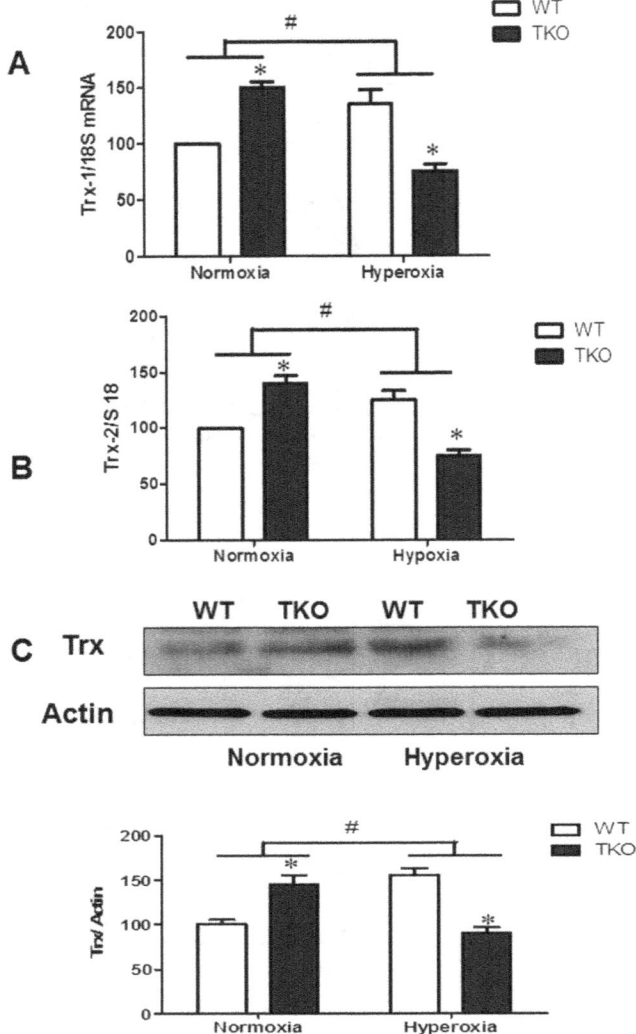

Figure 6. Deletion of TXNIP decreases thioredoxin levels under hyperoxic conditions. WT and TKO mice were subjected to 5 days hyperoxia (p7–12). Retinas were collected for protein and mRNA levels of thioredoxin (cytoplasmic Trx-1) (**A**) as well as mitochondrial Trx-2 (**B**) and total thioredoxin protein (**C**). A 2×2 analysis showed a significant interaction between gene (TKO) and oxygen levels (Hyperoxia) in both Trx-1 and Trx-2 as well as Trx total expression. Exposure of TKO pups to hyperoxia significantly decreased the anti-apoptotic thioredoxin on both mRNA as well as protein level compared to WT. (#P<0.05 Hyperoxia vs Normoxia, *P<0.05, TKO vs WT, n = 4–6).

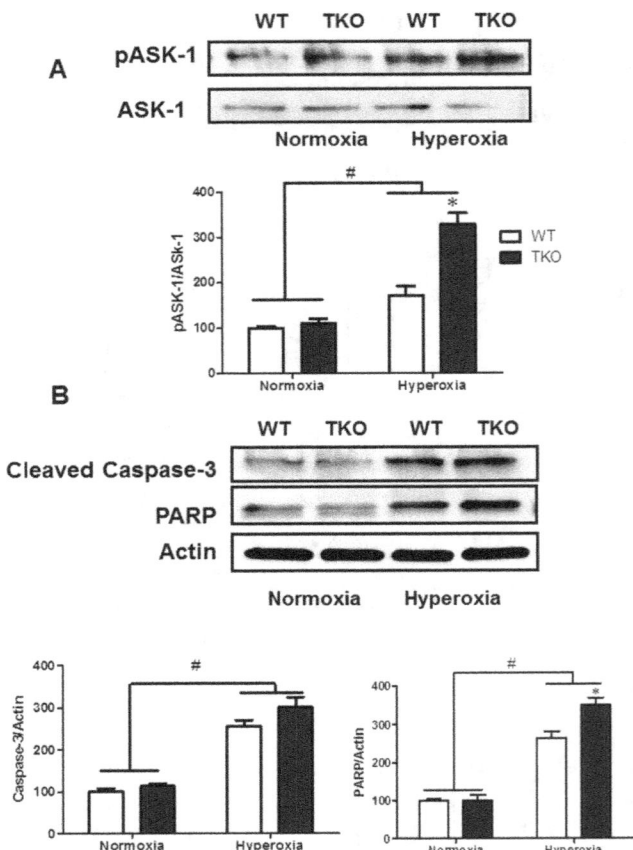

Figure 7. Deletion of TXNIP augments hyperoxia-induced ASK-1 activation and apoptotic markers. Wild type (WT) and TXNIP knockout (TKO) mice were subjected to 5 days hyperoxia (p7–12). Retinas were collected for protein ASK-1 (**A**) and apoptotic markers (Cleaved Caspase-3 and PARP) (**B**). A 2×2 analysis showed a significant interaction between gene (TKO) and oxygen levels (Hyperoxia) in activation of ASK-1. In parallen, Hyperoxia caused significant increase in cleaved caspase-3 and PARP expression compared to normoxia in WT and TKO. Reduced Trx levels were associated with a significant increase in activation of the apoptotic ASK-1, PARP and caspase-3 pathway. (#P<0.05 Hyperoxia vs Normoxia, *P<0.05, TKO vs WT, n = 4–6).

to hyperoxia versus hypoxia. Oxygen toxicity has been attributed to increases in oxidative and nitrative stress that can modulate levels of VEGF, the main survival factor of endothelial cells [33] or activation of VEGFR2 and its downstream signal [27,34]. We and others have demonstrated that preventing peroxynitrite formation and tyrosine nitration prevent capillary dropout in vivo [27,29-31] and retinal endothelial cell death in vitro [34-36]. Indeed, exposure of WT mice to hyperoxia induced capillary dropout and tyrosine nitration that was associated with decreases in phosphorylation of the survival protein Akt. These results confirmed our previous findings using same model where WT retinas showed similar VEGF and VEGFR2 activation yet, impaired Akt activation due to nitration of p85, the regulatory

subunit of the PI3-kinase [27,34]. In comparison to WT, retinas from TKO mice demonstrated similar VEGF level and minimal levels of tyrosine nitration and less VEGFR2/Akt activation in normoxia as previously characterized [23]. However, upon exposure to hyperoxia, retinas from TKO showed improved activation of VEGFR2 and Akt that became comparable to WT but still TKP showed greater capillary dropout area. These findings suggest that decreases in nitration and restoration of Akt survival signal were not enough to rescue TKO from oxygen toxicity.

Several studies showed that TXNIP is a regulator of cellular redox status and has emerged as a key component in the link between redox regulation and the pathogenesis of diseases (reviewed in [1–3]). While targeting TXNIP represents an attractive strategy toward achieving less oxidative stress, and inflammation and preventing neurotoxicity [17,37,38], its genetic deletion provides different insight. Silencing or reducing TXNIP expression should result in increase in Trx availability with subsequent beneficial action. Indeed, retinas from TKO showed higher Trx at the mRNA and protein expression that resulted in

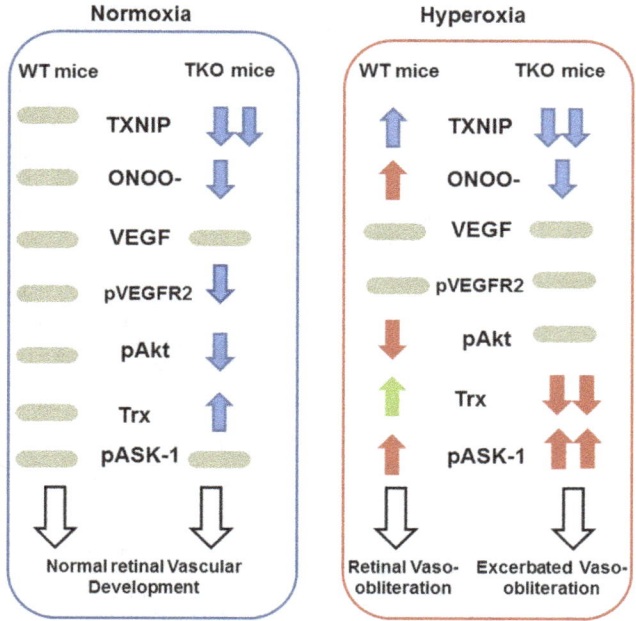

Figure 8. Representative diagram shows the impact of TXNIP deletion on retina vasculature under both normoxia and hyperoxia. Under normoxia, retinas from TXNIP-deficient mice showed similar VEGF levels, less peroxynitrite (ONOO-) levels, less VEGF receptor-2 (pVEGFR2) activation and upregulated thioredoxin (Trx) that collectively lead to normal vascular development in comparison to WT mice. Under hyperoxia, retinas from WT mice showed higher peroxynitrite formation, less survival Akt activation (pAkt) and upregulated proapoptotic signal of ASK-1 resulting in vaso-obliteration. Retinas from TKO although showed less peroxynitrite levels and maintained Akt activation, retinas experienced significant decreases in thioredoxin (Trx) that shift the balance of the ASK-1-Trx inhibitory complex and increases the activation of the proapoptotic ASK-1 pathway leading to exacerbated vasoobliteration compared to WT.

significant increases in reduced-glutathione and thioredoxin-reductase activity at normoxia as reported previously [8,23,24]. Nevertheless, upon exposure to hyperoxia and the highly oxidative milieu, expression of Trx-1, Trx-2 mRNA and protein expression was enhanced in WT but decreased in TKO resulting in activation of cell death signal. The decrease in Trx was consistent with the transcriptional and total protein levels for both cytoplasmic (Trx-1) and mitochondrial (Trx-2) levels. In agreement, previous reports demonstrated that exposure to hyperoxia upregulated the protective Trx1 in alveolar epithelial cells and brain of developing rats [39,40]. We believe that the TXNIP expression is important to maintain homeostasis of thioredoxin as anti-oxidant and anti-apoptotic and that genetic deletion of TXNIP will disturb cellular redox state. The finding that Trx compartmentalization under nitrosative/oxidative stress conditions is dependent on the expression levels of TXNIP [41] supports our hypothesis that TXNIP expression is important for Trx anti-apoptotic activity. Moreover, exposure to hyperoxia/oxidative stress can induce oxidation and inactivation of Trx-1 resulting in cell death as demonstrated in a lung model [42] and in a cellular model [43].

Recent studies by the same groups showed that shifting the antioxidant defense from thioredoxin system to glutathione is more beneficial in response to hyperoxia [44] and that glutathione/glutaredoxin system is essential for Trx reactivation after oxidation [43]. Of note, retinas from TKO experienced significant drop in reduced-GSH levels upon exposure to hyperoxia much more dramatic than WT mice (. 5.A)

We and others showed that Trx is not only an antioxidant, but it can also regulate cell survival by binding to and inhibiting the activity of apoptosis signal-regulating kinase 1 (ASK-1), a member of the MAP kinase kinase kinase family [14,16,45]. Trx can directly bind and inhibit ASK-1 or indirectly through inhibition of murine protein serine-threonine kinase 38 (MPK38), a member of the AMP-activated protein kinase-related serine/threonine kinase family that plays an important role in inducing ASK1-, TGF-β-, and p53-mediated apoptosis [45]. Here, we reported that hyperoxia caused reduction in Trx expression that was accompanied with significant increase in the activation of ASK-1. Increased activation of ASK-1 aggravated vascular cell death as confirmed with the increased expression of cleaved-caspase-3 and cleaved-PARP in TKO compared to WT. Our results lend further support to previous studies that demonstrated critical role of ASK-1 activation in hyperoxia-mediated injury [46,47].

In summary, we believe that TXNIP expression is essential for Trx system homeostasis. We and others showed that TXNIP is required for VEGF angiogenic signal via redox-dependent [23] and independent mechanisms [25]. Here, we provide new insights that TXNIP is indispensable for homeostasis of thioredoxin anti-apoptotic function in the developing retina. Genetic deletion of TXNIP, despite increased cellular antioxidant defense, significantly accelerated vascular cell death in response of hyperoxia. These findings highlight the importance of manipulating the antioxidant defense specially during the possible control of retinopathy of prematurity, a potentially blinding disorders that affect premature infants [48].

Supporting Information

Figure S1 TXNIP knockout mice have similar retinal vascular density comparable to WT. P7 retinas of both Wild type (WT) and TXNIP knockout (TKO) mice were fixed and stained with GS-IB4 conjugate-isolectin to quantify retinal vascular density. Images were processed via Image J software to be skeletonized to quantify vascular density. Our results showed no significant deference between TKO and WT in retinal vascular density during development at p7 compared to WT. (n = 12).

Acknowledgments

Authors are grateful to Dr. A.J. Lusis for providing TKO mice.

Author Contributions

Conceived and designed the experiments: ABE MAA. Performed the experiments: MAA SM. Analyzed the data: MAA SM ABE. Contributed reagents/materials/analysis tools: ABE AE. Wrote the paper: MAA ABE AE.

References

1. Lu J, Holmgren A (2012) Thioredoxin system in cell death progression. Antioxid Redox Signal 17: 1738–1747.

2. Yamawaki H, Berk BC (2005) Thioredoxin: a multifunctional antioxidant enzyme in kidney, heart and vessels. Curr Opin Nephrol Hypertens 14: 149–153.

3. Mahmood DF, Abderrazak A, El Hadri K, Simmet T, Rouis M (2013) The thioredoxin system as a therapeutic target in human health and disease. Antioxid Redox Signal 19: 1266–1303.

4. Tian L, Nie H, Zhang Y, Chen Y, Peng Z, et al. (2014) Recombinant human thioredoxin-1 promotes neurogenesis and facilitates cognitive recovery following cerebral ischemia in mice. Neuropharmacology 77: 453–464.

5. Adluri RS, Thirunavukkarasu M, Zhan L, Akita Y, Samuel SM, et al. (2011) Thioredoxin 1 enhances neovascularization and reduces ventricular remodeling during chronic myocardial infarction: a study using thioredoxin 1 transgenic mice. J Mol Cell Cardiol 50: 239–247.

6. Holmgren A (1995) Thioredoxin structure and mechanism: conformational changes on oxidation of the active-site sulfhydryls to a disulfide. Structure 3: 239–243.

7. Jung H, Choi I (2014) Thioredoxin-interacting protein, hematopoietic stem cells, and hematopoiesis. Curr Opin Hematol.

8. El-Azab MF, Baldowski BR, Mysona BA, Shanab AY, Mohamed IN, et al. (2014) Deletion of thioredoxin-interacting protein preserves retinal neuronal function by preventing inflammation and vascular injury. Br J Pharmacol 171: 1299–1313.

9. Perrone L, Devi TS, Hosoya K, Terasaki T, Singh LP (2009) Thioredoxin interacting protein (TXNIP) induces inflammation through chromatin modification in retinal capillary endothelial cells under diabetic conditions. J Cell Physiol 221: 262–272.

10. Dunn LL, Simpson PJ, Prosser HC, Lecce L, Yuen GS, et al. (2014) A critical role for thioredoxin-interacting protein in diabetes-related impairment of angiogenesis. Diabetes 63: 675–687.

11. Singh LP (2013) Thioredoxin Interacting Protein (TXNIP) and Pathogenesis of Diabetic Retinopathy. J Clin Exp Ophthalmol 4.

12. Huang C, Lin MZ, Cheng D, Braet F, Pollock CA, et al. (2014) Thioredoxin-interacting protein mediates dysfunction of tubular autophagy in diabetic kidneys through inhibiting autophagic flux. Lab Invest 94: 309–320.

13. Chen J, Cha-Molstad H, Szabo A, Shalev A (2009) Diabetes induces and calcium channel blockers prevent cardiac expression of proapoptotic thioredoxin-interacting protein. Am J Physiol Endocrinol Metab 296: E1133–1139.

14. Ichijo H, Nishida E, Irie K, ten Dijke P, Saitoh M, et al. (1997) Induction of apoptosis by ASK1, a mammalian MAPKKK that activates SAPK/JNK and p38 signaling pathways. Science 275: 90–94.

15. Yoshioka J, Schreiter ER, Lee RT (2006) Role of thioredoxin in cell growth through interactions with signaling molecules. Antioxid Redox Signal 8: 2143–2151.

16. Al-Gayyar MM, Abdelsaid MA, Matragoon S, Pillai BA, El-Remessy AB (2010) Neurovascular protective effect of FeTPPs in N-methyl-D-aspartate model: similarities to diabetes. Am J Pathol 177: 1187–1197.

17. Al-Gayyar MM, Abdelsaid MA, Matragoon S, Pillai BA, El-Remessy AB (2011) Thioredoxin interacting protein is a novel mediator of retinal inflammation and neurotoxicity. Br J Pharmacol 164: 170–180.

18. Ebrahimian T, Sairam MR, Schiffrin EL, Touyz RM (2008) Cardiac hypertrophy is associated with altered thioredoxin and ASK-signaling in a mouse model of menopause. Am J Physiol Heart Circ Physiol 295: H1481–1488.

19. Hsieh CC, Papaconstantinou J (2006) Thioredoxin-ASK1 complex levels regulate ROS-mediated p38 MAPK pathway activity in livers of aged and long-lived Snell dwarf mice. FASEB J 20: 259–268.

20. Zhang QY, Pan Y, Wang R, Kang LL, Xue QC, et al. (2014) Quercetin inhibits AMPK/TXNIP activation and reduces inflammatory lesions to improve insulin signaling defect in the hypothalamus of high fructose-fed rats. J Nutr Biochem 25: 420–428.

21. Devi TS, Hosoya K, Terasaki T, Singh LP (2013) Critical role of TXNIP in oxidative stress, DNA damage and retinal pericyte apoptosis under high glucose: implications for diabetic retinopathy. Exp Cell Res 319: 1001–1012.

22. Nivet-Antoine V, Cottart CH, Lemarechal H, Vamy M, Margaill I, et al. (2010) trans-Resveratrol downregulates Txnip overexpression occurring during liver ischemia-reperfusion. Biochimie 92: 1766–1771.

23. Abdelsaid MA, Matragoon S, El-Remessy AB (2013) Thioredoxin-Interacting Protein Expression Is Required for VEGF-Mediated Angiogenic Signal in Endothelial Cells. Antioxid Redox Signal 19: 2199–2212.

24. Hui ST, Andres AM, Miller AK, Spann NJ, Potter DW, et al. (2008) Txnip balances metabolic and growth signaling via PTEN disulfide reduction. Proc Natl Acad Sci U S A 105: 3921–3926.

25. Park SY, Shi X, Pang J, Yan C, Berk BC (2013) Thioredoxin-interacting protein mediates sustained VEGFR2 signaling in endothelial cells required for angiogenesis. Arterioscler Thromb Vasc Biol 33: 737–743.

26. Smith LE, Wesolowski E, McLellan A, Kostyk SK, D'Amato R, et al. (1994) Oxygen-induced retinopathy in the mouse. Invest Ophthalmol Vis Sci 35: 101–111.

27. Abdelsaid MA, Pillai BA, Matragoon S, Prakash R, Al-Shabrawey M, et al. (2010) Early intervention of tyrosine nitration prevents vaso-obliteration and neovascularization in ischemic retinopathy. J Pharmacol Exp Ther 332: 125–134.

28. Ferrara N (1996) Vascular endothelial growth factor. Eur J Cancer 32A: 2413–2422.

29. Brooks SE, Gu X, Samuel S, Marcus DM, Bartoli M, et al. (2001) Reduced severity of oxygen-induced retinopathy in eNOS-deficient mice. Invest Ophthalmol Vis Sci 42: 222–228.

30. Sennlaub F, Courtois Y, Goureau O (2002) Inducible nitric oxide synthase mediates retinal apoptosis in ischemic proliferative retinopathy. J Neurosci 22: 3987–3993.

31. Beauchamp MH, Sennlaub F, Speranza G, Gobeil F, Jr., Checchin D, et al. (2004) Redox-dependent effects of nitric oxide on microvascular integrity in oxygen-induced retinopathy. Free Radic Biol Med 37: 1885–1894.

32. Fujino G, Noguchi T, Matsuzawa A, Yamauchi S, Saitoh M, et al. (2007) Thioredoxin and TRAF family proteins regulate reactive oxygen species-dependent activation of ASK1 through reciprocal modulation of the N-terminal homophilic interaction of ASK1. Mol Cell Biol 27: 8152–8163.

33. Alon T, Hemo I, Itin A, Pe'er J, Stone J, et al. (1995) Vascular endothelial growth factor acts as a survival factor for newly formed retinal vessels and has implications for retinopathy of prematurity. Nat Med 1: 1024–1028.

34. El-Remessy AB, Bartoli M, Platt DH, Fulton D, Caldwell RB (2005) Oxidative stress inactivates VEGF survival signaling in retinal endothelial cells via PI 3-kinase tyrosine nitration. J Cell Sci 118: 243–252.

35. Gu X, El-Remessy AB, Brooks SE, Al-Shabrawey M, Tsai NT, et al. (2003) Hyperoxia induces retinal vascular endothelial cell apoptosis through formation of peroxynitrite. Am J Physiol Cell Physiol 285: C546–554.

36. de Bem AF, Fiuza B, Calcerrada P, Brito PM, Peluffo G, et al. (2013) Protective effect of diphenyl diselenide against peroxynitrite-mediated endothelial cell death: a comparison with ebselen. Nitric Oxide 31: 20–30.

37. Yoshihara E, Masaki S, Matsuo Y, Chen Z, Tian H, et al. (2014) Thioredoxin/Txnip: Redoxisome, as a Redox Switch for the Pathogenesis of Diseases. Front Immunol 4: 514.

38. Chen J, Jing G, Xu G, Shalev A (2014) Thioredoxin-Interacting Protein Stimulates its Own Expression via a Positive Feedback Loop. Mol Endocrinol: me20141041.

39. Shan R, Chang L, Li W, Liu W, Rong Z, et al. (2011) Effects of hyperoxia on cytoplasmic thioredoxin system in alveolar type epithelial cells of premature rats. J Huazhong Univ Sci Technolog Med Sci 31: 258–263.

40. Bendix I, Weichelt U, Strasser K, Serdar M, Endesfelder S, et al. (2012) Hyperoxia changes the balance of the thioredoxin/peroxiredoxin system in the neonatal rat brain. Brain Res 1484: 68–75.

41. Ogata FT, Batista WL, Sartori A, Gesteira TF, Masutani H, et al. (2013) Nitrosative/oxidative stress conditions regulate thioredoxin-interacting protein (TXNIP) expression and thioredoxin-1 (TRX-1) nuclear localization. PLoS One 8: e84588.

42. Tipple TE, Welty SE, Nelin LD, Hansen JM, Rogers LK (2009) Alterations of the thioredoxin system by hyperoxia: implications for alveolar development. Am J Respir Cell Mol Biol 41: 612–619.

43. Du Y, Zhang H, Zhang X, Lu J, Holmgren A (2013) Thioredoxin 1 is inactivated due to oxidation induced by peroxiredoxin under oxidative stress and reactivated by the glutaredoxin system. J Biol Chem 288: 32241–32247.

44. Britt RD, Jr., Velten M, Locy ML, Rogers LK, Tipple TE (2014) The thioredoxin reductase-1 inhibitor aurothioglucose attenuates lung injury and improves survival in a murine model of acute respiratory distress syndrome. Antioxid Redox Signal 20: 2681–2691.

45. Manoharan R, Seong HA, Ha H (2013) Thioredoxin inhibits MPK38-induced ASK1, TGF-beta, and p53 function in a phosphorylation-dependent manner. Free Radic Biol Med 63: 313–324.

46. Makena PS, Gorantla VK, Ghosh MC, Bezawada L, Kandasamy K, et al. (2012) Deletion of apoptosis signal-regulating kinase-1 prevents ventilator-induced lung injury in mice. Am J Respir Cell Mol Biol 46: 461–469.

47. Kolliputi N, Waxman AB (2009) IL-6 cytoprotection in hyperoxic acute lung injury occurs via suppressor of cytokine signaling-1-induced apoptosis signal-regulating kinase-1 degradation. Am J Respir Cell Mol Biol 40: 314–324.

48. Chen J, Smith LE (2007) Retinopathy of prematurity. Angiogenesis 10: 133–140.

16

The Activity-Integrated Method for Quality Assessment of Reduning Injection by On-Line DPPH-CE-DAD

Yan-xu Chang[1,2]*[9], Jiao Liu[1][9], Yang Bai[1], Jin Li[1], Er-wei Liu[1], Jun He[1], Xiu-cheng Jiao[1],
Zhen-zhong Wang[2], Xiu-Mei Gao[1], Bo-li Zhang[1], Wei Xiao[2]*

1 Tianjin State Key Laboratory of Modern Chinese Medicine, Tianjin University of Traditional Chinese Medicine, Tianjin, China, **2** State Key Laboratory of New-tech for Chinese Medicine Pharmaceutical Process, Kanion Pharmaceutical Co., Ltd, Lianyungang, China

Abstract

A sensitive on-line DPPH-CE-DAD method was developed and validated for both screening and determining the concentration of seven antioxidants of Reduning injection. The pH and concentrations of buffer solution, SDS, β-CD and organic modifier were studied for the detection of DPPH and seven antioxidants. By on-line mixing DPPH and sample solution, a DPPH-CE method for testing the antioxidant activity of the complex matrix was successfully established and used to screen the antioxidant components of Reduning injection. Then, antioxidant components including caffeic acid, isochlorogenic acid A, isochlorogenic acid B, isochlorogenic acid C, chlorogenic acid, neochlorogenic acid and cryptochlorogenic acid were quantified by the newly established CE–DAD method. Finally, the total antioxidant activity and the multiple active components were selected as markers to evaluate the quality of Reduning injection. The results demonstrated that the on-line DPPH-CE-DAD method was reagent-saving, rapid and feasible for on-line simultaneous determination of total pharmacological activity and contents of multi-components samples. It was also a powerful method for evaluating the quality control and mechanism of action of TCM injection.

Editor: Irina V. Lebedeva, Columbia University, United States of America

Funding: This research was supported National Natural Science Foundation of China (81374050) and Program for Innovative Research Team in Universities of Tianjin (TD12-5033) and Tianjin Research Program of Application Foundation Advanced Technology (12JCQNJC08800) and Stated Key Development Program for Basic Research of China (No. 973:2012CB723504). The funders had no role in study design, data collection and analysis, decision to publish, or preparation of the manuscript.

Competing Interests: There are no competing interests and financial disclosure that declare the affiliation(s) to Kanion Pharmaceutical Co., LTD, along with any other relevant declarations relating to employment, consultancy, patents, products in development or marketed products etc.

* Email: Tcmcyx@126.com (Y-xC) Xiaowei@163.com (WX)

❥ These authors contributed equally to this work.

Introduction

Traditional Chinese medicines (TCMs) have shown an increasing prospect in recent years as an alternative therapy. Considering that TCMs usually contain one or more substances, multi-ingredient methods for assessing the quality of TCMs are important in order to meet the clinical requirements of safety and efficacy. Some common methods of quality control of TCMs include High-Performance Liquid Chromatography (HPLC) [1–3], Ultra-High Performance Liquid Chromatography (UPLC) [4], Gas Chromatography-Mass Spectrometry (GC-MS) [5], HPLC-MS [6] and UPLC-MS [7]. These methods are limited in the determination of the chemical contents of TCMs and they fail to reflect the comprehensive pharmacological effect. Taking these problems into account, it has become imperative to develop a new method of linking pharmacological effects and quality control closely. In light of these, the dual-standard quality assessment, which was defined as a method to evaluate the quality using the total activity and contents of multi-ingredients, was proposed to evaluate the quality of TCMs.

Reduning injection made from extracts of *Gardenia jasminoides Ellis*, *Artemisia annua* L. and *Lonicera japonica* Thunb. is a widely used TCM preparation for the treatment of common cold,

cough, acute upper respiratory infection and acute bronchitis in the clinic [8]. The prospective clinical trial demonstrated that Reduning injection could be safely used in the treatment of severe hand, foot, and mouth disease (HFMD) [9]. Clinical studies reported that Reduning injection was safe and effective in curing pneumonia [8]. Reactive oxygen species (ROS) are believed to be crucial in the induction of lung damage caused by pneumonia, while therapeutic agents that could effectively scavenge ROS may prevent or reduce the deleterious effects of pneumonia. Reduning injection could increase the activity of superoxide dismutase (SOD) in lungs of rats treated with lipopolysaccharide [8]. Thus, Reduning injection has antioxidant properties. From available literatures, the components which possess antioxidant properties in Reduning injection have not yet been reported. Therefore, it is considered paramount to investigate its antioxidant properties.

Free radical species such as ABTS (2, 2-azino-bis (3-ethyl-benzthiazoline-6-sulfonic acid) and DPPH (1, 1-diphenyl-2-picryl-hydrazyl) were primarily selected to test antioxidant capabilities of ingredients or extracts of TCMs by spectrophotometry [10]. Recently, HPLC-DPPH [11] and TLC-DPPH [12] methods were reported to analyze the antioxidant activity of ingredients in extracts of TCMs. Although these analytical methods prevented interference from color pigments of analytes, they often involved usage of large amount of organic reagents which were considered harmful to environment. As a powerful analytical technique, capillary electrophoresis (CE) has gained much attention in separation science [13], since it owned the advantages such as simple preparation process, short analysis time, excellent efficiency and few organic reagents. Capillary electrophoresis with the Diode Array Detector (CE-DAD) method has been used to separate components and successfully applied to quality control of herbal medicines [14–16]. In this study, an on-line DPPH-CE-DAD method was developed and validated for the determination of the antioxidant activity of Reduning injection and its quality control.

To our knowledge, the on-line DPPH-CE-DAD method for the analysis of Reduning injection has not been reported yet in literature. In this study, CE was used to separate the chemical components of Reduning injection and then further used to develop the antioxidant activity-integrated method of this herbal preparation. The antioxidant activity-integrated method was used to evaluate the quality of different batches of the sample. The feasibility and precision of this on-line DPPH-CE-DAD method for quality control has been discussed in our report. The on-line DPPH-CE-DAD could not only determine total antioxidant activity of samples, but also screen active components rapidly from the complex Reduning injection. Furthermore CE-DAD could be used to analyze the contents of multiple active components in Reduning injection. This antioxidant activity-integrated method obtained by on-line DPPH-CE-DAD will become an advantageous tool for quality control of herbal medicines.

Materials and Methods

Chemicals and Reagents

Seven reference compounds including caffeic acid, isochlorogenic acid A, isochlorogenic acid B, isochlorogenic acid C, chlorogenic acid, neochlorogenic acid and cryptochlorogenic acid were purchased from Chengdu Must Bio. Sci. and Tec. Co. Ltd. (Chengdu, China). 25 batches of Reduning injections were offered from Jiangsu Kanion Pharmaceutical Co. Ltd. (Lianyungang, China). Deionized water used for sample preparations and buffer solutions was provided by a Milli-Q Academic ultra-pure water system (Millipore, Milford, MA, USA). Acetonitrile (ACN) and methanol were purchased from Merck (Germany). DPPH (2,2-Diphenyl-1-picrylhydrazyl) was purchased from Sigma (USA). All other chemicals were of reagent grade.

Apparatus and conditions

All experiments were performed on an Agilent CE system equipped with a Diode Array Detector (Waldbronn, Germany). Instrumental control and data analysis were operated by Agilent ChemStation software. Separations were performed in a 60.5 cm total length (52 cm to the detector) and 50 μm i.d. bare fused-silica capillary (Ruifeng, Heibei, China). New capillaries were flushed sequentially with 1.0 M NaOH, 0.1 M NaOH and deionized water (10 min each). Prior to every separation, the capillary was conditioned by rinsing with 0.1 M NaOH and deionized water followed by the background electrolyte (BGE) (3 min each). After the last run of each day, capillaries were washed orderly with 0.1 M NaOH (10 min) and deionized water (5 min). BGEs in vials were replenished for every injection to obtain the highest reproducibility of the migration times.

Preparation of standard solutions and samples

All standard phenolic acids and DPPH were individually dissolved with methanol. Fresh DPPH solution was prepared on each day of analysis at a concentration of 500 μg/mL and stored in the dark. The mixed standards were dissolved in 50% methanol at a stock concentration of 1 mg/mL. A 200 μL aliquot of Reduning injection was diluted independently to 10 mL with the diluent (methanol: water 50:50, v/v). Standard solutions were stored at 4°C. All solutions were filtered prior to use through a 0.22 μm nylon syringe filter.

Preparation of quality control samples

Quality control (QC) samples of caffeic acid, isochlorogenic acid A, isochlorogenic acid B, isochlorogenic acid C, chlorogenic acid, neochlorogenic acid and cryptochlorogenic acid were prepared at low, medium and high concentration levels by dissolving appropriate mixed standard solutions in 50% methanol, respectively.

Condition of on-line DPPH-CE-DAD method

The total antioxidant activities were tested by on-line DPPH-CE-DAD method. Sample solution was injected hydrodynamically into a bare capillary tube at 50 mbar pressure for 5 s and pure DPPH (500 μg/mL) solution was injected immediately after the sample injection in the same way as the sample. Subsequently, a 25 kV voltage was applied with positive polarity setting and the temperature of capillary was maintained at 22°C. The buffer solution for detecting the peak of DPPH to evaluate total antioxidant activity was 20 mM NaH_2PO_4 (pH 6.0) – 50 mM sodium dodecyl sulfate (SDS). The pH of BGEs for solution of DPPH was adjusted by 1.0 M NaOH. The wavelength of detection for DPPH was 517 nm, respectively.

Condition of CE-DAD method for determining multi-components

CE-DAD was used to separate and determine the multiple active components in Reduning injection. Sample solution was injected hydrodynamically at 50 mbar pressure for 5 s. A 25 kV voltage was applied with positive polarity setting and the temperature of capillary was maintained at 22°C. The aqueous BGE for sample separation and determination consisted of 20 mM NaH_2PO_4 (pH 4.2), 10 mM β-cyclodextrin (β-CD) and 5% ACN. The pH of BGEs for sample separation was adjusted by using 1% phosphate. The wavelength of detection for the sample separation was 325 nm.

Condition of DPPH-CE-DAD method for screening antioxidants of Reduning injection

The injection process is the same as the above-referred DPPH-CE-DAD. The chemical reaction of sample and DPPH was performed in the bare fused-silica capillary before separation of analytes and DPPH was carried out by capillary electrophoresis. DPPH is a radical-containing compound which is neutralized by antioxidant molecules. This reaction changes the absorbance properties of DPPH. Consequently, the magnitude of decrease of the DPPH peak seen in the electrophoretogram can be used to quantify the antioxidant activity of the TCM sample. The buffer solution for screening antioxidants of Reduning injection was 20 mM NaH_2PO_4 (pH 4.2), 10 mM β-cyclodextrin (β-CD) and 5% ACN. The wavelength of detection for antioxidants was 325 nm.

Results and Discussion

Optimization of on-line DPPH-CE-DAD

DPPH is a well-known radical which is used to determine the antioxidant activity. For the purpose of on-line evaluation of

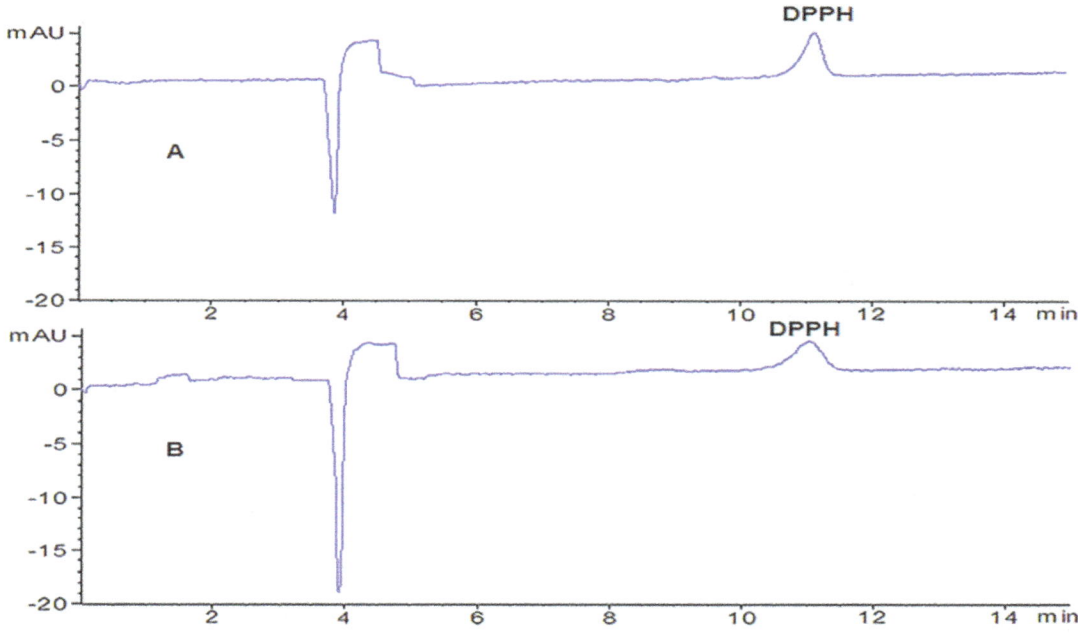

Figure 1. Capillary electropherograms of DPPH: (A) DPPH solution and (B) on-line mixed with Reduning injection (n = 3). Experimental conditions: 50 µm i.d. ×375 µm o.d. ×60.5 cm length (52 cm effective length), uncoated; 20 mM NaH₂PO₄ (pH 6.0)–50 mM SDS, voltage, 25 kV; temperature, 22°C; detection wavelength, 517 nm; pressure injection, 50 mbar for 5 s.

antioxidant activity of Reduning injection, there was a need to detect the DPPH peak in a DPPH solution and an on-line spiked Reduning injection mixture. Some parameters including the pH of buffer, the concentrations of buffer and SDS were optimized to develop a capillary electrophoresis method for detecting DPPH. Each optimized experiment was repeated three times.

The concentrations of phosphate buffer and SDS were maintained at 20 and 50 mM, the different pH values (5.5, 6.0, 6.5, 7.0, 7.5 and 8.0) were evaluated. The results showed that shorter migration time, no tailing, less peak width and more peak area were observed at pH 6.0 (**Figure S1A and Figure S2A**). Therefore, pH 6.0 was selected as the optimized pH of BGE.

The different concentrations of phosphate buffer (10, 20, 30, 40 and 50 mM) were tested for the research under constant instrumentation conditions (pH 6.0, 50 mM SDS, 25 kV, 22°C). The results showed that the electrophoretograms were similar at the different concentrations (20, 30, 40 and 50 mM). When the concentration of phosphate buffer was 10 mM, none of the compounds were detected. Meanwhile, high currents which lead to broadening of the peak because of excessive joule heating were observed at the different concentrations (30, 40 and 50 mM) (**Figure S1B and Figure S2B**). Thus, the optimum concentration of phosphate buffer was set at 20 mM.

Finally, the effect of different SDS concentrations (0, 10, 30, 50, 80, 100 mM) on the performance of DPPH separation was studied. At lower concentration (0 mM), the peak of DPPH was not detected. At higher concentrations (100 mM), the migration time became longer (**Figure S1C**). The least peak width was found when a suitable SDS concentration of 50 mM was used (**Figure S2C**). Thus, 50 mM SDS was chosen as the optimum in the following experiments. Therefore, 20 mM phosphate buffer (pH 6.0)–50 mM SDS were optimized to analyze DPPH which was incubated with the TCM in order to evaluate the antioxidant activities of TCM.

On-line determination of total antioxidant activity of Reduning injection

The established on-line DPPH-CE-DAD method was used to evaluate total antioxidant activity of Reduning injection. From the **Figure 1**, it can be observed that there was no inference with the peak of DPPH within shorter time (12 min). The peaks of DPPH with no interference in a DPPH solution and an on-line spiked Reduning injection mixture were obtained by DPPH-CE-DAD method. The experiment was repeated three times.

The relative percentage of inhibition of DPPH was calculated using the equation [Inhibition(%) = (P_0−P_1)/P_0×100%] where P_0 was peak area of DPPH (blank) and P_1 was peak area of DPPH (on-line spiked Reduning injection). The half maximal inhibitory concentration (IC$_{50}$) of DPPH was selected to assess the antioxidant activity of the different batches of Reduning injection. The IC$_{50}$ values of 25 batches of Reduning injections were listed in **Table 1** (IC$_{50}$ value was represented by the diluted times). The results showed that Reduning injection possessed antioxidant activity. Moreover, the average IC$_{50}$ value of 25 batches was 1049 (diluted times) and relative standard deviation (RSD) was less than 9%, which illustrated that the total antioxidant activity varied a little between each batch considering IC$_{50}$ values. It was concluded that on-line DPPH-CE-DAD could be a powerful method for evaluating total antioxidant activity of complex TCMs.

Comparison of on-line DPPH-CE-DAD, HPLC-DPPH and spectrophotometry Methods

The analytical performance of on-line DPPH-CE-DAD and HPLC-DPPH were compared in terms of chemical reagent consumption. Chandrasekar *et al.* had used a mobile phase composition of methanol and water (80:20, v/v) to establish an HPLC-DPPH method and analysis time was 10 min. Thus, 8 mL methanol (organic solvent) was needed to determine the antioxidant activity of one sample [11]. However, in the developed

Table 1. Contents of seven compounds in different samples and the result of total IC_{50} values (mg/mL) (n = 3).

Sample	caffeic acid	isochlorogenic acid A	isochlorogenic acid C	chlorogenic acid	isochlorogenic acid B	neochlorogenic acid	cryptochlorogenic acid	IC_{50}
091001	0.058	0.314	0.486	7.163	0.564	3.104	3.850	922.5
091002	0.057	0.313	0.483	7.143	0.557	3.114	3.785	890.5
091108	0.054	0.316	0.481	7.106	0.569	3.098	3.832	898.1
091103	0.056	0.313	0.484	7.366	0.575	3.086	3.905	1029
091109	0.057	0.312	0.481	7.062	0.575	3.141	3.919	955.1
091117	0.056	0.314	0.491	7.259	0.573	3.106	3.944	1006
081008	0.055	0.312	0.491	7.387	0.577	3.134	3.935	1047
091104	0.055	0.317	0.486	7.177	0.575	3.106	3.932	971.8
100301	0.057	0.321	0.490	7.483	0.582	3.195	3.908	1173
090908	0.056	0.313	0.492	7.343	0.576	3.169	3.934	1038
090911	0.056	0.313	0.487	7.396	0.568	3.164	3.955	1026
091012	0.057	0.312	0.481	7.233	0.575	3.146	3.929	1001
100206	0.055	0.313	0.488	7.382	0.565	3.194	3.959	1059
090913	0.057	0.317	0.492	7.394	0.572	3.203	3.962	1097
090912	0.057	0.314	0.488	7.422	0.575	3.215	4.000	1189
090907	0.057	0.317	0.492	7.379	0.567	3.200	3.918	1071
090906	0.057	0.316	0.487	7.408	0.576	3.192	3.958	1171
090909	0.056	0.316	0.484	7.505	0.568	3.194	3.930	1186
091206	0.057	0.317	0.488	7.259	0.575	3.164	3.899	1017
100210	0.055	0.316	0.483	7.363	0.577	3.175	3.946	1048
091115	0.056	0.315	0.487	7.396	0.589	3.222	3.966	1173
100101	0.055	0.313	0.491	7.420	0.575	3.179	3.969	1172
091013	0.054	0.311	0.483	7.363	0.570	3.126	3.899	1034
100123	0.056	0.323	0.491	7.319	0.570	3.123	3.900	1024
091014	0.057	0.311	0.464	7.433	0.573	3.205	3.908	1031

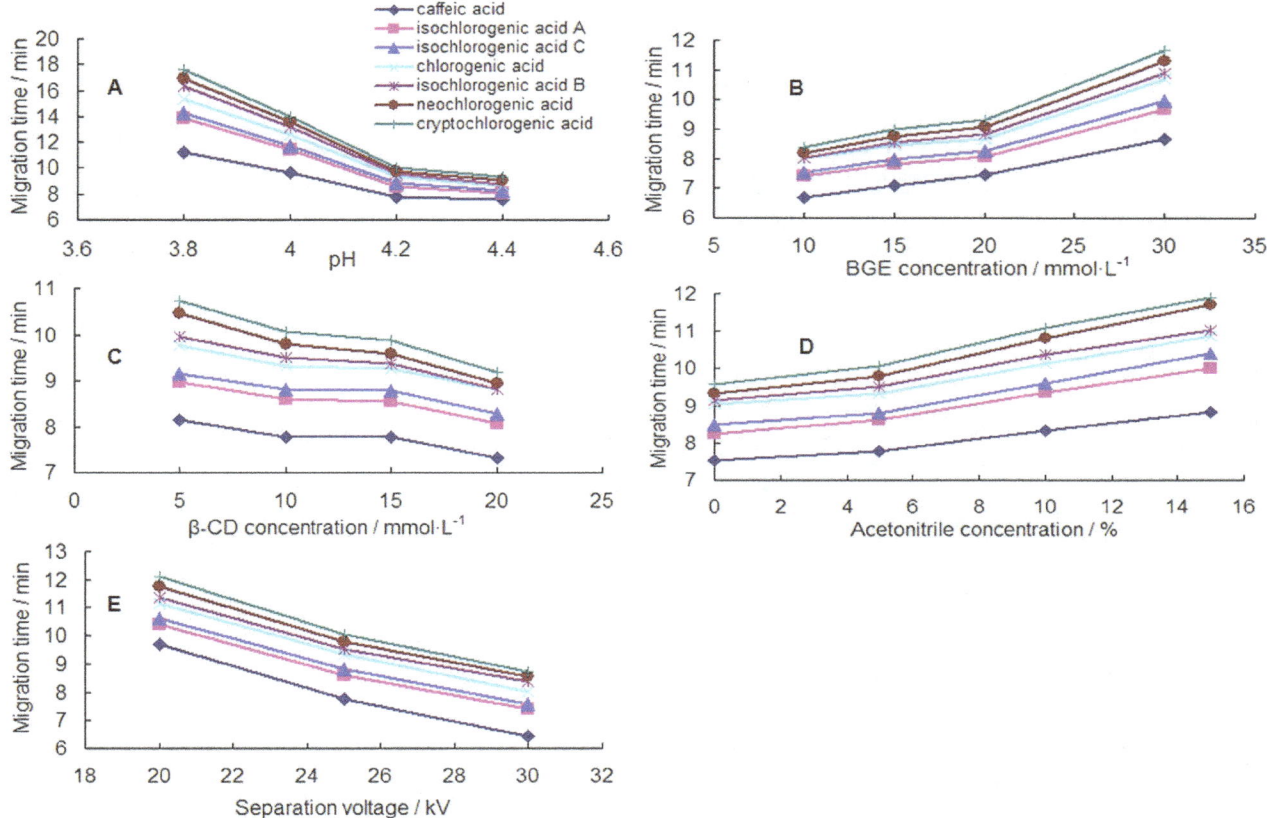

Figure 2. Effects of parameters on the migration time and resolution of seven peaks (n = 3). (A) pH of the phosphate buffer, (B) BGE concentration, (C) β-CD concentration, (D) acetonitrile concentration, (E) separation voltage.

DPPH-CE-DAD method, the BGE used during the Reduning injection screening contained 5% methanol and only 2.1 μL acetonitrile was used to analyze the antioxidant activity of one sample. Yamaguchi and his team had observed that the color pigments contained in analyzed samples could contribute to the observed difference in IC_{50} values when UV-vis direct spectrophotometric method was used. They also noted that UV-vis direct spectrophotometric method was nonspecific for DPPH [17]. Compared with UV-vis direct spectrophotometric method, the DPPH-CE-DAD method was based on the reduction in DPPH peak area and thus considered specific for DPPH. The superiority of DPPH-CE-DAD method was more evident from the detection of small changes in the DPPH absorbance reflected by the peak area even in the presence of mixtures of TCM extracts and colorants. Therefore, the newly developed on-line DPPH-CE-DAD method was a reagent-saving, rapid, feasible technique and it was also considered as a green technology with no harmful effects to the environment.

Optimization of CE-DAD method for determining multi-components

To achieve the optimum separation of all seven compounds, the most important parameters were optimized including the pH of buffer, the applied voltage and cassette temperature, the concentrations of buffer, β-cyclodextrin (β-CD) and organic modifier. Each optimized experiment was repeated three times.

Effect of pH

The pH which influenced peak efficiency and migration time was of key importance for the separation performance. By regulating the pH of BGE, the migration velocity of weak electrolyte and the velocity of the electroosmotic flow (EOF) changed [17]. The pH of the running buffer ranging from 3.8 to 4.4 were chosen for the study under constant instrumentation conditions (20 mM NaH_2PO_4, 10 mM β-CD, 5% ACN, 25 kV, 22°C). **Figure 2A** showed the effect of the pH on the selectivity of the separation and the resolutions (**Figure 3A**). At pH 4.2, appropriate migration was obtained. It was also found that migration time decreased with increasing pH of the buffer in the range of 3.8–4.4. Compared with the electrophoretograms at different pH, the resolution decreased when the buffer pH exceeded 4.2 (**Figure S3A**). Therefore, by comprehensive consideration of the separation and migration time, the optimized pH of BGE was 4.2.

Effect of BGE concentration

The effect of different concentrations of NaH_2PO_4 buffer (10, 15, 20 and 30 mM) on migration and resolution was investigated (**Figure 2B and Figure 3B**). As can be seen from **Figure 2B**, prolonged migration time was obtained with increase in the concentration of buffer at pH 4.2. There were bad resolutions among all the components in the electrophoretograms of concentration of buffer at 10 and 15 mM (**Figure S3B**, **S3C**). The improved resolutions were detected in the range of 20–30 mM. By comprehensive consideration of the separation and

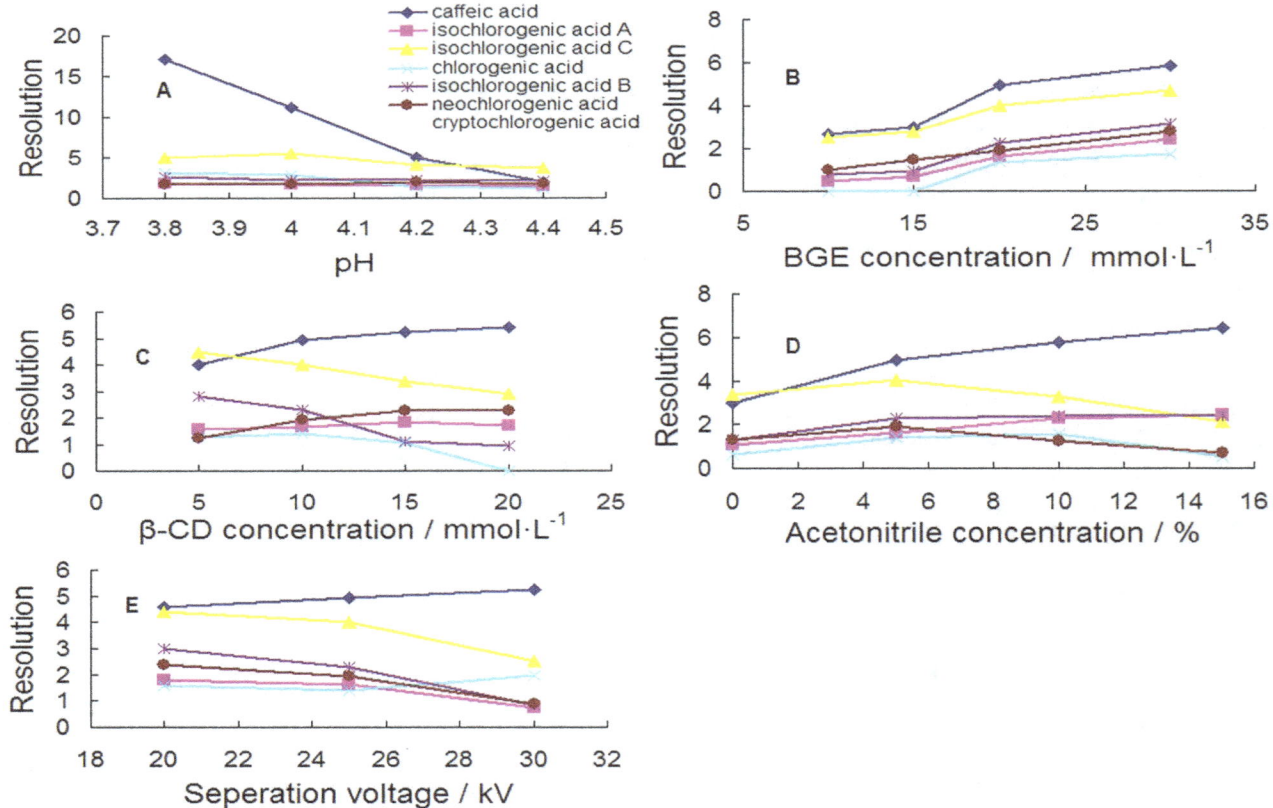

Figure 3. Effects of parameters on the resolution of seven peaks (n = 3). (A) pH of the phosphate buffer, (B) BGE concentration, (C) β-CD concentration, (D) acetonitrile concentration, (E) separation voltage.

migration time, 20 mM NaH$_2$PO$_4$ was selected as optimum concentration.

Effect of β-CD concentration

β-CD with a relatively hydrophobic cavity which forms inclusion complexes with analytes and an outer hydrophilic layer has been confirmed for the separation of the isomers. The formation of inclusion complexes between the enantiomers and the β-CD is strongly influenced not only by the hydrophobic interaction in the cavity but also by interaction between the hydroxyl groups (or other substituents) on the rim of CDs and substituents of the asymmetric center of the analytes [18]. In our lab, we studied the effect of different β-CD concentrations (5, 10, 15, 20 mM) on the separation performance. The results were shown as **Figure 2C and Figure 3C**, in spite of the shortened migration time with increasing β-CD concentration, it was necessary to mention a poor resolution at a concentration of 5 mM and a slightly broadened peak shape at 15 and 20 mM from electrophoretograms (**Figure S3D**). When β-CD was treated with 10 mM, all the compounds could achieve a good baseline separation and peak shape.

Effect of acetonitrile concentration

It is known that organic solvents added into a running buffer can enhance the solubility of the various substance and separation selectivity through changes in the physicochemical properties [18,19]. The most widely used organic modifier in MEKC is acetonitrile, which is particularly effective in reducing retention

factors and expanding the migration window [20]. Hence, acetonitrile ranging from 0% to 15% was chosen to be the organic modifier in this study. There was a poor resolution at a concentration of 0% acetonitrile in the electrophoretogram (**Figure S3E**). Although it caused an increased resolution (**Figure 3D**), prolonged migration times were observed at concentrations more than 5% from **Figure 2D**. Therefore, 5% acetonitrile was chosen as the optimum organic modifier.

Effect of separation voltage and temperature

The influence of voltages (20, 25, 30 kV) and cassette temperatures (20, 22, 25°C) on peak separation were examined. In despite of the decreasing migration time (**Figure 2E**), all peaks could not reach baseline separation in the electrophoretograms of voltage over 25 kV (**Figure S3F**). The best resolution was observed when a voltage of 25 kV was applied (**Figure 3E**). The suitable migration time and resolutions were similarly observed at 22°C. Therefore, 20 mM phosphate buffer (pH 4.2)–10 mM β-CD - 5% ACN, 25 kV and 22°C were selected to develop the CE-DAD method for determining multi-components of Reduning injection.

Method validation

Linearity, Limits of detection and quantification. The linearity of the method was evaluated by constructing six point calibration curves. Calibration graphs were constructed in the range of 1–100 μg/mL for caffeic acid, isochlorogenic acid A, isochlorogenic acid B and isochlorogenic acid C, 2–200 μg/mL

Table 2. The calibration curves, linearity ranges, LODs, LOQs and recoveries of seven compounds (n = 6).

Compounds	Regression equation	R^2	Linearity range (µg/mL)	LOD (µg/mL)	LOQ (µg/mL)	Recovery Average (%)	RSD (%)
caffeic acid	y = 1.0359x−0.1309	0.9999	1–100	0.25	0.80	104	2.5
isochlorogenic acid A	y = 0.7039x−0.9727	0.9986	1–100	0.25	0.80	101	3.4
isochlorogenic acid C	y = 0.924x−0.9422	0.9990	1–100	0.25	0.80	101	2.2
chlorogenic acid	y = 0.5662x−6.0336	0.9984	3–300	0.50	1.50	99.9	5.0
isochlorogenic acid B	y = 0.6993x−0.2756	0.9985	1–100	0.25	0.80	101	3.4
neochlorogenic acid	y = 0.6867x−3.5464	0.9983	2–200	0.50	1.50	98.1	4.6
cryptochlorogenic acid	y = 0.6389x−3.3038	0.9984	2–200	0.50	1.50	98.3	4.5

for neochlorogenic acid and cryptochlorogenic acid and 3–300 µg/mL for chlorogenic acid, respectively. Linear regression equations for all standard phenolic acids indicated good correlation with the correlation coefficients (R^2)>0.9983. The regression equations for caffeic acid, isochlorogenic acid A, isochlorogenic acid B, isochlorogenic acid C, chlorogenic acid, neochlorogenic acid and cryptochlorogenic acid were y = 1.0359x−0.1309, y = 0.7039x−0.9727, y = 0.6993x−0.2756, y = 0.924x−0.9422, y = 0.5662x−6.0336, y = 0.6867x−3.5464, y = 0.6389x−3.3038, respectively (**Table 2**). On the basis of signal-to-noise ratio of 3, limit of detection (LOD) of caffeic acid, isochlorogenic acid A, isochlorogenic acid B and isochlorogenic acid C were 0.25 µg/mL and those of other phenolic acids were 0.50 µg/mL. The limit of quantification (signal-to-noise ratio = 10) is the lowest amount of compound in a sample which can be quantitatively determined. The limits of quantification of caffeic acid, isochlorogenic acid A, isochlorogenic acid B and isochlorogenic acid C were 0.80 µg/mL and all other phenolic acids were 1.50 µg/mL, respectively (**Table 2**).

Precision and accuracy. Precision and accuracy were evaluated by assaying quality control samples containing seven phenolic acids at low, medium, and high concentrations and DPPH solutions (n = 6). The data of intra- and inter-day precision and accuracy were presented in **Table 3**. The intra- and inter-day accuracies of seven phenolic acids were within the range of 91.4%-108%. The RSDs for both intra- and inter-day were below 4.2%. The RSDs for both intra- and inter-day of DPPH were less than 3.0%. The results indicated that the method was accurate, reliable and reproducible.

Stability. The stability of seven phenolic acids during the sample storage and processing procedures was evaluated by analyzing quality control samples at low, medium, and high concentrations. The results of stability studies were presented in **Table 3**. After 24 h storage at 4°C, the accuracies of seven phenolic acids were within the range of 92.2%–107% and the RSDs were below 4.8%. The RSD of DPPH was 4.4%. The above results demonstrated that the developed method could be used to determine phenolic acids in the Reduning injection and DPPH.

Recovery. The recovery test was performed by spiking equivalent amount of seven investigated compounds into a certain amount Reduning injection sample. The original and resultant samples were analyzed using the method mentioned above. The recovery was calculated by the formula: Recovery (%) = (found amount − original amount)/spiked amount ×100%. The recoveries of seven phenolic acids of Reduning injections mixed with appropriate standard solutions for caffeic acid, isochlorogenic acid A, isochlorogenic acid B, isochlorogenic acid C, chlorogenic acid, neochlorogenic acid and cryptochlorogenic acid were assayed, respectively. The mean recoveries of seven phenolic acids determined ranged between 98.1–104% and the RSDs were below 5.0% (**Table 2**).

On-line screening of antioxidant of Reduning injection

To identify the active ingredients, the various diluted sample solutions and 500 µg/mL DPPH were injected into the on-line CE-DAD system at 50 mbar pressure for 5 s. The experiment was repeated three times. Compared with the electrophoretogram of the sample without DPPH, seven antioxidants were screened based on the decreased peak area (**Figure 4**). These components were identified by comparing with reference standards, including caffeic acid, isochlorogenic acid A, isochlorogenic acid B, isochlorogenic acid C, chlorogenic acid, neochlorogenic acid and cryptochlorogenic acid (**Figure 5**). Thus, these ingredients were the main antioxidant in Reduning injection. It was clearly proved

Table 3. Intra-day and Inter-day accuracy and precision, stability of seven compounds and DPPH (n = 6).

Compounds	Concentration (µg/mL)	Intra-day		Inter-day		Stability for 24 h	
		Accuracy (%)	RSD (%)	Accuracy (%)	RSD (%)	Remains (%)	RSD (%)
caffeic acid	2	99.1	4.2	94.6	4.8	94.5	4.8
	10	97.5	3.6	97.9	1.9	96.4	2.1
	50	106	1.2	101	3.7	102	3.2
isochlorogenic acid A	2	103	0.9	102	0.9	102	1.0
	10	96.4	2.2	96.8	0.4	95.1	1.2
	50	98.0	2.1	97.2	0.7	99.7	2.6
isochlorogenic acid C	2	107	2.1	107	0.07	108	1.0
	10	95.1	1.2	95.2	0.6	94.5	1.0
	50	101	1.8	103	1.6	102	1.8
chlorogenic acid	6	96.6	2.1	96.4	0.7	96.3	0.7
	30	94.5	0.2	93.9	0.7	93.7	1.1
	150	101	1.3	99.8	1.3	101	0.3
isochlorogenic acid B	2	106	2.2	107	0.4	108	1.5
	10	95.4	0.3	96.0	0.8	95.0	0.4
	50	100	2.6	100	0.2	101	1.5
neochlorogenic acid	4	96.2	2.4	95.6	2.7	97.6	1.3
	20	95.3	0.2	94.9	0.6	95.4	0.3
	100	102	1.8	101	1.8	102	0.6
cryptochlorogenic acid	4	92.1	2.1	92.2	0.2	91.4	0.6
	20	93.3	0.8	94.4	1.0	93.1	1.5
	100	101	1.1	99.9	1.5	100	1.2
DPPH	500	-	2.8	-	3.0	-	4.4

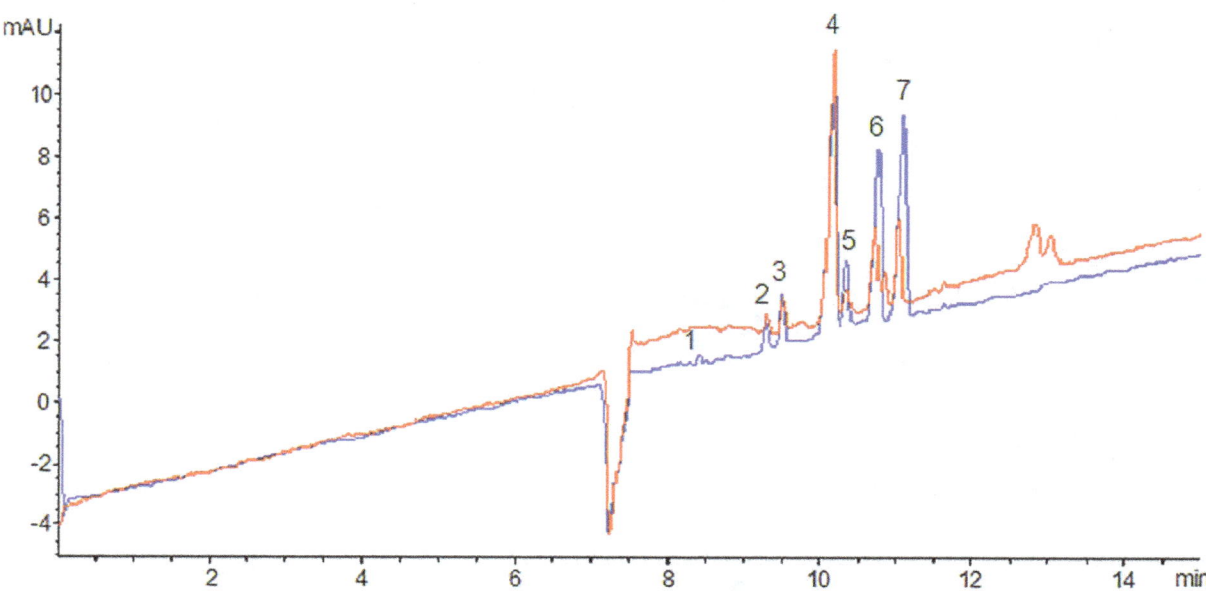

Figure 4. Capillary electropherograms of Reduning injection: Reduning injection (blue) and on-line mixed with DPPH (red) (n = 3). Peaks: 1 = caffeic acid, 2 = isochlorogenic acid A, 3 = isochlorogenic acid C, 4 = chlorogenic acid, 5 = isochlorogenic acid B, 6 = neochlorogenic acid, 7 = cryptochlorogenic acid. Experimental conditions: 50 µm i.d. ×375 µm o.d. ×60.5 cm length (52 cm effective length), uncoated; 20 mM NaH_2PO_4 (pH 4.2) - 10 mM β-CD - 5% (v/v) acetonitrile, voltage, 25 kV; temperature, 22°C; detection wavelength, 325 nm; pressure injection, 50 mbar for 5 s.

that the antioxidant activity of Reduning injection was the consequence of the effect of these screened compounds. The results observed were consistent with previous studies [21–23]. These seven active components were necessarily selected as markers for quality assessment of Reduning injection.

Method application

The developed method was applied for the determination of seven phenolic acids in 25 batches of Reduning injection and the results were presented in **Table 1**. The typical chromatographic profile of Reduning injection obtained under the above-mentioned

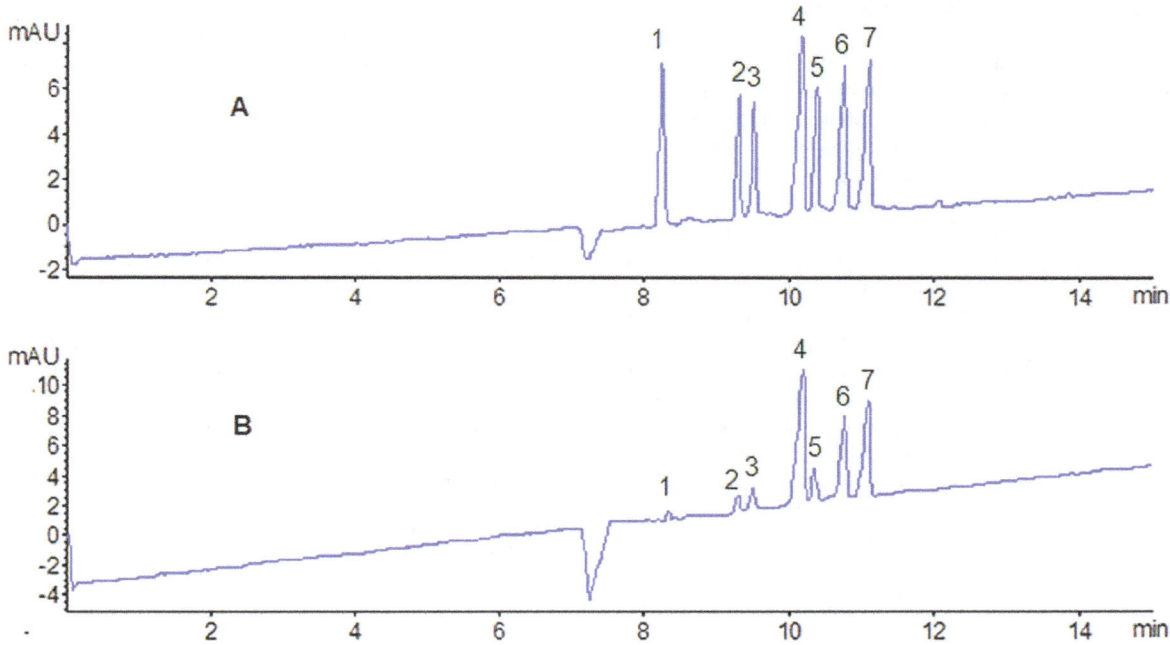

Figure 5. Capillary electropherograms of standard mixture of seven compounds (A) and Reduning injection (B) (n = 3). Peaks: 1 = caffeic acid, 2 = isochlorogenic acid A, 3 = isochlorogenic acid C, 4 = chlorogenic acid, 5 = isochlorogenic acid B, 6 = neochlorogenic acid, 7 = cryptochlorogenic acid. Experimental conditions: 50 µm i.d. ×375 µm o.d. ×60.5 cm length (52 cm effective length), uncoated; 20 mM NaH_2PO_4 (pH 4.2) - 10 mM β-CD - 5% (v/v) acetonitrile, voltage, 25 kV; temperature, 22°C; detection wavelength, 325 nm; pressure injection, 50 mbar for 5 s.

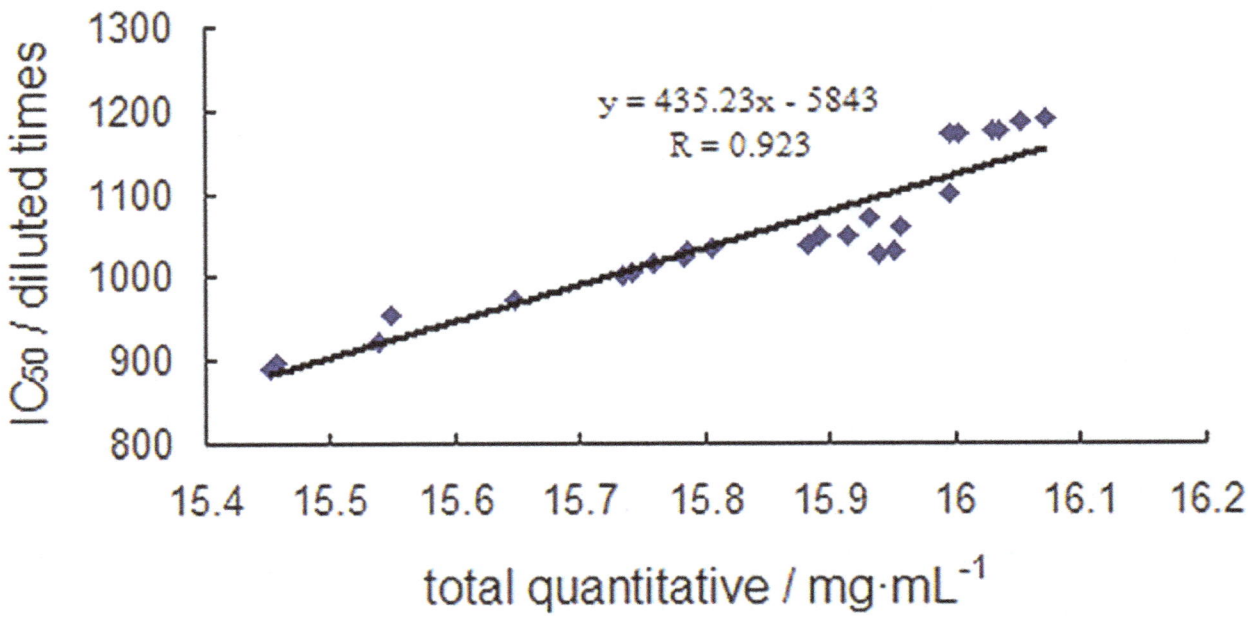

Figure 6. Relationship of the total quantitative and antioxidant activity of Reduning injection.

conditions was shown in **Figure 5B**. The ranges of quantitative determination for caffeic acid, isochlorogenic acid A, isochlorogenic acid B, isochlorogenic acid C, chlorogenic acid, neochlorogenic acid and cryptochlorogenic acid were 0.054–0.058 mg/ml, 0.311–0.323 mg/ml, 0.464–0.492 mg/ml, 7.062–7.505 mg/ml, 3.785–4.000 mg/ml, 0.557–0.589 mg/ml and 3.086–3.222 mg/ml, respectively. The average of total content of 25 batches was 15.836 mg/ml and the RSD was less than 3.0%. It showed that the seven compounds in 25 batches of Reduning injections varied slightly.

It is generally known that traditional quality control methods for herbal samples only present the chemical information and could not reflect the real and comprehensive pharmacological information of active constituents, which made these traditional methods less reliable. Therefore, a more reliable approach to evaluate the quality of samples could be the determination of both contents of multiple active components and their activity in contrast to these traditional methods. In our research, we attempted to investigate the relationship between total content and antioxidant activity to evaluate the quality of Reduning injection. As shown in **Figure 6**, there was a good correlation (R = 0.923) between total content of seven compounds and total antioxidant activity across 25 batches of Reduning injection. Therefore, it was important and suitable to choose seven active antioxidant compounds including caffeic acid, isochlorogenic acid A, isochlorogenic acid B, isochlorogenic acid C, chlorogenic acid, neochlorogenic acid and cryptochlorogenic acid as key quality markers to maintain batch-to-batch uniformity and efficacy of Reduning injection. Hence, the combination of total antioxidant activity with multiple active components for dual-standard quality assessment of Reduning injection was concluded to be reliable.

Conclusions

The dual-standard quality assessment has been proposed to evaluate the quality of Reduning injections with the total antioxidant activity and active multi-ingredients quantification.

The antioxidant activity-integrated CE-DAD method was developed for evaluating the total antioxidant activity and screening antioxidant compounds in Reduning injection. By mixing DPPH and sample solution, the on-line DPPH-CE-DAD method was successfully applied to assessing total antioxidant activity and screening the active antioxidant compounds in the samples. Seven components (caffeic acid, isochlorogenic acid A, isochlorogenic acid B, isochlorogenic acid C, chlorogenic acid, neochlorogenic acid and cryptochlorogenic acid) of Reduning injection were characterized as antioxidants and markers and simultaneously determined by CE-DAD method. The results demonstrated that it was reliable to combine total antioxidant activity with multiple active components contents to evaluate the quality of Reduning injection. The on-line DPPH-CE-DAD method was highly efficient due to use of few organic reagents and anti-interference of color pigments in comparison with other existing techniques such as UV-vis direct spectrophotometry, HPLC-DPPH and TLC-DPPH. Hence, it was considered useful in the study of antioxidant activities of other TCMs and valuable for reporting the antioxidant activity of unknown molecules in complex matrix. In contrast to the traditional method of quality control, the present improved method was more available and precise for linking closely total pharmacological activity and contents of multi-components to evaluate quality control of TCM injection.

Supporting Information

Figure S1 Effects of parameters on the migration time of DPPH: (A) pH of the phosphate buffer, (B) BGE concentration, (C) SDS concentration.

Figure S2 Effects of parameters on the peak width of DPPH: (A) pH of the phosphate buffer, (B) BGE concentration, (C) SDS concentration.

Figure S3   Some electrophoretograms of standard mixture of seven compounds with poor resolution in different experimental conditions: (A) pH = 4.4, 20 mM NaH$_2$PO$_4$, 10 mM β-CD, 5% (v/v) acetonitrile, 25 kV, (B) pH = 4.2, 10 mM NaH$_2$PO$_4$, 10 mM β-CD, 5% (v/v) acetonitrile, 25 kV, (C) pH = 4.2, 15 mM NaH$_2$PO$_4$, 10 mM β-CD, 5% (v/v) acetonitrile, 25 kV, (D) pH = 4.2, 10 mM NaH$_2$PO$_4$, 5 mM β-CD, 5% (v/v) acetonitrile, 25 kV, (E) pH = 4.2, 10 mM NaH$_2$PO$_4$, 10 mM β-CD, 0% (v/v) acetonitrile, 25 kV, (F) pH = 4.2, 10 mM NaH$_2$PO$_4$, 10 mM β-CD, 5% (v/v) acetonitrile, 30 kV.

Author Contributions

Conceived and designed the experiments: YC. Performed the experiments: J. Liu J. Li EL JH YB. Analyzed the data: YC J. Liu. Contributed reagents/materials/analysis tools: XJ XG BZ WX ZW. Wrote the paper: YC J. Liu.

References

1. Zhao L, An Q, Qin F, Xiong Z (2014) Simultaneous determination of six constituents in the fruit of Acanthopanax sessiliflorus (Rupr. et Maxim.) Seem. by HPLC-UV. Nat Prod. Res. 28:500–502.

2. Jian ZY, Wang WQ, Xu GF, Meng L, Hou JL (2013) Comprehensive quality evaluation of Chishao by HPLC. Nutr Hosp 28:1681–1687.

3. Tian S, Yu Q, Wang D, Upur H (2012) Development of a rapid resolution liquid chromatography-diode array detector method for the determination of three compounds in Ziziphora clinopodioides Lam from different origins of Xinjiang. Pharmacogn Mag 8: 280–284.

4. Naldi M, Fiori J, Gotti R, Périat A, Veuthey JL, et al. (2014) UHPLC determination of catechins for the quality control of green tea. J Pharm Biomed Anal 88: 307–314.

5. Hu YC, Kong WJ, Yang XH, Xie LW, Wen J, et al. (2013) GC-MS combined with chemometric techniques for the quality control and original discrimination of Curcumae longae rhizome: Analysis of essential oils. J Sep Sci. DOI: 10.1002/jssc.201301102.

6. Liang J, Wu WY, Sun GX, Wang DD, Hou JJ, et al. (2013) A dynamic multiple reaction monitoring method for the multiple components quantification of complex traditional Chinese medicine preparations: Niuhuang Shangqing pill as an example. J Chromatogr A 1294: 58–69.

7. Li TX, Hu L, Zhang MM, Sun J, Qiu Y, et al. (2014) A sensitive UPLC-MS/MS method for simultaneous determination of eleven bioactive components of Tong-Xie-Yao-Fang decoction in rat biological matrices. J Chromatogr B Analyt Technol Biomed Life Sci 944: 90–100.

8. Tang LP, Xiao W, Li YF, Li HB, Wang ZZ, et al. (2014) Anti-inflammatory effects of Reduning injection on lipopolysaccharide-induced acute lung injury of rats. Chin J Integr Med 20:1–9.

9. Li X, Zhang X, Ding J, Xu Y, Wei D, et al. (2014) Comparison between Chinese herbal medicines and conventional therapy in the treatment of severe hand, foot, and mouth disease: a randomized controlled trial. Evid Based Complement Alternat Med. DOI: 10.1155/2014/140764.

10. de Oliveira RG Jr, Souza GR, Guimarães AL, de Oliveira AP, Silva Morais AC, et al. (2013) Dried extracts of Encholirium spectabile (Bromeliaceae) present antioxidant and photoprotective activities in vitro. J Young Pharm 5: 102–105.

11. Chandrasekar D, Madhusudhana K, Ramakrishna S, Diwan PV (2006) Determination of DPPH free radical scavenging activity by reversed-phase HPLC: a sensitive screening method for polyherbal formulations. J Pharm Biomed Anal 40: 460–464.

12. Cieśla Ł, Kryszeń J, Stochmal A, Oleszek W, Waksmundzka-Hajnos M (2012) Approach to develop a standardized TLC-DPPH• test for assessing free radical scavenging properties of selected phenolic compounds. J Pharm Biomed Anal 70: 126–135.

13. Cao J, Qu HB, Cheng YY (2010) The use of novel ionic liquid-in-water microemulsion without the addition of organic solvents in a capillary electrophoretic system. Electrophoresis 31: 3492–3498.

14. Cao J, Li P, Yi L (2011) Ionic liquids coated multi-walled carbon nanotubes as a novel pseudostationary phase in electrokineticchromatography. J Chromatogr A 1218: 9428–9434.

15. Cao J, Li P, Chen J, Tan T, Dai HB (2013) Enhanced separation of Compound Xueshuantong capsule using functionalized carbon nanotubes with cationic surfactant solutions in MEEKC. Electrophoresis 34: 324–330.

16. Rodríguez J, Castañeda G, Contento AM, Muñoz L (2012) Direct and fast determination of paclitaxel, morphine and codeine in urine by micellar electrokinetic chromatography. J Chromatogr A 1231: 66–72.

17. Yamaguchi T, Takamura H, Matoba T, Terao J (1998) HPLC method for evaluation of the free radical-scavenging activity of foods by using 1,1-diphenyl-2-picrylhydrazyl. Biosci Biotechnol Biochem 62:1201–1204.

18. Orlandini S, Gotti R, Furlanetto S (2014) Multivariate optimization of capillary electrophoresis methods: A critical review. J Pharm Biomed Anal 87: 290–307.

19. Huie CW (2003), Effects of organic solvents on sample pretreatment and separation performances in capillary electrophoresis. Electrophoresis 24: 1508–1529.

20. Gottardo R, Bertaso A, Pascali J, Sorio D, Musile G, et al. (2012) Micellar electrokinetic chromatography: a new simple tool for the analysis of synthetic cannabinoids in herbal blends and for the rapid estimation of their log P values. J Chromatogr A 1267: 198–205.

21. Gašić U, Kečkeš S, Dabić D, Trifković J, Milojković-Opsenica D, et al. (2014) Phenolic profile and antioxidant activity of Serbian polyfloral honeys. Food Chem 145: 599–607.

22. Takeoka GR, Dao LT (2003) Antioxidant constituents of almond Prunus dulcis (Mill.) D.A. Webb] hulls. J Agric Food Chem 51: 496–501.

23. Iwai K, Kishimoto N, Kakino Y, Mochida K, Fujita T (2004) In vitro antioxidative effects and tyrosinase inhibitory activities of seven hydroxycinnamoyl derivatives in green coffee beans. J Agric Food Chem 52: 4893–4898.

17

The Association between Selenium and Other Micronutrients and Thyroid Cancer Incidence in the NIH-AARP Diet and Health Study

Thomas J. O'Grady[1]*, Cari M. Kitahara[2], A. Gregory DiRienzo[1], Margaret A. Gates[1]

1 University at Albany, School of Public Health, Rensselaer, New York, United States of America, **2** Division of Cancer Epidemiology and Genetics, National Cancer Institute, National Institutes of Health, Rockville, Maryland, United States of America

Abstract

Background: Selenium is an essential trace element that is important for thyroid hormone metabolism and has antioxidant properties which protect the thyroid gland from oxidative stress. The association of selenium, as well as intake of other micronutrients, with thyroid cancer is unclear.

Methods: We evaluated associations of dietary selenium, beta-carotene, calcium, vitamin D, vitamin C, vitamin E, folate, magnesium, and zinc intake with thyroid cancer risk in the National Institutes of Health – American Association of Retired Persons Diet and Health Study, a large prospective cohort of 566,398 men and women aged 50–71 years in 1995–1996. Multivariable-adjusted Cox proportional hazards regression was used to examine associations between dietary intake of micronutrients, assessed using a food frequency questionnaire, and thyroid cancer cases, ascertained by linkage to state cancer registries and the National Death Index.

Results: With the exception of vitamin C, which was associated with an increased risk of thyroid cancer ($HR_{Q5\ vs\ Q1}$, 1.34; 95% CI, 1.02–1.76; P_{trend}, <0.01), we observed no evidence of an association between quintile of selenium ($HR_{Q5\ vs\ Q1}$, 1.23; 95% CI, 0.92–1.65; P_{trend}, 0.26) or other micronutrient intake and thyroid cancer.

Conclusion: Our study does not suggest strong evidence for an association between dietary intake of selenium or other micronutrients and thyroid cancer risk. More studies are needed to clarify the role of selenium and other micronutrients in thyroid carcinogenesis.

Editor: Javier S. Castresana, University of Navarra, Spain

Funding: This work was supported in part by the Intramural Research Program of the National Cancer Institute, National Institutes of Health. The funders had no role in study design, data collection and analysis, decision to publish, or preparation of the manuscript. The authors received no other funding of any type for the preparation of this manuscript.

Competing Interests: The authors have declared that no competing interests exist.

* Email: togrady@albany.edu

Introduction

The past three decades have seen a rapid increase in thyroid cancer incidence in the United States (U.S.) and other countries [1–5]. Increased medical surveillance and diagnostic scrutiny are likely responsible for some, but not all, of this trend [6]. In addition to established risk factors such as ionizing radiation exposure and benign thyroid nodules [7], recent studies have focused on modifiable etiologic factors such as diet and obesity. Findings from these studies indicate that obesity and excessive weight gain during adulthood [8–10] and dietary nitrate and nitrite intake [11,12] are associated with increased thyroid cancer risk while eating various fruits and vegetables [13–17], having adequate iodine intake and consuming fish [18], and a Polynesian-style diet [19] may be protective. Still, there is no consensus as to what dietary factors contribute to or inhibit thyroid carcinogenesis, as

others have reported that a traditional Western diet [19,20] and fruit and vegetable consumption [21] are unassociated with thyroid cancer risk.

While there has been research into dietary patterns as a whole, less work has been done to assess specific dietary constituents such as micronutrients and their impact on thyroid cancer risk. Additional research into the association between micronutrients and thyroid cancer may help further the understanding of the biological mechanisms involved in thyroid carcinogenesis. Selenium is an essential trace element that is found at higher concentrations in the thyroid gland than in other organs [22]. Selenium is important in the metabolism of thyroid hormones triiodothyronine (T3) and thyroxine (T4). Specifically, type I 5'-deiodinase is a selenium-containing protein which assists in activating naturally occurring T4 (which has little biological activity) into biologically active T3. Additionally, selenium has antioxidant properties which may help protect the thyroid gland from H_2O_2 and reactive oxygen species [23–26]. While selenium has been shown to have an inverse association with other cancers [27] this relationship has not been investigated for thyroid cancer. A recent meta-analysis of the association between supplemental micronutrients and thyroid cancer showed both that there is limited research on the topic of micronutrients and thyroid cancer and that the results of these prior studies are largely inconclusive [28].

We therefore examined the association between dietary intake of micronutrients and thyroid cancer risk in the U.S. National Institutes of Health-American Association of Retired Persons (NIH-AARP) Diet and Health Study, a large prospective cohort of 566,398 men and women ages 50–71 years at baseline. The primary micronutrient of interest was selenium because of its established importance for proper function of the thyroid gland, its potential antioxidant properties, and the possible inverse association between selenium and other cancers [23,24,27–29]. To our knowledge this is the first prospective evaluation of dietary selenium intake in relation to thyroid cancer risk. Additionally, we evaluated the associations between dietary intake of beta-carotene, calcium, folate, magnesium, vitamin C, vitamin D, vitamin E, and zinc with thyroid cancer risk to expand upon recent studies investigating supplemental intake of these micronutrients and thyroid cancer [28].

Methods

Study Population

The NIH-AARP Diet and Health study began in 1995–1996 with the mailing of an extensive baseline questionnaire to 3.5 million AARP members aged 50–71 years old and residing in six U.S. states (California, Florida, Louisiana, North Carolina, New Jersey, and Pennsylvania) or two U.S. metropolitan areas (Atlanta, Georgia and Detroit, Michigan). Information ascertained by the questionnaire included usual dietary intake over the past twelve months, use of individual and multivitamin supplements, alcohol intake, smoking history, height and weight at baseline, and other factors. Of the 566,398 individuals who were deemed to have satisfactorily completed the baseline questionnaire we excluded, in the following order, participants with proxy respondents (n = 15,760), those who reported poor health or end stage renal disease (n = 9,134), participants with a previous diagnosis of cancer other than non-melanoma skin cancer (n = 53,195), those with extreme or missing values for total energy intake (n = 4,279) and individuals with extreme values for daily selenium intake (n = 1,223). The remaining analytic cohort included 482,807 participants (287,944 men and 194,863 women). The NIH-AARP Diet and Health Study was approved by the Special Studies Institutional Review Board of the U.S. National Cancer Institute and the Institutional Review Boards from all participating institutions approved the use of these data.

Cancer Ascertainment

Participants accrued time in the study from the date of baseline questionnaire completion to the date of any cancer other than non-melanoma skin cancer, the date when a person moved out of the registry ascertainment area, death, or December 31, 2006, whichever occurred first. Incident thyroid cancers (International Classification of Disease for Oncology, Third Edition (ICD-O-3), topography code C73) [30] were identified through December 31, 2006, via linkage of the NIH-AARP cohort membership to state cancer registries and the National Death Index [31]. The state

cancer registries are certified by the North American Association of Central Cancer Registries as being at least 90% complete within two years of the close of the diagnosis year [32]. A validation study comparing linkage to state cancer registries with self-report and subsequent medical record conformation of incident cancers estimated that 90% of all cancer cases identified in the NIH-AARP Diet and Health Study were valid [31]. Subtypes of thyroid cancer were defined by ICD-O-3 morphology codes (papillary, 8050, 8052, 8130, 8260, 8340–8344) and (follicular, 8290, 8330–8332, 8335).

Dietary Intake

The baseline questionnaire had a dietary component, which included questions about the frequency of dietary consumption during the past twelve months of 124 food items and the corresponding portion sizes of 100 of those food items. Intake frequency was recorded as one of ten categories ranging from "never" to "2+ times per day" for foods and "never" to "6+ times per day" for beverages. Additionally, each item included three possible portion size responses. The methods of Subar et al [33] along with national dietary data from the U.S. Department of Agriculture's 1994–1996 Continuing Survey of Food Intake by Individuals (CSFII) [34] were used to construct the food items, portion sizes, nutrient database, and pyramid food servings database. A recipe file was used by the Pyramid Servings Database to disaggregate food mixtures into component ingredients and assign the components to food groups. The FFQs used by the NIH-AARP Diet and Health Study have been validated using two 24-hour recalls in a subset of the cohort [35]. The daily micronutrient consumption of an individual was determined by multiplying the frequency of consumption of each line item by its micronutrient content (determined from CSFII) and summing over all line items.

Statistical Analysis

Cox proportional hazards regression [36] with person-years as the underlying time metric was used to estimate the cause-specific hazard ratios (HR) and corresponding 95% confidence intervals (CI) for thyroid cancer by quintiles of selenium and other micronutrient intake for thyroid cancer overall and the papillary and follicular subtypes. Quintiles of micronutrient intake were assessed first using the same cut points for men and women, resulting in an equal number of participants in each quintile when summed across men and women but not within sex. We then ran the analysis using separate, sex-specific, cut points for men and women. We assessed and verified, using cumulative sums of martingale residuals [37], that there was no violation of the proportional hazards assumption. We tested for linear trend by including the median value of each micronutrient category as a continuous variable in the model and assessing the significance of the Wald Chi-square p-value.

Micronutrient intake was adjusted for total energy intake using the nutrient residual method [38], which computes nutrient intake by first removing both calorie and micronutrient outliers, separately by sex, and then taking residuals from the regression model with total caloric intake as the independent variable and absolute nutrient intake as the dependent variable and adding a fixed constant (mean caloric intake by sex) of the study population.

Our minimal model was adjusted for age (continuous) and sex in the overall analysis and age in the sex specific analysis. Potential confounding variables were identified based on a review of the literature and previous studies on thyroid cancer using the NIH-AARP Diet and Health Study. Potential confounders were assessed using a backward elimination method in which we

Table 1. Baseline characteristics of the NIH-AARP Diet and Health Study cohort by quintiles of residual adjusted selenium intake.

	Male N = 287,944					Female N = 194,683				
N* =	Q1	Q2	Q3	Q4	Q5	Q1	Q2	Q3	Q4	Q5
	22,396	39,795	60,176	77,287	88,290	74,165	56,767	36,385	19,275	8,271
Number of thyroid cancer cases	19	25	49	68	96	119	106	66	29	15
Number of papillary thyroid cancer cases	12	14	33	49	56	83	77	49	21	12
Number of follicular thyroid cancer cases	4	8	8	14	23	21	14	14	5	2
Energy adjusted Selenium quintile median (mcg/day Residual Method)	7.1	7.6	8	8.4	8.9	7.1	7.6	8	8.4	8.9
Unadjusted Selenium quintile median (mcg/day)	47	68.4	86.6	108.6	150.1	47	68.4	86.6	108.6	150.1

Parameter	Characteristics	Q1	Q2	Q3	Q4	Q5	Q1	Q2	Q3	Q4	Q5
Age+		62.4 (5.4)	62.5 (5.3)	62.4 (5.3)	62.1 (5.3)	61.7 (5.4)	62.1 (5.4)	61.9 (5.4)	61.7 (5.4)	61.5 (5.4)	61.5 (5.4)
Race	White (%)	88.2	91.4	92.7	93.4	93.5	86.9	90.8	91.3	91.1	91.6
	Black (%)	5.6	3.4	2.5	2.2	2.1	7.8	4.4	4	3.9	3.8
	Other (%)	4.5	3.9	3.7	3.4	3.4	3.5	3.5	3.4	3.6	3.3
Currently Married (%)		79.6	83.9	85.6	86.4	85.7	41.7	45.9	46.9	46.4	43.9
BMI (kg/m^2)		26.8 (4.2)	26.8 (4.1)	27.0 (4.1)	27.3 (4.2)	27.7 (4.5)	26.3 (5.7)	26.8 (5.9)	27.2 (6.1)	27.5 (6.3)	27.8 (6.4)
Smoking History	Never Smoker (%)	26.7	28.9	29.6	30	29.6	44.9	45	44	41.8	39.4
	Former Smoker (%)	53.6	55	56.2	56.6	57.8	35.1	38.5	40.1	42.7	45.1
	Current Smoker (%)	15.3	12	10.3	9.7	8.8	16.3	13.3	12.5	11.9	11.1
Education College Grad/PostGrad (%)		37.6	40.7	43.8	45.9	48.4	26.5	30.9	32.7	34.1	34.4
20 minutes Physical activity 5 or more times per week (%)		21.1	21.3	20.9	21.0	22.5	16.2	15.9	16.1	16.8	19.3
Post- menopause (% Yes)		n/a	n/a	n/a	n/a	n/a	94.5	93.9	93.8	93.5	93.2
Hormone Use (% Yes Ever)		n/a	n/a	n/a	n/a	n/a	51.3	55.2	55.2	54.5	51.5
Currently taking hormones (% yes)		n/a	n/a	n/a	n/a	n/a	42.2	46.1	46.2	45.5	43.2
Dietary Intakes (Vitamins Residually Adjusted)	Calories	2,226.0 (1,018.8)	1,995.4 (875.0)	1,935.1 (800.6)	1,944.1 (781.5)	2,064.9 (812.5)	1,568.6 (677.7)	1,522.9 (616.0)	1,561.0 (623.4)	1,623.0 (647.0)	1,702.1 (648.5)
	Vitamin C (mg/day)	9.7 (2.6)	9.6 (2.1)	9.5 (1.9)	9.4 (1.7)	9.2 (1.7)	9.5 (2.0)	9.2 (1.6)	9.1 (1.6)	9.0 (1.6)	8.9 (1.7)
	Vitamin D (mcg/day)	1.2 (0.8)	1.5 (0.8)	1.6 (0.7)	1.7 (0.7)	1.8 (0.6)	1.2 (0.8)	1.4 (0.7)	1.4 (0.6)	1.5 (0.6)	1.7 (0.7)
	Vitamin E (mg/day)	2.2 (0.5)	2.3 (0.4)	2.4 (0.4)	2.4 (0.4)	2.4 (0.3)	2.1 (0.4)	2.2 (0.3)	2.2 (0.3)	2.2 (0.3)	2.2 (0.3)
	Beta-carotene (mcg/day)	9.5 (1.2)	9.7 (1.1)	9.8 (1.0)	9.9 (1.0)	10.0 (1.0)	9.9 (1.1)	10.1 (1.0)	10.1 (1.0)	10.2 (1.0)	10.2 (1.1)
	Calcium (mg/day)	7.3 (0.6)	7.5 (0.5)	7.5 (0.5)	7.6 (0.5)	7.5 (0.4)	7.3 (0.5)	7.4 (0.5)	7.4 (0.5)	7.3 (0.4)	7.3 (0.4)
	Folate (mg/day)	12.8 (1.7)	13.3 (1.2)	13.4 (1.1)	13.6 (1.1)	13.8 (1.0)	12.5 (1.3)	12.8 (1.0)	12.9 (1.0)	13.0 (1.0)	13.1 (1.0)

Table 1. Cont.

	Male N = 287,944					Female N = 194,683				
	Q1	Q2	Q3	Q4	Q5	Q1	Q2	Q3	Q4	Q5
Magnesium (mg/day)	10.8 (1.0)	11.2 (0.7)	11.3 (0.7)	11.4 (0.6)	11.5 (0.6)	10.6 (0.8)	10.8 (0.6)	10.8 (0.6)	10.9 (0.6)	11.0 (0.6)
Zinc (mg/day)	2.4 (0.4)	2.7 (0.4)	2.9 (0.3)	3.0 (0.3)	3.1 (0.3)	2.4 (0.4)	2.6 (0.3)	2.6 (0.3)	2.7 (0.3)	2.7 (0.3)

The same quintile cut points were used for men and women. In an analysis using sex-specific cut points, the results presented in the manuscript were unchanged.
Mean and (standard deviation).

removed the least significant covariate in the model and assessed whether this changed the main exposure HR by more than 10%. We assessed for confounding by age (continuous), sex, body mass index (BMI; <18.5, 18.5–24.99, 25–29.99, >30), total calories (continuous), education (high school or less, some college, college or post graduate, unknown), physical activity (<1–2 times per week, 3–4 times per week, 5+ times per week, unknown), race (White, Black, other, unknown), smoking status (never, current, former, unknown), marital status (yes, no, unknown), alcohol intake (≤1, 2, 3, 4+ drinks/day), and micronutrient intake (continuous intake of vitamin C, vitamin E, beta-carotene, and folate). Effect modification was assessed using the likelihood ratio test comparing a model with the cross-product terms to one without.

Finally, we tested the assumption that the risk of thyroid cancer was log-linearly associated with selenium and other micronutrient intake by comparing the linear model with a non-parametric regression curve obtained with restricted cubic splines [39]. We used a stepwise selection process to identify the number and location of knots for each micronutrient analyzed. The likelihood ratio test was used to fit the restricted cubic splines. SAS software version 9.3 (SAS Institute, Cary, NC) was used for all statistical analyses. All reported p-values are based on two-sided tests and an alpha level of 0.05.

Results

Over a total of 4,406,634 person-years of follow-up we identified 592 incident thyroid cancer cases (257 in men and 335 in women) of which 406 were of the papillary histologic subtype (164 in men and 242 in women) and 113 (57 in men and 56 in women) were of the follicular histologic subtype. Select characteristics of the study population are presented by quintile of dietary selenium intake (Table 1). Participants in the highest quintile of selenium intake were more likely to be married, never or former smokers, and college educated when compared to the lowest quintile of selenium intake. The mean and standard deviation for selenium intake was 94.0±42.9 mcg/day (other micronutrient mean and standard deviation intakes presented in Table 2). The five largest contributors to selenium intake were breads & rolls (13.9%), pasta (6.3%), tuna (4.8%), fish – not fried (4.1%), and eggs (4.0%). Contributors to vitamin C intake were orange & grapefruit juices (29.1%), oranges & tangelos (7.9%), broccoli (7.3%), other juices (5.1%), and grapefruit (4.5%). The dietary contributors of each micronutrient studied were similar for men and women. Table 2 provides a detailed summary of the major dietary contributors of each micronutrient in our study.

Table 3 presents the association between dietary micronutrient intake and thyroid cancer risk. After controlling for potential confounders there were no statistically significant associations between increasing quintile of selenium intake and risk of thyroid cancer ($HR_{Q5\ vs\ Q1}$, 1.23; 95% CI, 0.92–1.65; P_{trend}, 0.26). There was a statistically significant increase in risk of thyroid cancer for the highest versus lowest quintile of vitamin C intake in our multivariable model prior to adjusting for antioxidant intake ($HR_{Q5\ vs\ Q1}$, 1.34; 95% CI, 1.02–1.76; P_{trend}, <0.01), but not after ($HR_{Q5\ vs\ Q1}$, 1.35; 95% CI, 0.99–1.84; P_{trend}, 0.09). Although we observed evidence of a statistically significant positive linear trend for vitamin C, the HRs were highest for the fourth versus first quintile and were slightly attenuated for the highest quintile. For the remaining micronutrients in our study there was no clear evidence of a positive or negative association, or of a linear trend, between intake of any micronutrient and thyroid cancer risk.

Table 2. Top Five Primary Dietary Sources for Micronutrients in NIH-AARP Diet and Health Study for Men and Women Combined.

Micronutrient[+] Mean (SD) Intake	Primary Source (%)	Secondary Source (%)	Tertiary Source (%)	Fourth Source (%)	Fifth Source (%)
Selenium: 94±42.9	Bread/Rolls (13.9)	Pasta (6.7)	Tuna (4.8)	Fish - Not Fried (4.1)	Eggs (4.0)
Vitamin C: 157.0±103.7	Orange/Grapefruit Juice (29.1)	Orange/Tangelos (7.9)	Broccoli (7.3)	Other Juice (5.1)	Grapefruit (4.5)
Beta-carotene: 4,234.3±3,767.6	Carrots (36.8)	Sweet Potatoes (12.4)	Spinach/Greens (9.8)	Vegetable Medley (5.3)	Broccoli (2.9)
Calcium: 766.5±429.1	Milk –1 & 2% (11.2)	Milk - Skim (10.8)	Milk - In Cereal (9.0)	Bread/Rolls (5.4)	Cereal (3.7)
Folate: 411.4±182.2	Cereal (14.5)	Orange/Grapefruit Juice (10.8)	Lettuce (4.9)	Bread/Rolls (4.7)	Spinach/Greens (4.1)
Magnesium: 326.2±129.7	Coffee (11.3)	Bread/Rolls (5.6)	Cereal (5.4)	Orange/Grapefruit Juice (4.2)	Bananas (3.5)
Vitamin E: 8.8±4.6	Cereal (10.2)	Salad Dressing (7.5)	Oils (3.8)	Nuts/Seed -Whole (3.7)	Tomato Sauces w/Meat (3.6)
Zinc: 10.4±4.8	Cereal (11.0)	Beef - Steak (4.8)	Bread/Rolls (4.4)	Beef - Burger (4.2)	Beef - Meatball (3.3)

[+]Micronutrients measured as followed: Selenium (mcg/day), Betacarotene (mcg/day), Calcium (mg/day), Folate (mcg/day), Magnesium (mg/day), Vitamin C (mg/day), Vitamin D (mcg/day), Vitamin E (mg/day), Zinc (mg/day). Vitamin D food sources not available.

We also evaluated the associations between micronutrient intake and thyroid cancer separately by sex because of the higher incidence of thyroid cancer in women and suspected differences in the etiology of thyroid cancer by sex [40] (Tables S1 and S2). Although for some micronutrients, such as vitamin C, the strength and direction of the association appeared to differ by sex there were no statistically significant interactions by sex. Results restricted to papillary or follicular thyroid cancer, the two most common subtypes of thyroid cancer, were largely similar to the overall analysis (Tables S3–S8).

When we restricted the analysis to only participants with complete covariate information the results did not vary substantially from the presented data with an indicator variable for unknown values. There was also no indication that additionally adjusting for intake of the antioxidants vitamin C, vitamin E, beta-carotene, and folate in our multivariable model had an impact on the association with selenium or any other micronutrient (multivariable model 2 vs. 3). Finally there was no evidence that using sex-specific quintiles of micronutrient intake in men and women had a substantial impact on the results presented.

Discussion

Using a large prospective cohort study, we observed no evidence of an association between quintile of selenium intake and incidence of total thyroid cancer or the papillary and follicular subtypes. This was the first prospective study on this topic. We observed a suggestion of a positive linear relationship between increasing quintile of vitamin C and the risk of total thyroid cancer and the papillary and follicular subtypes, but no evidence of an association between thyroid cancer risk and calcium, folate, vitamin E, vitamin D, magnesium, or zinc intakes.

Selenium is a biologically important micronutrient shown to have a preventive effect for cancers other than thyroid [27]. Selenium assists in the production and proper function of thyroid hormones and has antioxidant properties [23–26]. Several limitations of our study may have contributed to the null results observed for selenium. Using an FFQ to determine dietary selenium intake may have resulted in exposure misclassification because the selenium content of soil varies largely by geographic region [41,42], resulting in a difference in the accumulation of

selenium in animals and plant products and in the selenium content of foods. An additional limitation of our study is that measurement error in dietary micronutrient intake, which was likely non-differential, may have attenuated the associations of interest. Use of bio-specimens in future studies would allow for more accurate measurement of an individual's selenium intake. Measuring plasma selenium is a good indicator of recent intake [43,44] although plasma selenium concentrations are not useful for determining long term intake and recent infections can influence plasma selenium levels [45]. Toenail clippings have been used to indicate long term selenium intake and are useful in investigations between selenium and chronic diseases [45,46].

Another limiting factor of using an FFQ in our study population was that the participants were not asked about every possible contributor to dietary selenium. For example, dietary consumption of Brazilian nuts, the food with the highest selenium content [47], was not assessed. Intake of fresh halibut and sardines, two other food sources with high selenium content were also not assessed. A study of the major food sources of antioxidants in U.S. adults [48] showed that Brazilian nuts, halibut, and sardines are not major dietary contributors of selenium in the U.S. diet and that the dietary contributors of selenium included in our study are representative of the major sources of selenium in the U.S. diet. Therefore, the impact of not having information on Brazilian nut consumption in our study was likely minimal. Further, the bio-availability of selenium from different food sources varies, and wheat, which was the highest contributor of dietary selenium in our study, has a high bio-availability [49].

In our analysis of other micronutrients and their association with thyroid cancer we saw evidence for an association with intake of vitamin C only. Higher intake of vitamin C appeared to be positively associated with thyroid cancer risk. Biologically, vitamin C has been shown to improve and mediate abnormalities seen in the levels of thyroid hormones and thyroid stimulating hormone in the serum of humans [50] and rats [51]. Previous studies have indicated both a positive and negative association between increased vitamin C intake and thyroid cancer [28]. A recent case-control study by Jung et al. indicated that vitamin C intake and citrus consumption in controls was higher than in individuals with either benign or malignant thyroid cancer, although the difference was not statistically significant [13]. A positive

Table 3. Hazard Ratios (HRs) and corresponding 95% confidence intervals (CIs) for total thyroid cancer by quintile of micronutrient intake in men and women combined in The NIH-AARP Diet and Health Study.

Selenium	Q1	Q2	Q3	Q4	Q5	P trend
Median Intake	7.05	7.64	8.03	8.41	8.93	
Number of Cases	138	131	115	97	111	
Age-adjusted HR[1] (95% CI)	1.00 (ref)	0.96 (0.75, 1.22)	0.85 (0.66, 1.09)	0.72 (0.56, 0.94)	0.83 (0.65, 1.07)	0.03
Multivariable HR[2] (95% CI)	1.00 (ref)	1.00 (0.79, 1.28)	0.99 (0.76, 1.29)	1.00 (0.75, 1.33)	1.23 (0.92, 1.65)	0.26
Multivariable HR[3] (95% CI)	1.00 (ref)	1.07 (0.83, 1.36)	1.07 (0.81, 1.40)	1.10 (0.82, 1.48)	1.35 (0.99, 1.84)	0.09

Vitamin C	Q1	Q2	Q3	Q4	Q5	P trend
Median Intake	7	8.41	9.36	10.28	11.67	
Number of Cases	36	43	46	67	63	
Age-adjusted HR[1] (95% CI)	1.00 (ref)	1.19 (0.77, 1.86)	1.27 (0.82, 1.96)	1.80 (1.20, 2.70)	1.57 (1.04, 2.36)	0.01
Multivariable HR[2] (95% CI)	1.00 (ref)	1.03 (0.78, 1.36)	1.07 (0.81, 1.42)	1.43 (1.09, 1.86)	1.34 (1.02, 1.76)	<0.01
Multivariable HR[3] (95% CI)	1.00 (ref)	1.03 (0.77, 1.37)	1.11 (0.82, 1.49)	1.47 (1.09, 1.98)	1.46 (1.05, 2.04)	<0.01

Betacarotene	Q1	Q2	Q3	Q4	Q5	P trend
Median Intake	8.67	9.38	9.89	10.43	11.3	
Number of Cases	111	116	107	123	132	
Age-adjusted HR[1] (95% CI)	1.00 (ref)	1.03 (0.80, 1.34)	0.95 (0.73, 1.24)	1.09 (0.84, 1.40)	1.16 (0.90, 1.50)	0.2
Multivariable HR[2] (95% CI)	1.00 (ref)	1.00 (0.77, 1.30)	0.87 (0.67, 1.15)	1.01 (0.78, 1.32)	1.07 (0.83, 1.39)	0.6
Multivariable HR[3] (95% CI)	1.00 (ref)	0.93 (0.71, 1.22)	0.78 (0.59, 1.04)	0.87 (0.66, 1.16)	0.89 (0.65, 1.20)	0.38

Calcium	Q1	Q2	Q3	Q4	Q5	P trend
Median Intake	8.67	9.38	9.89	10.43	11.3	
Number of Cases	111	116	107	123	132	
Age-adjusted HR[1] (95% CI)	1.00 (ref)	1.03 (0.80, 1.34)	0.95(0.73, 1.24)	1.09 (0.84, 1.40)	1.16(0.90, 1.50)	0.2
Multivariable HR[2] (95% CI)	1.00 (ref)	0.94 (0.71, 1.23)	0.92 (0.69, 1.23)	1.15 (0.85, 1.56)	1.08 (0.74, 1.58)	0.63
Multivariable HR[3] (95% CI)	1.00 (ref)	0.92 (0.69, 1.21)	0.90 (0.66, 1.21)	1.11 (0.81, 1.53)	1.00 (0.67, 1.48)	0.94

Folate	Q1	Q2	Q3	Q4	Q5	P trend
Median Intake	11.7	12.58	13.17	13.78	14.72	
Number of Cases	121	115	129	110	117	
Age-adjusted HR[1] (95% CI)	1.00 (ref)	0.94 (0.73, 1.22)	1.06 (0.83, 1.36)	0.90 (0.70, 1.17)	0.96 (0.75, 1.24)	0.7
Multivariable HR[2] (95% CI)	1.00 (ref)	0.94 (0.73, 1.22)	1.17 (0.91, 1.51)	1.04 (0.79, 1.36)	1.27 (0.97, 1.67)	0.06
Multivariable HR[3] (95% CI)	1.00 (ref)	0.94 (0.73, 1.22)	1.17 (0.91, 1.51)	1.04 (0.79, 1.36)	1.27 (0.97, 1.67)	0.06

Vitamin E	Q1	Q2	Q3	Q4	Q5	P trend
Median Intake	1.85	2.09	2.26	2.43	2.71	
Number of Cases	125	133	126	108	97	
Age-adjusted HR[1] (95% CI)	1.00 (ref)	1.06 (0.83, 1.35)	1.01 (0.79, 1.29)	0.87 (0.67, 1.12)	0.78 (0.60, 1.02)	0.03
Multivariable HR[2] (95% CI)	1.00 (ref)	1.09 (0.84, 1.39)	1.12 (0.87, 1.45)	1.05 (0.80, 1.37)	0.99 (0.75, 1.31)	0.92
Multivariable HR[3] (95% CI)	1.00 (ref)	1.07 (0.83, 1.38)	1.11 (0.85, 1.44)	1.02 (0.77, 1.35)	0.93 (0.69, 1.25)	0.58

Vitamin D	Q1	Q2	Q3	Q4	Q5	P trend
Median Intake	0.58	1.14	1.51	1.89	2.46	
Number of Cases	114	116	135	117	109	
Age-adjusted HR[1] (95% CI)	1.00 (ref)	1.04 (0.80, 1.35)	1.19 (0.91, 1.57)	0.99 (0.73, 1.34)	0.90 (0.62, 1.31)	0.77
Multivariable HR[2] (95% CI)	1.00 (ref)	1.10 (0.84, 1.44)	1.24 (0.94, 1.65)	1.09 (0.79, 1.49)	1.03 (0.70, 1.50)	0.72
Multivariable HR[3] (95% CI)	1.00 (ref)	1.08 (0.82, 1.43)	1.25 (0.93, 1.66)	1.08 (0.78, 1.49)	1.03 (0.70, 1.53)	0.71

Magnesium	Q1	Q2	Q3	Q4	Q5	P trend
Median Intake	10.1	10.72	11.11	11.49	12.03	
Number of Cases	145	115	116	113	100	
Age-adjusted HR[1] (95% CI)	1.00 (ref)	0.79 (0.62, 1.01)	0.80 (0.63, 1.02)	0.79 (0.61, 1.00)	0.70 (0.54, 0.90)	<0.01

Table 3. Cont.

Magnesium	Q1	Q2	Q3	Q4	Q5	P trend
Multivariable HR[2] (95% CI)	1.00 (ref)	0.85 (0.66, 1.09)	0.96 (0.74, 1.23)	1.03 (0.79, 1.34)	1.00 (0.76, 1.33)	0.66
Multivariable HR[3] (95% CI)	1.00 (ref)	0.79 (0.61, 1.03)	0.86 (0.66, 1.14)	0.88 (0.66, 1.19)	0.80 (0.57, 1.13)	0.33
Zinc	**Q1**	**Q2**	**Q3**	**Q4**	**Q5**	**P trend**
Median Intake	2.24	2.54	2.75	2.95	3.24	
Number of Cases	146	138	106	104	98	
Age-adjusted HR[1] (95% CI)	1.00 (ref)	0.95 (0.75, 1.20)	0.74 (0.57, 0.94)	0.73 (0.57, 0.94)	0.69 (0.54, 0.90)	<0.01
Multivariable HR[2] (95% CI)	1.00 (ref)	0.96 (0.76, 1.22)	0.85 (0.66, 1.11)	0.95 (0.72, 1.25)	0.98 (0.73, 1.32)	0.96
Multivariable HR[3] (95% CI)	1.00 (ref)	0.94 (0.74, 1.21)	0.82 (0.62, 1.08)	0.87 (0.64, 1.18)	0.89 (0.63, 1.25)	0.36

[1]Adjusted for entry age.
[2]Adjusted for entry age, sex (overall), calories, smoking status, race, education, BMI, and physical activity.
[3]Additionally adjusted for vitamin C, vitamin E, beta-carotene, and folate.

association between citrus consumption and thyroid cancer was also seen by Xiao et al. and appeared to be driven by the intake of orange and grapefruit juice [52]. Orange and grapefruit juices were the primary contributors of vitamin C intake in our study. Xiao et al. suggest some individuals may have included artificially flavored drinks in their report of orange and grapefruit juice consumption, leading to misclassification and potentially a false positive association.

It is also possible that residual confounding may have contributed to our observed association between increasing quintile of vitamin C intake and thyroid cancer risk. Participants in the highest quintile of vitamin C intake had higher education, which has been associated with healthcare utilization and increased rates of thyroid cancer diagnosis [53,54]. Additionally, individuals in the highest quintile of vitamin C intake had greater physical activity and lower caloric intake, which are characteristics associated with a health conscious lifestyle. A healthier lifestyle, again, may correspond to greater healthcare utilization and increased opportunity for thyroid cancer detection. Although we did control for these factors in our analysis it is possible that they were measured imperfectly or that we were unable to fully capture aspects of a healthy lifestyle or health consciousness that could influence detection. It seems unlikely, however, that increased detection would occur only in men and impact only the results for vitamin C.

Our study has several strengths. While previous studies have utilized case-control designs this study is the first to examine the association between selenium, and other micronutrient, intake and thyroid cancer using a large prospective cohort design, greatly reducing the possibility of recall and selection biases [55]. Furthermore, there was generally complete follow-up of study participants to ascertain outcomes and a relatively large number of thyroid cancer cases, which allowed us to analyze thyroid cancer by histologic subtype. Additionally we had covariate information that allowed us to adjust for potential confounders in our multivariable analysis of this study population. Using quintiles of selenium intake allowed for a natural comparison group, as the lowest unadjusted quintile of intake for selenium (less than 47 mcg/day) was the only group which fell below the recommended daily intake for men and women of 55 mcg/day [56].

It is possible that exposure to essential micronutrients earlier in life, or over the course of an individual's lifetime, may be more important in determining thyroid cancer risk. Although the NIH-AARP study utilized follow-up questionnaires that asked about

diet at different points in the participant's life, the follow-up questions were not designed to assess selenium and therefore were not used in this analysis. Future studies with information on dietary intake over the span of a person's life could add valuable information to dietary studies on thyroid cancer risk. Finally, because the FFQ in this study was not designed to measure iodine consumption, we did not have information on daily iodine intake, which is important for thyroid function and may be a potential important confounding variable. However, iodine deficiency and the subsequent impact of not controlling for iodine intake on our study results is likely minimal, as iodine fortification of salt and other foods has occurred in the United States since the 1920's [57].

In conclusion we observed no evidence of an association between thyroid cancer and quintile of selenium intake. Given the unexpected positive association observed for vitamin C, further evaluation of the association between vitamin C and thyroid cancer in other prospective cohorts is warranted. Valuable follow-up studies of selenium and thyroid cancer risk could evaluate selenium intake and intake of other micronutrients more objectively, such as by use of biomarkers.

Supporting Information

Table S1 Hazard Ratios (HRs) and corresponding 95% confidence intervals (CIs) for total thyroid cancer by quintile of micronutrient intake among men in The NIH-AARP Diet and Health Study.

Table S2 Hazard Ratios (HRs) and corresponding 95% confidence intervals (CIs) for total thyroid cancer by quintile of micronutrient intake among women in The NIH-AARP Diet and Health Study.

Table S3 Hazard Ratios (HRs) and corresponding 95% confidence intervals (CIs) for papillary thyroid cancer by quintile of micronutrient intake among men and women combined in The NIH-AARP Diet and Health Study.

Table S4 Hazard Ratios (HRs) and corresponding 95% confidence intervals (CIs) for papillary thyroid cancer

by quintile of micronutrient intake among men in The NIH-AARP Diet and Health Study.

Table S5 Hazard Ratios (HRs) and corresponding 95% confidence intervals (CIs) for papillary thyroid cancer by quintile of micronutrient intake among women in The NIH-AARP Diet and Health Study.

Table S6 Hazard Ratios (HRs) and corresponding 95% confidence intervals (CIs) for total thyroid cancer by quintile of micronutrient intake among men in The NIH-AARP Diet and Health Study.

Table S7 Hazard Ratios (HRs) and corresponding 95% confidence intervals (CIs) for follicular thyroid cancer

by quintile of micronutrient intake among men in The NIH-AARP Diet and Health Study.

Table S8 Hazard Ratios (HRs) and corresponding 95% confidence intervals (CIs) for follicular thyroid cancer by quintile of micronutrient intake among women in The NIH-AARP Diet and Health Study.

Author Contributions

Conceived and designed the experiments: TJO CMK MAG. Performed the experiments: TJO. Analyzed the data: TJO AGD. Contributed to the writing of the manuscript: TJO. Revised the manuscript critically for important intellectual content: CMK AGD MAG.

References

1. Kilfoy BA, Zheng T, Holford TR, Han X, Ward MH, et al. (2009) International patterns and trends in thyroid cancer incidence, 1973–2002. Cancer Causes Control 20: 525–531.
2. Colonna M, Grosclaude P, Remontet L, Schvartz C, Mace-Lesech J, et al. (2002) Incidence of thyroid cancer in adults recorded by French cancer registries (1978–1997). Eur J Cancer 38: 1762–1768.
3. dos Santos Silva I, Swerdlow AJ (1993) Thyroid cancer epidemiology in England and Wales: time trends and geographical distribution. Br J Cancer 67: 330–340.
4. Enewold L, Zhu K, Ron E, Marrogi AJ, Stojadinovic A, et al. (2009) Rising thyroid cancer incidence in the United States by demographic and tumor characteristics, 1980–2005. Cancer Epidemiol Biomarkers Prev 18: 784–791.
5. Montanaro F, Pury P, Bordoni A, Lutz JM, Swiss Cancer Registries N (2006) Unexpected additional increase in the incidence of thyroid cancer among a recent birth cohort in Switzerland. Eur J Cancer Prev 15: 178–186.
6. Cramer JD, Fu P, Harth KC, Margevicius S, Wilhelm SM (2010) Analysis of the rising incidence of thyroid cancer using the Surveillance, Epidemiology and End Results national cancer data registry. Surgery 148: 1147–1152; discussion 1152–1143.
7. Ron E, Schneider A (2006) Thyroid Cancer. In Shottenfeld, D.; Fraumeni, JF., Jr, editors. Cancer Epidemiology and Prevention. New York: Oxford University Press.
8. Leitzmann MF, Brenner A, Moore SC, Koebnick C, Park Y, et al. (2010) Prospective study of body mass index, physical activity and thyroid cancer. Int J Cancer 126: 2947–2956.
9. Kitahara CM, Platz EA, Freeman LE, Hsing AW, Linet MS, et al. (2011) Obesity and thyroid cancer risk among U.S. men and women: a pooled analysis of five prospective studies. Cancer Epidemiol Biomarkers Prev 20: 464–472.
10. Kitahara CM, Platz EA, Park Y, Hollenbeck AR, Schatzkin A, et al. (2012) Body fat distribution, weight change during adulthood, and thyroid cancer risk in the NIH-AARP Diet and Health Study. Int J Cancer 130: 1411–1419.
11. Ward MH, Kilfoy BA, Weyer PJ, Anderson KE, Folsom AR, et al. (2010) Nitrate intake and the risk of thyroid cancer and thyroid disease. Epidemiology 21: 389–395.
12. Kilfoy BA, Zhang Y, Park Y, Holford TR, Schatzkin A, et al. (2011) Dietary nitrate and nitrite and the risk of thyroid cancer in the NIH-AARP Diet and Health Study. Int J Cancer 129: 160–172.
13. Jung SK, Kim K, Tae K, Kong G, Kim MK (2013) The effect of raw vegetable and fruit intake on thyroid cancer risk among women: a case-control study in South Korea. Br J Nutr 109: 118–128.
14. Dal Maso L, Bosetti C, La Vecchia C, Franceschi S (2009) Risk factors for thyroid cancer: an epidemiological review focused on nutritional factors. Cancer Causes Control 20: 75–86.
15. Bosetti C, Negri E, Kolonel L, Ron E, Franceschi S, et al. (2002) A pooled analysis of case-control studies of thyroid cancer. VII. Cruciferous and other vegetables (International). Cancer Causes Control 13: 765–775.
16. Markaki I, Linos D, Linos A (2003) The influence of dietary patterns on the development of thyroid cancer. Eur J Cancer 39: 1912–1919.
17. Galanti MR, Hansson L, Bergstrom R, Wolk A, Hjartaker A, et al. (1997) Diet and the risk of papillary and follicular thyroid carcinoma: a population-based case-control study in Sweden and Norway. Cancer Causes Control 8: 205–214.
18. Clero E, Doyon F, Chungue V, Rachedi F, Boissin JL, et al. (2012) Dietary iodine and thyroid cancer risk in French Polynesia: a case-control study. Thyroid 22: 422–429.
19. Clero E, Doyon F, Chungue V, Rachedi F, Boissin JL, et al. (2012) Dietary patterns, goitrogenic food, and thyroid cancer: a case-control study in French Polynesia. Nutr Cancer 64: 929–936.
20. Mack WJ, Preston-Martin S, Bernstein L, Qian D (2002) Lifestyle and other risk factors for thyroid cancer in Los Angeles County females. Ann Epidemiol 12: 395–401.
21. George SM, Park Y, Leitzmann MF, Freedman ND, Dowling EC, et al. (2009) Fruit and vegetable intake and risk of cancer: a prospective cohort study. Am J Clin Nutr 89: 347–353.
22. Kohrle J (1999) The trace element selenium and the thyroid gland. Biochimie 81: 527–533.
23. Hard GC (1998) Recent developments in the investigation of thyroid regulation and thyroid carcinogenesis. Environ Health Perspect 106: 427–436.
24. Dora JM, Machado WE, Rheinheimer J, Crispim D, Maia AL (2010) Association of the type 2 deiodinase Thr92Ala polymorphism with type 2 diabetes: case-control study and meta-analysis. Eur J Endocrinol 163: 427–434.
25. Kohrle J (2013) Pathophysiological relevance of selenium. J Endocrinol Invest 36: 1–7.
26. Beckett GJ, Arthur JR (2005) Selenium and endocrine systems. J Endocrinol 184: 455–465.
27. Vinceti M, Dennert G, Crespi CM, Zwahlen M, Brinkman M, et al. (2014) Selenium for preventing cancer. Cochrane Database Syst Rev 3: CD005195.
28. Zhang LR, Sawka AM, Adams L, Hatfield N, Hung RJ (2013) Vitamin and mineral supplements and thyroid cancer: a systematic review. Eur J Cancer Prev 22: 158–168.
29. Drutel A, Archambeaud F, Caron P (2013) Selenium and the thyroid gland: more good news for clinicians. Clin Endocrinol (Oxf) 78: 155–164.
30. Fritz A, Percy C, Jack A, Shanmugaratnam K, Parkin DM, et al. (2000) International Classification of Diseases for Oncology (ICD-O), 3rd Edition. WHO, Geneva.
31. Michaud DS, Midthune D, Hermansen S, Leitzmann M, Harlan L, et al. (2005) Comparison of Cancer Registry Case Ascertainment with SEER Estimates and Self-reporting in a Subset of NIH-AARP Diet and Health Study. Journal of Registry Management 32: 70–75.
32. North American Association of Central Cancer Registries (NAACCR). North American Association of Central Disease Registries. Standards for Cancer Registries Vol. 3. Standards for completeness, quality, analysis, and management of data. 2008.
33. Subar AF, Midthune D, Kulldorff M, Brown CC, Thompson FE, et al. (2000) Evaluation of alternative approaches to assign nutrient values to food groups in food frequency questionnaires. Am J Epidemiol 152: 279–286.
34. Friday JB (2006) MyPyramid Equivalents Database for USDA Survey Food Codes, 1994–2002 USDA, ARS, Community Nutrition Research Group.
35. Thompson FE, Kipnis V, Midthune D, Freedman LS, Carroll RJ, et al. (2008) Performance of a food-frequency questionnaire in the US NIH-AARP (National Institutes of Health-American Association of Retired Persons) Diet and Health Study. Public Health Nutr 11: 183–195.
36. Cox DR (1972) Regression Models and Life-Tables. Journal of the Royal Statistical Society 34: 187–220.
37. Lin DY, Wei LJ, Ying Z (1993) Checking the Cox model with cumulative sums of martingale-based residuals. Biometrika 80: 557–572.
38. Willett W, Stampfer MJ (1986) Total energy intake: implications for epidemiologic analyses. Am J Epidemiol 124: 17–27.
39. Durrleman S, Simon R (1989) Flexible regression models with cubic splines. Stat Med 8: 551–561.
40. Rahbari R, Zhang L, Kebebew E (2010) Thyroid cancer gender disparity. Future Oncol 6: 1771–1779.
41. Sunde RA. Selenium. In: Bowman B RR, eds. Present Knowledge in Nutrition. 9th ed. Washington, DC: International Life Sciences Institute; 2006: 480–97.
42. Sunde RA. Selenium. In: Ross AC CB, Cousins RJ, Tucker KL, Ziegler TR, eds. Modern Nutrition in Health and Disease. 11th ed. Philadelphia, PA: Lippincott Williams & Wilkins; 2012: 225–37.
43. Duffield AJ, Thomson CD, Hill KE, Williams S (1999) An estimation of selenium requirements for New Zealanders. Am J Clin Nutr 70: 896–903.

44. Persson-Moschos M, Alfthan G, Akesson B (1998) Plasma selenoprotein P levels of healthy males in different selenium status after oral supplementation with different forms of selenium. Eur J Clin Nutr 52: 363–367.

45. Longnecker MP, Stampfer MJ, Morris JS, Spate V, Baskett C, et al. (1993) A 1-y trial of the effect of high-selenium bread on selenium concentrations in blood and toenails. Am J Clin Nutr 57: 408–413.

46. van den Brandt PA, Goldbohm RA, van 't Veer P, Bode P, Dorant E, et al. (1993) A prospective cohort study on toenail selenium levels and risk of gastrointestinal cancer. J Natl Cancer Inst 85: 224–229.

47. U.S. Department of Agriculture, Agricultural Research Service. USDA National Nutrient Database for Standard Reference, Release 25. Nutrient Data Laboratory Home Page, 2012.

48. Chun OK, Floegel A, Chung SJ, Chung CE, Song WO, et al. (2010) Estimation of antioxidant intakes from diet and supplements in U.S. adults. J Nutr 140: 317–324.

49. Mutanen M (1986) Bioavailability of selenium. Ann Clin Res 18: 48–54.

50. Jubiz W, Ramirez M (2014) Effect of vitamin C on the absorption of levothyroxine in patients with hypothyroidism and gastritis. J Clin Endocrinol Metab: jc20134360.

51. Ambali SF, Orieji C, Abubakar WO, Shittu M, Kawu MU (2011) Ameliorative effect of vitamin C on alterations in thyroid hormones concentrations induced by subchronic coadministration of chlorpyrifos and lead in wistar rats. J Thyroid Res 2011: 214924.

52. Xiao Q, Park Y, Hollenbeck AR, Kitahara CM (2014) Dietary Flavonoid Intake and Thyroid Cancer Risk in the NIH-AARP Diet and Health Study. Cancer Epidemiol Biomarkers Prev.

53. Sprague BL, Warren Andersen S, Trentham-Dietz A (2008) Thyroid cancer incidence and socioeconomic indicators of health care access. Cancer Causes Control 19: 585–593.

54. Roche LM, Niu X, Pawlish KS, Henry KA (2011) Thyroid cancer incidence in New Jersey: time trend, birth cohort and socioeconomic status analysis (1979–2006). J Environ Public Health 2011: 850105.

55. Key TJ, Schatzkin A, Willett WC, Allen NE, Spencer EA, et al. (2004) Diet, nutrition and the prevention of cancer. Public Health Nutr 7: 187–200.

56. (2000) Institute of Medicine, Food and Nutrition Board. Dietary Reference Intakes: Vitamin C, Vitamin E, Selenium, and Carotenoids. National Academy Press, Washington, DC.

57. Leung AM, Braverman LE, Pearce EN (2012) History of U.S. iodine fortification and supplementation. Nutrients 4: 1740–1746.

18

Bi-Functional Biobased Packing of the Cassava Starch, Glycerol, Licuri Nanocellulose and Red Propolis

Samantha Serra Costa[1]*, Janice Izabel Druzian[2], Bruna Aparecida Souza Machado[3], Carolina Oliveira de Souza[2], Alaíse Gil Guimarães[2]

1 Faculty of Pharmacy, Federal University of Bahia, Rua Barão de Geremoabo, s/n, Ondina, CEP 40171-970, Salvador, Bahia, Brasil, 2 Faculty of Pharmacy, Department of Bromatological Analysis, Federal University of Bahia, Rua Barão de Geremoabo, s/n, Ondina, CEP 40171-970, Salvador, Bahia, Brasil, 3 Faculty of Technology, SENAI/CIMATEC, Serviço Nacional de Aprendizagem Industrial - SENAI, Av. Orlando Gomes, n° 1845, Piatã, CEP 41650-010, Salvador, Bahia, Brasil

Abstract

The aim of this study was to characterize and determine the bi-functional efficacy of active packaging films produced with starch (4%) and glycerol (1.0%), reinforced with cellulose nanocrystals (0–1%) and activated with alcoholic extracts of red propolis (0.4 to 1.0%). The cellulose nanocrystals used in this study were extracted from licuri leaves. The films were characterized using moisture, water-activity analyses and water vapor-permeability tests and were tested regarding their total phenolic compounds and mechanical properties. The antimicrobial and antioxidant efficacy of the films were evaluated by monitoring the use of the active films for packaging cheese curds and butter, respectively. The cellulose nanocrystals increased the mechanical strength of the films and reduced the water permeability and water activity. The active film had an antimicrobial effect on coagulase-positive staphylococci in cheese curds and reduced the oxidation of butter during storage.

Editor: Guo-Qiang Chen, Tsinghua University, China

Funding: The research was funded by CAPES (Coordination of Improvement of Higher Education Personnel), through design nanobiotec. The funders had no role in study design, data collection and analysis, decision to publish, or preparation of the manuscript.

Competing Interests: The authors have declared that no competing interests exist.

* Email: manthasc@yahoo.com.br

Introduction

The food industry seeks to develop products with quality and safety to meet a consumer market increasingly demanding. Thus, besides the application of good hygiene and sanitation practices, the packaging of the product in adequate to protect and conserve food during periods of storage and marketing packaging is required, assuring the consumer to purchase a healthy product [1].

Among the major types of containers for food and are flexible plastic films, which have excellent mechanical and gas barrier properties and water vapour [2,3].

It has been estimated that 2% of all plastics eventually reach the environment, thus contributing considerably to a currently acute ecological problem. [4] In this context, packaging is seen as a major contributor to the environmental impact, since over a third of current plastics production is used to make them, and too because its relativity short life cycle. [4,5] In addition, recent increases in the cost of raw petroleum have led to a dramatic increase in the cost of plastics. Hence, plastic pollution have driven the development of biobased packing with biodegradable polymers derived from renewable resources, that are kinds of environmentally-friendly materials, which can be degraded into carbon dioxide and water by microorganisms in natural environment [2,3,4].

Lately, numerous studies [3,4,6,7] have been undertaken to provide alternative packaging made from renewable agricultural materials, such as starch. The starch based films are transparent, non-toxic, have moderately low permeability to oxygen and moisture barrier. [6,7] However, the low mechanical resistance and high sensitivity to water restricts the use of these films, especially in foods with high moisture.

In this context, research has been conducted in an attempt to incorporate into the biodegradable matrices, organic nanoparticles such as cellulose nanocrystals (CNCs) or nanocellulose, which can improve the mechanical properties of the films. [2,7] These nanoparticles are obtained from different sources of natural fibers such as cotton, wood, sisal, bamboo, coconut and sugarcane bagasse, sugarcane, among others [7].

Besides the possibility of becoming active biodegradable films with nanoparticles, there is a tendency to activate them bioactive compounds, resulting in active films reinforced mechanically retaining the biodegradability. [4,7] The antimicrobial films, and antioxidants, for example, can control the development of specific species of microorganisms and enhance the oxidative stability of the packaged product. [8,9,10,11,12].

Most active packaging marketed have synthetic additives, as butilhidroxitolueno (BHT) and polihiroxialcanoatos (BHA), which despite their proven effectiveness, have restrictions on their use, as these components can migrate into food, posing a potential health risk. Thus, recently, has increased demand for natural bioactive ingredients, when incorporated into the packaging can present antioxidant and antimicrobial effects about packaged product [6,8,9,10].

Among these substances, propolis has great potential to be added to the packaging, since its high antimicrobial and antioxidant activity has been confirmed in several studies [8,9].

In this context, the aim of this study was to investigate the effect of incorporation of extract of red propolis (ERP) and licuri cellulose nanocrystals (CNCs) on the mechanical, barrier and bifunctional properties of the composites films based glycerol and cassava starch.

Materials and Methods

2.1. Material

Licuri leaves used for extraction of CNCs were collected from farms (Cruz das Almas, Bahia, Brazil, located about 12 degrees south latitude and 43 degrees west longitude). The red propolis was donated by apiaries (Aracaju, Sergipe, Brazil). No specific permissions were required for these locations/activities. The studies did not involve endangered or protected species and provide. Cassava starch was donated by Cargill Agrícola S.A. (Porto Ferreira, Bahia, Brazil). Polyethylene was donated by Braskem S.A. (Camaçari, Bahia, Brazil). The cheese and butter were purchased at a local market (Salvador, Bahia, Brazil).

2.2. Licuri Cellulose Nanocrystals (CNC$_S$) Preparation

Sulfuric acid hydrolysis of the leaves *licuri* was performed as described previously in the literature and in our previous works, with minor modifications. [13,14,15] Briefly, leaves *licuri* were ground and subjected to four washes using NaOH (80°C/4 hours). After bleached treatment of the cellulose, using NaOH and NaClO2 (sodium chlorite), the resulting material was ground until a fine particulate was obtained. Then, 10.0 g of cellulose was added to 160.0 mL of 64 wt% sulfuric acid under strong mechanical stirring. Hydrolysis was performed at 50°C for about 20 min. After hydrolysis, the dispersion was diluted twofold in water, and the suspension was washed using three repeated centrifuge cycles. The last washing was conducted using dialysis against deionized water until the dispersion reached a pH of 5–6. A stable suspension of CNCs was obtained through sonication for approximately 5 min.

2.3. Extract of the Propolis (ERP) Preparation

To prepare the extract, propolis was crushed and ground, adding grain alcohol 70%. The mixture was kept in a water bath temperature of 70°C under constant stirring for a period of 30 minutes. The extract was centrifuged at 4400 rpm, the temperature of 10°C for 10 minutes. The supernatant was filtered and then stored in amber vial at a temperature of 10°C.

2.4. Films Preparation

Sample films were prepared by casting a mixture of cassava starch (4 g) and glycerol (1 g) in a suspension containing 100 mL of distilled water. [3] The films were prepared by adding the aqueous dispersion of CNCs and ERP in the desired quantity. A series of cassava starch films plasticized with glycerol were prepared with CNC contents of 0–1.0 wt%, and ERP contents of 0.4–1.0 wt%, according to a central composite experimental design. The samples were heated to 70°C under constant stirring and with dehydration under renewable circulated air (35±5°C) in Petri plastic dishes. Samples were stored at 23°C and 75% relative humidity (RH) for 10 days prior to testing [6,16].

2.5. Licuri Nanocellulose (CNCs) and Films Characterization

Transmission electron microscopy (TEM) was used to characterize the CNCs. TEM images were taken using a FEI Tecnai G2-Spirit with a 120-kV acceleration voltage. Diluted water suspensions of the nanowhiskers (0.01% wt/v) were deposited on a carbon–formvar-coated copper grid (300-mesh). The samples were subsequently stained with a 2% uranyl acetate solution. The length and width of the CNCs were obtained from several TEM images (50 images) [7].

Water activity (Aw) of the films was measured with a Decagon, Aqualab Lite. Pure water (Aw of 1.000±0.001%) and LiCl (Aw of 0.500±0.015%) were used as calibration standards. Preconditioned samples (4 cm^2) were cut from the center of the films and evaluated in triplicate.

The moisture contents of films were determined by measuring their weight loss upon drying in an oven at 105°C until they reached a constant weight (dry sample weight). Samples were analyzed at least in triplicate, and results are expressed as percentage (g/100g) moisture contents of samples.

Mechanical properties of the films were obtained using an Emic Universal Testing Instrument (Model DL2000), operated as specified in the ASTM standard method D882-00. [17] Film strips of 8 cm×2.5 cm (length×width) were cut from each preconditioned sample and mounted between the grips of the machine. The initial grip separation and crosshead speed were set to 50 mm and 12.5 mm/min, respectively. At least 10 replicates of each specimen were averaged together. Values were analyzed of elastic modulus (E), tensile strength (σ) and elongation at break (ε) in films [17].

The water vapour permeability (WVP) of the films gravimetrically according to the methodology proposed by ASTM method E96-80 has been determined [18].

The determinations of total phenolic compounds (TPC) of the samples of the films were determined by spectrophotometry. The result was expressed as gallic acid equivalents (AG/g mg) calculated by curve constructed with varying acid concentrations 25–300 mg/mL (R2 = 0.99491) [19].

2.6. Antimicrobial Efficacy of the Films

To evaluate the antimicrobial activity of the films was used cheese rennet type. The cheese was sliced and submitted the incidence of UV light in a laminar flow chamber for 15 minutes, in order to reduce the initial microbial load. The samples were packed with active film containing highest concentration of both additives (F9-0.70% of ERP and 0.50% of CNCs), and stored in the refrigerator at a temperature of 7±2°C, along with the controls: samples packed with cassava starch films (without propolis; C1), with low-density polyethylene (LDPE; C2), and without any package (C3). The polyethylene film (0.089±0.011 mm) were extruded using a twin screw extruder. Samples were analyzed periodically for counting Staphylococci Coagulase positive (CSCP) for up to 28 days [20].

The formulation containing highest concentration of both additives (F9-0.70% of ERP and 0.50% of CNCs), was used to the evaluate the antimicrobial and antioxidant efficacy of films by presenting better mechanical characteristics and barrier.

2.7. Antioxidant Efficacy of the Films

The oxidative stability from a packaged product was carried out using peroxide value (PV) content at 0, 7, 15, 30, 45 and 60 days. Butter samples packed with cassava starch films (without propolis; C1), with low-density polyethylene (LDPE; C2), and without any package (C3) were used as controls. The PV was determined by titration according to Association of Analytical Communities (AOCS, 1997) methodology [6,21].

2.8. Statistical Analysis

A central composite experimental design was used to evaluate the influence of concentration differences of the ERP and CNCs incorporated in bio-based films. ERP (%, w/w; X1) and CNCs (%, w/w; X2) were chosen as independent variables. The real and coded values of these variables are shown in Table 1. The Aw, TPC, WVP, E, σ and ε were used as dependent variables (Y).

A central composite experimental design with four axial points (R = 1.41) and three replications at the center points (for a total of 11 experiments) was employed (Table 1). The second-degree polynomials were calculated to estimate the response of the dependent variable (Stat, Inc., Minneapolis, MN) using eq 1:

$$Y = b_0 + b_1 X_1 + b_2 X_2 + b_{11} X_1 + B_{22} X_2 + b_{12} X_1 X_2$$

where Y is the predicted response, X1 and X2 are the independent variables, b0 is the offset term, b1 and b2 are the linear effects, b11 and b22 are the squared effects, and b12 is the interaction term. A film without ERP and CNCs (C 0) and the other without ERP (C 1) were used as controls for comparison of results.

The results of the evaluation of the antimicrobial efficacy of the antioxidant and films were analyzed by ANOVA using a statistical program StatSoft v.7 (StatSoft, Inc., Tulsa, Okla., USA). The Tukey test was used to evaluate average differences (at a 95% confidence interval).

Results and Discussion

Figure 1 is a TEM image of the cellulose nanocrystals from licuri leaves. Nanoparticles can be observed in the image and are represented by both aggregate and individual fibrils that are commonly found in TEM images of cellulose nanocrystals extracted from natural fibers. [11,15] The *licuri* fiber used to extract the nanocellulose was composed of approximately 68% cellulose, 15.9% lignin and 8% hemicellulose.

The nanocellulose was generated in an aqueous suspension at a concentration of 0.00117 g/ml, with a yield of 33% relative to the

mass of cellulose pulp used and 5.7% relative to the ground dry weight of the *licuri* leaves.

The isolated cellulose nanocrystals had a mean length of 157 ± 24 nm and a mean diameter of 5.7 ± 1.6 nm, resulting in an aspect ratio (L/D) of 27. Notably, there are no data in the literature on the nanocellulose extracted from licuri, but these values are in the range for nanocrystals from other lignocellulosic fibers that have great potential for use as reinforcement in biodegradable films. [10,14] *Licuri* is a palm tree that originates in the semi-arid region of Brazil. This species is an abundant source of lignocellulosic fibers with high cellulose content and low cost, adding value to the supply chain, which is mainly familial.

When incorporated in different concentrations as additives in cassava starch and glycerol films, licuri cellulose nanocrystals (CNCs) and alcoholic aqueous extract of red propolis (ERP) significantly influenced the mechanical properties, barrier quality and simultaneous bi-functionality of the films.

The Aw values of the various film formulations ranged from 0.438 to 0.494, depending on the concentration of incorporated CNCs and ERP additives (Table 2). The Pareto chart and response surface plot (Figure 2) indicate that the cellulose nanocrystals had a significant effect ($p < 0.05$) on the water activity (Aw) of the films and show a linear inverse relationship between the increase in CNCs and decrease in the Aw (R2 = 0.9706, Figure 3A), as confirmed by the equation generated for the model (Equation 2).

$$Aw = 0.493 - 0.047 Y \left(R^2 = 0.97 \right) \qquad (2)$$

Statistical analyses indicated that the addition of ERP did not significantly interfere ($p > 0.05$) with the Aw values (Figure 2).

There was an 11% reduction in the Aw values of the films with 0.5% CNCs (C1 - cassava starch films without propolis) compared to the control films (C0 - cassava starch films without propolis and cellulose nanocrystals), indicating that CNCs reduce water availability within the matrix of the nanocomposite material and have a relevant role in controlling water availability due to the increase in the number of hydrogen bonds with the polymeric matrix and plasticizer [7].

Table 1. Values of the independents variables added to films as experimental design.

| Film formulations | Coded values | | Real values (%) | |
	ERP (X₁)	CNCs (X₂)	ERP (X₁)	CNCs (X₂)
F1	−1.00	−1.00	0.50	0.15
F2	−1.00	+1.00	0.50	0.85
F3	+1.00	−1.00	0.90	0.15
F4	+1.00	+1.00	0.90	0.85
F5	−1.41	0.00	0.40	0.50
F6	+1.41	0.00	1.00	0.50
F7	0.00	−1.41	0.70	0.00
F8	0.00	+1.41	0.70	1.00
F9[a]	0.00	0.00	0.70	0.50
F10[a]	0.00	0.00	0.70	0.50
F11[a]	0.00	0.00	0.70	0.50

[a]*central points;* ERP = extract of red propolis; CNCs = *licuri* cellulose nanocrystals.

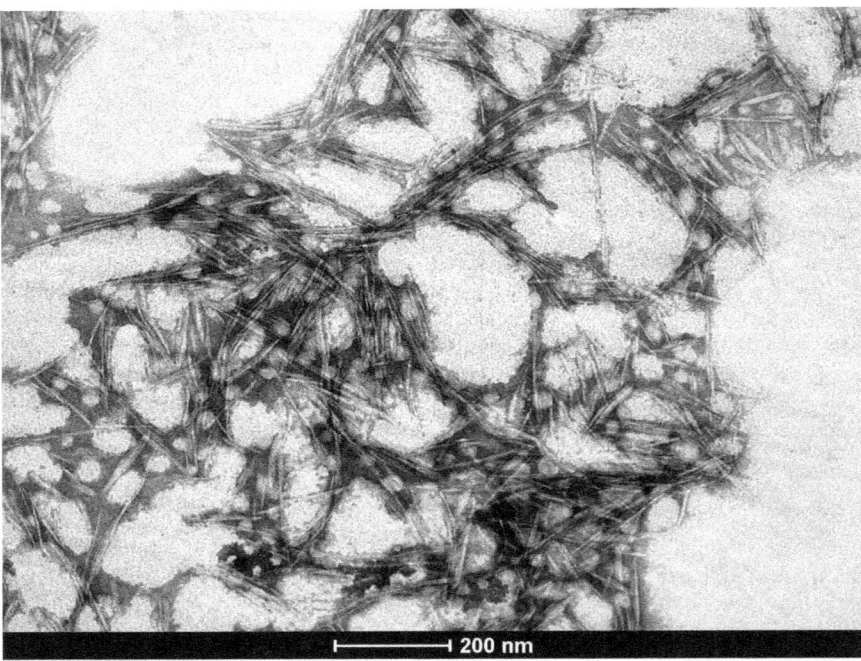

Figure 1. TEM image of licuri cellulose nanocrystals (CNCs).

Considering that water migrates from areas of high Aw to areas of low Aw, it is essential that films for food-product packaging have low values for this parameter because such films can reduce the amount of water available for the growth of microorganisms or unwanted chemical reactions during product storage.

The interference of the cellulose nanocrystals was also observed in the moisture and water vapor-permeability (WVP) results (Table 2). The formulations that had the highest concentrations of CNCs also had the lowest WVP rates and moisture. There was a 55% reduction in the WVP for the films with 0.5% CNCs (C1) compared to the control film (C0). Regarding moisture reduction was 4%.

The moisture is an important parameter to measure, is the basis for several characteristics of the films, including the mechanical properties and water vapour permeability. The moisture of the films ranged 61.28–98.33%, with differences between the formulations, however, there was a very low correlation coefficient of this parameter with the water vapour permeability ($r^2 = 0.007$), elongation ($r^2 = 0.40$) and tensile strength ($r^2 = 0.0005$), indicating that this parameter has little influence on the properties of the films analyzed.

The WVP values and the CNC concentrations of the films showed an inverse proportional and linear relationship, similar to the Aw values, with $R^2 = 0.9706$ (Figure 3A). At any concentration, the cellulose nanocrystals in the films are intended to reduce the WVP by providing a physical barrier, reducing the free spaces in the polymer matrix, restricting the passage of vapors and inhibiting water absorption. The degree of crystallinity of the cellulose nanoparticles and the strong hydrogen bonds formed within the matrix restrict access and hence the permeability of water through the film [14,15,22].

The correlation between the concentration of the ERP of the films and the WVP was not significant (p>0.05), although other studies [23,24] have shown a reduction in the WVP of polymer matrices with the addition of natural products rich in phenolic compounds. These compounds can form hydrogen bonds with the

reactive groups in the polymer matrix, lowering the rate of water permeation by limiting the availability of hydrogen groups to form hydrophilic bonds with water [23].

Comparing the obtained films with polyethylene films that are commonly utilized, it is noted that biodegradable films showed higher elastic modulus and tensile strength greater than the polyethylene film proving to the mechanical reinforcement by the addition of nanocrystals (Table 2).

With respect to total phenolic compounds (TPC) derived from the addition of ERP to the films, the Pareto chart and response surface (Figure 4) show that the concentrations of ERP and CNCs significantly influence (p<0.05) this independent variable. As expected, as the concentration of ERP in the films increases, there is an almost linear and directly proportional increase in the TPC, indicating that there is a good incorporation of the propolis and its active compounds in the nanocomposite. The equation for this model (Equation 3) indicates that the TPC content in the films does not depend on the interaction of the independent variables studied but is only influenced by the variables in an isolated manner.

$$TPC = 21.59 + 117.11X - 46.33X2 + 4.14Y - 11.50Y2 (R^2 = 0.94) \quad (3)$$

The presence of CNCs also played an important role in the mechanical properties of the films. The values obtained for the mechanical properties of the films did not fit well with the adopted mathematical model; however, the results indicate that the elastic modulus and tensile strength values for the films increased 176% and 111%, respectively, with the use of 0.5% CNC (C1). These results demonstrate that cellulose nanocrystals can be used to reinforce biodegradable matrices, even at low concentrations.

The tensile strength values can be defined as the maximum resistance provided by the films when subjected to tension, and

Table 2. Properties of the films formulations with different contents of ERP and CNCs.

Film Formulations	Aw	M (%)	TPC (mg AG/g of film)	WVP×10^{-8} (gH$_2$O μg/m^2·h·mmHg)	Mechanical properties		
					E (Mpa)	σ (Mpa)	ε (%)
C0	0.522±0.002	11,01±0,97	0.00	4.16±0.07	107.20±9.43	3.70±0.34	68.00±5.67
C1	0.462±0.002	10,61±0,67	0.00	1.86±0.06	296.20±14.86	7.80±0.45	22.80±5.22
C2	-	-	-	1.00-1.50	102.0-240.0	6.9-16.0	100.0-800.0
F1	0.487±0.004	10,23±1,09	68.42±1.09	2.43±0.02	217.80±32.51	5.40±0.89	52.80±6.10
F2	0.441±0.003	9,55±0,98	64.53±0.41	1.77±0.04	257.20±44.06	6.20±0.84	46.80±5.02
F3	0.488±0.004	10,08±0,73	86.32±0.48	2.47±0.01	190.80±42.76	4.40±0.55	54.00±7.35
F4	0.439±0.006	10,43±1,49	83.68±0.39	1.77±0.03	264.00±38.12	6.00±1.41	48.80±8.90
F5	0.466±0.003	11,46±0,91	61.28±0.56	2.42±0.03	221.40±19.51	5.20±0.84	52.00±8.60
F6	0.471±0.009	10,75±0,67	98.33±0.44	1.95±0.03	237.00±34.66	5.40±0.55	50.00±7.35
F7	0.494±0.004	10,09±1,07	83.24±0.97	3.10±0.07	122.40±8.73	4.00±0.71	66.00±6.93
F8	0.438±0.002	9,38±0,55	79.44±0.71	1.71±0.05	290.20±58.52	8.00±0.71	26.40±6.54
F9[a]	0.464±0.001	11,27±0,76	81.15±0.78	2.06±0.03	246.60±32.44	7.20±0.84	33.20±7.69
F10[a]	0.464±0.001	11,44±1,53	82.26±0.15	2.11±0.01	247.00±34.73	7.20±1.64	33.60±8.05
F11[a]	0.471±0.004	11,60±1,38	82.18±0.59	2.07±0.03	246.20±15.35	7.20±1.30	33.60±9.21

[a]central points; C0 = cassava starch films without propolis and cellulose nanocrystals; C1 = cassava starch films without propolis; C2 = polyethylene film [35]; Aw = water activity; M = moisture; TPC = total phenolic compounds; WVP = water vapor permeability; E = elastic modulus; σ = tensile strength; ε = elongation at break; ERP = extract of the red propolis; CNCs = *licuri* cellulose nanocrystals.

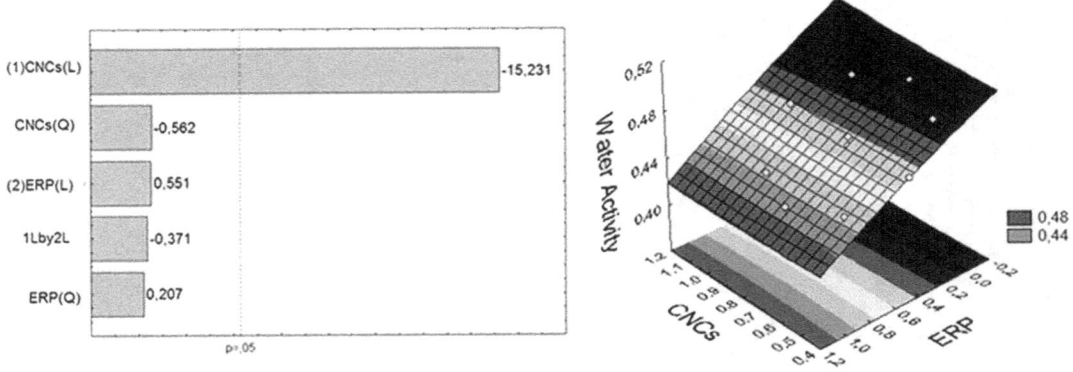

CNCs: Licuri Nanocellulose; ERP: Extract of the Propolis; (L): Linear variable; (Q): Quadratic variable; 1Lby2L: Variable interaction.

Figure 2. Pareto graph (A) and response surface (B) generated between CNCs, Water Activity and ERP.

larger values may be related to the interaction between the polymer chains and the cellulose nanocrystals, forming a structure that is more resistant to tension [15].

The addition of ERP alone did not affect the mechanical properties of the films: similar results were obtained for the F 7

(without ERP) and control films (C 0) (Table 2). However, the results for all the formulations indicated that the addition of ERP combined with CNCs, even at low concentrations, has an opposite effect than the CNCs alone. In the formulations with the same CNC content and different concentrations of ERP (F1 and F3),

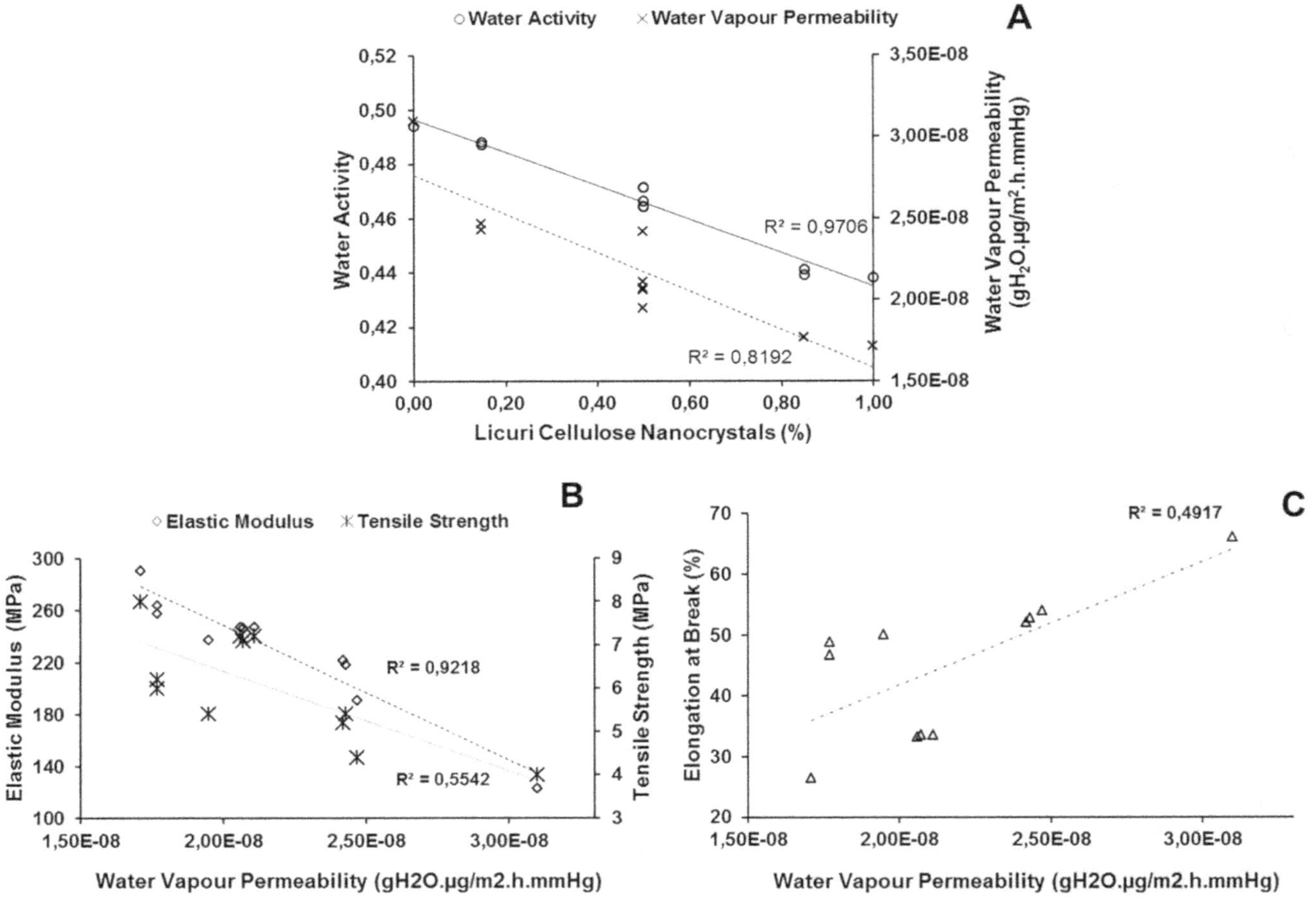

Figure 3. Linear correlation between the CNCs and the physico-chemical properties (A) and WVP and mechanical properties (B and C) of the films.

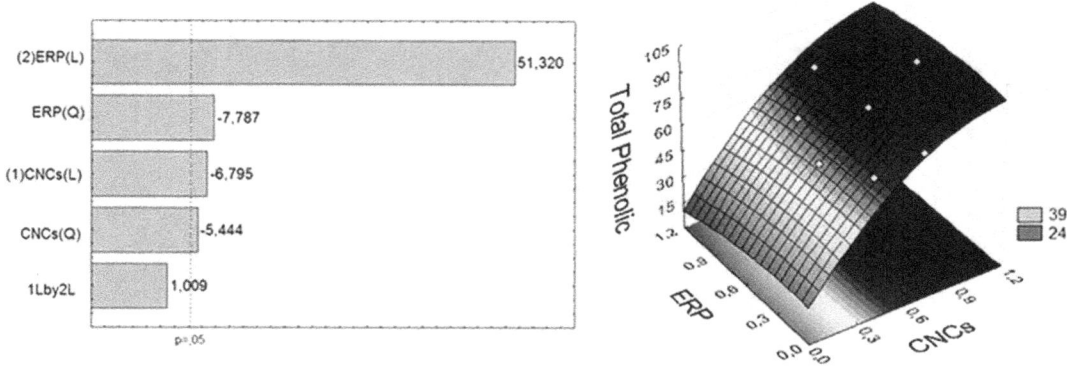

CNCs: Licuri Nanocellulose; ERP: Extract of the Propolis; (L): Linear variable; (Q): Quadratic variable;
1Lby2L: Variable interaction.

Figure 4. Pareto graph (A) and response surface (B) generated between CNCs, Total Phenolic and ERP.

increases in the proportion of ERP further reduce the elastic modulus and tensile strength values of the films.

A reduction in the strength of films with an increase in the concentration of oils and extracts from natural products has been reported in other studies [25,26] and is related to the replacement of polymer components by natural components, resulting in a less-resistant matrix [25] Phenolic derivatives derived from propolis extracts result in weakened interactions between the other components of the nanocomposite.

The CNCs also had an effect on the elongation at break of the films. With the addition of 0.5% CNCs to the films, there was a 66% reduction in the deformation of the films (C1). The addition of nanocrystals favors the formation of a more compact structure, which increases stiffness and reduces the flexibility of the films. [7] This behavior has also been reported in other studies and can be used as an indication of good interactions between the different film components, [22] forming a continuous network of nanocrystals, connected via strong hydrogen bonds [27].

Figure 3B and 3C shows the linear correlation between the WVP values and mechanical properties of the films. There is a linear and inverse proportional correlation between the WVP and elastic modulus of the films ($R^2 = 0.9218$). The increased permeability of the films significantly reduced the film's stiffness and hence its ability to stretch, i.e., its modulus of elasticity. This correlation is extremely important because it indicates that the addition of CNCs can linearly reduce the WVP and increase the elastic modulus of the films, making the film more competitive with standard products. The film (F9) containing intermediate levels of ERP (0.70%) and CNCs (0.50%) was selected for food packaging and to evaluate the bi-functional efficacy of the films because it had the combination of values for Aw, WVP, TPC and mechanical properties that together provided the most satisfactory film characteristics.

3.1. Antimicrobial Efficacy of the Films

Figure 5A shows the number of CFUs of coagulase-positive staphylococci in cheese packed using films C1, C2, C3 and F9 during 28 days of storage. The active film F9 containing 0.70% and 0.50% of ERP and CNCs, respectively, showed an approximately 1 log reduction in the number of these microorganisms at 4, 12, 20 and 28 days compared to all the controls. As expected, the exposed cheese (3 C) showed the highest number of CFUs at all the periods analyzed. There were no significant

differences (p>0.05) between the number of CFUs in the cheese packed in the polyethylene film (C 2) and the starch film without propolis (C 1), demonstrating that these films play equivalent roles in protecting the product from contamination by coagulase-positive staphylococci (Figure 5A, Table 3). This result reflects the addition of nanocellulose to the biodegradable film, which reduced the Aw and WVP of the film, which are extremely important properties in controlling microbial growth in foods.

Despite the increase in the number of CFUs of coagulase-positive staphylococci during the shelf life of the packaged product, active films played an active role in inhibiting the growth of these microorganisms, given that the number of CFUs was low over the analysis period, indicating that the film has an antimicrobial effect on the microorganisms in the matrix studied. The high initial load of staphylococci in cheese (4.5E+02 CFU/g) can be reduced because of the antimicrobial efficacy of the film. Moreover, the intrinsic factors (fat, protein, water, pH and preservatives) and extrinsic factors (temperature, vacuum packaging and microbial characteristics) of the food can also influence the sensitivity of the bacteria and reduce the effectiveness of the films [28].

The number of CFUs indicates that the ERP in the active film had inhibitory activity against the microorganism evaluated, which reached its lowest number of CFUs over the entire storage period.

The antimicrobial activity of alcoholic extracts of propolis on Gram-positive bacteria has been reported in several studies [24,25,29] and can be attributed to the presence of phenolic compounds that inhibit bacterial growth via inhibition of the bacterial RNA polymerase and disruption of the bacterial cell membrane and cytoplasm. [8,9,30] In products with a low initial microbial load, the antimicrobial effect of the active films is greater and more easily observed because for these products, the packaging should only provide an inhibitory effect by slowing bacterial growth, whereas in products with high microbial loads, active films must have both an inhibitory and bactericidal effect to eliminate pre-existing microorganisms.

The action of antimicrobial films can occur via the diffusion of active compounds into the food or may only be the result of direct contact between the food and the surface of the film, without diffusion of active compounds. [31] Importantly, the interaction between polymer groups and the active compounds of the embedded agent can reduce or even prevent the diffusion of active compounds into the system, reducing the effectiveness of the compounds. Thus, it is evident that the inhibitory effect of the

Table 3. Effect Antimicrobial and antioxidant of active film during products storage.

Film formulations	Count of coagulase-positive staphylococci (CFUs) in cheese, during 28 days of storage (days)							
	0	4	8	12	16	20	24	28
C1	4,5E+02[a]	1,1E+03[a]	6,5E+03[b]	1,7E+04[b]	4,6E+04[b]	8,7E+04[b]	1,5E+05[b]	4,1E+05[b]
C2	4,5E+02[a]	1,8E+03[a]	7,4E+03[b]	1,9E+04[b]	5,1E+04[b]	9,8E+04[b]	1,9E+05[b]	4,5E+05[b]
C3	4,5E+02[a]	2,1E+03[a]	8,3E+03[a]	3,2E+04[a]	7,4E+04[a]	1,2E+05[a]	4,7E+05[a]	5,9E+05[a]
F9	4,5E+02[a]	7,5E+02[b]	1,1E+03[c]	4,7E+03[c]	2,3E+04[c]	4,8E+04[c]	7,2E+04[c]	8,4E+04[c]

Film formulations	Peroxide index (PI) of the butter (meq/Kg) during storage (days)					
	0	7	15	30	45	60
C1	0.35[a]	0.71[a]	0.81[b]	0.95[b]	0.90[a]	0.80[a]
C2	0.35[a]	0.75[a]	0.89[a,b]	1.04[a,b]	0.87[a]	0.68[a,b]
C3	0.35[a]	0.94[a]	1.14[a]	1.22[a]	0.91[a]	0.58[a,b]
F9	0.35[a]	0.40[b]	0.46[c]	0.54[c]	0.58[b]	0.52[b]

Means with the same letters in the same columns presented no statistical difference (p>0.05) according to Tukey's test.
C1 = film without propolis; C2 = film of low-density polyethylene; C3 = product without any package; F9 = active film (starch, glycerol and 0.5% licuri cellulose nanocrystals).

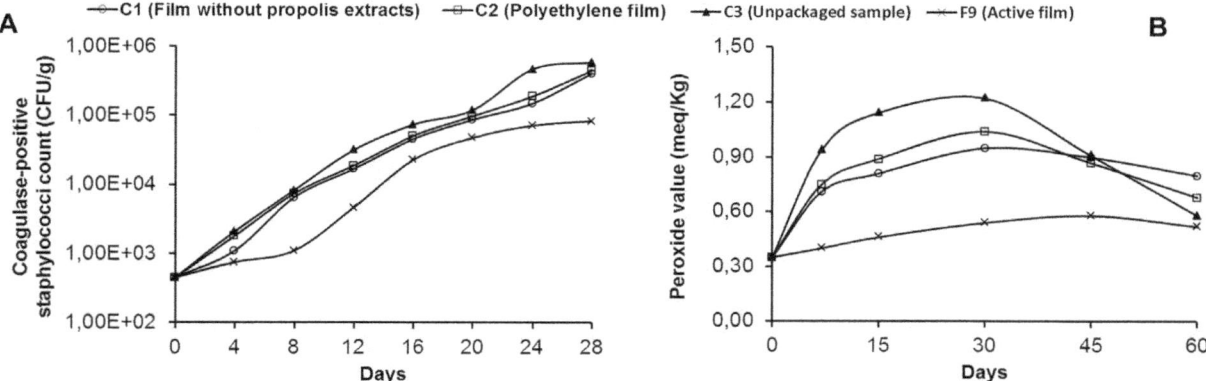

Figure 5. Antimicrobial efficacy (A) and antioxidant (B) the active film during products storage.

bioactive compounds also appears to be associated with the macromolecules used in the films' formulation, and it is therefore essential to understand the macromolecule's properties and how they interact with the active compounds [32].

Similar results were found in a study that evaluated the antimicrobial activity of starch films activated with oregano oil in storage of fresh beef. The authors observed that films with addition of 1.5% of oregano oil fell 2 log reduction counts of mesophilic and pseudomonas [33].

3.2. Antioxidant Efficacy of the Films

The antioxidant efficacy of the film containing 0.70% ERP and 0.50% CNCs (F 9) was evaluated for butter storage using the nanocomposite material. The active film with propolis significantly slowed the oxidation process of butter ($p < 0.05$) compared to controls over the 60-day storage period (Figure 5B, Table 3). The results in this investigation are in agreement with studies of Souza et al. [6] who reported that the PV of palm oil packaged with starch film containing mango and acerola pulp added as antioxidants were lower than those of the control during storage. At 60 days of storage, the PV of the butter packaged in the active film (F 9) increased 49%. The PV of the butter packaged in the film without propolis extracts (C 1) and in low-density polyethylene (C 2) increased 129% and 94%, respectively, after 30 days of storage. At this time, the PV of the butter packaged in the F 9 film was approximately 3 times lower, showing a large reduction in the induction period for oxidation.

The PV of the samples packed in the control films C 1 and C 2 were not significantly different ($p > 0.05$) from the unpackaged sample (C 3) at 7, 45 and 60 days. These results suggest that these films did not provide any protection to the product during storage (Figure 5B).

Notably, in the first 30 days of storage, the butter wrapped in the control film (C 2 - starch, glycerol and nanocellulose) had a lower PV than the sample stored in a polyethylene film (C 1), indicating that the nanocomposite film without propolis extract was more effective in controlling oxidation of the product than the synthetic film. These results suggest that the mechanical reinforcement achieved by the addition of nanocellulose to the film reduces both the film's permeability to water vapor and also its permeability to gases, which would have a direct effect on the oxidation of the stored product. These results are in agreement with some studies that have shown that oxygen can permeate the film and react preferentially with compounds present in the film formulation, allowing the packaged product to be preserved for a longer period of time. For films containing yerba mate extract and mango pulp used to pack palm oil, the authors reported a significant maximum reduction of 25,40% (between formulations) and 87,40% (between formulations and control) ($p < 0.05$) for the oxidation of oil after 45 days of storage [34].

The main purpose of using antioxidants in packaging for food lipids is to delay a significant accumulation of free radicals and thus improve oxidative stability. The results of this study suggest that the protection of packaged products against oxidation can be attributed to the concentration-dependent radical scavenging activity of antioxidant compounds present in film-forming dispersions (Table 3). This outcome occurred because antioxidants act by inhibiting or interrupting the mechanism of lipid auto-oxidation of free radicals. The protective effect against lipid oxidation is likely due to a physical and synergistic process mostly from the total polyphenols and total flavonoids presents in propolis.

Conclusions

The results indicate that biodegradable films derived from starch and glycerol with the addition of cellulose nanocrystals extracted from licuri and alcoholic extracts of red propolis can be a competitive alternative in reducing the use of synthetic packaging films in the storage of some foods. These films may have a simultaneous antimicrobial and antioxidant bi-functional effect. The highly ordered structure of cellulose nanocrystals increases the mechanical strength of the films and reduces the moisture, water activity and water-vapor permeability, which are both critical properties in food packaging. The active film containing 0.70% and 0.50% of the additives ERP and CNCs, respectively, limited the growth of coagulase-positive staphylococci in cheese and slowed the oxidation of butter. Additional studies are therefore needed to evaluate the effect of these films on other food matrices, in which the storage conditions and the characteristics of products and their possible interactions with the films should demonstrate the true effectiveness of the active packaging. The potential of bi-functional films for use in medical and clinical areas and for cosmetics and pharmaceutical products should also be evaluated, increasing the potential applications for these nanocomposite materials.

Author Contributions

Conceived and designed the experiments: SSC BASM COS JID AGG. Performed the experiments: SSC BASM COS JID AGG. Analyzed the data: SSC JID AGG. Contributed reagents/materials/analysis tools: SSC

JID AGG. Contributed to the writing of the manuscript: SSC BASM COS
JID AGG. Obtaining cellulose nanocrystals: SSC BASM COS.

References

1. Mohamed AA, Ali SI, El-Braz FK (2013) Antioxidant and Antibacterial Activities of Crude Extracts and Essential Oils of Syzygium cumini Leaves. Plos one 8(4): 1–7.

2. Habibi Y, Lucia LA, Rojas OJ (2010) Cellulose nanocrystals: chemistry, self-assembly, and applications. Chem Rev 110: 3479–3500.

3. Veiga-Santos P, Oliveira LM, Cereda MP, Scamparini ARP (2007) Sucrose and inverted sugar as plasticizer. Effect on cassava starch–gelatin film mechanical properties, hydrophilicity and water activity. Food Chem 103: 255–262.

4. Wenbin G, Jian T, Chao Y, Cunjiang SM, Weitao G, et al. (2012) Introduction of Environmentally Degradable Parameters to Evaluate the Biodegradability of Biodegradable Polymers. Plos One 7: 5.

5. Luz JMR, Paes SA, Nunes MD, Silva MCS, Kasuya MCM (2013) Degradation of Oxo-Biodegradable Plastic by Pleurotus ostreatus. Plos one 8: 8.

6. Souza CO, Silva LT, Silva JR, Lopez JA, Veiga-Santos P, et al. (2011) Mango and Acerola Pulps as Antioxidant Additives in Cassava Starch Bio-based Film. Journal of Agricultural and Food Chemistry 1: 1.

7. Silva JBA, Pereira FV, Druzian JI (2012) Cassava Starch-Based Films Plasticized with Sucrose and Inverted Sugar and Reinforced with Cellulose Nanocrystals. Journal of Food Science 77: 14–19.

8. Taketoshi H, Shigemi T, Shozo O, Mee-Ra R, Kenji I (2012) Artepillin C, a Major Ingredient of Brazilian Propolis, Induces a Pungent Taste by Activating TRPA1 Channels. Plos One 7: 11.

9. Tanja P, Tomaz P, Lea D, Polona J (2013) Fractionation of Phenolic Compounds Extracted from Propolis and Their Activity in the Yeast Saccharomyces cerevisiae. Plos One 8: 2.

10. Amal AM, Sami IA, Farouk KE (2013) Antioxidante and Antibacterial Activities of Crude Extracts and Essential Oils of Syzygium cumini Leaves. Plos one 8(4): 1–7.

11. Holt BL, Stoyanov SD, Pelan E, Paunov VN (2010) Novel anisotropic materials from functionalized colloidal cellulose and cellulose derivatives. J Mater Chem 20: 10058–10070.

12. Kechichian V, Ditchfield C, Veiga-Santos P, Tadini CC (2010) Natural antimicrobial ingredients incorporated in biodegradable films based on cassava starch. LWT Food Scienc and Technology 43: 1088–1094.

13. Beck-Candanedo S, Roman M, Gray DG (2005) Effect of Reaction Conditions on the Properties and Behaviour of Wood Cellulose Nanocrystal Suspensions. Biomacromolecules 6: 1048–1054.

14. Mesquita J, Patricio P, Donnici C, Petri D, de Oliveira L, et al. (2011) Hybrid layer-by-layer assembly based on animal and vegetable structural materials: multilayered films of collagen and cellulose nanowhiskers. Soft Matter 7: 4405–4413.

15. Rosa MF, Medeiros ES, Malmonge JA, Gregorski KS, Wood DF, et al. (2010) Cellulose nanowhiskers from coconut husk fibers: Effect of preparation conditions on their termal and morphological behavior. Carbohydrate Polymers 81: 83–92.

16. Pelissari FM, Grossmann MVE, Yamashita F, Pineda EAG (2009) Antimicrobial, mechanical, and barrier properties of cassava starchechitosan films incorporated with orégano essential oil. Journal of Agricultural and Food Chemistry 57: 7499–7504.

17. American Society for Testing and Materials (ASTM) (2001) Standard Test Method for tensile properties of thin plastic sheeting. ASTM D882-00.

18. American Society for Testing and Materials (ASTM) (2005) Standard Test for Water Vapor Transmission of Materials. E96/E96M.

19. Cavalcante DRR, Oliveira PS, Góis SM, Soares AF, Cardoso JC, et al. (2011) Effect of green propolis on oral epithelial dysplasia in rats. Braz J Otorhinolaryngol 77: 278–84.

20. American Public Health Association (APHA) (2001) Compendium of Methods for the Microbiological Examination of Foods 4: 676.

21. American Oil Chemists Society (AOCS) (1997) Official Methods of Analysis, Cd 8b-90.

22. Mathlouthi M (2001) Water content, water activity, water structure and the stability of foodstuffs. Food Control 12(7): 409–17.

23. Siripatrawan U, Harte BR (2010) Physical properties and antioxidant activity of an active film from chitosan incorporated with green tea extract. Food Hydrocolloids 24: 770–775.

24. Rohn S, Rawel HM, Kroll J (2004) Antioxidant activity of protein-bound quercetin. Journal of Agricultural and Food Chemistry 52: 4725–4729.

25. Gontard N, Guilbert S, Cuq JL (1992) Edible wheat gluten films: influence of the main processes variables on films properties using response surface methodology. Journal of Food Science 57: 190–195.

26. Azeredo HMC, Mattoso LHC, Wood D, Williams TG, Avena-Bustillos RJ, et al. (2009) Nanocomposite edible films from mango puree reinforced with cellulose nanofibers. Journal of Food Science 74(5): 31–35.

27. Samir MASA, Aloin F, Dufresne A (2005) Review of Recent Research into Cellulosic Whiskers, Their Properties and Their Application in Nanocomposite Field. Biomacromolecules 6: 612.

28. Gill AO, Delaquis P, Russo P, Holley RA (2002) Evaluation ofantilisterial action of cilantro oil on vacuum packed ham. International. Journal of Food Microbiology 1: 83–92.

29. Park YK, Ikegaki M (1998) Preparation of water and ethanolic extracts of propolis and evaluation of the preparations. Biosci Biotechnol Biochem 62: 2230–2232.

30. Massaro FC, Brooks PR, Wallace HM, Nsengiyumva V, Narokai L, et al. (2013) Effect of Australian Propolis from Stingless Bees (Tetragonula carbonaria) on Pre-Contracted Human and Porcine Isolated Arteries. Plos one 8: 1–10.

31. Santos FA, Bastos EMA, Uzeda M, Carvalho MAR, Farias LM (2002) Antibacterial activity of Brazilian propolis and fractions against oral anaerobic bacteria. Journal of Ethnopharmacology 80: 1–7.

32. Cagri A, Ustunol Z, Ryser ET (2001) Antimicrobial, mechanical, and moisture barrier properties of low pH whey protein-based edible films containg p-Amminobenzoic or sorbic acid. Journal of Food Science 66: 865–870.

33. Zinoviadou KG, Koutsoumanis KP, Biliaderis CG (2009) Physico-chemical properties of whey protein isolate films containing oregano oil and their antimicrobial action against spoilage flora of fresh beef. Meat Science 82: 338–345.

34. Reis LCB, Souza CO, Silva JBA, Martins AC, Nunes IL, et al. (2014) Active biocomposites of cassava starch: the effect of yerba mate extract and mango pulp as antioxidant additives on the properties and the stability of a packaged product. Food and Bioproduct Processing: DOI:10.1016/j.fbp.2014.05.004.

35. Coutinho FMB, Mello IL, Maria LCS (2003) Polietileno: principais tipos, propriedades e aplicações. Polímeros 13: 1.

19

Response of *Daphnia's* Antioxidant System to Spatial Heterogeneity in Cyanobacteria Concentrations in a Lowland Reservoir

Adrianna Wojtal-Frankiewicz[1]*, Joanna Bernasińska[2], Piotr Frankiewicz[1], Krzysztof Gwoździński[2], Tomasz Jurczak[1]

1 Department of Applied Ecology, University of Lodz, Lodz, Poland, 2 Department of Molecular Biophysics, University of Lodz, Lodz, Poland

Abstract

Many species and clones of *Daphnia* inhabit ecosystems with permanent algal blooms, and they can develop tolerance to cyanobacterial toxins. In the current study, we examined the spatial differences in the response of *Daphnia longispina* to the toxic *Microcystis aeruginosa* in a lowland eutrophic dam reservoir between June (before blooms) and September (during blooms). The reservoir showed a distinct spatial pattern in cyanobacteria abundance resulting from the wind direction: the station closest to the dam was characterised by persistently high *Microcystis* biomass, whereas the upstream stations had a significantly lower biomass of *Microcystis*. Microcystin concentrations were closely correlated with the cyanobacteria abundance ($r = 0.93$). The density of daphniids did not differ among the stations. The main objective of this study was to investigate how the distribution of toxic *Microcystis* blooms affects the antioxidant system of *Daphnia*. We examined catalase (CAT) activity, the level of the low molecular weight antioxidant glutathione (GSH), glutathione S-transferase (GST) activity and oxidative stress parameters, such as lipid peroxidation (LPO). We found that the higher the abundance (and toxicity) of the cyanobacteria, the lower the values of the antioxidant parameters. The CAT activity and LPO level were always significantly lower at the station with the highest *M. aeruginosa* biomass, which indicated the low oxidative stress of *D. longispina* at the site with the potentially high toxic thread. However, the low concentration of GSH and the highest activity of GST indicated the occurrence of detoxification processes at this site. These results demonstrate that daphniids that have coexisted with a high biomass of toxic cyanobacteria have effective mechanisms that protect them against the toxic effects of microcystins. We also conclude that *Daphnia*'s resistance capacity to *Microcystis* toxins may differ within an ecosystem, depending on the bloom's spatial distribution.

Editor: Franck Chauvat, CEA-Saclay, France

Funding: This study was supported by the Polish Ministry of Science and Higher Education (Grant No. 3988/B/P01/2010/39). The funders had no role in study design, data collection and analysis, decision to publish, or preparation of the manuscript.

Competing Interests: The authors have declared that no competing interests exist.

* Email: adwoj@biol.uni.lodz.pl

Introduction

Planktivorous zooplankton are one of the groups most affected by the mass development of toxic cyanobacteria in inland waters [1]. Specifically, the large-bodied, efficient grazer *Daphnia* usually exhibits slower growth rates and decreased survival and reproduction in the presence of cyanobacteria [2–4]. However, in recent years, it has been observed that the sensitivity of *Daphnia* to cyanobacteria depends on the species and even varies among clones [5–7]. An increasing number of publications have shown that *Daphnia* populations can evolve mechanisms that allow them to coexist with toxic cyanobacteria [8–10,6]. Such resistance results from genetic changes that result in the local co-adaptation of *Daphnia* to cyanobacterial toxins [11,12]. The sensitivity of daphniids to cyanotoxins is most striking in species or clones that are isolated from distinctive habitat types, ecosystems with different trophy and abundances of toxic strains of cyanobacteria [13,5,14]. Little is known about how *Daphnia* sp. respond to spatial differences in cyanobacteria abundance within an ecosystem. Instead, previous research has focused on asynchrony in the zooplankton – the spatial distribution of cyanobacteria and the formation of the "refuge sites" that allow large grazers to persist during blooms [15,16].

The main objective of this study was to investigate how the antioxidant system in *Daphnia longispina* (O. F. Müller) responds to the spatial distribution of toxic *Microcystis aeruginosa* (Kutzing) blooms within a lowland reservoir. Our previous research indicated that daphniids that had coexisted with high concentrations of microcystins in the environment had effective mechanisms to protect them against the accumulation and toxic effect of these metabolites [17]. On the basis of those results, we hypothesise that the oxidative stress of *D. longispina* in the sites with high toxic cyanobacteria abundance will be relatively low compared to sites with less biomass of *Microcystis*.

Materials and Methods

Ethics statement

No specific permits were required for the field studies described herein. There was no activity involving endangered or protected species in this study.

Study Site

Sulejow Reservoir is a 39-year-old lowland dam reservoir situated on km 138.9 of the Pilica River (the Vistula River catchment) in central Poland (Fig. 1). Its maximum length is 15.5 km, and its maximum width is 2.1 km. At its maximum capacity (75×10^6 m^3), the reservoir covers 1980 ha, with a mean depth of 3.3 m and a maximum depth of 11 m. The mean water retention time in the reservoir is 30 days [18]. The reservoir subcatchment is mainly covered by agricultural land (50% arable land, 13% meadows and pastures, 1% orchards) and forests (31%). The length of the shoreline is approximately 54 km. The Sulejow Reservoir is a eutrophic ecosystem with annual cyanobacterial blooms [17,18,19]. The dominant species of bloom-forming cyanobacteria is *Microcystis aeruginosa*, which produces micro-cystin-LR, -YR and RR [20,21]. The genus was determined by verifying the accuracy of the amplification products obtained for the 16S rRNA gene and mcyA gene, which are specific to the *Microcystis* genera. In all the analysed samples, we found homology (99–100%) for *M. aeruginosa* NIES-843 [22].

Studies were conducted at three sampling stations in the Sulejow Reservoir in 2012: Tresta (TR), Bronisławów (BR) and Zarzęcin (ZA). The TR station is located in the lower section near the dam. The BR is located in the middle section of the reservoir in front of the former water pump station, and the ZA station is located in the upper part of the reservoir near the backwater (Fig. 1). The fieldwork was conducted in four periods: before cyanobacterial blooms (beginning of June) and during blooms (July–September). The sampling dates were established according to the results of monitoring of the Sulejow Reservoir, which has been performed weekly for eighteen years from April to October by the Department of Applied Ecology in the University of Lodz. Thus, the sampling was conducted on June 4th, July 2nd, August 21st and September 26th. However, in June we performed the fieldwork only on two sites at opposite ends of the reservoir, Tresta and Zarzęcin, due to the clear water phase in the reservoir and homogenous physical, chemical and biological conditions at BR and ZA (data not shown).

To confirm the pattern observed in 2012, additional samples were collected at TR, BR and ZA on 11th September 2014.

Plankton collection, preparation and identification

Zooplankton samples were collected from a 0.5-m depth using a 64-μm mesh. A net with a diameter of 0.25 m was dragged by a boat for 10 min with a speed of 0.5 m s^{-1}, which resulted in sieving approximately 15,000 L of water. During the most intensive cyanobacterial blooms (in August and September), zooplankton were collected using a 5-L sampler (with many repetitions) at a 4-m depth due to the high concentration of *Microcystis aeruginosa* in the surface water layer. This procedure was feasible because regular mixing of the reservoir's water caused unification of the physical and chemical conditions in the water column, which was controlled by the YSI Professional Plus multisensors during sampling (data not shown).

In the laboratory, the samples were placed into a 2-L glass separator and repeatedly washed by water to separate the zooplankton from the phytoplankton. Next, individuals of *Daphnia* sp. were selected under a Nikon 102 microscope (magnification of ×40–60) with a pipette and placed into a vessel with distilled water. Then, *Daphnia* sp. individuals were gravitationally filtrated (without using a pump) on Whatman GF/C filter paper. The animals were gently removed from the filters with a needle and separated into nine Eppendorf test tubes (approximately 500–600 daphniids per each tube). The test tubes were frozen at −70°C before the chemical analyses.

To identify the zooplankton species and determine their density, 35 L of water was filtered using a 64 μm mesh net, and the samples were concentrated to 10 ml and preserved in a 4% Lugol's solution. Zooplankton taxa were distinguished under a Nikon 115 microscope (magnification of ×100–200). The *Daphnia* species collected were morphologically identified using Benzie [23].

Cyanobacteria species composition (qualitative analyses) was examined in 10 L^{-1} water samples taken on each sampling occasion. The water samples for phytoplankton estimation were preserved in a Lugol's solution and sedimented in the laboratory. Algae were counted using a Fusch-Rosenthal counting cell. At least 400 cells or colonies were counted to reduce the error to less than 10% (P = 0.05). The collected material was morphologically analysed according to Starmach [24], Komarek [25], and Komarek and Anagnostidis [26].

Estimation of cyanobacteria abundance by chlorophyll *a* concentration

The concentration of chlorophyll *a* (μg L^{-1}) was measured immediately after sampling in a 1-L integrated water sample using a bbe Algae Online Analyser (AOA, Version 1.5 E1, bbe-Moldaenke company Kiel, Germany). The measurement principle of the bbe AOA is based on the determination of the fluorescence spectrum and the fluorescence kinetics of the algae. By analysing the interaction of chlorophyll *a* with other pigments, AOA discriminates four main groups of algae (green algae, cyanobacteria, diatoms and cryptophytes). This method is recognised as a reliable on-line analysis for chlorophyll *a* measurements [27] and as a useful tool for monitoring the phytoplankton community composition, particularly as an early warning system for the detection of harmful algal blooms [28]. The AOA fluorometer has been previously tested as an early warning method for cyanobacterial blooms in the Sulejow Reservoir. The results of these studies demonstrated a significant positive correlation (r = 0.68, n = 46, P<0.05) between cyanobacterial biovolume, as determined by cell counts, and cyanobacterial chlorophyll *a*, as measured by the AOA [29].

Microcystin concentrations – sample preparation and analyses

Samples of water containing cyanobacteria were stored in dark glass bottles of 1-L volume and transported to the laboratory for analyses immediately after sampling.

The microcystins (MCs) were analysed in two fractions: the first was dissolved in water (extracellular) and the second in cell-bound form in suspended matter (intracellular). For the analysis of microcystins in suspended matter, the water samples were filtered on Whatman GF/C filter paper and dried to estimate the dry biomass of cyanobacteria. The filters were then stored at −20°C until future use. Microcystins in the suspended material were extracted in 75% aqueous methanol. The samples were sonicated for 30 s in a Misonix ultrasonicator (Farmingdale, NY, USA) equipped with an ultrasonic probe (100 W, 19 mm diameter with "spike") and a liquid processor XL. The extracts were then centrifuged twice at $11000 \times g$ for 10 min at 4°C in an Eppendorf

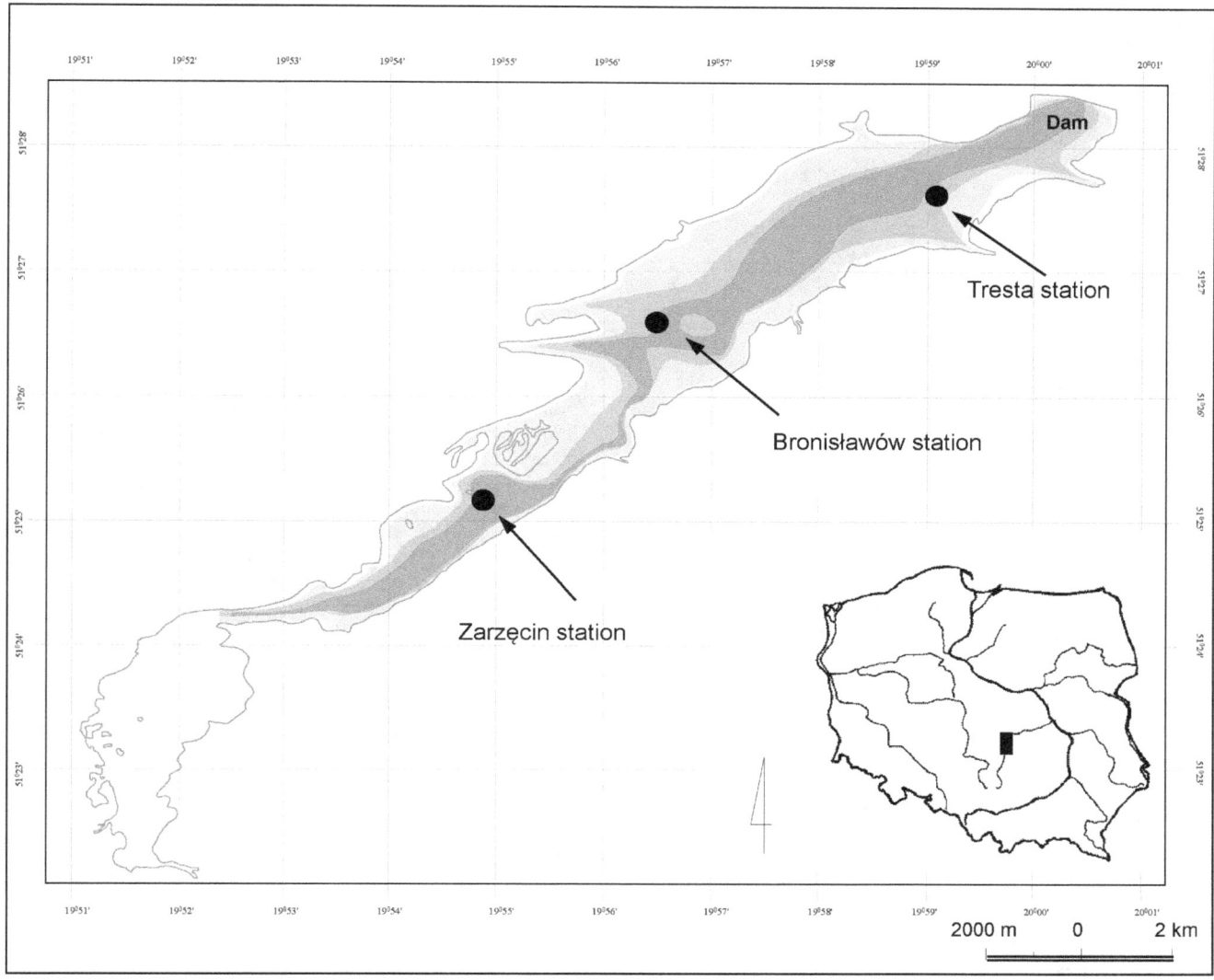

Figure 1. Study site. Map of the Sulejow Reservoir with sampling sites.

5804 centrifuge (Hamburg, Germany). The supernatants were collected and evaporated in a SC110A Speedvac Plus (Thermo-Savant, Holbrook, NY, USA). Before the High Performance Liquid Chromatography (HPLC) analysis, the samples were redissolved in 1 ml of 75% aqueous methanol and filtered through a Gelman GHP Acrodisc 13-mm syringe filter with a 0.45-μm GHP membrane and a minispike outlet (East Hills, NY, USA).

For dissolved microcystins, 1-L samples of filtered water were concentrated in Baker C_{18} solid phase extraction (SPE) cartridges (Deventer, Netherlands; sorbent mass: 500 mg). The microcystins were eluted from the C_{18} cartridges with 3 ml of 90% aqueous methanol containing 0.1% trifluoroacetic acid (TFA). The eluates were then evaporated to dryness, and the samples were redissolved in 1 ml of 75% aqueous methanol before HPLC analysis.

The samples were analysed using an Agilent 1100 series (Waldbronn, Germany) HPLC comprising a quaternary pump, an autosampler, a thermostated column compartment and a diode-array detector. Chromatographic separation was achieved on a Merck Purospher Star RP-18e column (55×4 mm; 3 μm) with a C_{18} guard column (4×4 mm). A determination of

microcystins by HPLC-DAD was performed using a gradient mobile phase of H_2O +0.05% TFA (eluent A) and acetonitrile (ACN) +0.05% TFA (eluent B) and diode-array detection at 200–300 nm. The linear gradient conditions were as follows: 25% B at 0.0 min, 70% B at 5.0 min, 70% B at 6.0 min and 25% B at 6.1 min. The sample volume was 20 μl, the flow rate was 1 ml min^{-1} and the column temperature 40°C. The microcystins in the cyanobacterial extracts were identified using the microcystin standards MC-LR, MC-RR and MC-YR, with their characteristic absorption spectra and retention times. Microcystins MC-LR, MC-RR and MC-YR are the main microcystins detected in Polish freshwaters [30]. Three microcystin standards from seven commercially available products from Calbiochem (La Jolla, CA, USA) were used.

Determination of thiobarbituric reactive substances

Daphnia sp. individuals (wet mass 0.06–0.16 g) were placed at a 10% w/v ratio into a 100-mM sodium phosphate buffer with a pH 7.4 with 100 mM KCl and 1 mM EDTA on ice.

Table 1. Density of *Daphnia longispina* in the sites of the Sulejow Reservoir.

Date	TR	BR	ZA
04/06/2012	32	36	31
02/07/2012	85	73	76
21/08/2012	42	37	41
26/09/2012	25	31	29
11/09/2014	53	48	59

Density of *Daphnia longispina* [ind dm^{-3}] in the three studied sites: Tresta (TR), Bronisławów (BR) and Zarzęcin (ZA). The sampling was conducted monthly between June and September of 2012, and on September of 2014.

Homogenisation using a CAT X-120 knife homogeniser was performed on ice at 2000 rpm for 2 min, and the homogenates were then centrifuged at 10,000×g for 10 min (4°C). The supernatants were immediately used for a cellular lipid peroxidation (LPO) estimation. Lipid peroxidation was measured using a thiobarbituric acid reactive substances (TBARS) assay [31] with modifications [32]. The concentration of the thiobarbituric acid reactive substances is an index of lipid peroxidation and oxidative stress. The assay was monitored for the appearance of conjugated complexes of thiobarbituric acid, mainly malondialdehyde (MDA), which is the end product of LP at 532 nm. The MDA levels were calculated using the MDA extinction coefficient, 156 L mmol^{-1} cm^{-1}, and expressed as nanomoles per milligram of protein (nmol/mg protein).

Determination of glutathione content

Daphnia sp. individuals (wet mass 0.06–0.16 g) were placed in a 10% w/v ratio in homogenisation buffer (154 mM KCl, 5 mM diethylenetriaminepentaacetic acid (DTPA) and 0.1 M potassium phosphate buffer, pH 6.8) and homogenised on ice at 2000 rpm for 2 min. An aliquot was removed for protein determination using the Lowry method [33]. The remainder of the homogenate was mixed with a cold acid buffer containing 40 mM HCl, 10 mM DTPA, 20 mM ascorbic acid and 10% trichloroacetic acid (TCA) at a ratio of 1:1. The suspension was centrifuged at 13,000×g for 10 min (4°C). The supernatants were immediately used for further determination of the glutathione (GSH) content.

The GSH was estimated according to the fluorescence assay of Senft et al. [34] using o-phthalaldehyde (OPA). In this method, OPA, which has a low fluorescence background, reacts only with GSH to generate a strong fluorescence so that GSH can be specifically quantified. The OPA-derived fluorescence was measured at 365 nm excitation and 430 nm emission. The glutathione concentration was calculated using the calibration curve for different concentrations of reduced glutathione, as a standard, and expressed as nanomoles per milligram protein (nmol/mg protein).

Determination of glutathione S-transferase activity

Glutathione S-transferase (GST) activity was determined in September 2014 using the method of Habig et al. [35] by evaluating the increase in absorbance at 340 nm due to the formation of the conjugate of the 1-chloro-2,4-dinitrobenzene (CDNB) substrate in the presence of reduced glutathione (GSH). The reaction mixture was prepared by mixing 1.5 ml sodium phosphate buffer 0.1 M pH 6.5, 0.2 ml GSH 9.2 mM, 0.02 ml CDNB 0.1 M and 0.1 ml of the sample. The absorbance was measured spectrophotometrically at 340 nm and +25°C.

Determination of catalase activity

Daphnia sp. individuals (wet mass 0.06–0.16 g) were placed in a 10% w/v ratio in 100 mM sodium phosphate buffer, pH 7.4, with 100 mM KCl and 1 mM EDTA on ice. Homogenisation was performed on ice at 2000 rpm for 2 min, and the homogenates were then centrifuged at 10,000×g for 10 min (4°C). The supernatants were immediately used for the determination of catalase activity. The activity of catalase (CAT) was measured spectrophotometrically according to Aebi [36]. The method is based on the decomposition of hydrogen peroxide as indicated by decreased absorbance at 240 nm. The assay mixture consisted of homogenate, 50 mM potassium phosphate buffer pH 7.0 and 0.1% hydrogen peroxide in the final volume of 3 ml. The results of this enzymatic assay were reported in units of CAT activity per milligram of protein (U/mg protein), where 1 U of CAT is defined as the amount of enzyme decomposing 1 μmol of H_2O_2 per minute.

Determination of the protein concentration

The protein concentration was evaluated using the spectrophotometric method [33] with Folin's reagent. The amount of protein in each sample was estimated using the calibration curve for different concentrations of bovine albumin as a standard.

Statistical methods

All the statistical analyses were conducted in Statistica 9.0 (StatSoft, Inc). To test for the effects of the season and sites on the analysed parameters, we used a two-way ANOVA with the months and sites as categorical factors and the *Microcystis* biomass and oxidative stress parameters in *Daphnia* sp. tissues as the dependent factors.

To test for the effect of sites on the analysed parameters within a given month, we applied a one-way ANOVA with the sites as categorical factors and the *Microcystis* biomass and the oxidative stress parameters in *Daphnia* sp. tissues as the dependent factors.

To determine the relationship between *Microcystis* biomass and microcystin concentration, a Pearson correlation coefficient r was calculated.

Results

Identification of *Daphnia* species and determination of their densities

Two species of *Daphnia* were identified in the samples from the Sulejow Reservoir: *D. longispina* (O. F. Müller) and *D. cucullata* (Sars). Throughout the sampling season, the numerical share of *D. longispina* in the samples was on average 90%±7.12%, whereas the numerical share of *D. cucullata* was on average only 10%±7.12%. Thus, we isolated the species from the samples

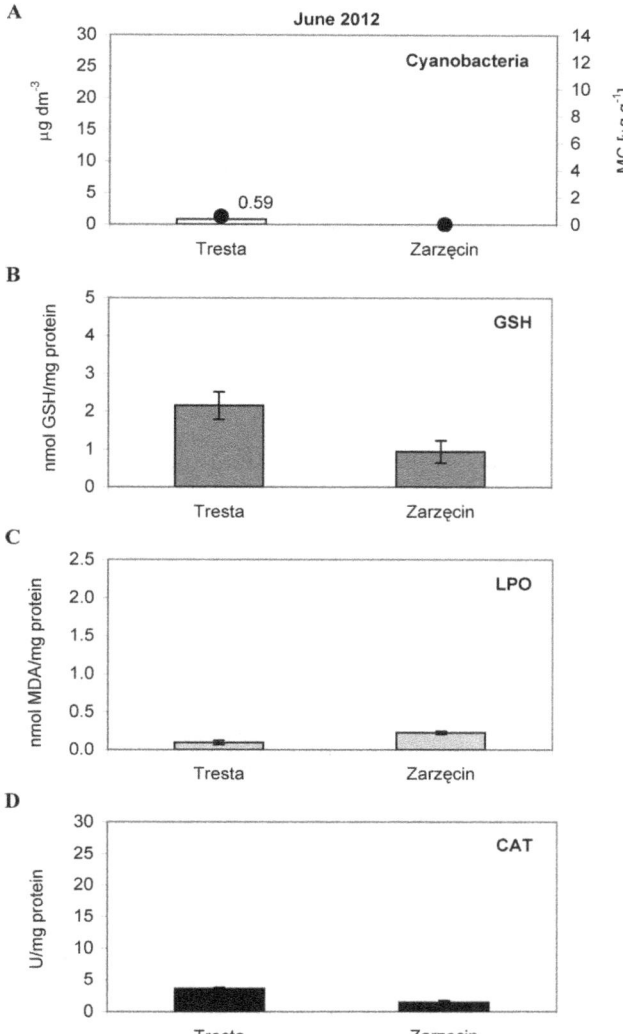

Figure 2. The cyanobacteria abundance and activity of the antioxidant parameters in *Daphnia* tissues, measured in June 2012. (A) The bars represent an average of three replicates (±S.D.) of cyanobacteria abundance (μg dm^{-3}); the dark dots indicate the total concentration of microcystins LR and RR (μg g^{-1}), (B) The bars represent an average of seven replicates (±S.D.) of glutathione (GSH) concentrations (nmol/mg protein), (C) The bars represent an average of seven replicates (±S.D.) of lipid peroxidation (LPO) (nmol/mg protein) and (D) The bars represent an average of six replicates (±S.D.) of catalase (CAT) activity (U/mg protein) in *Daphnia* tissues from the Sulejow Reservoir in 2012. The same letters above the bars indicate that the values did not differ significantly. Each panel of the figure includes the one-way ANOVA test results. Details concerning GSH, LPO and CAT data are presented in Tables S1, S2 and S3.

and analysed the dominant species. The densities of *D. longispina* did not differ between the stations. However, the densities did vary by season, which resulted from typical annual dynamics of *Daphnia* populations. The details are presented in Table 1.

Assessment of Cyanobacteria abundance by measurement of chlorophyll *a*

A qualitative analysis confirmed that the species of cyanobacteria occurring in all samples was *Microcystis aeruginosa*. In June, the chlorophyll *a* concentration for cyanobacteria in the Sulejow

Reservoir was detected only at the TR station (Fig. 2A). The value was very low and averaged 0.80 μg L^{-1}. During the summer, the chlorophyll *a* concentration increased, consistently reaching its highest values at the TR station (Table 2). The differences between months were also significant, and the test showed the following relationship between values of *Microcystin* abundance: July < August < September (see Table 2 for details of two-way ANOVA results).

In July, the chlorophyll *a* concentrations for cyanobacteria were 12.27 μg L^{-1} at the TR station and 1.08 μg L^{-1} at the BR station (Fig. 3A). There were no cyanobacteria present at the ZA station. In August, the concentration of chlorophyll *a* for cyanobacteria was 8.75 μg L^{-1} at TR, 6.89 μg L^{-1} at BR, and 3.46 μg L^{-1} at ZA (Fig. 4A). During September, a substantial bloom of *M. aeruginosa* was observed at the TR station (23.77 μg L^{-1}). At the other sites, the concentration of chlorophyll *a* for cyanobacteria was significantly lower, only 5.16 μg L^{-1} at BR and 0.23 μg L^{-1} at ZA (Fig. 5A). The analysis of the interaction plot indicated that there was a significant effect of the months × sites interaction, resulting mostly from high *Microcystis* abundance at the Tresta station in July and September.

In September 2014, the concentration of chlorophyll *a* for cyanobacteria was 12.21 μg L^{-1} at TR, 6.43 μg L^{-1} at BR, and 3.70 μg L^{-1} at ZA (Fig. 6A).

A one-way ANOVA indicated that the average values of *M. aeruginosa* abundance differed significantly between the sites during all the sampling periods (details on Figs. 2A–6A).

Microcystin concentrations

In 2012, the microcystins LR and RR in samples from the Sulejow Reservoir were identified. An HPLC analysis detected the presence of microcystins in the cells (Fig. 2A–5A) but not in the water of the studied ecosystem. In June, the concentrations of MC-RR in the cells were 0.35 μg g^{-1}, and the concentrations of MC-LR were 0.24 μg g^{-1} at the TR station. In July, the microcystins were still detected only at the TR station, and their concentration increased to 1.06 μg g^{-1} (MC-RR) and 0.88 μg g^{-1} (MC-LR). In August, both MC-RR and MC-LR were found at all three stations: MC-RR = 1.67 μg g^{-1} and MC-LR = 1.18 μg g^{-1} at TR; MC-RR = 1.53 μg g^{-1} and MC-LR = 0.91 μg g^{-1} at BR, and MC-RR = 0.48 μg g^{-1} and MC-LR = 0.24 μg g^{-1} at ZA. The highest values of microcystins at the TR station were detected in September (MC-RR = 10.66 μg g^{-1} and MC-LR = 2.58 μg g^{-1}). At the BR station, MC-RR reached 1.02 μg g^{-1} and MC-LR reached 0.73 μg g^{-1}. We did not find microcystins at ZA (Fig. 5A).

In September 2014, we identified the microcystins LR, YR and RR in the samples. The highest values of microcystins at the TR station were MC-RR = 0.86 μg g^{-1}, MC-YR = 0.20 μg g^{-1} and MC-LR = 0.41 μg g^{-1}. At the BR station, MC-RR reached 0.19 μg g^{-1}, MC-YR reached 0.05 μg g^{-1} and MC-LR reached 0.10 μg g^{-1}. At the ZA station, only MC-RR was identified in an amount of 0.06 μg g^{-1} (Fig. 6A).

The combined microcystin concentration was closely correlated with *M. aeruginosa* abundance (r = 0.93) within all the sampling months.

Activity of the antioxidant system in *Daphnia* tissues

The average values of GSH, LPO, CAT differed significantly among animals from the three sites in each month studied (one-way ANOVA, details on Figs. 2B, C, D–5B, C, D). Significant differences for all the studied parameters were also observed in 2014 (one-way ANOVA, Fig. 6 B, C, D, E).

Table 2. The two-way ANOVA test results.

Overall effect	df	F-ratio	P value	Tukey, posthoc
CYANOBACTERIA				
Months	2,18	73.31	<0.001	Jul <Aug <Sep
Sites	2,18	530.42	<0.001	TR> BR> ZA
Months x sites	4,18	90.93	<0.001	
GSH				
Months	2,54	75.39	<0.001	Aug <Sep <Jul
Sites	2,54	49.14	<0.001	TR <ZA <BR
Months x sites	4,54	24.01	<0.001	
LPO				
Months	2,54	67.53	<0.001	Aug <Sep <Jul
Sites	2,54	75.68	<0.001	TR <BR <ZA
Months x sites	4,54	32.30	<0.001	
CAT				
Months	2,45	828.23	<0.001	Aug <Sep <Jul
Sites	2,45	349.63	<0.001	TR <ZA <BR
Months x sites	4,45	180.54	<0.001	

Results of the two-way ANOVA test for the effects of month and site on the cyanobacteria abundance and antioxidant parameters measured in *Daphnia* tissues in 2012. GSH – glutathione, LPO – lipid peroxidation, CAT – catalase. Analysis do not include data from June.

In June, the glutathione concentrations in the tissues of daphniids were 2.15 nmol GSH/mg protein at TR and 0.94 at ZA (Fig. 2B). In July, the GSH values reached 2.39 (TR), 3.82 (BR) and 3.33 nmol GSH/mg protein at ZA (Fig. 3B). Lower concentrations of glutathione were found in August: 1.66 (TR), 1.75 (BR) and 2.17 (ZA) nmol GSH/mg protein (Fig. 4B). In September, the highest GSH concentration was detected at BR (3.91 nmol GSH/mg protein), and it was comparable to those found at TR and ZA (2.24 and 1.94 nmol GSH/mg protein, respectively) (Fig. 5B). The lipid peroxidation of daphniid cells was the lowest in June (0.10 nmol MDA/mg protein at TR and 0.23 nmol MDA/mg protein at ZA) and highest in July (0.87 (TR), 0.89 (BR) and 1.83 nmol MDA/mg protein at ZA (Figs. 2C, 3C). In August, the lipid peroxidation reached a medium level: 0.34 (TR), 0.53 (BR) and 0.93 nmol MDA/mg protein at ZA (Fig. 4C). More diverse values of LPO were observed in September (0.28 at TR, 1.23 at BR and 0.72 nmol MDA/mg protein at ZA) (Fig. 5C). Similar to the LPO, the catalase activity was the lowest in June (3.58 at TR and 1.42 U/mg protein at ZA) and the highest in July (7.57 (TR), 24.20 (BR) and 22.81 U/mg protein at ZA) (Figs. 2D, 3D). In August (Fig. 4D), the values of CAT were as follows: 5.22 (TR), 6.47 (BR) and 8.45 U/mg protein (ZA). In September, the catalase activity reached 6.16 (TR), 12.31 (BR) and 4.81 U/mg protein at ZA (Fig. 5D).

The two-way ANOVA indicated a significant seasonal difference in the values of all the studied antioxidant parameters in *Daphnia* (Table 2). The values of GSH were the highest in June and lowest in September. The same pattern was observed in the case of the lipid peroxidation values and catalase activity. Differences between sites were considerable in all the months studied. However, as there were only two sites sampled in June, a two-way ANOVA was used for the July, August and September data analyses. Statistical analysis showed that values of GSH, LPO and CAT were always significantly higher in the TR station than in the BR and ZA stations. The month × site interactions in the antioxidant system parameters were significant and resulted from

high LPO values at BR in September and low CAT values at both ZA in September and TR in July. In the case of glutathione, the month × site interaction was also significant and was largely an effect of the low GSH concentration both at ZA in September and BR in August (Table 2).

The results from September 2014 showed the same pattern of antioxidant system parameters activity as in the 2012 season. The lowest concentration of GSH (0.35 nmol GSH/mg protein) was observed in TR. The GSH concentrations at the other stations amounted to the following: 1.91 nmol GSH/mg protein in BR and 2.47 nmol GSH/mg protein in ZA (Fig. 6B). The GST activity was the highest in TR (83.46 U/mg protein) compared to BR (72.94 U/mg protein) and ZA (60.09 U/mg protein) (Fig. 6C). The values of LPO were lower in TR and ZA (0.087 and 0.074 nmol MDA/mg protein, respectively) than in BR (0.126 nmol MDA/mg protein) (Fig. 6D). The CAT activity was the lowest in TR (8.36 U/mg protein) and the highest in BR (17.81 U/mg protein) (Fig. 6E).

Discussion

The spatial distribution of *M. aeruginosa* was not homogenous in the Sulejow Reservoir. The lacustrine part of the reservoir, which is below ZA, is characterised by physical and chemical parameters favouring cyanobacterial development: relatively stable (flood-free) hydrological conditions with retention times of up to 60 days and a high supply of nutrients from the catchment area [37]. In addition, the presence of numerous bays, i.e., shallow and wind-protected areas in this part of the reservoir, may be crucial for the recruitment of cyanobacteria from sediments (Izydorczyk, unpublished data). Moreover, the Sulejow Reservoir is geographically oriented from the southwest to northeast, whereas the winds in this area blow predominantly from the west and southwest. This means that wind moves blooms towards the dam [38], and therefore the TR station is characterised by the highest cyanobacterial concentrations, as confirmed by our results.

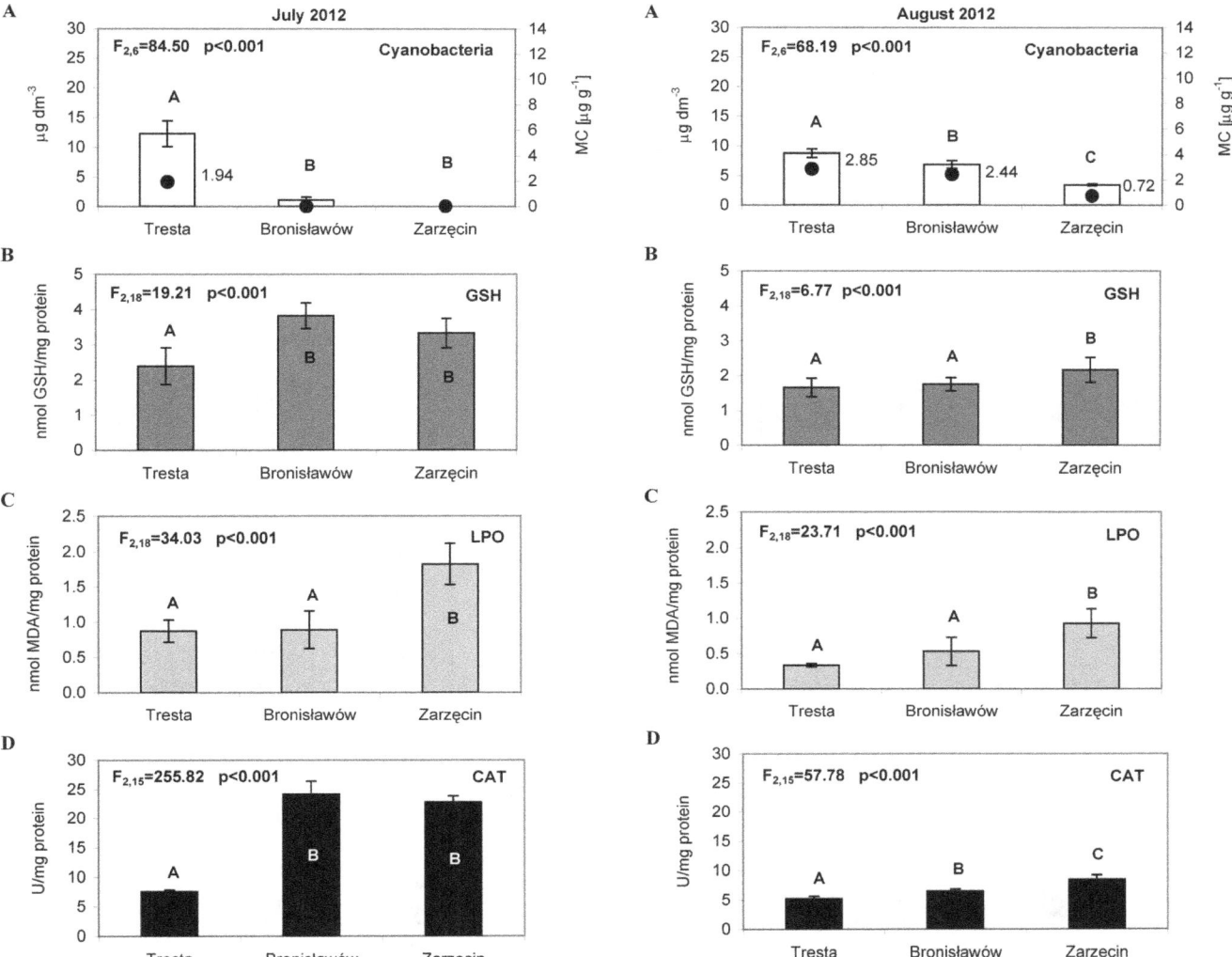

Figure 3. The cyanobacteria abundance and activity of the antioxidant parameters in *Daphnia* tissues, measured in July 2012. (A) The bars represent an average of three replicates (±S.D.) of cyanobacteria abundance (µg dm^{-3}); the dark dots indicate the total concentration of microcystins LR and RR (µg g^{-1}), (B) The bars represent an average of seven replicates (±S.D.) of glutathione (GSH) concentrations (nmol/mg protein), (C) The bars represent an average of seven replicates (±S.D.) of lipid peroxidation (LPO) (nmol/mg protein) and (D) The bars represent an average of six replicates (±S.D.) of catalase (CAT) activity (U/mg protein) in *Daphnia* tissues from the Sulejow Reservoir in 2012. The same letters above the bars indicate that the values did not differ significantly. Each panel of the figure includes the one-way ANOVA test results. Details concerning GSH, LPO and CAT data are presented in Tables S1, S2 and S3.

Figure 4. The cyanobacteria abundance and activity of the antioxidant parameters in *Daphnia* tissues, measured in August 2012. (A) The bars represent an average of three replicates (±S.D.) of cyanobacteria abundance (µg dm^{-3}); the dark dots indicate the total concentration of microcystins LR and RR (µg g^{-1}), (B) The bars represent an average of seven replicates (±S.D.) of glutathione (GSH) concentrations (nmol/mg protein), (C) The bars represent an average of seven replicates (±S.D.) of lipid peroxidation (LPO) (nmol/mg protein) and (D) The bars represent an average of six replicates (±S.D.) of catalase (CAT) activity (U/mg protein) in *Daphnia* tissues from the Sulejow Reservoir in 2012. The same letters above the bars indicate that the values did not differ significantly. Each panel of the figure includes the one-way ANOVA test results. Details concerning GSH, LPO and CAT data are presented in Tables S1, S2 and S3.

In 2012, the microcystin concentrations were closely correlated with cyanobacteria abundance, leading to spatial differences in the toxicity. However, the densities of *D. longispina* were comparable at all of the studied stations in a given month (Table 1), which indicates that the distribution of daphniids was not related to differences in the distribution of cyanobacteria. Nevertheless, we expected differences in the antioxidant system activity of *D. longispina* depending on the intensity of blooms. Indeed, the similar results achieved in both 2012 and 2014 indicate the occurrence of a relationship between the spatial diversity of

cyanobacteria abundance and the effectiveness of the system that protects *Daphnia* against oxidative stress.

Microcystins, the main group of cyanotoxins, can induce oxidative stress in the cells of aquatic animals [39], which is related to the production of reactive oxygen species (ROS) and leads to an increase in lipid peroxidation [40,41]. It is worth noting that lipid peroxidation is considered to be the major mechanism by which oxyradicals can cause tissue damage, impair cellular function and disrupt the physicochemical properties of cell membranes [42]. In our study, the oxidative stress of *D. longispina* caused by MCs was determined by the TBARS assay, reacting

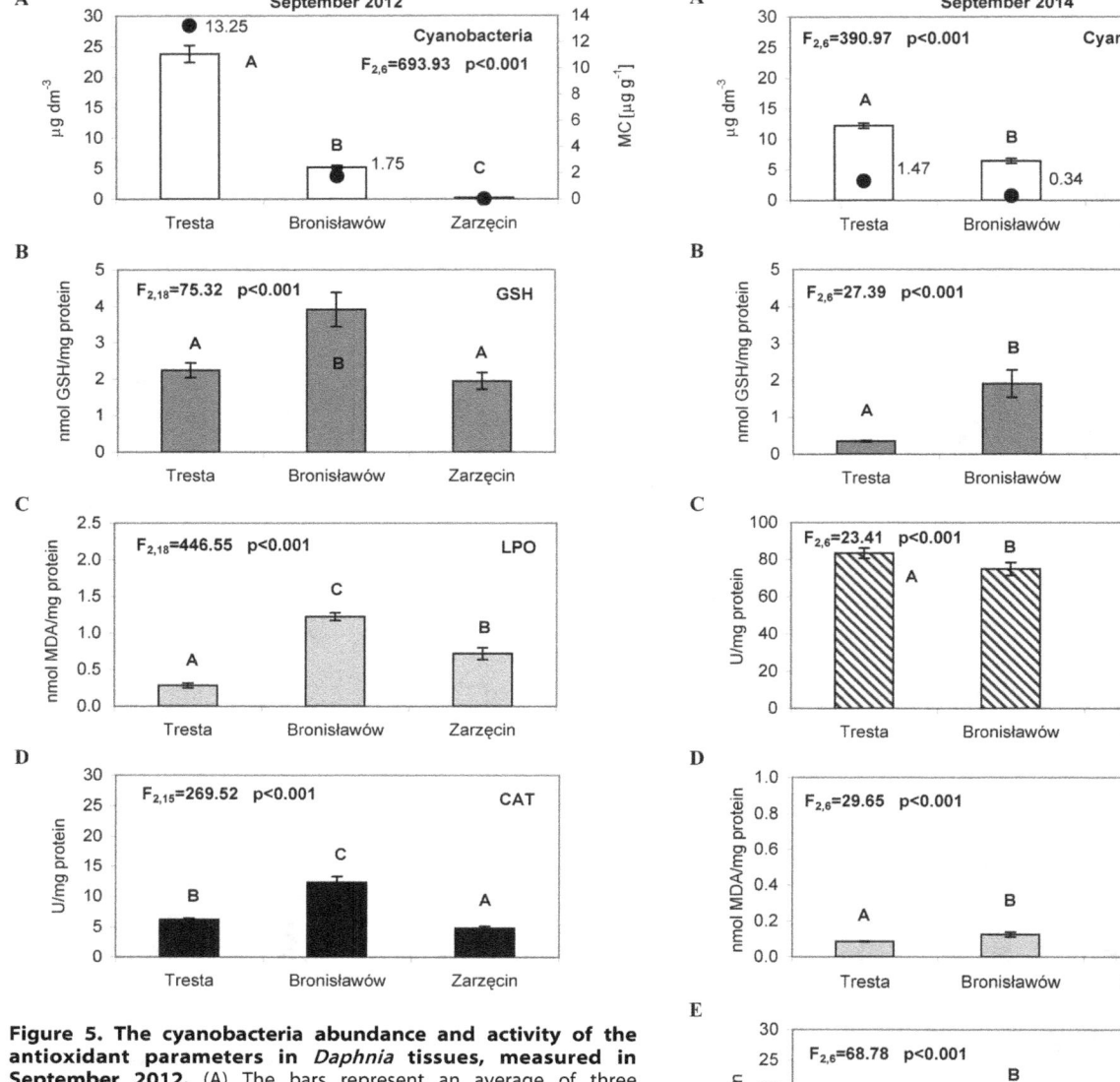

Figure 5. The cyanobacteria abundance and activity of the antioxidant parameters in *Daphnia* tissues, measured in September 2012. (A) The bars represent an average of three replicates (±S.D.) of cyanobacteria abundance (µg dm^{-3}); the dark dots indicate the total concentration of microcystins LR and RR (µg g^{-1}), (B) The bars represent an average of seven replicates (±S.D.) of glutathione (GSH) concentrations (nmol/mg protein), (C) The bars represent an average of seven replicates (±S.D.) of lipid peroxidation (LPO) (nmol/mg protein) and (D) The bars represent an average of six replicates (±S.D.) of catalase (CAT) activity (U/mg protein) in *Daphnia* tissues from the Sulejow Reservoir in 2012. The same letters above the bars indicate that the values did not differ significantly. Each panel of the figure includes the one-way ANOVA test results. Details concerning GSH, LPO and CAT data are presented in Tables S1, S2 and S3.

Figure 6. The cyanobacteria abundance and activity of the antioxidant parameters in *Daphnia* tissues, measured in September 2014. (A) The bars represent an average of three replicates (±S.D.) of cyanobacteria abundance (µg dm^{-3}); the dark dots indicate the total concentration of microcystins LR, YR and RR (µg g^{-1}), (B) The bars represent an average of three replicates (±S.D.) of glutathione (GSH) concentrations (nmol/mg protein), (C) The bars represent an average of three replicates (±S.D.) of glutathione S-transferase (GST) activity (U/mg protein) (D) The bars represent an average of three replicates (±S.D.) of lipid peroxidation (LPO) (nmol/mg protein) and (D) The bars represent an average of three replicates (±S.D.) of catalase (CAT) activity (U/mg protein) in *Daphnia* tissues from the Sulejow Reservoir in 2014. The same letters above the bars indicate that the values did not differ significantly. Each panel of the figure includes the one-way ANOVA test results. Details concerning GSH, GST, LPO and CAT data are presented in Tables S1, S2, S3 and S4.

mainly with malondialdehyde (MDA) as the principal product from lipid peroxidation. Intriguingly, the results indicated the lowest level of LPO at the TR site, which was characterised by the highest *M. aeruginosa* abundance and toxicity in all studied months (Fig. 2C–5C). The high levels of LPO (also corresponding to high CAT activity) occurred at sites with a minimum biomass of cyanobacteria – mainly at ZA in July and August (Fig. 3A, C, D–4A, C, D). A possible explanation may be that daphniids from upper sites were less susceptible to cyanotoxins and thus had a less resistant antioxidant system, which responded more rapidly to biotic and abiotic environmental factors. The CAT activity increased and hence mitigated oxidative damage, which is why we observed elevated values of CAT when LPO reached high levels (Fig. 3C, D–5C, D). This would confirm the finding that, when faced with an increasing production of ROS resulting from the effects of cyanotoxins, organisms usually up-regulate antioxidant defences such as catalase activity [43]. However, the low CAT activity at TR indicates that the main defence mechanism of Daphnia in the presence of toxic blooms was instead the process of microcystin detoxification. This may also be supported by the low glutathione concentration at TR, as glutathione can be used for the production of MC-GSH conjugates [17]. The additional measurement of the glutathione S-transferase activity in September 2014 confirmed this conclusion. The high GST activity corresponded with low GSH concentration at the TR station, which indicates the most intensive detoxification at the site with the highest toxic thread (Fig. 6C, D). The chemical conjugation of MCs with glutathione is recognised as the first step of detoxification in aquatic organisms exposed to cyanobacterial toxins because production of conjugates reduces the toxicity of MCs and facilitates their excretion by organisms [44]. This ability of glutathione confirms its particularly important function in diminishing oxidative stress [45].

Our results corroborate the conclusions of Lemaire et al. [12] that Daphnia do not develop generalised responses against Microcystis but rather specifically adapt to local assemblages of toxic cyanobacteria strains. Additionally, our study indicates that such adaptation also appears within an ecosystem with a different spatial distribution of blooms. These new facets of the Daphnia-Microcystis interaction may be crucial for the stabilisation of the top-down effect of grazers in the trophic structure, for the limitation of microcystin accumulation by Daphnia and thus for the reduction of the contribution of daphniids to the transfer of toxins to higher trophic levels [6,9,46].

References

1. Lampert W (1982) Further studies on the inhibitory effect of the toxic blue-green Microcystis aeruginosa on the filtering rate of zooplankton. Arch Hydrobiol 95: 207–220.

2. DeMott WR (1999) Foraging strategies and growth inhibition in five daphniids feeding on mixtures of a toxic cyanobacterium and a green alga. Freshwat Biol 42: 263–274.

3. Sarnelle O (1993) Herbivore effects on phytoplankton succession in a eutrophic lake. Ecol Monogr 63: 129–149.

4. Dao TS, Do-Hong LC, Wiegand C (2010) Chronic effects of cyanobacterial toxins on Daphnia magna and their offspring. Toxicon 55: 1244–1254.

5. Blanchette ML, Haney JF (2002) The effect of toxic Microcystis aeruginosa on four different populations of Daphnia. UNH Center for Freshwater Biology Research 4: 1–10.

6. Sarnelle O, Wilson AE (2005) Local adaptation of Daphnia pulicaria to toxic cyanobacteria. Limnol Oceanogr 50: 1565–1570.

7. Tillmanns AR, Wilson AE, Pick FR, Sarnelle O (2008) Meta-analysis of cyanobacterial effects on zooplankton population growth rate: species-specific responses. Fund Appl Limnol 171: 285–295.

8. Hairston NG Jr, Lampert W, Cáceres CE, Holtmeier CL, Weider LJ, et al. (1999) Rapid evolution revealed by dormant eggs. Nature 401: 446.

9. Hairston NG Jr, Holtmeier CL, Lampert W, Weider LJ, Post DM, et al. (2001) Natural selection for grazer resistance to toxic cyanobacteria: evolution of phenotypic plasticity? Evolution 55: 2203–2214.

10. Gustafsson S, Hansson L-A (2004) Development of tolerance against toxic cyanobacteria in Daphnia. Aquat Ecol 38: 37–44.

11. Sarnelle O, Gustafsson S, Hansson L-A (2010) Effects of cyanobacteria on fitness components of the herbivore Daphnia. J Plankton Res 32: 471–477.

12. Lemaire V, Brusciotti S, van Gremberghe I, Vyverman W, Vanoverbeke J, et al. (2012) Genotype x genotype interactions between the toxic cyanobacterium Microcystis and its grazer, the waterflea Daphnia. Evolutionary Applications 5: 168–182.

13. Repka S (1997) Effects of food type on the life history of Daphnia clones from lakes differing in trophic state. I. Daphnia galeata feeding on Scenedesmus and Oscillatoria. Freshwat Biol 38: 675–683.

14. Jiang X, Zhang L, Liang H, Li Q, Zhao Y, et al. (2013) Resistance variation within a Daphnia pulex population against toxic cyanobacteria. J Plankton Res 35: 1177–1187.

15. Roy S (2008) Spatial Interaction Among Nontoxic Phytoplankton, Toxic Phytoplankton and Zooplankton: Emergence in Space and Time. J Biol Phys 34: 459–474.

16. Reichwaldt ES, Song H, Ghadouani A (2013) Effects of the Distribution of a Toxic Microcystis Bloom on the Small Scale Patchiness of Zooplankton. PLoS ONE 8(6): e66674. doi:10.1371/journal.pone.0066674.

17. Wojtal-Frankiewicz A, Bernasińska J, Jurczak T, Gwoździński K, Frankiewicz P, et al. (2013) Microcystin assimilation and detoxification by Daphnia spp. in two ecosystems of different cyanotoxin concentrations. Journal of Limnology 72: 154–171.

18. Wagner I, Izydorczyk K, Kiedrzynska E, Mankiewicz-Boczek J, Jurczak T, et al. (2009) Ecohydrological system solution for enhancement of ecosystem services: the Pilica river demonstration project. Ecohydrol Hydrobiol 9: 13–39.

19. Gągała I, Izydorczyk K, Jurczak T, Pawełczyk J, Dziadek J, et al. (2013) Role of Environmental Factors and Toxic Genotypes in The Regulation of Microcystins-Producing Cyanobacterial Blooms. Microbial Ecology doi: 10.1007/s00248-013-0303-3.

20. Tarczyńska M, Romanowska-Duda Z, Jurczak T, Zalewski M (2001) Toxic cyanobacterial blooms in drinking water reservoir – causes, consequences and management strategy. Water Sci Technol: Water Supply 1: 237–246.

21. Jurczak T, Tarczyńska M, Izydorczyk K, Mankiewicz J, Zalewski M, et al. (2005) Elimination of microcystins by water treatment process – examples from Sulejow Reservoir, Poland. Water Res 39: 2394–2406.

22. Gągała I (2013) Impact of the environmental factors with particular emphasis on the role of bacteria in dynamics of occurrence and toxicity of freshwater cyanobacteria - PhD Dissertation, University of Lodz, Poland.

23. Benzie JAH (2005) The genus Daphnia (including Daphniopsis) (Anomopoda: Daphniidae). Guides to the Identification of the Macroinvertebrates of the Continental Waters of the World. Coordination editor: H.J.F. Dumont. Kenobi Productions, Ghent; Backhuys Publishers, Leiden.

24. Starmach K (1966) Cyanophyta – Sinice. Flora słodkowodna Polski, 2. PWN, Warszawa (in Polish).

25. Komarek J (1991) A review of water-bloom forming Microcystis species, with regard to populations from Japan. Arch Hydrobiol, Suppl Algological Studies 65: 115–127.

Supporting Information

Table S1   The data represent three/seven replicates (1–7), mean and standard deviation (SD) of glutathione concentration (nmol/mg protein) in Daphnia tissues from the Sulejow Reservoir.

Table S2   The data represent three/seven replicates (1–7), mean and standard deviation (SD) of lipid peroxidation (nmol/mg protein) in Daphnia tissues from the Sulejow Reservoir.

Table S3   The data represent three/six replicates (1–6), mean and standard deviation (SD) of catalase activity (U/mg protein) in Daphnia tissues from the Sulejow Reservoir.

Table S4   The data represent three replicates (1–3), mean and standard deviation (SD) of glutathione S-transferase activity (U/mg protein) in Daphnia tissues from the Sulejow Reservoir.

Author Contributions

Conceived and designed the experiments: AWF JB PF KG TJ. Performed the experiments: AWF JB PF KG TJ. Analyzed the data: AWF JB PF KG TJ. Contributed reagents/materials/analysis tools: AWF JB PF KG TJ. Wrote the paper: AWF PF.

26. Komarek J, Anagnostidis K (1999) Cyanoprokaryota. Teil: Chroococcales. In: H. Ettl, J. Gartner, H. Heyning. and D. Mollenhauer (eds.), Süsswasserflora von Mitteleuropa 19/1. Gustav Fischer Verlag, Jena.

27. Cagnard O, Boudin I, Lemoigne I, Cartnick K (2006) Assessment of emerging optic sensors (fluoroprobes) for algae on-line monitoring. In: Water Quality Technology Conference (WQTC) Proceedings, pp. 1–10. American Water Works Association, Denverss.

28. Richardson TL, Lawrenz E, Pinckney JL, Guajardo RC, Walker EA, et al. (2010) Spectral fluorometric characterization of phytoplankton community composition using the Algae Online Analyser. Water Res 44: 2461–2472.

29. Izydorczyk K, Carpentier C, Mrówczyński J, Wagenvoort A, Jurczak T, et al. (2009) Establishment of an Alert Level Framework for cyanobacteria in drinking water resources by using the Algae Online Analyser for monitoring cyjanobacterial chlorophyll a. Water Res 43: 989–996.

30. Jurczak T, Tarczyńska M, Karlsson K, Meriluoto J (2004) Characterization and diversity of cyanobacterial hepatotoxins (microcystins) in blooms from Polish freshwaters identified by Liquid Chromatography-Electrospray Ionisation Mass Spectrometry. Chromatographia 59: 571–578.

31. Stocks R, Dormandy TL (1971) The autooxidation of human red cell lipids induced by hydrogen peroxide. Br J Haematol 20: 95–111.

32. Rice-Evans CA, Diplock AT, Symons MCR (1991) Techniques in free radicals research, pp. 147–149. In: R.H. Burdon and P.H. van Knippenberg (eds.), Laboratory techniques in biochemistry and molecular biology. Elsevier, Amsterdam.

33. Lowry OH, Rosenbrough NJ, Farr AL, Randal RJ (1951) Protein measurement with Folin phenol reagent. J Biol Chem 193: 265–275.

34. Senft AP, Dalton TP, Shertzer HG (2000) Determining glutatione and glutatione disulfide using the fluorescence probe o-phthalaldehyde. Anal Biochem 280: 80–86.

35. Habig WH, Pabst MJ, Jokoby WB (1974) Glutathione S-transferase. The first enzyme step in mercapturic acid formation. J Biol Chem 249: 7130–7139.

36. Aebi H (1984) Catalase in vitro. Methods Enzymol 105: 121–126.

37. Izydorczyk K, Skowron A, Wojtal A, Jurczak T (2008) The stream inlet to a shallow bay of a drinking water reservoir, a "hot-spot" for Microcystis blooms initiation. International Review of Hydrobiology 93: 257–268.

38. Izydorczyk K, Tarczyńska M (2005) The influence of wind on cyanobacterial bloom development in shallow, lowland reservoir in central Poland. Ecohydrology & Hydrobiology 5: 195–203.

39. Campos A, Vasconcelos V (2010) Molecular mechanisms of microcystin toxicity in animal cells. Int J Mol Sci 11: 268–287.

40. Wiegand C, Pflugmacher S (2005) Ecotoxicological effects of selected cyanobacterial secondary metabolites a short review. Toxicol Appl Pharmacol 203: 201–218.

41. Amado LL, Monserrat JM (2010) Oxidative stress generation by microcystins in aquatic animals: Why and how. Environ Int 36: 226–235.

42. Rikans LE, Hornbrook KR (1997) Lipid peroxidation, antioxidant protection and aging. Biochim Biophys Acta 1362: 116–127.

43. Livingstone DR (2003) Oxidative stress in aquatic organisms in relation to pollution and aquaculture. Rev Med Vet 154: 427–430.

44. Pflugmacher S, Wiegand C, Oberemm A, Beattie KA, Krause E, et al. (1998) Identification of an enzymatically formed glutathione conjugate of the cyanobacterial hepatotoxin microcystin-LR: the first step of detoxication. Biochim Biophys Acta 1425: 527–533.

45. Wiegand C, Peuthert A, Pflugmacher S, Carmeli S (2002) Effects of microcin SF608 and microcystin-LR, two cyanobacterial compounds produced by Microcystis sp., on aquatic organisms. Environ Toxicol 17: 400–406.

46. von Elert E, Zitt A, Schwarzenberger A (2012) Inducible tolerance to dietary protease inhibitors in Daphnia magna. The Journal of Experimental Biology 215: 2051–2059.

20

Lactobacillus casei-01 Facilitates the Ameliorative Effects of Proanthocyanidins Extracted from Lotus Seedpod on Learning and Memory Impairment in Scopolamine-Induced Amnesia Mice

Juan Xiao[2], Shuyi Li[3], Yong Sui[1], Qian Wu[1], Xiaopeng Li[1], Bijun Xie[1], Mingwei Zhang[2], Zhida Sun[1]*

1 College of Food Science and Technology, Huazhong Agricultural University, 1 Shizishan Street, Hongshan District, Wuhan 430070, Hubei, China, **2** Key Laboratory of Functional Foods, Ministry of Agriculture/ Guangdong Key Laboratory of Agricultural Products Processing, Sericultural & Agri-Food Research Institute Guangdong Academy of Agricultural Sciences, 133 Yiheng Road, Dongguan Zhuang, Tianhe District, Guangzhou 510610, Guangdong, China, **3** College of Food Science and Engineering, Wuhan Polytechnic University, Xuefu South Road, Changqing Garden, Dongxihu District, Wuhan 430023, Hubei, China

Abstract

Learning and memory abilities are associated with alterations in gut function. The two-way proanthocyanidins-microbiota interaction *in vivo* enhances the physiological activities of proanthocyanidins and promotes the regulation of gut function. Proanthocyanidins extracted from lotus seedpod (LSPC) have shown the memory-enhancing ability. However, there has been no literature about whether *Lactobacillus casei-01* (LC) enhances the ameliorative effects of LSPC on learning and memory abilities. In this study, learning and memory abilities of scopolamine-induced amnesia mice were evaluated by Y-maze test after 20-day administration of LC (10^9 cfu/kg body weight (BW)), LSPC (low dose was 60 mg/kg BW (L-LSPC) and high dose was 90 mg/kg BW (H-LSPC)), or LSPC and LC combinations (L-LSPC+LC and H-LSPC+LC). Alterations in antioxidant defense ability and oxidative damage of brain, serum and colon, and brain cholinergic system were investigated as the possible mechanisms. As a result, the error times of H-LSPC+LC group were reduced by 41.59% and 68.75% relative to those of H-LSPC and LC groups respectively. LSPC and LC combinations ameliorated scopolamine-induced memory impairment by improving total antioxidant capacity (TAOC) level, glutathione peroxidase (GSH-Px) and total superoxide dismutase (T-SOD) activities of brain, serum and colon, suppressing malondialdehyde (MDA) level of brain, serum and colon, and inhibiting brain acetylcholinesterase (AchE), myeloperoxidase, total nitric oxide synthase and neural nitric oxide synthase (nNOS) activities, and nNOS mRNA level. Moreover, LC facilitated the ameliorative effects of H-LSPC on GSH-Px activity of colon, TAOC level, GSH-Px activity and ratio of T-SOD to MDA of brain and serum, and the inhibitory effects of H-LSPC on serum MDA level, brain nNOS mRNA level and AchE activity. These results indicated that LC promoted the memory-enhancing effect of LSPC in scopolamine-induced amnesia mice.

Editor: Ranji Cui, Jilin University, China

Funding: This work was supported by National Natural Science Foundation of China (NSFC, No. 31071633) and the Innovation Capacity Construction Special Project of Main Scientific Research Organization of Guangdong Province (2011). The funders had no role in study design, data collection and analysis, decision to publish, or preparation of the manuscript.

Competing Interests: The authors have declared that no competing interests exist.

* Email: sunzhida@sina.com

Introduction

Accumulating data indicate that imbalance of gut bacteria not only contributes to gut dysfunction [1], chronic metabolic disorders [2] and aging [3], but also plays an important role in memory and cognition dysfunction [1,4]. Recent researches have reported that learning and memory abilities in mice are associated with diet-induced alterations in gut bacteria [1,5]. *C rodentium* infection or high-fat diet resulted in memory impairment via disturbing the balance of gut bacteria, which were reversed by regulation of gut microbiota and colonic inflammatory [1,6]. Moreover, McCarthy investigated the concomitant symptoms of dementia patients randomly in 20 health districts of England, and found that 85% of the patients suffered from gut dysfunction [7]. Relieving symptoms of gut dysfunction such as constipation was beneficial to the attenuation of memory and cognition dysfunction

in dementia patients [8]. Therefore, modulation of gut microbiota and gut function may be the potential strategies for reversing learning and memory deficits and related diseases [9–11].

Lotus seedpod, a part of *Nelumbo nucifera* Gaertn, contains an abundance of proanthocyanidins. It has been demonstrated that the main components in LSPC are monomers, dimers, trimers and tetramers of proanthocyanidins, in which dimers are the most component, and catechin and epicatechin are the base units [12]. LSPC have been proven to possess a potent antioxidant activity [12], and ameliorate memory deficits and oxidative damage in scopolamine-induced amnesia mice, SAMP8 and cognitively impaired aged rats [13–15]. However, there is no literature about the gut-regulation effect of LSPC. Previous studies have reported that proanthocyanidins B1 and B2, and proanthocyanidins in red wine modulate intestinal function by regulation of microbiota [16,17]. Moreover, the oligomer and polymer of proanthocyanidins are poorly absorbed compared to the monomers in small intestine, and then accumulate to colon, where they are metabolized by the microbiota to more bioactive compounds with potent physiological effects [18,19]. Importantly, proanthocyanidins and their metabolites in colon play an important role in the maintenance of intestinal health by inhibiting the growth of pathogenic bacteria, and keeping the growth of probiotics such as *Lactobacillus* [20]. Thus, the gut-modulation effect of B-type proanthocyanidins and the two-way proanthocyanidins-microbiota interaction suggest that LSPC may have a gut-modulation effect.

It is generally accepted that probiotics contribute to the balance of gut microecology and gut health [1,6]. Recently, accumulating evidence has revealed the memory-enhancing effect of probiotics. Davari and colleagues have reported that administration of mixed probiotics pronouncedly ameliorates spatial memory impairment in diabetic rats by recovering basic synaptic transmission and hippocampal long-term potentiation, and inhibiting oxidative damage [21]. *Lactobacillus pentosus var. plantarum* C29 attenuates scopolamine-induced amnesia by inducing brain BDNF and p-CREB expressions [22]. Moreover, several researches have shown that probiotics prevent memory deficits by modulating gut function. The study of Gareau and colleagues has revealed that combination of *Lactobacillus rhamnosus* and *Lactobacillus helveticus* improves spatial memory impairment in *C rodentium*-infected mice by regulation of gut microbiota and inhibition of colonic inflammatory and epithelial cell hyperplasia [1]. Ameliorative effects of *Lactobacillus helveticus* on memory deficits in IL-$10^{-/-}$ mice with high-fat diet are closely associated with the modulation of gut microbiota, colonic inflammation and cytokine expression [6].

It has been proven that both probiotics and LSPC have the memory-enhancing effect and gut-modulation effect. And there is a two-way interaction between probiotics and LSPC [19,20]. However, there has been no literature about whether probiotics enhance the ameliorative effects of LSPC on learning and memory ability. Our lab found that oligomeric proanthocyanidins from Litchi pericarp did not change the growth of *Lactobacillus casei-01* at concentrations of 0.25 and 0.5 mg/mL *in vitro*, and were decomposed into many kinds of phenolic acids with more potent antioxidant ability than their parent proanthocyanidins [23]. These results indicate that *Lactobacillus casei-01 in vivo* may enhance the biotransformation of LSPC, and that may further increase the ameliorative effects of LSPC on intestinal function and learning and memory ability.

The purpose of this study was to investigate whether *Lactobacillus casei-01* enhanced the ameliorative effects of LSPC on learning and memory ability in scopolamine-induced amnesia mice. Alterations in antioxidant defense ability and oxidative damage of brain, serum and colon, and brain cholinergic system were investigated as the possible mechanisms.

Materials and Methods

Chemical and reagents

Mature lotus seedpods of *Nelumbo Nucifera* Gaertn (cultivar: Number 2 Wuhan plant) were harvested from Honghu District ($113°7'–114°5'E$, $29°39'–30°2'N$) in Hubei province, China, in late July, 2011, and arrived in the laboratory within 24 h postharvest, which were kept at $-20°C$ prior to extraction. Lotus seedpod was identified by Prof. Xueming Ni from Department of Botany, Wuhan Plant Institute of the Chinese Academy of Science. *Lactobacillus casei-01* was purchased from Chr. Hansen Company (Beijing, China).

Kits for determination of glutathione peroxidase (GSH-Px, Cat. No. A005), total superoxide dismutase (T-SOD, Cat. A001-1), acetylcholinesterase (AchE, Cat. No. A024), total nitric oxide synthase (TNOS, Cat. No. A014-1), neural nitric oxide synthase (nNOS, Cat. No. A014-1), inducible nitric oxide synthase (iNOS, Cat. No. A014-1) and myeloperoxidase (MPO, Cat. No. A044) activities, levels of malondialdehyde (MDA, Cat. No. A003-1), total antioxidant capacity (TAOC, Cat. No. A015) and protein (Cat. No. A045-2) were purchased from Nianjing Jiancheng Bioengineering Institute, Nanjing, China. Scopolamine hydrobromide injection was purchased from Xuzhou RYEN Pharma.CO., LTD, Xuzhou, China. Piracetam was purchased from Hubei Huazhong Pharma.CO., LTD, Hubei, China.

Preparation of LSPC and strain

Frozen lotus seedpods were extracted to obtain LSPC as described previous [24]. LSPC were stored at $-20°C$ and the purity of 98.7% was measured on the basis of comparison with a calibration curve of grape seed procyanidin extract by the method reported by Porter [25]. Electrospray ionization mass spectrometry analysis has revealed that monomers, dimers, trimers and tetramers of proanthocyanidins are the main ingredients of LSPC, and the base units of LSPC are (+)-catechin and (−)-epicatechin [12]. The percentage of catechin, epicatechin, dimers, trimers and tetramers in LSPC is 10.9%, 9.1%, 53.6%, 19.5% and 1.9%, respectively. *Lactobacillus casei-01* powder was incubated in MRS fluid nutrient medium at $37°C$ for 18 h to activate. Then, the viable bacteria were inoculated (1%) in the same medium, which was incubated at $37°C$ for 24 h to reach the strain concentration at about 10^7 cfu/mL. The strain solution (50 mL) was put in a centrifuge tube, and centrifuged at 5590 g for 10 min at $4°C$. The precipitate was dissolved in distilled water (50 mL) and centrifuged at 5590 g for 10 min at $4°C$, which was repeated two times to remove the MRS. The precipitate was the strain ($5×10^8$ cfu in each tube) and stored at $4°C$.

Animals

Male Kunming mice ($20±2$ g) were purchased from Wuhan University Research Center for Animal Experiment, China. Animals were housed five per cage, kept under a controlled temperature $22±1°C$, humidity 55–60% and 12-h light/12-h dark cycle throughout the experiment. A normal solid diet and water were available *ad libitum*. The diet was provided by the Wuhan University Research Center for Animal Experiment, China.

Ethics statement

All experimental procedures involving animals followed the Guiding Principles in the Care and Use of Animals, and were approved by the ethics committee of Wuhan General Hospital of Guangzhou Military Command (SYXK (Hubei) 2008-0007), and Huazhong Agricultural University, Hubei Province, China. We made all efforts to minimize suffering. The animals were killed by cervical dislocation under anesthesia. Mature lotus seedpods used in this study were industrial crops and obtained from private land, for future permissions should be contacted with Zong Zhang (+86 13972369298). No specific permissions were required for Honghu District, Hubei province, China. The study did not involve any endangered or protected species.

Treatments

After acclimatization for 3 days, the animals were chosen by Y-maze test. Then, a total of 80 male Kunming mice were randomly divided into eight groups with ten mice in each group: control

(CON), vehicle scopolamine control (SCOP), positive drug control (Piracetam), *Lactobacillus casei-01* (LC), low and high dose of LSPC (L-LSPC and H-LSPC), L-LSPC and LC combination (L-LSPC+LC), H-LSPC and LC combination (H-LSPC+LC) groups (Table 1). They were given the respective drugs at a dose of 0.1 mL/10 g body weight (BW) by oral gavage once daily for 20 days. Groups of CON, SCOP, Piracetam, LC, L-LSPC and H-LSPC were given distilled water, distilled water, piracetam (400 mg/kg BW), LC (10^9 cfu/kg BW), LSPC (60 mg/kg BW) and LSPC (90 mg/kg BW), respectively. L-LSPC+LC and H-LSPC+LC were given a mixture of LSPC (60 mg/kg BW) and LC (10^9 cfu/kg BW), and a mixture of LSPC (90 mg/kg BW) and LC (10^9 cfu/kg BW), respectively (Table 1).

Solutions of Piracetam (40 mg/mL), L-LSPC (6 mg/mL LSPC) and H-LSPC (9 mg/mL LSPC) were prepared in distilled water. Five mL of distilled water, L-LSPC and H-LSPC solutions were mixed with the strains in three tubes, respectively, as the solutions for the groups of LC, L-LSPC+LC and H-LSPC+LC. Doses of LC, LSPC, and LSPC and LC combinations were based on the early experiments results *in vivo* and in *vitro*. All solutions were freshly prepared and administrations began at 9:00 a.m. At the 20th day, animals except CON group received scopolamine (3 mg/kg BW, i.p.) to induce memory impairment 30 min before a training course of Y-maze test [26]. CON group received 0.9% saline (i.p.) in the same way as mentioned above (Table 1).

Behavioral procedures

Learning and memory abilities of the mice were tested at the end of the treatment experiment by Y-maze test as described previous [24]. Briefly, each mouse was placed at the end of one arm to adapt for 5 min without electric shock. Then electric shock made the mouse move to the safe arm, where the mouse adjusted to the situation for 30 s; and then another electric shock was switched following the sequence (ABCCAB) to compel the mouse to move to the safe arm. Such trial was repeated 35 times during a training course and repeated 10 times after 24 h during the test course. During the experiment, it would be an incorrect response if the mouse did not move to the safe arm directly within 30 s. The error times in 10 tests was taken as the learning and memory ability. The more the incorrect responses, the weaker the learning and memory ability of the mouse was.

Determination of MDA and TAOC levels, and GSH-Px, T-SOD, MPO, TNOS, nNOS, iNOS and AchE activities

After behavioral test, all mice were anesthetized by an intraperitoneal injection of sodium thiopental (40 mg/kg BW). Blood of each mouse was obtained from the tail veins, afterwards, mice were sacrificed by cervical dislocation under anesthesia. The brain was immediately removed and washed with ice-cold normal saline. The colon was immediately dissected and its content was removed. Left brain and the colon were weighed and stored at −80°C, and then homogenized with a phosphate buffer (50 mmol/L, pH 7.0) containing 0.1 mmol/L EDTA before use. Brain homogenate was divided into two parts. One part was centrifuged at 2054 g for 10 min at 4°C and the supernatant was used for MDA, TAOC, GSH-Px, T-SOD, TNOS, nNOS, iNOS and AchE tests. Another part was used for determination of MPO activity directly. Blood and colon homogenate were centrifuged at 2054 g for 10 min at 4°C to obtain serum and colon supernatant for MDA, TAOC, GSH-Px and T-SOD tests. All parameters were determined using the respective kits according to the manufacturer's specifications.

Quantitative real-time PCR

For nNOS, total RNA of right brain (four in each group) was isolated using Trizol reagent (15596-026, Invitrogen) according to the manufacturer's instruction, and reverse-transcribed using a RevertAid First Strand cDNA Synthesis Kit (K-1622, Fermentas) at 42°C for 30 min followed by 70°C for 5 min. Real-time polymerase chain reaction was executed with Thunderbird SYBR qPCR Mix (QPS-201, Toyobo Biologics), using the real-time thermocycler (Hangzhou Bioer Technology Co., LTD). The cycle conditions were as follows: denaturation (95°C for 60 s), cycles (40 times), renaturation (temperature declines to 58°C in 15 s), stretch (temperature increases to 72°C in 20 s, and then keeps at 72°C for 20 s). The dissociation curve of each gene was performed and analyzed using the SLAN Quantitative Real-Time PCR detection system (Shanghai Hongshi Medical Technology Co., Ltd), and the result verified the specificity of the product. Each sample was performed in triplicate, and normalized to β-actin. The relative expression levels of the genes were calculated by the $2^{-\Delta\Delta CT}$ method as previously described [27]. The sequences of the primers (Invitrogen) for the genes were as follows. nNOS: forward 5'-GCTTCAGGAATATGAGGAATGG-3', reverse 5'-TGATG-GAATAGTAGCGAGGTTGT-3'; β-actin: forward 5'-CTGA-GAGGGAAATCGTGCGT-3', reverse 5'-CCACAGGATTC-CATACCCAAGA-3'.

Table 1. Groups and treatments.

Groups (abbreviations)	Treatments[a] (oral gavage, 1st–20th day)	Induce memory impairment[b] (intraperitoneal, the 20th day)
Control (CON)	Distilled water	Saline
Vehicle scopolamine control (SCOP)	Distilled water	3 mg/kg BW scopolamine
Positive drug control (Piracetam)	400 mg/kg BW[c] Piracetam	3 mg/kg BW scopolamine
Lactobacillus casei-01 (LC)	10^9 cfu/kg BW LC	3 mg/kg BW scopolamine
Low dose of LSPC (L-LSPC)	60 mg/kg BW LSPC	3 mg/kg BW scopolamine
High dose of LSPC (H-LSPC)	90 mg/kg BW LSPC	3 mg/kg BW scopolamine
L-LSPC and LC combination (L-LSPC+LC)	60 mg/kg BW LSPC and 10^9 cfu/kg BW LC combination	3 mg/kg BW scopolamine
H-LSPC and LC combination (H-LSPC+LC)	90 mg/kg BW LSPC and 10^9 cfu/kg BW LC combination	3 mg/kg BW scopolamine

[a]: Administrated once daily by oral gavage from 1st–20th day;
[b]: Administrated intraperitoneal 30 min before a training course of Y-maze test at the 20th day;
[c]: body weight.

Statistical analysis

Data were expressed as mean ± standard deviation (SD). All data were analyzed using a one-way ANOVA, followed by Duncan *post hoc test*. Statistical analyses were performed by the SPSS 16.0 software and P<0.05 was regarded as statistical significance.

Results

Effects of combined LSPC and LC on behavioral performance in Y-maze test

Effects of combined LSPC and LC on behavioral performance in Y-maze test are shown in Fig. 1. SCOP group exhibited obviously increased error times in Y-maze test in comparison with CON group (P<0.05). However, error times were significantly reduced by treatments with Piracetam, LC, L-LSPC, H-LSPC and LSPC and LC combinations as compared to SCOP group (all P<0.05). Importantly, H-LSPC, L-LSPC+LC and H-LSPC+LC groups exhibited the decreased error times with respect to CON, Piracetam and LC groups (all P<0.05). Moreover, error times of H-LSPC+LC group were reduced by 41.59% relative to those of H-LSPC group without significant difference.

Effects of combined LSPC and LC on brain antioxidant defense capacity and oxidative damage

MDA and TAOC levels, and activities of GSH-Px, T-SOD and MPO in brain were measured to investigate the changes of brain oxidative damage and antioxidant defense capacity. As shown in Table 2, scopolamine resulted in noticeable increases in MDA level and MPO activity, and marked declines in TAOC level, T-SOD and GSH-Px activities (all P<0.05). Compared to SCOP group, LC treatment declined MDA level, and raised TAOC level and GSH-Px activity remarkably (all P<0.05), but had no

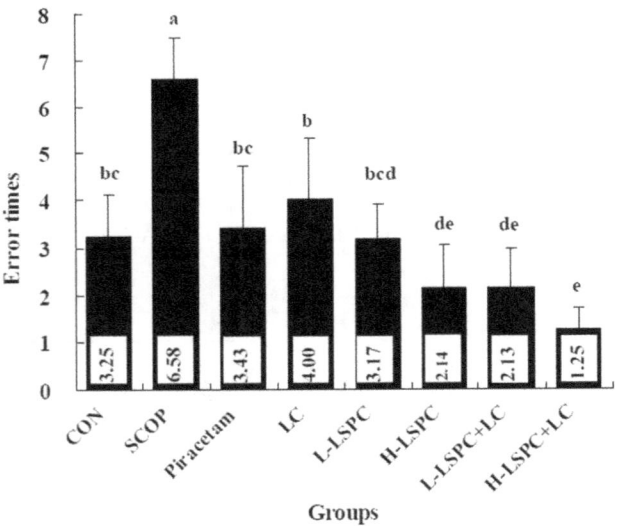

Figure 1. Effects of combined proanthocyanidins extracted from lotus seedpod (LSPC) and *Lactobacillus casei-01* (LC) on scopolamine-induced memory impairment in Y-maze test. Groups without any same letters above the bars signify statistically significant differences (P<0.05). CON, SCOP, Piracetam and LC is control, vehicle scopolamine control, positive drug control and *Lactobacillus casei-01* group, respectively. L-LSPC and H-LSPC is low and high dose of LSPC group, respectively. L-LSPC+LC is L-LSPC and LC combination group. H-LSPC+LC is H-LSPC and LC combination group.

significant effect on T-SOD and MPO activities. However, treatments with Piracetam, L-LSPC, H-LSPC, L-LSPC+LC and H-LSPC+LC significantly reversed scopolamine-induced changes of MDA and TAOC levels, and T-SOD, GSH-Px and MPO activities, to the level of equal or superior to CON group. Moreover, compared to LC group, both L-LSPC and H-LSPC groups showed the obviously reduced MDA level and MPO activity, and the markedly raised activities of T-SOD and GSH-Px (all P<0.05). Importantly, MDA and TAOC levels, and T-SOD, GSH-Px and MPO activities in H-LSPC+LC group were significantly better than those in LC group (all P<0.05). H-LSPC+LC treatment exhibited higher TAOC level and GSH-Px activity than H-LSPC treatment (P<0.05 and P<0.05). Additionally, as indexes of oxidative stress, ratio of T-SOD to MDA (T-SOD/MDA) and ratio of GSH-Px to T-SOD (GSH-Px/T-SOD) were also calculated. Scopolamine led to marked declines in T-SOD/MDA and GSH-Px/T-SOD (P<0.05 and P<0.05). However, all treatments normalized GSH-Px/T-SOD. T-SOD/MDA was enhanced in all treatments groups relative to SCOP group (all P<0.05). Treatments with L-LSPC and H-LSPC+LC significantly raised T-SOD/MDA in comparison with CON group (P<0.05 and P<0.05). Effect of H-LSPC+LC treatment on improving T-SOD/MDA was markedly superior to those of both H-LSPC and LC treatments (P<0.05 and P<0.05). Furthermore, compared to Piracetam group, H-LSPC+LC group exhibited significant increases in TAOC level, GSH-Px activity and T-SOD/MDA, and a significant decrease in MPO activity (all P<0.05).

Based on the results, LC, LSPC, and their combinations had the ability to improve brain antioxidant defense capacity and suppress oxidative damage, and the abilities of LSPC, and combined LSPC and LC were more potent than LC. LC enhanced the ameliorative effects of H-LSPC on brain antioxidant defense ability and oxidative damage.

Effects of combined LSPC and LC on brain AchE activity

ANOVA indicated that significantly increased brain AchE activity in both SCOP and LC groups was observed when compared with CON group (P<0.05 and P<0.05) (Table 2). Brain AchE activity in Piracetam, L-LSPC, H-LSPC and the combination groups was markedly weaker than that in SCOP group (all P<0.05). Importantly, H-LSPC+LC treatment reduced brain AchE activity significantly as compared to CON, Piracetam, LC, H-LSPC and L-LSPC+LC groups (all P<0.05). LC facilitated the ameliorative effects of H-LSPC on brain cholinergic activity.

Effects of combined LSPC and LC on serum antioxidant defense capacity and oxidative damage

As shown in Table 3, scopolamine resulted in an obvious enhancement of serum MDA level, and significant reductions of serum TAOC level, GSH-Px and T-SOD activities, and T-SOD/MDA relative to CON group (all P<0.05). LC treatment notably enhanced GSH-Px and T-SOD activities (P<0.05 and P<0.05) compared with SCOP group. Piracetam, L-LSPC, H-LSPC and L-LSPC+LC treatments normalized the variations of serum MDA and TAOC levels, GSH-Px and T-SOD activities and T-SOD/MDA induced by scopolamine. No significant differences in these five parameters were found among Piracetam, L-LSPC and H-LSPC groups. Serum TAOC level of H-LSPC group was significantly higher than that of LC group (P<0.05). L-LSPC+LC treatment showed a marked increase in T-SOD/MDA relative to both LC and L-LSPC treatments (P<0.05 and P<0.05). It was worthy of note that serum MDA, TAOC levels and GSH-Px activity as well as T-SOD/MDA in H-LSPC+LC group were better than those in CON, Piracetam, LC, L-LSPC, H-LSPC and

Table 2. Effects of combined proanthocyanidins extracted from lotus seedpod (LSPC) and *Lactobacillus casei-01* (LC) on brain malondialdehyde (MDA) and total antioxidant capacity (TAOC) levels, total superoxide dismutase (T-SOD), glutathione peroxidase (GSH-Px), myeloperoxidase (MPO) and acetylcholinesterase (AchE) activities, ratio of T-SOD to MDA (T-SOD/MDA) and ratio of GSH-Px to T-SOD (GSH-Px/T-SOD) in scopolamine-induced amnesia mice.

Group	MDA (nmol/mgprot)	T-SOD (U/mgprot)	GSH-Px (U/mgprot)	TAOC (U/mgprot)	MPO (U/g)	AchE (U/mgprot)	T-SOD/MDA	GSH-Px/T-SOD
CON	1.72±0.13[bc]	113.36±7.74[bc]	59.51±3.77[c]	1.16±0.11[bc]	0.12±0.01[bc]	0.47±0.05[b]	66.04±5.62[cd]	0.53±0.04[b]
SCOP	2.07±0.22[a]	102.72±2.31[a]	42.15±4.20[a]	0.84±0.10[a]	0.13±0.01[a]	0.59±0.07[a]	50.12±5.02[a]	0.41±0.02[a]
Piracetam	1.63±0.15[cd]	119.98±5.05[cd]	62.86±4.04[c]	1.11±0.06[b]	0.12±0.01[bc]	0.48±0.09[b]	73.83±5.30[de]	0.53±0.04[b]
LC	1.87±0.14[b]	106.46±10.79[ab]	55.59±5.10[b]	1.15±0.11[bc]	0.13±0.02[ab]	0.55±0.04[a]	56.84±2.82[b]	0.53±0.06[b]
L-LSPC	1.61±0.13[cd]	119.86±7.96[cd]	61.19±2.74[c]	1.10±0.10[b]	0.10±0.01[d]	0.44±0.05[bc]	75.01±7.42[ef]	0.51±0.03[b]
H-LSPC	1.68±0.16[cd]	118.43±8.00[cd]	60.92±3.20[c]	1.19±0.13[bc]	0.10±0.01[d]	0.47±0.04[b]	71.16±7.23[cde]	0.52±0.05[b]
L-LSPC+LC	1.60±0.21[cd]	113.89±6.55[bc]	62.46±5.12[c]	1.15±0.10[bc]	0.11±0.01[cd]	0.48±0.08[b]	72.02±10.01[cde]	0.55±0.06[b]
H-LSPC+LC	1.51±0.12[d]	124.03±3.79[d]	66.63±4.21[d]	1.32±0.11[d]	0.10±0.01[d]	0.38±0.04[c]	82.55±5.26[fg]	0.54±0.03[b]

Notes: Means in the same column with different superscript are significantly different ($P<0.05$), while sharing any same letters signify insignificant differences. CON, SCOP, Piracetam and LC is control, vehicle scopolamine control, positive drug control and *Lactobacillus casei-01* group, respectively. L-LSPC and H-LSPC is low and high dose of LSPC group, respectively. L-LSPC+LC is L-LSPC and LC combination group. H-LSPC+LC is H-LSPC and LC combination group.

L-LSPC+LC groups (all $P<0.05$). These results revealed that LC promoted the ameliorative effects of H-LSPC on serum antioxidant defense ability and oxidative damage.

Effects of combined LSPC and LC on colon antioxidant defense capacity and oxidative damage

As shown in Table 4, obviously raised colon MDA level, and reduced colon GSH-Px activity, T-SOD/MDA and GSH-Px/T-SOD in SCOP group were observed when compared with CON group (all $P<0.05$). LC treatment significantly lowered MDA level and increased T-SOD/MDA compared to SCOP group ($P<0.05$ and $P<0.05$). Moreover, treatments with Piracetam, L-LSPC, H-LSPC, L-LSPC+LC and H-LSPC+LC markedly reversed scopolamine-induced changes in colon MDA and TAOC levels, GSH-Px activity, T-SOD/MDA and GSH-Px/T-SOD, to the level of equal or superior to CON group. By comparison with SCOP group, H-LSPC+LC treatment significantly enhanced T-SOD activity ($P<0.05$). Furthermore, both L-LSPC and H-LSPC groups showed notable increases in TAOC level, GSH-Px activity and GSH-Px/T-SOD versus LC group (all $P<0.05$). Effects of H-LSPC+LC treatment on improving T-SOD and GSH-Px activities, TAOC level, T-SOD/MDA and GSH-Px/T-SOD were superior to those of LC treatment (all $P<0.05$), and only GSH-Px activity in H-LSPC+LC group was higher than that in H-LSPC group ($P<0.05$). Moreover, compared to Piracetam group, H-LSPC+LC group showed significant increases in GSH-Px activity and GSH-Px/T-SOD ($P<0.05$ and $P<0.05$). These results suggested that ameliorative effects of LSPC, and combined LSPC and LC on colon antioxidant defense ability and oxidative stress were stronger than LC. LC was helpful for H-LSPC to promote colon antioxidant defense ability.

Effects of combined LSPC and LC on brain TNOS, iNOS and nNOS activities and nNOS mRNA level

As shown in Table 5, brain TNOS and nNOS activities of SCOP group were markedly increased ($P<0.05$ and $P<0.05$) relative to those of CON group. Between SCOP and LC groups, there were no significant differences in brain TNOS, iNOS and nNOS activities. However, treatments with Piracetam, L-LSPC, H-LSPC, L-LSPC+LC and H-LSPC+LC normalized the raised TNOS and nNOS activities induced by scopolamine, and no remarkable difference was found among these groups. Importantly, brain iNOS activity was notable diminished by H-LSPC+LC treatment compared with SCOP, Piracetam, L-LSPC and L-LSPC+LC groups (all $P<0.05$).

The variations in brain mRNA level of nNOS were parallel to the alterations in brain nNOS activity in all the experimental groups. As shown in Fig. 2, SCOP group showed significantly higher nNOS mRNA level than CON group. LC group revealed a significant decrease in nNOS mRNA level relative to SCOP group ($P<0.05$), but showed significantly higher nNOS mRNA level than CON group ($P<0.05$). However, nNOS mRNA level in Piracetam, L-LSPC, H-LSPC, L-LSPC+LC and H-LSPC+LC groups was remarkably lower than that in both SCOP and LC groups (all $P<0.05$), and comparative with that in CON group. Furthermore, H-LSPC+LC group exhibited a significant decline in nNOS mRNA level relative to Piracetam, LC, L-LSPC, H-LSPC and L-LSPC+LC groups (all $P<0.05$).

Discussion

The present study provides initial evidence that supports the hypothesis that LC facilitates the ameliorative effects of LSPC on learning and memory impairment in scopolamine-induced amne-

Table 3. Effects of combined proanthocyanidins extracted from lotus seedpod (LSPC) and *Lactobacillus casei-01* (LC) on serum malondialdehyde (MDA) and total antioxidant capacity (TAOC) levels, total superoxide dismutase (T-SOD) and glutathione peroxidase (GSH-Px) activities, ratio of T-SOD to MDA (T-SOD/MDA) and ratio of GSH-Px to T-SOD (GSH-Px/T-SOD) in scopolamine-induced amnesia mice.

Group	MDA (nmol/mL)	T-SOD (U/mL)	GSH-Px (U/mL)	TAOC (U/mL)	T-SOD/MDA	GSH-Px/T-SOD
CON	24.43±9.10[bc]	196.05±15.62[bcd]	581.28±29.53[b]	5.29±0.51[bc]	8.97±2.90[bcd]	2.98±0.25
SCOP	33.87±9.12[a]	170.85±20.13[a]	529.51±33.92[a]	4.65±0.35[a]	5.40±1.64[a]	3.14±0.45
Piracetam	23.45±7.73[bc]	201.59±18.35[bcd]	604.52±17.45[b]	5.19±0.28[bc]	9.38±2.70[cd]	3.02±0.25
LC	29.22±6.63[ab]	188.01±12.95[b]	602.07±24.45[b]	5.03±0.20[ab]	6.86±2.13[ab]	3.22±0.31
L-LSPC	25.32±6.91[bc]	189.67±19.60[bc]	585.59±31.48[b]	5.18±0.21[bc]	7.80±1.44[bc]	3.13±0.46
H-LSPC	23.03±5.60[bc]	195.42±11.30[bcd]	599.62±29.08[b]	5.41±0.23[c]	8.91±2.02[bcd]	3.08±0.25
L-LSPC+LC	19.11±5.89[c]	206.80±22.87[cd]	599.66±19.67[b]	6.24±0.75[cd]	12.06±4.75[d]	2.93±0.32
H-LSPC+LC	11.54±3.28[d]	210.88±10.84[d]	627.46±22.79[c]	6.40±0.68[d]	19.38±4.49[e]	2.98±0.17

Notes: Means in the same column with different superscript are significantly different ($P<0.05$), while sharing any same letters signify insignificant differences. CON, SCOP, Piracetam and LC is control, vehicle scopolamine control, positive drug control and *Lactobacillus casei-01* group, respectively. L-LSPC and H-LSPC is low and high dose of LSPC group, respectively. L-LSPC+LC is L-LSPC and LC combination group. H-LSPC+LC is H-LSPC and LC combination group.

sia mice. In our study, H-LSPC+LC group had better behavioral performance in Y-maze test than H-LSPC and LC groups. Moreover, LC promoted the memory-enhancing effect of LSPC by improving antioxidant defense ability of brain, serum and colon, ameliorating brain cholinergic activity, and suppressing oxidative damage of serum and brain as well as brain nNOS mRNA level.

Animals with scopolamine-induced memory impairment have been generally employed to appraise the possible memory-improving activity of herbal and other agents [24,28]. In this study, scopolamine resulted in elevated error times in Y-maze test compared with CON group, which verified the serious memory deterioration in mice [28]. Piracetam, a clinic medicine, has shown the memory-enhancing effect in many animal model systems [29]. Compared with Piracetam group, LC and L-LSPC groups showed comparative effects on attenuating scopolamine-induced memory impairment in Y-maze test. However, H-LSPC, and LSPC and LC combinations exhibited stronger memory-enhancing effects than piracetam, which was suggested by markedly decreased error times of these groups compared to Piracetam group. The memory-enhancing effects of LC and LSPC were consistent with previous studies [6,13,14,15,22]. Noteworthily, error times of H-LSPC+LC group were reduced by 41.59% and 68.75% relative to those of H-LSPC and LC groups respectively, which indicated that LC was potentially helpful for H-LSPC to attenuate scopolamine-induced memory impairment.

Previous researches have found that learning and memory impairment is closely associated with oxidative stress [14,15,30]. As a muscarinic receptor antagonist, scopolamine results in memory deficits in animals by leading to cholinergic neurotransmission deficits and oxidative damage [28,31,32]. In this study, scopolamine resulted in noticeable declines in TAOC level, T-SOD and GSH-Px activities, T-SOD/MDA of brain and serum, and brain GSH-Px/T-SOD, and obvious increases in MDA level of brain and serum, and brain MPO activity. It is well known that TAOC reflects the capacity of nonenzymatic antioxidant defense system, and GSH-Px and SOD are two important enzymes of antioxidant defense system to eliminate reactive oxygen species (ROS) [33]. MDA, a byproduct of lipid peroxidation, is considered as the biomarker of oxidative stress [34]. The increases in GSH-Px/T-SOD and T-SOD/MDA indicate that more antioxidant enzymes are activated and ROS are efficiently scavenged by them, and thus oxidative damage is decreased [35,36]. The results indicated that marked decreases in antioxidant defense ability and increases in oxidative damage of mice brain and serum were closely associated with scopolamine-induced memory deficits [28,32]. LC treatment brought a certain increase in antioxidant defense ability of brain and serum, which may be due to the antioxidant abilities of *Lactobacilli* strains [37,38]. LC treatment only significantly reversed raised MDA level of brain, revealing the weak effect of LC on inhibiting oxidative damage. Moreover, LSPC and LSPC and LC combinations reversed scopolamine-induced variations of brain and serum, verifying that enhancement of antioxidant defense ability and inhibition of oxidative damage were effective ways to improving memory [14,30]. Previous studies have demonstrated that LSPC have a potent antioxidant activity *in vitro* [12], and an inhibitory effect on oxidative stress in brain and serum [14,15]. Especially, LSPC groups showed comparative MDA level and T-SOD/MDA of brain and serum with Piracetam group, and remarkably lower MPO activity of brain than Piracetam group, which prompted that LSPC were the effective inhibitor of oxidative damage. Noteworthily, effects of H-LSPC+LC treatment on lowering serum MDA level and raising T-SOD/MDA, TAOC level and GSH-Px activity of brain and serum were

Table 4. Effects of combined proanthocyanidins extracted from lotus seedpod (LSPC) and *Lactobacillus casei-01* (LC) on colon malondialdehyde (MDA) and total antioxidant capacity (TAOC) levels, total superoxide dismutase (T-SOD) and glutathione peroxidase (GSH-Px) activities, ratio of T-SOD to MDA (T-SOD/MDA) and ratio of GSH-Px to T-SOD (GSH-Px/T-SOD) in scopolamine-induced amnesia mice.

Group	MDA (nmol/mgprot)	T-SOD (U/mgprot)	GSH-Px (U/mgprot)	TAOC (U/mgprot)	T-SOD/MDA	GSH-Px/T-SOD
CON	0.74±0.10[b]	58.91±5.49[abc]	304.34±26.27[b]	0.79±0.12[ab]	81.35±15.05[bc]	5.18±0.39[bc]
SCOP	0.89±0.14[a]	55.18±3.20[a]	250.40±21.40[a]	0.65±0.17[a]	63.31±12.00[a]	4.57±0.60[a]
Piracetam	0.74±0.09[b]	58.57±2.52[abc]	305.56±14.75[b]	0.98±0.12[bc]	80.44±9.48[bc]	5.23±0.37[bc]
LC	0.75±0.16[b]	55.47±5.21[a]	269.80±35.94[a]	0.70±0.15[ab]	76.57±15.98[b]	4.89±0.66[ab]
L-LSPC	0.76±0.11[b]	58.73±5.46[abc]	345.84±42.01[cd]	0.97±0.05[c]	78.98±11.64[bc]	5.88±0.44[d]
H-LSPC	0.75±0.08[b]	58.61±3.76[abc]	333.71±45.68[bc]	1.04±0.09[c]	79.21±9.01[bc]	5.71±0.80[cd]
L-LSPC+LC	0.70±0.12[b]	57.52±3.92[ab]	320.15±27.06[bc]	0.93±0.20[bc]	84.71±16.29[bc]	5.59±0.61[cd]
H-LSPC+LC	0.69±0.12[b]	62.08±2.77[c]	365.92±43.16[d]	1.29±0.31[c]	92.19±14.27[c]	5.89±0.54[d]

Notes: Means in the same column with different superscript are significantly different (P<0.05), while sharing any same letters signify insignificant differences. CON, SCOP, Piracetam and LC is control, vehicle scopolamine control, positive drug control and *Lactobacillus casei-01* group, respectively. L-LSPC and H-LSPC is low and high dose of LSPC group, respectively. L-LSPC+LC is L-LSPC and LC combination group. H-LSPC+LC is H-LSPC and LC combination group.

stronger than those of Piracetam, H-LSPC and LC treatments, suggesting that LC facilitated the ameliorative effects of H-LSPC on antioxidant defense ability and oxidative damage of brain and serum.

Learning and memory impairment is associated with gut dysfunction [1,5,6]. Oxidative damage of colonic mucosa is one of main symptoms of intestinal dysfunction in F344 rats, and inhibition of oxidative damage is an important approach to modulate intestinal function and carcinogenesis [17]. This study measured colon MDA and TAOC levels, and activities of T-SOD and GSH-Px to evaluate the changes of gut function. Obviously raised MDA level, and lessened GSH-Px activity, T-SOD/MDA and GSH-Px/T-SOD in colon of SCOP mice revealed the increases of oxidative damage and the decreases of antioxidant defense ability in colon, and further indicated mild colon dysfunction induced by scopolamine. Previous studies have reported that *Lactobacilli* strains have antioxidant abilities [37,38], and can modulate gut function [1,6]. In this study, LC treatment showed the effective inhibition of colon oxidative damage, as indicated by markedly diminished MDA level and

enhanced T-SOD/MDA. Thus, inhibition of oxidative damage was an approach to modulate gut function [17], and the antioxidant ability of LC may partly contribute to the amelioration of gut dysfunction. This study showed that LSPC, and LSPC and LC combinations had potent abilities of suppressing oxidative damage and improving antioxidant defense of colon. Importantly, GSH-Px activity of H-LSPC+LC group was superior to that of Piracetam, LC and H-LSPC groups, suggesting that LC facilitated the ameliorative effects of H-LSPC on antioxidant defense ability of colon. These results also indicated that improving gut dysfunction may contribute to the amelioration of learning and memory ability [1,5,6].

Base on the above results, LC facilitated the ameliorative effects of H-LSPC on memory impairment by improving antioxidant defense ability of brain, serum and colon, and inhibiting oxidative damage of serum and brain. LSPC had a potent antioxidant activity [12], and parts of monomers, dimers, and trimers of proanthocyanidins in LSPC were absorbed to blood and then distributed into the brain to exert the antioxidant activity [39–41], therefore, oxidative damage and antioxidant defense ability of

Table 5. Effects of combined proanthocyanidins extracted from lotus seedpod (LSPC) and *Lactobacillus casei-01* (LC) on brain total nitric oxide synthase (TNOS), nitric oxide synthase (iNOS) and neural nitric oxide synthase (nNOS) activities in scopolamine-induced amnesia mice.

Group	TNOS (U/mgprot)	iNOS (U/mgprot)	nNOS (U/mgprot)
CON	0.48±0.06[b]	0.14±0.02[ab]	0.34±0.08[b]
SCOP	0.60±0.05[a]	0.15±0.02[a]	0.45±0.05[a]
Piracetam	0.48±0.04[b]	0.15±0.02[a]	0.35±0.07[b]
LC	0.60±0.08[a]	0.14±0.02[ab]	0.46±0.07[a]
L-LSPC	0.48±0.06[b]	0.15±0.01[a]	0.34±0.07[b]
H-LSPC	0.45±0.09[b]	0.14±0.04[ab]	0.30±0.09[b]
L-LSPC+LC	0.50±0.04[b]	0.16±0.02[a]	0.35±0.04[b]
H-LSPC+LC	0.47±0.06[b]	0.12±0.03[b]	0.36±0.06[b]

Notes: Means in the same column with different superscript are significantly different (P<0.05), while sharing any same letters signify insignificant differences. CON, SCOP, Piracetam and LC is control, vehicle scopolamine control, positive drug control and *Lactobacillus casei-01* group, respectively. L-LSPC and H-LSPC is low and high dose of LSPC group, respectively. L-LSPC+LC is L-LSPC and LC combination group. H-LSPC+LC is H-LSPC and LC combination group.

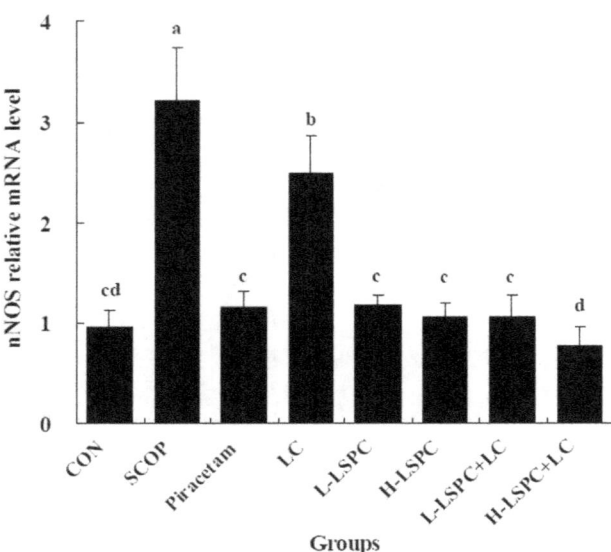

Figure 2. Effects of combined proanthocyanidins extracted from lotus seedpod (LSPC) and *Lactobacillus casei-01* (LC) on the mRNA level of neural nitric oxide synthase (nNOS). Groups without any same letters above the bars signify statistically significant differences (P<0.05). CON, SCOP, Piracetam and LC is control, vehicle scopolamine control, positive drug control and *Lactobacillus casei-01* group, respectively. L-LSPC and H-LSPC is low and high dose of LSPC group, respectively. L-LSPC+LC is L-LSPC and LC combination group. H-LSPC+LC is H-LSPC and LC combination group.

brain and serum were ameliorated. Previous researches proved the gut-regulation effect and antioxidant ability of *Lactobacilli* strains [37,38], thus in this study, LC showed a certain capacity of inhibiting oxidative damage and enhancing antioxidant defense ability. Most ingredients of LSPC were not absorbed directly and then accumulated in colon [18,19]. In colon, LSPC were metabolized by the colonic microbiota and generated phenolic acids, oligomeric proanthocyanidins and their isomers, and conjugated lactones [42,43], which showed stronger physiological and biological activities than their parent proanthocyanidins [19,44]. Proanthocyanidins metabolites by biotransformation of

LC *in vitro* showed more potent antioxidant ability than their parent proanthocyanidins [23], thus the metabolites of the combination groups exerted stronger antioxidant ability than LC and LSPC groups *in vivo*. Additionally, proanthocyanidins metabolites in colon kept the growth of probiotics such as *Lactobacillus* and inhibited the growth of pathogenic bacteria [20]. Consequently, in this study, the metabolites of the combination groups exerted stronger effects on improving gut function by modulating gut bacteria and decreasing oxidative damage. Ultimately, LSPC and LC combination exhibited a potent effect on ameliorating scopolamine-induced memory impairment.

Conclusions

In conclusion, LC, LSPC, and LSPC and LC combinations exhibited the ameliorative effects on scopolamine-induced memory impairment in mice. The ameliorative effects of LSPC, and LSPC and LC combinations were more effective than LC. The mechanisms involved in improving memory deficits of LSPC and LC combinations were associated with the improvement of antioxidant defense ability of brain, serum and colon, inhibition of oxidative damage of brain, serum and colon, suppression of brain nNOS activity and mRNA level, and amelioration of brain cholinergic activity. LC promoted the memory-enhancing effect of LSPC by improving antioxidant defense ability of brain, serum and colon, ameliorating brain cholinergic activity, and suppressing oxidative damage of serum and brain as well as brain nNOS mRNA level. These findings suggest LSPC and LC combination may provide a viable therapy in the treatment of memory impairment in aging process and some related diseases such as AD.

Acknowledgments

We are extremely grateful to Yiliang Zhu, Xiaokun Wang and Bo Diao for the technical assistance. They are members of Wuhan General Hospital of Guangzhou Military Command.

Author Contributions

Conceived and designed the experiments: JX ZDS. Performed the experiments: JX SYL YS QW XPL. Analyzed the data: JX BJX. Contributed reagents/materials/analysis tools: JX BJX ZDS. Wrote the paper: JX ZDS BJX MWZ.

References

1. Gareau MG, Wine E, Rodrigues DM, Cho JH, Whary MT, et al. (2011) Bacterial infection causes stress-induced memory dysfunction in mice. Gut 60: 307–317.
2. Zhao L, Shen J (2010) Whole-body systems approaches for gut microbiota-targeted, preventive healthcare. Journal of Biotechnology 149: 183–190.
3. Biagi E, Candela M, Fairweather-Tait S, Franceschi C, Brigidi P (2012) Ageing of the human metaorganism: the microbial counterpart. Age 34: 247–267.
4. Bajaj JS, Ridlon JM, Hylemon PB, Thacker LR, Heuman DM, et al. (2012) Linkage of gut microbiome with cognition in hepatic encephalopathy. American Journal of Physiology-Gastrointestinal and Liver Physiology 302: G168–G175.
5. Li W, Dowd SE, Scurlock B, Acosta-Martinez V, Lyte M (2009) Memory and learning behavior in mice is temporally associated with diet-induced alterations in gut bacteria. Physiology Behavior 96: 557–567.
6. Ohland CL, Kish L, Bell H, Thiesen A, Hotte N, et al. (2013) Effects of *Lactobacillus helveticus* on murine behavior are dependent on diet and genotype and correlate with alterations in the gut microbiome. Psychoneuroendocrinology 38: 1738–1747.
7. McCarthy M, Addington-hall J, Altmann D (1997) The experience of dying with dementia: a retrospective study. International Journal of Geriatric Psychiatry 12: 404–409.
8. Moujalli S, Chang E, Andrew S, Halcomb E (2008) Effectiveness of strategies to manage constipation in individuals of residential care facilities with dementia. The JBI Database of Systematic Reviews and Implementation Reports 6: S35–58.
9. Collins SM, Surette M, Bercik P (2012) The interplay between the intestinal microbiota and the brain. Nature Reviews Microbiology 10: 735–742.
10. Bhattacharjee S, Lukiw WJ (2013) Alzheimer's disease and the microbiome. Frontiers in Cellular Neuroscience 7: 1–3.
11. Cryan JF, Dinan TG (2012) Mind-altering microorganisms: the impact of the gut microbiota on brain and behaviour. Nature Reviews Neuroscience 13: 701–712.
12. Ling ZQ, Xie BJ, Yang EL (2005) Isolation, characterization, and determination of antioxidative activity of oligomeric proanthocyanidins from the seedpod of *Nelumbo nucifera* Gaertn. Journal of Agricultural and Food Chemistry 53: 2441–2445.
13. Xu JQ, Rong S, Xie BJ, Sun ZD, Zhang L, et al. (2009) Procyanidins extracted from the lotus seedpod ameliorate scopolamine-induced memory impairment in mice. Phytotherapy Research 23: 1742–1747.
14. Xu JQ, Rong S, Xie BJ, Sun ZD, Zhang L, et al. (2009) Rejuvenation of antioxidant and cholinergic systems contributes to the effect of procyanidins extracted from the lotus seedpod ameliorating memory impairment in cognitively impaired aged rats. European Neuropsychopharmacology 19: 851–860.
15. Gong YS, Liu LG., Xie BJ, Liao YC, Yang EL, et al. (2008) Ameliorative effects of lotus seedpod proanthocyanidins on cognitive deficits and oxidative damage in senescence-accelerated mice. Behavioural Brain Research 194: 100–107.
16. Bustos I, García-Cayuela T, Hernández-Ledesma B, Peláez C, Requena T, et al. (2012) Effect of flavan-3-ols on the adhesion of potential probiotic Lactobacilli to intestinal cells. Journal of Agricultural and Food Chemistry 60: 9082–9088.

17. Dolara P, Luceri C, Filippo CD, Femia AP, Giovannelli L, et al. (2005) Red wine polyphenols influence carcinogenesis, intestinal microflora, oxidative damage and gene expression profiles of colonic mucosa in F344 rats. Mutation Research 591: 237–246.

18. Roowi S, Stalmach A, Mullen W, Lean MEJ, Edwards CA, et al. (2010) Green tea flavan-3-ols: colonic degradation and urinary excretion of catabolites by humans. Journal of Agricultural and Food Chemistry 58: 1296–1304.

19. Monagas M, Urpi-Sarda M, Sánchez-Patán F, Llorach R, Garrido I, et al. (2010) Insights into the metabolism and microbial biotransformation of dietary flavan-3-ols and the bioactivity of their metabolites. Food Function 1: 233–253.

20. Lee HC, Jenner AM, Low CS, Lee YK (2006) Effect of tea phenolics and their aromatic fecal bacterial metabolites on intestinal microbiota. Research in Microbiology 157: 876–884.

21. Davari S, Talaei SA, Alaei H (2013) Probiotics treatment improves diabetes-induced impairment of synaptic activity and cognitive function: behavioral and electrophysiological proofs for microbiome-gut-brain axis. Neuroscience 240: 287–296.

22. Jung IH, Jung MA, Kim EJ, Han MJ, Kim DH (2012) *Lactobacillus pentosus var. plantarum* C29 protects scopolamine-induced memory deficit in mice. Journal of applied microbiology 113: 1498–1506.

23. Li SY, Chen L, Yang T, Wu Q, Lv ZJ, et al. (2013) Increasing antioxidant activity of procyanidin extracts from the pericarp of Litchi chinensis processing waste by two probiotic bacteria bioconversions. Journal of Agricultural and Food Chemistry 61: 2506–2512.

24. Xiao J, Sui Y, Li SY, Wu Q, Yang T, et al. (2013) Combination of proanthocyanidins extracted from lotus seedpod and L-cysteine ameliorate memory impairment induced by alcohol and scopolamine in mice. European Food Research and Technology 236: 671–679.

25. Porter LJ, Hrstich LN, Chan BG (1985) The conversion of procyanidins and prodelphinidins to cyanidin and delphinidin. Phytochemistry 25: 223–230.

26. Xu SY, Bian LR, Cheng X (2001) Behavioral pharmacological experiment. In: Zhang JT, editor. Pharmacological experimental methodology. Beijing: People's Health press. pp. 826–829.

27. Livak KJ, Schmittgen TD (2001) Analysis of relative gene expression data using real-time quantitative PCR and the $2^{-\Delta\Delta CT}$ Method. Methods 25: 402–408.

28. Kwon S-H, Lee H-K, Kim J-A, Hong S-I, Kim H-C, et al. (2010) Neuroprotective effects of chlorogenic acid on scopolamine-induced amnesia via anti-acetylcholinesterase and anti-oxidative activities in mice. European Journal of Pharmacology 649: 210–217.

29. Chen J, Long Y, Han M, Wang T, Chen Q, et al. (2008) Water-soluble derivative of propolis mitigates scopolamine-induced learning and memory impairment in mice. Pharmacology, Biochemistry and Behavior 90: 441–446.

30. Devasagayam TPA, Tilak JC, Boloor KK, Sane KS, Ghaskadbi SS, et al. (2004) Free radicals and antioxidants in human health: current status and future prospects. JAPI 52: 794–804.

31. Kopelman MD, Corn TH (1988) Cholinergic blockade as a model of cholinergic depletion: a comparison of the memory deficits with those of Alzheimer-type dementia and the alcoholic korsakoff syndrome. Brain 111: 1079–1110.

32. Jeong EJ, Lee KY, Kim SH, Sung SH, Kim YC (2008) Cognitive-enhancing and antioxidant activities of iridoid glycosides from *Scrophularia buergeriana* in scopolamine-treated mice. European Journal of Pharmacology 588: 78–84.

33. Reiter RJ (1995) Oxidative processes and antioxidative defense mechanisms in the aging brain. The FASEB Journal 9: 526–533.

34. Maes M, Galecki P, Chang YS, Berk M (2011) A review on the oxidative and nitrosative stress (O&NS) pathways in major depression and their possible contribution to the (neuro)degenerative processes in that illness. Progress in Neuro-Psychopharmacology Biological Psychiatry 35: 676–692.

35. Formigari A, Irato P, Santon A (2007) Zinc, antioxidant systems and metallothionein in metal mediated-apoptosis: biochemical and cytochemical aspects. Comparative Biochemistry and Physiology Part C: Toxicology Pharmacology 146: 443–459.

36. Škrha J, Šindelka G, Kvasnička J, Hilgertova J (1999) Insulin action and fibrinolysis influenced by vitamin E in obese type 2 diabetes mellitus. Diabetes Research and Clinical Practice 44: 27–33.

37. Laura P, Saleta S, Monica CA (2007) Comparative study of the preventative effects exerted by two probiotics, *lactobacillus reuteri* and *lactobacillus fermentum*, in the trinitrobenzenesulfonic acid model of rat colitis. British Journal of Nutrition 97: 96–103.

38. Kullisaar T, Zilmer M, Mikelsaar M, Vihalemm T, Annuk H, et al. (2002) Two antioxidative lactobacilli strains as promising probiotics. International Journal of Food Microbiology 72: 215–224.

39. Gonthier M-P, Donovan JL, Texier O, Felgines C, Remesy C, et al. (2003) Metabolism of dietary procyanidins in rats. Free Radical Biology and Medicine 35: 837–844.

40. Kohri T, Matsumoto N, Yamakawa M, Suzuki M, Nanjo F, et al. (2001) Metabolic fate of (−)-[4-3H] epigallocatechin gallate in rats after oral administration. Journal of Agricultural and Food Chemistry 49: 4102–4112.

41. Van PH, Lucero MJ, Yeo GW, Stecker K, Heivand N, et al. (2007) Plant-derived flavanol (−)epicatechin enhances angiogenesis and retention of spatial memory in mice. The Journal of Neuroscience 27: 5869–5878.

42. Groenewoud G, Hundt HKL (1986) The microbial metabolism of condensed (+)-catechins by rat-caecal microflora. Xenobiotica 16: 99–107.

43. Li SY, Sui Y, Xiao J, Wu Q, Hu B, et al. (2013) Absorption and urinary excretion of A-type procyanidin oligomers from Litchi chinensis pericarp in rats by selected ion monitoring liquid chromatography–mass spectrometry. Food Chemistry 138: 1536–1542.

44. Baba S, Osakabe N, Natsuxne M, Terao J (2002) Absorption and urinary excretion of proeyanidin B2 [epicatechin-(4beta-8)-epicatechin] in rats. Free Radical Biology and Medicine 33: 142–148.

21

An Extract of Pomegranate Fruit and Galangal Rhizome Increases the Numbers of Motile Sperm: A Prospective, Randomised, Controlled, Double-Blinded Trial

Maja D. K. Fedder[1], Henrik B. Jakobsen[2], Ina Giversen[2], Lars P. Christensen[3], Erik T. Parner[4], Jens Fedder[1,5]*

1 Laboratory of Reproductive Biology, Scientific Unit, Regional Hospital of Horsens, Horsens, Denmark, **2** Nerthus ApS, Lejre, Denmark, **3** Department of Chemical Engineering, Biotechnology and Environmental Technology, Faculty of Engineering, University of Southern Denmark, Odense, Denmark, **4** Department of Public Health, Section for Biostatistics, University of Aarhus, Aarhus, Denmark, **5** Centre of Andrology, Fertility Clinic, Department D, Odense University Hospital, Odense, Denmark

Abstract

Pomegranate fruit (*Punica granatum*) and galangal (*Alpinia galanga*) have separately been shown to stimulate spermatogenesis and to increase sperm counts and motility in rodents. Within traditional medicine, pomegranate fruit has long been used to increase fertility, however studies on the effect on spermatogenesis in humans have never been published. With this study we investigated whether oral intake of tablets containing standardised amounts of extract of pomegranate fruit and powder of greater galangal rhizome (Punalpin) would increase the total number of motile spermatozoa. The study was designed as a prospective, randomized, controlled, double-blinded trial. Enrolment was based on the mean total number of motile spermatozoa of two ejaculates. The participants delivered an ejaculate after 4–8 days of tablet intake and two ejaculates just before they stopped taking the tablets. Seventy adult men with a semen quality not meeting the standards for commercial application at Nordic Cryobank, but without azoospermia, were included in the study. Participants were randomized to take tablets containing extract of pomegranate fruit (standardised with respect to punicalagin A+B, punicalin and ellagic acid) and freeze-dried rhizome of greater galangal (standardised with respect to 1′S-1′-acetoxychavicol acetate) or placebo on a daily basis for three months. Sixty-six participants completed the intervention (active treatment: n = 34; placebo: n = 32). After the intervention the total number of motile spermatozoa was increased in participants treated with plant extracts compared with the placebo group (p = 0.026). After three months of active treatment, the average total number of motile sperm increased by 62% (from 23.4 to 37.8 millions), while for the placebo group, the number of motile sperm increased by 20%. Sperm morphology was not affected by the treatment. Our findings may help subfertile men to gain an improved amount of motile ejaculated sperm by taking tablets containing preparations of pomegranate fruit extract and rhizome of greater galangal.

Trial Registration: ClinicalTrials.gov NCT01357044

Editor: Samuel Kim, University of Kansas Medical Center, United States of America

Funding: The study was funded by Centre for Science and Research of Alternative Treatment (VIFAB), who played no other role in the study. Grant number 1002282. VIFAB does not exist anymore. Nerthus ApS is owned by authors Ina Giversen (IG) and Henrik Byrial Jakobsen (HBJ). Nerthus ApS did not have any additional role in the study design, data collection and analysis, decision to publish, or preparation of the manuscript. The specific roles of these authors are articulated in the "author contributions" section.

Competing Interests: Two of the authors (IG and HBJ) are commercializing the product; however, their company, Nerthus APS, has not contributed financially to the study. The full names of the two filed patent applications are: Dry preparation of Alpinia galanga or Alpinia conchigera with high content of 1′S-1′-acetoxychavicol acetate:(Application no. PCT/EP2014/061880); and Composition for enhancing semen quality in a male subject: (Application no. PCT/EP2014/061851). The product name of the food supplement: Punalpin®. There are no further patents, products in development, or marketed products to declare.

* Email: fedder@dadlnet.dk

Introduction

In the Western world, a decline in sperm quality during the last 40 years has been suggested [1,2], although the decrease may have been stabilized during the last decade [3]. Only 23% of Danish men have optimal semen quality and 15% are at high risk of needing fertility treatment in order to biologically father a child [3]. It is evident that any improvement in poor quality semen brings about an increased chance for obtaining pregnancy, also with regard to assisted reproductive technology. Intracytoplasmic sperm injection is often the only alternative offered to couples, where the male suffers from poor semen quality. This makes randomised controlled trials of supplements, drugs or diet advices, aimed to improve semen quality, important in order to create other solutions helpful for males with poor semen quality.

Pomegranate (*Punica granatum*) and greater galangal (*Alpinia galanga*) belong to a limited group of well-known edible plants, which show promising effects *in vivo* on rodent sperm quality without causing undesirable side effects [4–7]. The pomegranate fruit consists of a leathery pericarp containing numerous seeds, each covered by a juicy, translucent aril. Both pericarp and aril juice contains hydrolysable tannins, with the pericarp being the far most concentrated source [8]. In the pericarp, the ellagitannins punicalagin A and B predominates, while punicalins, which are partial hydrolysis products of punicalagins [9], constitute a minor fraction. The ellagitannins are only found in trace amounts in the aril juice [8]. Instead, other hydrolysable tannins prevail in the juice together with another important group of polyphenols, the anthocyanins [10]. Ellagitannins and, to a lesser degree, anthocyanins possess pronounced antioxidative activity in vitro [8]. Juice from P. granatum fruit has been shown to stimulate spermatogenesis, to increase sperm motility, and to decrease the percentage of abnormal sperm in rats after seven weeks of gavage [4]. An increase in sperm motility and concentration together with a decrease in abnormal sperm rate was observed after 6 weeks of oral administration of pomegranate juice extract to rats [5]. Further, oral administration of pomegranate pericarp extract has been shown effective in reversing induced spermatogenic disruption in rats [6]. The ginger-like rhizome of A. galanga contains a wide variety of semivolatile and volatile compounds, the dominating substances being a group of phenylpropanoids with 1'S-1'-acetoxychavicol acetate (ACA) as the most abundant. The phenylpropanoids have shown potent antioxidative activity in vitro [11], while rhizome extracts of A. galanga have shown anti-inflammatory effect in vivo [12]. Additionally, extracts of the rhizome of A. galanga have been show to increase sperm counts and motility in mice [7] and to increase serum testosterone levels in rats [13].

In this study it was investigated whether oral consumption of tablets with a standardised content of P. granatum extract and A. galanga powder would increase the total number of motile spermatozoa (TMSC) and sperm morphology, defined by strict criteria, in adult men with reduced semen quality. Since this is the first published clinical study investigating the effects of P. granatum and A. galanga on human semen quality, we chose to focus on the most important (number and motility) [14] and basic (morphology) semen parameters [15].

Materials and Methods

The protocol for this trial and supporting CONSORT checklist are available as supporting information; see Checklist S1 and Protocol S1.

Study design

The hypothesis was formulated prior to the study, and the power was estimated with the assumptions that the difference between the study and placebo group would be 5 million TMSC and SD = 2.3 million TMSC. If a significance level of 5% and a statistical power of 80% were given, at least 18 participants should be included in each group. However, due to the high level of uncertainty of the assumptions in this power estimation, nearly twice as many participants were enrolled.

The study was a prospective, randomised, placebo-controlled, double-blinded trial. The participants were randomised 1:1 to daily treatment with tablets containing extract of pomegranate fruit (standardised with respect to punicalagin A+B, punicalin and ellagic acid) and freeze-dried rhizome of greater galangal (standardised with respect to 1'S-1'-acetoxychavicol acetate (ACA)) or placebo tablets, for 90 days. In addition to the two ejaculates delivered prior to intervention (baseline = mean of the two ejaculates), the participants delivered an ejaculate after 4–8 days of tablet intake and two ejaculates at the end of the study. Upon termination of the study, the participants completed a questionnaire in order to determine the occurrence of any negative or positive side effects.

Study participants and semen collection

Participants were recruited over a one-year period by advertising in Nordic Cryobank (sperm bank) and local newspapers. Inclusion criteria were: Healthy adult men ≥18 years of age with a semen quality not meeting the standards for commercial application at Nordic Cryobank (i.e. >200 mio. TMSC in raw semen). Potential sperm donors, who fulfilled the requirements, completed a questionnaire about former illness and medicine consumption to ensure that the reason for inferior semen quality was not due to obvious medical reasons, such as cryptorchidism or genital tract infection.

Enrolment was based on two ejaculates delivered by masturbation within a time span of 7–14 days and with 3–7 days of abstinence before each sample delivery. The ejaculates were kept at or above room temperature (max 30°C) and analysed within one hour from ejaculation. Participants with azoospermia were excluded from the study. Seventy participants were enrolled, but four chose to withdraw due to logistic difficulties.

Interventions

The daily active treatment consisted of 4 tablets with extract of P. granatum and 4 tablets with A. galanga (Punalpin, Nerthus ApS, Lejre, Denmark), two of each kind taken in the morning, and two of each kind in the evening. The placebo tablets were produced in two variations to visually match the two kinds of active treatment. Four of each kind of placebo tablets were taken daily, two of each kind in the morning and two of each kind in the evening. The participants of both groups received all the tablets at the beginning of the study in two separate containers. The daily active dose provided 1000 mg P. granatum extract and 764 mg A. galanga preparation, corresponding to 500–1000 mL high quality pomegranate fruit juice (based on the amount of characteristic ellagitannins) and approximately 1 g fresh greater galangal rhizome (based on the amount of ACA). For P. granatum this corresponded to: 106 mg punicalagin A (27 mg/tablet), 278 mg punicalagin B (69 mg/tablet), 4,7 mg punicalin (1,2 mg/tablet) and 9,6 mg ellagic acid (2,4 mg/tablet). For A. galanga, this corresponded to: 16 mg ACA (3.9 mg/tablet).

Initial screening of a range of commercially available extracts of P. granatum and A. galanga for their content of characteristic bioactive compounds showed that most extracts of P. granatum had either (i) very low levels of punicalagin A and B, punicalin and ellagic acid or (ii) low levels of punicalagin A and B and punicalin, but high levels of ellagic acid, indicating that the former compounds had been degraded to ellagic acid. This is in accordance with the study of Zhang et al. (2009) [16] who screened 27 commercial extracts of P. granatum and concluded that only 5 reflected the typical tannin profile of P. granatum while five extracts had no detectable tannins or ellagic acid. Our screening showed that the ethanolic fruit (pomace) extract P40P (Polinat, Las Palmas, Spain) had a fingerprint profile of punicalagin A and B, punicalin and ellagic acid, which was very similar to that of high quality pomegranate juice (data not shown) and was therefore selected for the present trial. P40P has been standardised to contain at least 35% punicalagin A+B and at least 40% punicosides (the total amount of punicalagins, punicalin, ellagic acid glycosides and ellagic acid).

Our screening of a number of commercial extracts of A. galanga revealed that the content of ACA was either very low or not detectable (data not shown). Therefore, we developed the following method for preparing an extract of *A. galanga* with high levels of ACA combined with very low levels of microbial counts: Fresh rhizomes of *A. galanga* imported from Thailand were purchased (Europa Frugt ApS, Valby, Denmark). The rhizomes were stored at approximately 5°C during transport. Upon arrival the stems were removed and the rhizomes were split longitudinally and transferred to a −24°C freezing chamber. After a few days of freezing the *A. galanga* rhizomes were freeze-dried. The water content in the rhizomes were 3–5% following drying. The rhizomes were transferred to gas tight sealed bags (VMPET12/Adhesive/LLDPE100, total 114 μm, WJ Packaging solutions, Canada) in nitrogen atmosphere (0–3% O_2) for 14 days. Subsequently the foil bags with rhizomes were placed in a heating oven for 3 hours at 78±2°C in order to further decrease the number of microorganisms. Prior to incorporation in tablets, the dry rhizomes were grated on a standard Co-mill pulveriser.

Tablets were produced containing either 191 mg of the above-mentioned dried powder of *A. galanga* or 250 mg of *P. granatum* extract. The *P. granatum* tablets additionally contained sodium bicarbonate, silicon dioxide, microcrystalline cellulose, syloid AL1, magnesium stearate and, for coating, propylene glycol, titanium dioxide, talc, and pharmacoat 615.

The *A. galanga* tablets contained talc, microcrystalline cellulose, syloid AL1, magnesium stearate, plus coating as just described added shellac and copper complexes of chlorophylls for colouring purpose. The placebo tablets contained 461 mg microcrystalline cellulose, syloid AL 1 and magnesium stearate plus coating, and the placebo tablets matching the *A. galanga* tablets also added copper complexes of chlorophylls.

The levels of marker compounds in the tablets were measured 1, 5, 12 and 21 months after production in order to monitor any degradation.

Outcome

The primary outcome was change in semen quality during the 3 month long treatment period, from baseline to follow-up, expressed as the total (progressive and non-progressive) motile sperm count (TMSC). The TMSC was defined as: Ejaculate volume × spermatozoa concentration × percentage of motile spermatozoa [15].

The secondary outcome was change in number of morphologically normal sperm based on WHO's strict sperm morphology criteria [15]. The outcomes were calculated as the difference between the follow-up value and the baseline value.

Chemical analysis of tablets

The level of ACA in the tablets was determined by gas chromatography-flame ionization detection (GC-FID) and gas chromatography-mass spectrometry (GC-MS), while levels of punicalagin A and B, punicalin, and ellagic acid were determined by analytical high-performance liquid chromatography with diode-array detection (HPLC-DAD).

For GC sample preparation, 15 tablets were ground in a mortar and 5.00 g of the ground tablets were mixed in 50 mL 99.9% ethanol in a 100 mL flask with screw cap. The mixture was subjected to a sonication bath for 20 min. Approximately 1 mL of this solution was filtered through a HPLC filter (Whatman Puradisc 13 syringe filters (0.2 μm), Sigma-Aldrich, Germany) into GC-vials before GC-FID and GC-MS analysis, respectively. GC-FID analysis was performed on an Agilent 6890 GC series (Agilent Technologies, Hørsholm, Denmark) and a Thermo Scientific DSQ II Single Quadrupole GC/MS system (Thermo Scientific, CA, USA) was used for GC-MS analysis. Separations were

performed on a ZB-Wax plus column (60 m×0.25 mm internal diameter (i.d.), df = 0.25 μm liquid phase, Phenomenex Denmark, Værløse, Denmark) using the following oven temperature program: 80°C for 1 min, from 80°C to 220°C at 5°C/min, followed by a constant temperature at 220°C for 20 min. Helium was applied as carrier gas with a flow rate of 1.2 mL. The injection volume was 1 μL with a split ratio of 1:50 (injector temperature 200°C). ACA was identified by GC-MS (70 eV) based by comparison of retention time (R_t) and mass spectrum with an authentic standard of ACA (PhytoLab GmbH & Co. KG, Vestenbergsgreuth, Germany) and quantified by GC-FID (FID operating at 300°C) based on a calibration curve of ACA covering the concentration range occurring in the samples. All analyses were performed in duplicates with a relative standard deviation of less than 1%.

For HPLC sample preparation, 15 tablets were ground in a mortar and 30 mg of the ground tablets were mixed with 20 mL distilled water in a 50 mL flask with screw cap. The mixture was subjected to a sonication bath for 20 min. The dissolved tablet sample was filtered through a HPLC filter (0.2 μm) before HPLC analysis. Analytical HPLC-DAD analysis was carried out on an Agilent 1200 Series analytical HPLC system (Hørsholm, Denmark) equipped with a DAD operating from 190 to 600 nm. Polyphenols were monitored at 258 nm, and UV spectra were recorded between 190 and 600 nm. Separations were performed on a Luna 3 μm C18(2) 100 Å, LC end-capped column (150 mm×4.6 mm internal diameter, 3 μm particle size, Phenomenex Denmark, Værløse, Denmark). The mobile phases consisted of 0.2% phosphoric acid in water as solvent A and 100% acetonitrile as solvent B, and the polyphenols were separated using the following solvent gradient: 0 min (5% B), 7 min (30% B), 8 min (70% B), 13 min (70% B), 15 min (5% B), and 17 min (5% B). Solvents were degassed before use. The flow rate was 1.0 mL/min, the column temperature was 40°C, and the injection volume was 10 μL. Stock standard solutions of punicalagin A and B (ChromaDex, Boulder, CO, USA), punicalin (Stanford Chemicals, Irvine, CA, USA), and ellagic acid (Sigma-Aldrich, Munich, Germany), respectively, were prepared in water:methanol (80:20, v:v) at a concentration of 1 mg/mL. Water was used for further dilution of the stock standard solutions. The polyphenol standards were used for identification (R_t on HPLC and UV spectra) and quantification (calibration curves covering the concentration range occurring in the samples). All analyses were performed in duplicates with a relative standard deviation of less than 5%.

Analysis of sperm motility

Number of motile sperm in each semen sample was determined using a Makler counting chamber (Sefi Medical Instruments Ltd., Haifa, Israel) by two laboratory technicians both blinded to the study. Each semen sample was allowed to liquefy and was analysed within one hour from delivery. Conditions during analysis were standardised to room temperature (22°C) and a preheated Leica microscope. The sperm in a minimum of 10 grid squares were counted, and when present, a minimum of 200 sperm per sample were counted in total. Intra-observer coefficient of variation was assessed in intervals and never found to exceed the accepted difference between independent replica counts [15]. In the cases of replica counts, the first reading was included in the study.

Analysis of sperm morphology

Number of morphologically normal sperm was determined by blinded strict morphology analysis of ethanol-fixated Papanicolaou stained semen smears. The staining procedure was performed as recommended by the WHO 5th edition [15]. On each smear, if present, 400 sperm were counted and rated either abnormal or normal based on WHO's strict sperm morphology criteria [15]. All the morphology analyses were performed by the same person and within one month in order to minimize rating variations.

Randomisation

Randomisation of the participants was performed by a computer-based block-randomisation sequence formed with allocation 1:1 and block size 10, meaning that for each 10 participants enrolled an equal number of participants are randomized to each group. Tablets with active compounds and placebo had identical appearances. The tablets were stored in separate plastic containers with consecutive numeration matching the participant randomisation number. Participants meeting the inclusion criteria received the tablets upon returning a written consent. The randomisation code was kept unknown to anyone except the one person, who managed the randomisation, and not revealed until the study was finished and all primary statistical calculations were performed.

Statistical analyses

The primary analysis was a comparison of the mean increase in TMSC between the treatment group and placebo group. The mean increase in TMSC was compared using an unequal variance t-test, taking into account possible variance heterogeneity between the two groups and deviation from normal distribution [17]. As a confirming analysis, all 95% confidence intervals and p-values were calculated by the bootstrap method [18]. In a secondary analysis adjustments were made in a linear regression model for age and BMI, both dichotomized by the median value. In a sensitivity analysis the adjustment for age and BMI were also performed using restricted cubic spline with each 2 knots in the linear regression model. For all comparative analyses, $p < 0.05$ was considered statistically significant.

Approvals

The study was approved by the Danish Data Protection Agency (J.nr. 2012-41-5441) and the Scientific Ethics Committee of Middle Jutland, Denmark (M-20100247) and registered in ClinicalTrials.gov (ID NCT01357044). All participants provided informed written consent.

Results

Study participants

From May 2011 to September 2012, 70 adult men (mean age: 29 years) were included and randomised, and 66 received active treatment (n = 32) or placebo (n = 34), Figure 1. The mean age for the group receiving active treatment was 30 ± 7 years, and for the group receiving placebo treatment the mean age was 28 ± 6 years. Body mass index (BMI) and life style factors did not differ significantly between the two groups (Table 1). In each group one participant had only one testicle (the other one presumably removed due to cryptorchidism), and two participants in each group reported previous infection with chlamydia. Fifty-three of the participants delivered a baseline semen sample with at least one abnormal semen parameter (volume, concentration, motility) according to the new standards from 2010 (≥ 1.5 mL and ≥ 15 million/mL and $\geq 40\%$ motile spermatozoa) [15]. One study participant in the treatment group had only one follow-up value.

Total motile sperm count (TMSC)

Treatment with standardised amounts of *P. granatum* extract and *A. galanga* powder for three months induced a 62% increase in the average total number of motile sperm (from 23.4 millions to 37.8 millions) which was a significantly higher increase than in the placebo group (20%: from 19.9 millions to 23.9 millions), p = 0.026 (Table 2). Bootstrap analysis showed very similar results (p = 0.021). Figure 2 shows the individual plots of the differences between follow-up and baseline TMSC. There was no increase in the number of motile sperm after one week of tablet intake for either of the groups (Table 2). When data were adjusted for age and BMI (using division at the median), the difference in increase in TMSC remained statistically significant (Table 2). Similar results were obtained when adjusting for age or BMI using restricted cubic spline (data not shown).

Although defined as normal semen quality, it has been shown that time to pregnancy increases, when the sperm concentration does not exceed 40×10^6/mL [19]. Subanalyses of our dataset show a clear trend of larger increase in TMSC in the active treatment group compared to the placebo group when looking at participants with very low baseline TMSC. The statistical significance is maintained when looking at the subgroup of participants with baseline TMSC at or below 40 million (active treatment: n = 27, placebo: n = 28) (p = 0.02), however due to the number of participants in the study, the statistical power is lost when baseline TMSC is further lowered.

To be mentioned as secondary results, the power was not sufficient to reveal a possible effect on motility (per cent motile sperm) alone (p = 0.2), but the concentration of motile spermatozoa also increased significantly in the active treatment group compared with placebo (p = 0.030; 3.8 mio./mL in active treatment vs. 1.1 mio./mL in placebo).

Two of the participants in the treatment group were experiencing influenza-like symptoms during the study period. For both these participants, TMSC did not improve.

Sperm morphology

Strict sperm morphology assessment did not reveal statistically significant difference between the treatment and the placebo groups (Table 3). Very similar results were obtained using an unequal variance t-test (Table 3) and the bootstrap method (data not shown).

Compliance and safety

Microbiological activity in both the *P. granatum* extract and the *A. galanga* powder were below the EU standards for herbal medicine, even prior to the applied heating procedure of *A. galanga*, which further decreased the microbiologic counts (data not shown).

Counting of the remaining tablets in all bottles at the end of the study showed that more than 85% of the participants had taken more than 80% of the tablets. Thirty-eight (57.6%) of the participants returned a questionnaire regarding side effects and state of health during the study period. Possibly unrelated to the treatment, three of the responders (two in the treatment group and one in the placebo group) reported influenza-like symptoms. One participant from the active treatment group reported side effects in the form of irregular defecation, diarrhoea, and increased flatulence, but nevertheless wished to complete the treatment.

Level of marker compounds

The amount of ACA in the tablets was reduced during the period in which monitoring took place. The initial level of ACA was reduced by approx. 50% 21 months after tablet production (Figure 3). The levels of punicalagin A and B, punincalin and ellagic acid in pomegranate tablets were unchanged during the same period (data not shown). Since the recruitment, and thus the starting points for the participants, was forthcoming, the first

CONSORT 2010 Flow Diagram

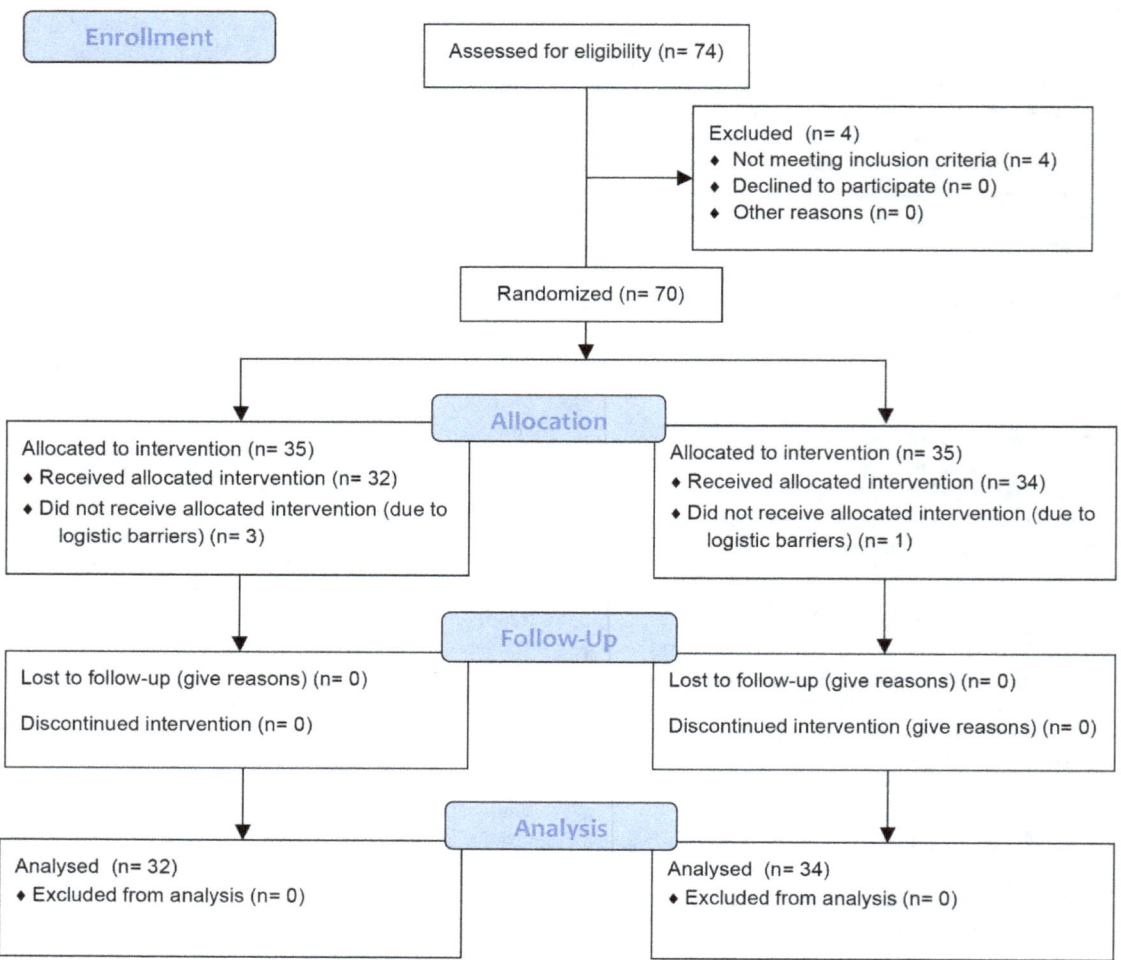

Figure 1. Flow diagram. Consort diagram showing the flow of the study participants.

Table 1. Demographics and life style factors.

	Active treatment	**Placebo**
	(n = 32)	**(n = 34)**
Age (years)	30.6±7.3 (CI: 28.1–33.2)	28.1±6.1 (CI: 26.1–30.2)
BMI	25.8±3.8 (CI: 24.5–27.1)	25.8±4.7 (CI: 24.2–27.4)
Smoking, n	5 (15.6%)	5 (14.7%)
Coffee drinking (≥4 cups/day), n	9 (28.1%)	3 (8.8%)
Alcohol drinking (≥14 units/week), n	2 (6.3%)	2 (5.9%)
Sauna regularly (monthly), n	2 (6.3%)	1 (2.9%)

Demographic data and life style factors of the participants showing a similar pattern for the two study groups. Participants were asked to state basic life style (smoking, coffee drinking, alcohol drinking and how often they attended sauna). Age at the time of baseline ejaculate delivery was calculated from social security numbers, and BMI was calculated from given weights and heights.

Table 2. Total motile sperm counts (TMSCs).

	Treatment		Crude		Adjusted	
	Active	**Placebo**				
	(n = 32)	**(n = 34)**	**Difference**	**P-value***	**Difference**	**P-value***
	Mean ±SD (95% CI)	**Mean ±SD (95% CI)**				
Baseline	23.4±25.1 (14.3; 32.4)	19.9±22.7 (12.0; 27.8)	3.5 (−8.3; 15.3)	0.56	4.4 (−8.2; 17.0)	0.49
Initiating treatment	23.6±25.4 (14.4; 32.8)	20.1±22.9 (12.1; 28.1)	3.5 (−8.4; 15.4)	0.56	4.7 (−8.0; 17.3)	0.46
Follow-up	37.8±39.5 (23.6; 52.1)	23.9±27.8 (14.2; 33.6)	14.0 (−3.0; 30.9)	0.10	14.2 (−3.8; 32.2)	0.12
Initiating treatment - baseline	0.2±6.0 (−1.9; 2.4)	0.2±9.9 (−3.3; 3.6)	0.1 (−4.0; 4.1)	0.98	0.3 (−3.9; 4.5)	0.89
Follow-up - baseline	14.5±21.3 (6.8; 22.1)	4.0±15.2 (−1.3; 9.3)	10.5 (1.3; 1 9.7)	0.026	9.8 (0.2; 19.5)	0.047

TMSCs for the treatment groups receiving the combination of *P. granatum* fruit extract and *A. galanga* rhizome powder or placebo. The results are shown unadjusted and adjusted for age and BMI.
*The difference in TMSC between the two groups was analysed by an unequal variance t-test. In the adjusted analyses the participants were divided in two groups at the median according to age and BMI.

recruited participants were taking tablets with a higher content of ACA in the *A. galanga* composition. For the first 34 participants included, the average increase in normal sperm counts was 74% (p = 0.045) (data not shown), possibly, but not certainly, reflecting the higher amount of ACA.

Discussion

This study revealed that the combination of *P. granatum* fruit extract and *A. galanga* rhizome powder (Punalpin) significantly increased the total number of motile sperm in men with a semen quality, which is reduced or low within the normal range, after three months of daily treatment.

Maturing sperm move from the testis to the epididymis, in which they spend in average six days prior to release during ejaculation [20]. Since no difference in the number of motile sperm was seen after the first week of intake of standardised extract of *P. granatum* and *A. galanga* powder, we propose that the combination exerts its positive effect during spermatogenesis rather than on the epididymal environment. This is in concordance with the rodent study of Türk *et al.*, who concluded that pomegranate juice stimulates spermatogenesis, increases sperm motility and decreases the number of abnormal sperm in rats in a dose-dependent manner [4]. However, the mechanism behind the observed improvement in semen quality still has to be clarified. One factor that may contribute considerably to the observed effects is the high levels of antioxidants present in both plants. A few studies have demonstrated a significant reduction in enzymatic and non-enzymatic antioxidant activity in seminal plasma of infertile men compared with fertile men and a positive influence of intake of antioxidants on the semen quality in humans [21,22]. Additionally, the consumption of pomegranate juice or fruit has resulted in significant rises in plasma antioxidant capacity [23,24]. Lastly, although not proved as causative factors, a few non-intervention studies have demonstrated statistical association between intake of fruits and greens and decreased risk of reduced semen quality [25,26].

Strengths and limitations

The present study was solidly designed. However, the relatively wide range in baseline values among the participants constitutes a weakness of the study. With a more homogenous study population, the 95% confidence intervals of the changes in TMSC would be expected to narrow down, followed by stronger statistical results.

It is also a weakness of the study that pregnancy rate was not an outcome parameter, but the participants were not recruited on the basis of a wish for pregnancy.

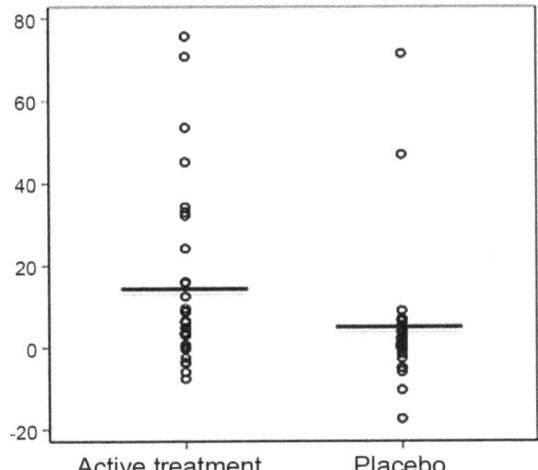

Δ motile sperm (mio.)

Figure 2. Differences in total motile sperm counts (TMSCs). Plot of the differences in TMSCs from baseline to follow-up for each participant in the active treatment receiving the combination of *P. granatum* fruit extract and *A. galanga* rhizome powder and the placebo group. Follow-up – baseline (Y-axis) represents the calculated differences in TMSC between the follow-up TMSCs following 90 days of administration of either the dry preparation of *A. galanga* and the *P. granatum* extract or the placebo and the corresponding TMSCs prior to administration (baseline). Horizontal bars indicate mean.

Table 3. Sperm morphology.

	Treatment			
	Active	Placebo		
	(n = 32)	(n = 34)	Difference	P-value*
	%±SD (95% CI)	%±SD (95% CI)		
Baseline	5.2±3.2 (4.0; 6.3)	4.9±3.0 (3.8; 5.9)	0.3 (−1.3; 1.9)	0.70
Follow-up	5.4±3.4 (4.2; 6.7)	5.0±2.8 (4.0; 5.9)	0.5 (−1.1; 2.0)	0.55
Follow-up − baseline	0.3±1.8 (−0.4; 0.9)	0.1±1.8 (−0.6; 0.7)	0.2 (−0.7; 1.1)	0.71

Morphologically normal sperm for the treatment groups receiving the combination of *P. granatum* fruit extract and *A. galanga* rhizome powder or placebo.
*The difference in number of morphologically normal sperm between the two groups was analysed by an unequal variance t-test.

Lastly, information on reactive oxygen species level and antioxidant capacity would have been highly relevant.

Comparison with other studies

Since this is, to our knowledge, the first study to address the effect of *P. granatum* and *A. galanga* on sperm quality in men, the magnitude of effect can only be compared with clinical human studies of other supplements. These studies mainly assess the effects of antioxidants.

The polyphenols in pomegranate exhibit significant antioxidant activity *in vitro* [10]. Mansour *et al.* administrated extract of *P. granatum* to rats and suggested that the observed decrease in the level of malondialdehyde, a by-product of lipid peroxidation, could be related to the accompanied decrease in abnormal sperm rate and increase in sperm motility, i.e. an antioxidant effect protecting the sperm cell membrane from lipid peroxidation [5]. Although numerous studies have addressed the question of whether oral intake of antioxidants is indeed beneficial for semen quality [22], several authors of recent studies and reviews conclude that more large well-designed studies are needed in this area. However, despite lack of consensus between studies and lack of evidence for improvement in the most important outcome parameter, pregnancy rate, the majority of studies find a positive effect of various antioxidants on semen quality, mainly measured by sperm concentration and motility [27–29]. Recently, an oral supplement of various antioxidants was shown to decrease the number of immotile sperm and to significantly increase the TMSC in men undergoing fertility treatment [30]. In another study, daily intake

of sesame was shown to increase sperm count and motility [31]. In both these studies, the interval of semen analysis from baseline to follow-up was at least two months, leaving no information on whether the supplements had an effect on epididymal sperm. Nevertheless, both studies referred to the antioxidant effect as the major reason for sperm quality improvement. In the study on the effects of sesame on sperm quality, sperm morphology was also assessed, but in accordance with our present findings there was no improvement in sperm morphology after three months of taking the sesame supplement [31].

Several studies have shown that oxidative stress in the seminal fluid, i.e., excess levels of reactive oxygen species (ROS) compared with the antioxidant levels, causes decreased sperm quality via two mechanisms: 1) Some ROS act as free radicals and cause damage to the sperm cell membrane thereby decreasing sperm motility and 2) the free radicals may cause damage to the sperm DNA [22]. Both *P. granatum* and *A. galanga* contain significant levels of antioxidants that may fully or partly explain the effect observed in this study on sperm quality. The antioxidant activity of *P. granatum* might be due to the abundant polyphenols present in the fruit as their chemical structure indicates that they can act as direct and indirect antioxidants. The bioavailability of polyphenols is, however, of major concern when talking about the direct antioxidant effect of these compounds because the concentration in the systemic system is often too low to have any significant direct antioxidant effect *in vivo*. Therefore, the preventive effects of polyphenols against, e.g., oxidative stress are probably related to their ability to activate the endogenous antioxidant defence system; hence acting as indirect antioxidants. Polyphenols containing *ortho*-phenol groups as those found in pomegranate fruit can be metabolised by phase 2 enzymes to electrophilic *ortho*-quinones, which is a prerequisite for activating the endogenous antioxidant defence system [32,33]. The antioxidant properties of the most abundant polyphenols in *P. granatum* are therefore expected to be due to their indirect antioxidant activity.

Hence, the fact that the TMSC did not increase after approximately one week of tablet intake in the present study does not prove against an antioxidant effect of the combination of *P. granatum* fruit extract and *A. galanga* rhizome powder. It rather suggests a possible effect of indirect antioxidants. In addition, *P. granatum* polyphenols may also exert modulatory effects in cells through selective actions on different components of the intracellular signalling cascades such as hormones and regulatory proteins that are vital for cellular functions such as growth and proliferation [34,35]. This may also explain the observed effects of the *P.*

Daily dose of ACA (mg)

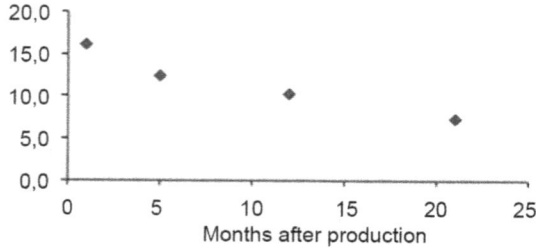

Figure 3. ACA content in tablets. Amount of 1′S-1′-acetoxychavicol acetate (ACA) in the *A. galanga* tablets within 21 months of production date.

granatum fruit extract and *A. galanga* rhizome powder combination.

As for the polyphenols in *P. granatum*, the major bioactive constituent of *A. galanga*, the 1'S-1'-acetoxychavicol acetate (ACA), also acts as an indirect antioxidant, although the chemical structure is unrelated to the *P. granatum* polyphenols. ACA is also a strong electrophile [36], which seems to explain many of the reported bioactivities of this natural product. ACA has been shown to enhance phase 2 enzyme activity and to regulate the glutathione metabolism by activating the endogenous antioxidant defence system [37–40] essential for its antitumor and chemopreventive effects [36,41–46].

Conclusion and future studies

This study indicates that intake of tablets containing preparations of extract of *P. granatum* fruit pomace and the rhizome of *A. galanga* (Punalpin) may help men in gaining an improved amount of motile ejaculated sperm.

Running chemical analyses of the tablets confirmed the presence of several potentially bioactive compounds, e.g. antioxidants, however at present we do not know, which of these compounds may be responsible for the increase in TMSC. Furthermore, it has yet to be investigated whether the combined effect of *P. granatum* and *A. galanga* is synergistic or additive. The mechanism by which the combination of *P. granatum* and *A. galanga* causes increase in the number of motile sperm should be investigated in future studies, where also pregnancy rate and time to pregnancy should be outcome measures. Such a study will expectedly justify the recommendation of the commercial product in clinical practice.

Supporting Information

Checklist S1   Supporting CONSORT checklist.

Protocol S1   Original detailed protocol for the study approved by the local Ethics Committee.

Background Data S1   The raw background data.

Acknowledgments

Part of this study was presented at the Annual meeting of the European Society of Human Reproduction in London July 7th–10th 2013.

The authors thank laboratory technicians Trine Matthäi from Laboratory of Reproductive Biology, Horsens Hospital, and Thomas Ebbesen from Nordic Cryobank, Aarhus, Denmark, for performing semen analyses. The study was funded by Centre for Science and Research of Alternative Treatment (VIFAB), who played no other role in the study.

Author Contributions

Conceived and designed the experiments: IG HBJ JF. Performed the experiments: JF MDKF LPC. Analyzed the data: ETP IG HBJ LPC JF MDKF. Contributed reagents/materials/analysis tools: IG HBJ. Wrote the paper: JF MDKF IG HBJ LPC ETP.

References

1. Carlsen E, Giwercman A, Keiding N, Skakkebaek NE (1992) Evidence for Decreasing Quality of Semen During Past 50 Years. BMJ 305: 609–613.
2. Swan SH, Elkin EP, Fenster L (1997) Have sperm densities declined? A reanalysis of global trend data. Environ Health Perspect. 105: 1228–1232.
3. Jørgensen N, Joensen UN, Jensen TK, Jensen MB, Almstrup K, et al. (2012) Human semen quality in the new millennium: a prospective cross-sectional population-based study of 4867 men. BMJ Open 2: e000990.
4. Türk G, Sönmez M, Aydin M, Yüce A, Gür S, et al. (2008) Effects of pomegranate juice consumption on sperm quality, spermatogenic cell density, antioxidant activity and testosterone level in male rats. Clin Nutr 27: 289–296.
5. Mansour SW, Sangi S, Harsha S, Khaleel MA, Ibrahim AR (2013) Sensibility of male rats fertility against olive oil, *Nigella sativa* oil and pomegranate extract. Asian Pac J Trop Biomed 3: 563–568.
6. Leiva P, Rubio J, Peralta F, Gonzales GF (2011) Effect of *Punica granatum* (pomegranate) on sperm production in male rats treated with lead acetate. Toxicol Mech Methods 21: 495–502.
7. Qureshi S, Shah AH, Ageel AM (1992) Toxicity studies on Alpinia galanga and Curcuma longa. Planta Med 58: 124–127.
8. Tzulker R, Glazer I, Bar-Ilan I, Holland D, Aviram M, et al. (2007) Antioxidant activity, polyhenol content and related compounds in different fruit juices and homogenates prepared from 29 different pomegranate accessions. J Agri Food Chem 55: 9559–9570.
9. Mayer W, Görner A, Andrä K (1977) Punicalagin und Punicalin, zwei Gerbstoffe aus den Schalen der Granatäpfel. Justus Liebigs Annalen der Chemie 11–12: 1976–1986.
10. Gil MI, Tomas-Barberán FA, Hess-Pierce B, Holcroft DM, Kader AA (2000) Antioxidant Activity of Pomegranate Juice and Its Relationship with Phenolic Composition and Processing. J Agric Food Chem 48: 4581–4589.
11. Kubota K, Ueda Y, Yasuda M, Masuda A (2001) Occurrence and antioxidative activity of 1'-acetoxychavicol acetate and its related compounds in the rhizomes of Alpinia galanga during cooking. Food Flavors and Chemistry - Advances of the New Millennium 274: 601–607.
12. Satish R, Dhananjayan R (2003) Evaluation of anti-inflammatory potential of rhizome of Alpinia galanga Linn. Biomedicine 23: 91–96.
13. Islam MW, Zakaria MNM, Radhakrishnan R, Liu XM, Ismail A, et al. (2000) Galangal (*Alpinia galanga* Willed.) and Black seeds (*Nigella sativa* Linn.) and sexual stimulation in male mice. J Pharm Pharmacol 52(Suppl.): 278–278.
14. Ayala C, Steinberger E, Smith DP (1996) The Influence of Semen Analysis Parameters on the Fertility Potential of Infertile Couples. J Androl 17: 718–725.
15. World Health Organization (2010) Laboratory manual for the examination and processing of human semen, 5th edition 2010. Cambridge University Press, Cambridge, UK.
16. Zhang Y, Wang D, Lee RP, Henning SM, Heber D (2009) Absence of Pomegranate Ellagitannins in the Majority of Commercial Pomegranate Extracts: Implications for Standardization and Quality Control. J Agric Food Chem 57: 7395–7400.
17. Ruxton GD (2006) The unequal variance t-test is an underused alternative to Student's t-test and the Mann–Whitney U test. Behav Ecol 17: 688–690.
18. James C, Bithell J (2000) "Bootstrap Confidence Intervals: When, Which, What? A Practical Guide for Medical Statisticians." Stat Med 19: 1141–1164.
19. Bonde JP, Ernst E, Jensen TK, Hjollund NH, Kolstad H, et al. (1998) Relation between semen quality and fertility: a population-based study of 430 first-pregnancy planners. Lancet 352: 1172–1177.
20. Turner TT (1995) On the Epididymis and its Role in the Development of the Fertile Ejaculate. J Androl 16: 292–298.
21. Fedder J (1996) Nonsperm cells in human semen: with special reference to seminal leukocytes and their possible influence on fertility. Arch Androl 36: 41–65.
22. Tremellen K (2008) Oxidative Stress and male infertility – a clinical perspective. Hum Reprod Update 14: 243–258.
23. Guo CJ, Wei JY, Yang J, Xu J, Pang W, et al. (2008) Pomegranate juice is potentially better than apple juice in improving antioxidant function in elderly subjects. Nutrition Research 28: 72–77.
24. Hajimahmoodi M, Oveisi MR, Sadeghi N, Jannat B, Nateghi M (2009) Antioxidant capacity of plasma after pomegranate intake in human volunteer. Acta Medica Iranica 47: 125–132.
25. Eslamian G, Amirjannati N, Rashidkhani B, Sadeghi MR, Hekmatdoost A (2012) Intake of food groups and idiopathic asthenozoospermia: a case-control study. Hum Reprod 27: 3328–3336.
26. Zareba P, Colaci DS, Afeiche M, Gaskins AJ, Jørgensen N, et al. (2013) Semen quality in relation to antioxidant intake in e healthy male population. Fertil Steril 100: 1572–1579.
27. Lombardo F, Sansone A, Romanelli F, Paoli D, Gandini L, et al. (2011) The role of antioxidant therapy in the treatment of male infertility: an overview. Asian J Androl 13: 690–696.
28. Zini A, Al-Hathal N (2011) Antioxidant therapy in male infertility: Fact or fiction? Asian J Androl 13: 374–381.
29. Gharagozloo P, Aitken RJ (2011) The role of sperm oxidative stress in male infertility and the significance of oral antioxidant therapy. Hum Reprod 26: 1628–1640.
30. Wirleitner B, Vanderzwalmen P, Stecher A, Spitzer D, Schuff M, et al. (2012) Dietary Supplementation of Antioxidants Improves Semen Quality of IVF Patients in Terms of Motility, Sperm Count, and Nuclear Vacuolization. Int J Vitam Nutr Res 82: 391–398.
31. Khani B, Bidgoli SR, Hassani H (2013) Effect of sesame on sperm quality of infertile men. J Res Med Sci 18: 184–187.

32. Dinkova-Kostova AT, Talalay P (2008) Direct and indirect antioxidant properties of inducers of cytoprotective proteins. Mol Nutr Food Res 52: S128–138.

33. Birringer M (2011) Hormetics: dietary triggers of an adaptive stress response. Pharm Res 28: 2680–2694.

34. Crozier A, Jaganath IB, Clifford MN (2009) Dietary phenolics: chemistry, bioavailability and effects on health. Nat Prod Rep 26: 1001–43.

35. Satpathy S, Patra A, Purohit AP (2013) Estrogenic activity of *Punica granatum* L. peel extract. Asian Pacific Journal of Reproduction 2: 19–24.

36. Miyauchi M, Nishikawa A, Furukawa F, Nakamura H, Son HY, et al. (2000) Inhibitory effects of 1'-acetoxychavicol acetate on N-nitrosobis(2-oxopropyl)-amine-induced initiation of cholangiocarcinogenesis in Syrian hamsters. Jpn J Cancer Res 91: 477–481.

37. Nakamura Y, Miyoshi N (2010) Electrophiles in foods: the current status of isothiocyanates and their chemical biology. Biosci Biotechnol Biochem 74: 242–255.

38. Yaku K, Matsui-Yuasa I, Azuma H, Kojima-Yuasa A (2011) 1'-Acetoxychavicol acetate enhances the phase II enzyme activities via the increase in intranuclear Nrf2 level and cytosolic p21 level. Amer J Chinese Med 39: 789–802.

39. Yaku K, Matsui-Yuasa I, Konishi Y, Kojima-Yuasa A (2013) AMPK synergizes with the combined treatment of 1'-acetoxychavicol acetate and sodium butyrate to upregulate phase II detoxifying enzyme activities. Mol Nutr Food Res 57: 1198–1208.

40. Higashida M, Xu S, Kojima-Yuasa A, Kennedy DO, Murakami A, et al. (2009) 1'-Acetoxychavicol acetate-induced cytotoxicity is accompanied by a rapid and drastic modulation of glutathione metabolism. Amino Acids 36: 107–113.

41. Ando S, Matsuda H, Morikawa T, Yoshikawa M (2005) 1'S-1'-Acetoxychavicol acetate as a new type inhibitor of interferon-b production in lipopolysaccharide-activated mouse peritoneal macrophages. Bioorg Med Chem 13: 3289–3294.

42. Ito K, Nakazato T, Ji Xian M, Yamada T, Hozumi N, et al. (2005) 1'-Acetoxychavicol acetate is a novel nuclear factor κB inhibitor with significant activity against multiple myeloma in vitro and in vivo. Cancer Res 65: 4417–4425.

43. Lee CC, Houghton P (2005) Cytotoxicity of plants from Malaysia and Thailand used traditionally to treat cancer. J Ethnopharmacol 100: 237–243.

44. Moffatt J, Hashimoto M, Kojima A, Kennedy DO, Murakami A, et al. (2000) Apoptosis induced by 1'-acetoxychavicol acetate in Ehrlich ascites tumor cells is associated with modulation of polyamine metabolism and caspase-3 activation. Carcinogenesis 21: 2151–2157.

45. Murakami A, Ohura S, Nakamura Y, Koshimizu K, Ohigashi H (1996) 1'-Acetoxychavicol acetate, a superoxide anion generation inhibitor, potently inhibits tumor promotion by 12-O-tetradecanoylphorbol-13-acetate in ICR mouse skin. Oncology 53: 386–391.

46. Ohnishi M, Tanaka T, Makita H, Kawamori T, Mori H, et al. (1996) Chemopreventive effect of a xanthine oxidase inhibitor, 1'-acetoxychavicol acetate, on rat oral carcinogenesis. Jpn J Cancer Res 87: 349–356.

22

Antioxidant and Antiproliferative Activities of Twenty-Four *Vitis vinifera* Grapes

Zhenchang Liang[1,2], Lailiang Cheng[3], Gan-Yuan Zhong[4]*, Rui Hai Liu[2]*

1 Institute of Botany, the Chinese Academy of Sciences, Beijing, China, **2** Department of Food Science, Cornell University, Ithaca, New York, United States of America, **3** Department of Horticulture, Cornell University, Ithaca, New York, United States of America, **4** USDA -ARS Grape Genetics Research Unit, Geneva, New York, United States of America

Abstract

Grapes are rich in phytochemicals with many proven health benefits. Phenolic profiles, antioxidant and antiproliferative activities of twenty-four selected *Vitis vinifera* grape cultivars were investigated in this study. Large ranges of variation were found in these cultivars for the contents of total phenolics (95.3 to 686.5 mg/100 g) and flavonoids (94.7 to 1055 mg/100 g) and antioxidant activities (oxygen radical absorbance capacity 378.7 to 3386.0 mg of Trolox equivalents/100 g and peroxylradical scavenging capacity14.2 to 557 mg of vitamin C equivalents/100 g), cellular antioxidant activities (3.9 to 139.9 μmol of quercetin equivalents/100 g without PBS wash and 1.4 to 95.8 μmol of quercetin equivalents /100 g with PBS wash) and antiproliferative activities (25 to 82% at the concentrations of 100 mg/mL extracts).The total antioxidant activities were significantly correlated with the total phenolics and flavonoids. However, no significant correlations were found between antiproliferative activities and total phenolics or total flavonoids content. Wine grapes and color grapes showed much higher levels of phytochemicals and antioxidant activities than table grapes and green/yellow grapes. Several germplasm accessions with much high contents of phenolics and flavonoids, and total antioxidant activity were identified. These germplasm can be valuable sources of genes for breeding grape cultivars with better nutritional qualities of wine and table grapes in the future.

Editor: Gianfranco Pintus, University of Sassari, Italy

Funding: The authors are grateful to Mr. Bernard Prins of the USDA-ARS National Clonal Germplasm Repository for Fruit and Nut Crops in Davis, California and Dr. Christopher Owens of the USDA-ARS Grape Genetics Research Unit in Geneva, New York for providing V. vinifera berry samples. This work was supported by a Special Cooperative Agreement between USDA-ARS and Cornell University. USDA-ARS is an equal opportunity provider and employer. This project was supported by the United States Department of Agriculture-Agricultural Research Service Project 1910-21220-004-00D through a Specific Cooperative Agreement with Cornell University. The funders had no role in study design, data collection and analysis, decision to publish, or preparation of the manuscript.

Competing Interests: The authors have declared that no competing interests exist.

* Email: rl23@cornell.edu (RHL); ganyuan.zhong@ars.usda.gov (GYZ)

Introduction

Fruits, as an important part of diet, are beneficial for human health.The Dietary Guidelines for Americans provide three reasons to eat more fruits and vegetables [1]. First, most fruits and vegetables are major contributors of a number of nutrients that are under-consumed in the United States, including folate, magnesium, potassium, dietary fiber, and vitamins A and C. Second, consumption of fruits and vegetables is associated with reduced risk of many chronic diseases. Specifically, moderate evidence indicates that intake of at least 2.5 cups of vegetables and fruits perday is associated with a reduced risk of cardiovascular disease, including heart attack and stroke. Some fruits and vegetables may be protective against certain types of cancer. Third, most fruits and vegetables, when prepared without added fats or sugars, are relatively low in calories. Consumption of fruits and vegetables instead of higher calorie foods can help adults and children achieve and maintain healthy weight. Currently, very few Americans consume diets that meet Dietary Guideline recommendations of fruits and vegetables. In fact, most Americans only consume about 42% of goal intake for fruits. For improving public health and reducing the risk of developing chronic diseases, the Dietary Guidelines for Americans recommend that most people should eat at least 9 servings of fruits and vegetables per day based on a 2000 kcal diet [1].

Grapes (*Vitis sp.*) are one of the most important fruit crops worldwide. They are consumed as fresh fruits as well as wine, juice and other processed products. There are about 60 grape species in the genus of *Vitis*, and the species *V. vinifera*, or European grapes, is most widely cultivated. There are thousands of *V. vinifera* cultivars extant in the world. To obtain an assessment of the full range of phytochemicals in grapes, we recently analyzed 36 phenolic compounds in the berry samples of 344 representative *V. vinifera* cultivars [2] and supported previous findings that grapes are rich in phytochemicals [3,4]. Demonstration of health benefits of these phytochemicals in grapes has been actively pursued. It has been reported that grape extracts exhibited antioxidant activities, including scavenging of free radicals, inhibition of lipid oxidation, and reduction of hydroperoxide formation [4,5], and inhibit cardiovascular diseases and certain types of cancers, reducing plasma oxidative stress, and slowing aging [6,7,8].While there were many reports on the assessment of phytochemical compounds and their antioxidant activities of grapes, these studies were limited to only a few cultivars [9–15].

In this study, we determined the phytochemical profiles of 24 *V. vinifera* grape cultivars and analyzed their total antioxidant and antiproliferative activities. These 24 cultivars were selected on the basis of a previous study in which we demonstrated that these cultivars contained a wide range of the composition and content of phenolic compounds [2]. The contribution of flavonoids to total phenolics and correlative relationships among phenolics, flavonoids, antioxidant activities and antiproliferative activities were also investigated.

Materials and Methods

Chemicals and reagents

Methanol (MeOH), ethanol (EtOH), acetone, hexane, ethyl acetate, hydrochloric acid (HCl), acetic acid (HAC), potassium chloride (KCl), sodium acetate (NaAC), sodium carbonate (NaCO3), sodium hydroxide (NaOH), potassium phosphate monobasic (KH_2PO4), and potassium phosphate dibasic (K_2HPO4) were of analytical grade and purchased from Mallinckrodt Chemicals (Phillipsburg, NJ). Ascorbic acid, 2′, 7′-dichlorofluorescin diaacetate (DCFH-DA), fluorescein disodium salt, sodium borohydride ($NaBH_4$, reagent grade), chloranil (analytical grade), vanillin (analytical grade), quercetin dehydrated, catechin hydrated, Folin-Ciocalteu reagent, 6-hydroxy-2, 5, 7, 8-tetramethylchroman-2-carboxylic acid (Trolox), trifluoroacetic acid (TFA, chromatographic grade), and acetonitrile (chromatographic grade) were purchased from Sigma-Aldrich, Inc. (St. Louis, MO). Tetrahydrofuran (THF, analytical grade), dimethyl sulfoxideand aluminum chloride ($AlCl_3 \cdot 6H_2O$, analytical grade) were purchased from Fisher Scientific (Fair Lawn, NJ). Gallic acid was obtained from ICN Biomedicals, Inc. (Aurora, OH). 2, 2′-Azobis (2-amidinopropane) dihydrochloride (ABAP) was purchased from Wako Chemicals USA, Inc. (Richmond, VA). HepG2 liver cancer cells were obtained from the American Type Culture Collection (ATCC) (Rockville, MD). Williams' Medium E (WME) and Hanks' Balanced Salt Solution (HBSS) were purchased from Gibco Life Technologies (Grand Island, NY). Fetal bovine serum (FBS) was obtained from Atlanta Biologicals (Lawrenceville, GA).

Plant materials

Berry samples of 24 *V. vinifera* grape cultivars were collected from the United States Department of Agriculture-Agricultural Research Service (USDA-ARS) *Vitis* clonal repository in Davis, California (Table 1). The grape berries were harvested upon ripening from the 2010 vintages. All the vines received standard fertilization, irrigation, pruning, and insect and disease control. About 100 of representative berries were collected from each individual cultivar. Berry color was scored as green-yellow, pink, red, dark-red violet, blue black or red-black and compared with those recorded in the Germplasm Resources Information Network [16] when a cultivar was introduced. The number of berries was counted and the berry weight was recorded for each sample before being frozen and stored at −80°C for further processing. The frozen berries were then crushed using a mortar and pestle. After removing all the seeds, flesh and peel tissues were ground in an IKA A11 mill (IKA Works, Inc, NC, USA) into powder while frozen using liquid nitrogen. All data collected for each grape cultivar were reported as mean ± SD with at least three replications.

Extraction of total phenolic compounds

Total phenolics were extracted from ground berry samples using the method previously reported from our laboratory [17]. Briefly, 10 g of grape powder was homogenized with 30 mL of 80% chilled acetone/water using a Virtis High Speed Homogenizer for 3 min on the ice. After centrifugation at 3000 g for10 min at 4°C, the supernatant was collected, and 30 mL of fresh extraction solution were added and the extraction was repeated twice. All extracts were transferred into a round bottom flask, and were evaporated using a rotary evaporator at 45°C until the weight of the evaporated filtrate was less than10% of the weight of the original filtrate. The extracts were brought to a final volume of 10 mL with 70% methanol. All extracts were stored at −40°C until use. All extractions were analyzed in triplicates.

Determination of total phenolic contents

Total phenolic content was determined using the Folin-Ciocalteu colorimetric method [18], which was modified by our laboratory [19]. Briefly, all extracts were diluted 1:20 with distilled water to obtain readings within the standard curve ranges of 0.0–600.0 μg of gallic acid/mL. The grape extracts were oxidized by the Folin-Ciocalteu reagent and the reaction was acted in sodium carbonate. The absorbance was measured at 760 nm, after 90 min at room temperature by a MRX II Dynex plate reader (Dynex Technologies Inc., Chanilly, VA). The absorbance values were then compared with those of standards with known gallic acid concentrations. All values were expressed as the mean (milligrams of gallic acid equivalents per 100 g of fresh sample) ± SD of three replications.

Determination of total flavonoid contents

The total flavonoid content was determined using the sodium borohydride/chloranil-based (SBC) assay developed by our laboratory [20] with modifications [21]. Briefly, 0.2 mL extracts of tested samples were added into test tubes (15×150 mm), dried under nitrogen gas, and reconstituted in 1 mL of terahydrofuran/ethanol (THF/EtOH, 1:1, v/v). Catechin standards (0.1–10.0 mM) were prepared fresh before use in 1 mL of THF/EtOH (1:1, v/v). Then 1 mL of 50 mM $NaBH_4$ solution and 0.5 mL of 74.6 mM $AlCl_3$ solution were added into each test tube with 1 mL of sample solution or 1 mL of catechin standard solution. The test tubes were shaken in an orbital shaker at room temperature for 30 min. An additional 0.5 mL of 50.0 mM $NaBH_4$ solution was added into each test tube with shaking continued for another 30 min at room temperature. Two millilitersof cold 0.8 M acetic acid solution was added into each test tube, and the solution was kept in the dark for 15 min after a thorough mix. One milliliter of 20.0 mM chloranil was added into each tube, which was heated at 95 °C with shaking for 60 min. The reaction solution was cooled using tap water, and the final volume was brought to 4 mL using methanol. One milliliter of 1052 mM vanillin was added into each tube and mixed. Then 2 mL of 12 M HCl was added to each tube, and the reaction solution was kept in the dark for 15 min after a thorough mix. Aliquots of final reaction solutions (200 μL) were added into each well of a 96-well plate, and the absorbance was measured at 490 nm using a MRX Microplate Reader with Revelation workstation (Dynex Technologies, Inc., Chantilly, VA). Total flavonoid content was expressed as milligrams of catechin equivalents per 100 g of fresh weight of sample. Data were reported as mean ± SD with at least triplicates.

Oxygen radical absorbance capacity (ORAC) assay

The antioxidant activity was determined using oxygen radical absorbance capacity (ORAC) assay [22] and modified in our laboratory [23]. Phenolic extract dilutions were prepared with 75 mM phosphate buffer (pH 7.4). The assay was performed in black-walled 96-well plates (Corning Scientific, Corning, NY). The outside wells of the plate were not used as there was much more

Table 1. Description of *V. vinifera* grape samples used in the study and total phenolics (µmol of Gallic acid equiv /100 g FW) and flavonoids (µmol of Catechinequiv /100 g FW) content, ratio of total flavonoids to total phenolics (F/P, %), ORAC (mg of Trolox equivalents/100 g FW) and PSC (mg of vitamin C equivalents/100 g FW) values, CAA with PBS washing and no PBS washing (µmol of QE/100 g of grape) values.

Accession	Cultivar	Primary use	Color	Phenolics	Flavonoids	F/P	ORAC	PSC	CAA with wash	CAA with no wash
DVIT2630	Agwam	Wine grape	Red	102.7±3.0	128.6±8.6	73.5±5.2	547.3±50.5	73.5±7.5	1.4±0.2	5.3±0.3
DVIT0632	Araclinos	Wine grape	Blue-black	347.9±7.1	408.1±11.2	68.9±2.7	1171.3±58.3	232.1±19.9	24.6±1.0	33.9±2.1
DVIT0677	Cabernet Sauvignon	Wine grape	Red-black	448.3±5.2	603.1±34.5	79.0±5.1	1394.7±54.3	226.0±18.0	22.0±2.6	43.6±5.3
DVIT0696	Corbeau	Wine grape	Red-black	582.8±11.7	687.1±43.5	69.0±3.5	2320.0±129.0	420.5±30.8	30.7±1.2	46.1±3.3
DVIT0384	Coudsi	Table grape	Green-yellow	159.5±4.8	201.7±14.4	74.0±3.2	710.3±63.6	31.2±1.4	9.8±1.2	14.7±1.5
DVIT0314	Demir Kara	Table grape	Red-black	345.0±13.2	341.0±13.3	58.0±1.0	1400.7±130.0	167.3±6.6	25.5±2.5	59.7±3.0
DVIT2705	Dornfelder	Table grape	Blue-black	391.6±9.5	530.0±36.4	79.3±4.7	1479.7±149.0	404.5±18.1	49.6±5.7	54.6±3.2
DVIT2642	Madam Matijas	Table grape	Green-yellow	201.5±5.4	146.4±7.8	42.6±2.4	566.0±67.0	73.0±6.5	10.2±0.9	12.2±0.1
DVIT0099	Mathilde	Table grape	Red	201.9±9.1	170.8±11.3	49.6±5.1	630.7±14.0	109.3±10.8	3.4±0.3	8.0±1.0
DVIT0825	Melon	Wine grape	Green-yellow	466.8±9.9	587.5±16.0	73.1±2.9	1651.7±138.0	103.8±10.9	11.9±0.8	42.5±0.6
DVIT1042	Mermark	Wine grape	Red-black	664.4±18.2	568.0±21.7	50.1±0.8	2422.7±221.0	244.0±15.6	14.1±1.3	47.5±1.5
DVIT0465	Muscat of Alexandria	Table grape	Green-yellow	95.3±1.8	94.7±3.0	58.3±1.2	392.0±34.9	14.2±0.6	1.8±0.2	3.9±0.1
DVIT2646	Pearl of Zola	Table grape	Green-yellow	291.2±1.5	230.0±4.1	46.3±0.7	733.0±65.2	58.0±3.5	17.0±1.2	26.4±0.9
DVIT2700	Pirobelle	Table grape	Blue-black	207.8±10.5	210.6±15.2	59.4±2.5	571.0±37.7	117.6±3.7	21.2±0.3	23.1±0.7
DVIT1119	Plavac Mali	Wine grape	Blue-black	351.1±11.0	502.6±31.8	84.2±6.5	1186.0±122.0	138.2±8.8	33.0±1.8	34.3±1.5
DVIT0503	Ribier	Table grape	Blue-black	294.3±7.9	251.9±10.1	50.2±2.6	909.3±93.5	46.9±6.0	19.9±2.5	21.5±2.6
DVIT0934	Robola	Wine grape	Green-yellow	384.4±13.6	425.2±27.4	65.1±5.2	1298.3±135.0	91.5±7.3	14.5±0.5	47.4±0.2
DVIT0937	Royalty	Wine grape	Red-black	686.5±1.9	1054.6±78.1	90.0±6.8	3386.3±221.0	557.0±50.0	95.8±5.0	139.9±9.2
DVIT0940	Ruby Cabernet	Wine grape	Red-black	486.7±13.2	758.2±48.1	91.3±3.9	1752.3±174.0	183.7±16.3	73.3±7.2	79.8±4.7
DVIT1060	Saint Macaire	Wine grape	Red-black	368.2±17.6	301.2±19.1	47.9±5.1	1402.7±158.0	285.0±26.6	16.9±1.0	50.0±2.9
DVIT0944	Salvador	Wine grape	Red-black	407.5±8.9	485.6±30.8	69.8±3.4	2504.7±271.0	376.2±36.7	81.1±7.7	91.0±8.8
DVIT0535	Thompson Seedless	Table grape	Green-yellow	151.1±6.6	201.4±7.3	78.3±3.2	378.7±24.8	24.1±2.2	7.0±0.6	10.5±0.9
DVIT1061	Touriga	Wine grape	Red-black	482.5±20.8	331.5±19.8	40.3±3.6	1960.3±85.6	302.0±11.0	21.1±0.1	54.4±2.3
DVIT0605	Yourutico	Table grape	Blue-black	419.1±6.0	534.6±29.3	74.7±3.8	1365.0±74.7	134.9±7.3	11.4±0.9	45.9±3.6

variation from them than from the inner wells. Each well contained 20 μL extracts or 20 μLTrolox standard (range 6.25–50 μM), and 200 μL fluoroscein (final concentration 0.96 μM), which were incubated at 37°C for 20 min. After incubation, 20 μL of 119 mM ABAP was added to each well. Fluorescence intensity was measured using Fluoroskan Ascent FL plate-reader (Thermo Labsystems, Franklin, MA) at excitation of 485 nm and emission of 520 nm for 35 cycles every 5 min. ORAC values were expressed as milligrams of TE/100 g FW. Data were reported as mean ± SD with triplicates.

Peroxylradicalscavenging capacity (PSC) assay

The antioxidant activity of extracts was determined by using a PSC assay developed in our lab [24] with modifications [25]. Just prior to use in the reaction, 107 μL of 2.48 mM dichlorofluorescein diacetate was hydrolyzed to dichlorofluorescein with 893 μL of 1.0 mM KOH for 5 min in a vial to remove the diacetate moiety and then diluted with 7 mL of 75 mM phosphate buffer (pH 7.4). ABAP (200 mM) was prepared fresh in the buffer and was kept at 4°C between runs. In an assay, 100 μL of extracts was diluted in 75 mM phosphate buffer (pH 7.4) and then transferred into reaction cells on a 96-well plate, and 100 μL of dichlorofluorescein was added. The 96-well plate was loaded into a Fluoroskan Ascent fluorescence spectrophotometer (Thermo Labsystems, Franklin, MA), and the solution in each cell was mixed by shaking at 1200 rpm for 20 s. The reaction was then initiated by adding 50 μL of ABAP from the autodispenser on the equipment. Each set of dilutions for a replicate and control was analyzed three times in adjacent columns. The reaction was carried out at 37°C, and fluorescence was monitored at 485 nm excitation and 538 nm emission with the fluorescence spectrophotometer. The buffer was used for control reactions. Data were acquired with Ascent software, version 2.6 (Thermo Labsystems). Data were reported as mean ± SD with three replicates.

Cell Culture

HepG2 cells (The American Type Culture Collection, ATCC, Rockville,MD) were grown in growth medium (WME supplemented with 5% FBS, 10 mMHepes, 2 mM L-glutamine, 5 μg/mL insulin, 0.05 μg/mL hydrocortisone, 50 units/mL penicillin, 50 μg/mL streptomycin, and 100 μg/mL gentamicin) and were maintained at 37°C and 5% CO_2 as described previously [26,27]. Cells used in this study were between passages or generations18 and 28 at which the cells were at the most stable physiological status.

Cytotoxicity

Cytotoxicity was measured using the methylene blue assay developed by our laboratory (Felice et al., 2009) with modifications [28]. Briefly, HepG2 cells were seeded at 4×10^4/well on a 96-well plate in 100 μL of growth medium and incubated for 24 h at 37°C. The medium was removed, and the cells were washed with PBS. Treatments of fruit extracts or antioxidant compounds in 100 μL of treatment medium (WME supplemented with 2 mM L-glutamine and 10 mM Hepes) were applied to the cells, and the plates were incubated at 37°C for 24 h. The treatment medium was removed, and the cells were washed with PBS. A volume of 50 μL/well methylene blue staining solution (98% HBSS, 0.67% glutaraldehyde, 0.6% methylene blue) was applied to each well, and the plate was incubated at 37°C for 1 h. The dye was removed, and the plate was immersed in fresh deionized water six times, or until the water became clear. The water was tapped out of the wells, and the plate was allowed to air-dry briefly before 100 μL of elution solution (49% PBS, 50% ethanol, 1% acetic

acid) was added to each well. The microplate was placed on a bench-top shaker for 20 min to allow uniform elution. The absorbance was read at 570 nm with blank subtraction using the MRX II DYNEX spectrophotometer (DynexInc., Chantilly, VA). Concentrations of fruit extracts that decreased the absorbance by >10% when compared to the control were considered to be cytotoxic (Felice et al., 2009).

Cellular Antioxidant Activity (CAA) of grape extracts

The CAA was determined using the method previously developed in our laboratory [29]. Briefly, HepG2 cells were seeded at a density of 6×10^4/well on a 96-well microplate in 100 μL of growth medium/well. Twenty-four hours after seeding, the growth medium was removed, and the wells were washed with PBS. Wells were then treated in triplicate for 1 h with 100 μL of treatment medium containing tested fruit extracts plus 25 μM DCFH-DA. When a PBS wash was utilized, wells were washed with 100 μL of PBS. When no PBS wash was done between treatments, the activity may have been higher, but the CV also tended to be higher. This was likely due to the interaction of the samples and oxidants with other factors in the residual medium on the cells. Washing the cells with PBS removed most of the interfering medium components and increased the consistency of the results. Then 600 μM ABAP was applied to the cells in 100 μL of HBSS, and the 96- well microplate was placed into a Fluoroskan Ascent FL plate reader (ThermoLabsystems, Franklin, MA) at 37°C. Emission at 538 nm was measured after excitation at 485 nm every 5 min for 1 h.

After blank subtraction and subtraction of initial fluorescence values, the area under the curve for fluorescence versus time was integrated to calculate the CAA value at each concentration of fruit as [30]

$$CAA(unit) = 1 - (\int SA / \int CA)$$

Where $\int SA$ is the integrated area under the sample fluorescence versus time curve and $\int CA$ is the integrated area from the control curve. The median effective dose (EC_{50}) was determined for the fruits from the median effect plot of $\log(fa/fu)$ versus $\log(dose)$, where fa is the fraction affected (CAA unit) and fu is the fraction unaffected (1 - CAA unit) by the treatment. The EC_{50} values were stated as mean ± SD for triplicate sets of data obtained from the same experiment. EC_{50} values were converted to CAA values, expressed as micromoles of quercetin equivalents (QE) per 100 g of grape, using the mean EC_{50} value for quercetin from five separate experiments.

Measurement of inhibition of HepG2 cell proliferation

The antiproliferative activity of different grape extracts was assessed by measurement of the inhibition of HepG2 (The American Type Culture Collection, ATCC, Rockville, MD) human cancer cell proliferation. Antiproliferative activities were determined by the colorimetric methylene blue method developed by our laboratory [31] with modifications [28,32]. Human hepatocellular carcinoma HepG2 cells were seeded at a density of 2.5×10^4/well on a 96-well microplate in 100 μL of growth medium/well. The outside wells of the plate were not used as there was much more variation from them than those from the inner wells. After 4 h of incubation, the growth medium was removed and media containing various concentrations (10, 20, 40, 60, 80 and 100 mg/mL) of grape extracts were added to the cells. Control cultures received the extraction solution minus the grape extracts, and blank wells contained 100 μL of growth medium without cells. Cell proliferation (percent) was determined at 96 h from the absorbance reading at 570 nm for each concentration when compared to the control, using at least three replications for each sample. The effective median dose (EC_{25} and EC_{50}) was determined and expressed as milligrams of grape extracts per milliliter ± SD.

Statistical analysis

Data from this study were reported as mean ± SD with at least three replicates for each sample extract. All graphical representations were performed using Sigmaplot 10.0 for Windows (SPSS, USA), and all data were analyzed using SPSS (Statistics for Social Science) 13.0 for Windows. Results were subjected to ANOVA, and differences between means were located using Tukey B multiple comparison test. Correlations between various parameters were also investigated. Significance was determined at $P < 0.05$.

Results

Total phenolic content

Table 1 shows the total phenolic contents of 24 V. vinifera grape cultivars. Among all the grape cultivars analyzed, Royalty (red flesh cultivar) and Mermark had the highest total phenolic content (686.5±3.2 and 664.4±31.5 mg of gallic acid equivalents/100 g FW, respectively), and were significantly higher than other grape cultivars ($P < 0.05$). Agwam and Muscat of Alexandria had the lowest total phenolic contents ($P < 0.05$). There were 7-fold differences in the total phenolic contents between the highest (Royalty) and lowest (Muscat of Alexandria) ranked cultivars. Most green-yellow grape cultivars had lower total phenolic contents with no more than 300 mg of gallic acid equivalents/100 g FW except for Melon and Robola. On the contrary, colored grape cultivars had higher total phenolic contents and were more than 300 mg of gallic acid equivalents/100 g FW except for Ribier and two red cultivars of Mathilde and Agwam. All wine grape cultivars had higher total phenolic contents with more than 347.9 mg of gallic acid equivalents/100 g FW, but Agwam was an exception. All table grape cultivars had low phenolic contents except for Yourutico and Dornfelder.

Total flavonoid content

Total flavonoid contents of the 24 grape cultivars are presented in Table 1. The Royalty grape presented the highest flavonoid content and was significantly higher than other cultivars ($P < 0.05$). Compared to most other cultivars, Pearl of Zola, Pirobelle, Coudsi, Thompson Seedless, Mathilde, Madam Matijas, and Agwam had relatively lower levels of total flavonoid contents. Muscat of Alexandria had the lowest flavonoid content. There were almost 11-fold differences in total flavonoid contents between the highest and lowest ranked cultivars. Melon and Robola had higher total flavonoid contents than other green-yellow cultivars in which no significant differences were detected. In general, colored cultivars had higher total flavonoid contents than green-yellow cultivars ($P < 0.05$). Compared to wine grape cultivars, all table grape cultivars had lower flavonoid contents (less than 250 mg of catechin equivalents/100 g FW) except for Yourutico, Dornfelder and Demir Kara. Total flavonoid contents of wine grape cultivars were higher than 300 mg of catechin equivalents/100 g FW except for Agwam.

Contribution of total flavonoids to total phenolics

The contributions of total flavonoids to total phenolics, calculated on the basis of micromoles of the total flavonoids and total phenolics, are presented in Table 1. The contributions of total flavonoids to the total phenolics in the 24 grape cultivars ranged from 40.3 to 91.3% with a mean 65.5%. The contributions of flavonoids to total phenolics in Ruby Cabernet and Royalty (91.3 and 90% of the total phenolics, respectively) were the highest among all cultivars. In contrast, the total flavonoid contents were no more than 50% of the total phenolic contents in Mathilde, Saint Macaire, Pearl of Zola, Madam Matijas and Touriga.

Oxygen radical absorbance capacity (ORAC)

The ORAC values for the 24 cultivars, expressed as mg of Trolox equivalents per 100 g of fresh grape, are shown in Table 1. Variation of ORAC values ranged from 378.7 (Thompson Seedless) to 3386.3 (Royalty) mg of Trolox equivalents/100 g FW, and the mean was 1339 mg. Royalty had the highest ORAC values, significantly higher than other cultivars ($P < 0.05$). The other red flesh cultivars Salvador, Mermark and Corbeau also had relatively high ORAC values ($P < 0.05$). Most green yellow cultivars had ORAC value no more than 733.0 mg of TE/100 g FW except for Melon and Robola. Most colored cultivars had ORAC values more than 909.0 mg of TE/100 g FW except for one blue-black cultivar (Pirobelle) and two red cultivars (Mathilde and Agwam). All table grape cultivars had lower ORAC values (lower than 1000 mg of TE/100 g FW) except for Dornfelder, Demir Kara and Yourutico. All ORAC values of wine grape cultivars were higher than 1170.0 mg of TE/100 g FW except for Agwam.

Peroxyl radical scavenging capacity (PSC)

Total antioxidant activities measured by peroxyl radical scavenging capacity (PSC) for the 24 cultivars are presented in Table 1 with the PSC values expressed as mg of vitamin C equivalents/100 g FW. Royalty had the highest PSC values, and was significantly higher than other cultivars ($P < 0.05$). Corbeau, Dornfelder and the other red flesh cultivar Salvador also had high PSC values ($P < 0.05$). PSC values of all green yellow cultivars were lower than 100 mg of vitamin C equivalents/100 g FW except for Melon. In addition, most PSC values of colored cultivars were higher than 109 mg of vitamin C equivalents/100 g FW except for a red cultivar (Agwam) and a blue black cultivar (Ribier). Dornfelder had the highest PSC value in all table grapes, followed by Demir Kara, Yourutico, Pirobelle and Mathilde. The other PSC values of table grapes were no more than 73 mg of vitamin C equivalents/100 g FW. Most wine grapes had high PSC values except for Agwam, Robola and Melon.

Cellular antioxidant activities

CAA values for 24 grape cultivars using the no PBS wash protocol are shown in Table 1. CAA values ranged from 3.9 to 139.9 μmol of QE/100 g of grape, and the mean was 41.5 μmol of QE/100 g of grape. Royalty had the highest CAA value ($P < 0.05$), followed by Salvador and Ruby Cabernet ($P < 0.05$). Demir Kara, Dornfelder, Touriga, Saint Macaire, Mermark, Robola, Corbeau and Yourutico, Cabernet Sauvignon, Melon had intermediate high CAA activities, but were not significantly different from each other ($P > 0.05$). Then the rankings of CAA activities were followed by Plavac Mali, Araclinos, Pearl of Zola, Pirobelle, Ribier and Coudsi. Madam Matijas, Thompson Seedless, Mathilde, Agwam and Muscat Alexandria had the lowest CAA values among the 24 grape cultivars tested.

In using the PBS wash protocol, CAA values ranged from 1.4 to 95.8 μmol of QE/100 g of grape, and the mean was 25.7 μmol of QE/100 g of grape (Table 2). Royalty had the greatest cellular antioxidant activity among the cultivars tested ($P<0.05$). Salvador ranked second, and Ruby Cabernet and Dornfelder were third and fourth in CAA activities, respectively ($P<0.05$). Plavac Mali and Corbeau and other cultivars were in descending order in their CAA activities. Muscat of Alexandria and Agwam had the lowest CAA values. In both PBS wash and no PBS wash protocols, the CAA values were significantly higher in wine grapes than in table grapes. This was also true in colored grapes when compared with green-yellow grapes ($P<0.05$, Table 2).

Correlationsamong total phenolics, total flavonoids and antioxidant activities

Correlations among the total phenolics, total flavonoids, ORAC values and PSC values were analyzed (Table 3). Significantly positive correlations were found between total phenolics and total flavonoids ($r^2 = 0.888$), and between ORAC and PSC values ($r^2 = 0.863$). The ORAC values had significantly positive correlations with total phenolics ($r^2 = 0.924$) and total flavonoids ($r^2 = 0.862$). In comparison, the PSC values also had significantly positive correlations with total phenolics ($r^2 = 0.757$) and total flavonoids ($r^2 = 0.737$), but the correlation coefficients were lower than those observed for the ORAC values.

The relationships of total phenolics content, total flavonoids content, ORAC values, and PSC values with CAA values were also determined. CAA values were significantly positively correlated with total phenolics ($r^2 = 0.574$), total flavonoids ($r^2 = 0.741$), ORAC values ($r^2 = 0.736$) and PSC values ($r^2 = 0.739$) when no PBS wash protocol was used ($P<0.05$). While significantly positive correlation relationships were also found between CAA values and total phenolics, total flavonoids, ORAC values and PSC values using the PBS wash protocol ($P<0.05$),the coefficients of correlation were lower than those with no PBS wash.

Inhibition of human cancer cell proliferation

The effects of extracts from the 24 grape cultivars against the growth of HepG2 human liver cancer cells *in vitro* are presented in Figure 1. Overall, grape extracts had potent antiproliferative activities against HepG2 human liver cancer cell growth in dose-dependent manners. The inhibition of cancer cell proliferation by the 24 grape extracts ranged from 25 to 82% at the concentration of 100 mg/mL extract applied (Figure 1). Robola, Plavac Mali, Royalty, Saint Macaire, Touriga, Coudsi, Ruby Cabernet,and Demir Kara exhibited relatively higher antiproliferative activities towards HepG$_2$ cells (Figure 1and Table 4). The EC$_{25}$ and EC$_{50}$of the antiproliferative activities of different grape cultivars are presented in Table 4. Lower EC$_{25}$ and EC$_{50}$values represent

higher antiproliferative activities. There were no significant differences in antiproliferative activities between wine and table grapes, but in general, the antiproliferative activities of red grapes were higher than those of yellow-green grapes. There were no detected cytotoxicities towards HepG2 cells *in vitro* at the concentration of 150 mg/mL of grape extracts in all 24 grape cultivars tested, indicating that antiproliferative activities were not caused by cytoxicity [32].

Discussion

Regular consumption of fruits and vegetables has been linked to reduced risk of developing chronic diseases [33,34]. Grapes and their products are rich in phytochemicals, which are proven to possess many health benefits. Grape phytochemicals include anthocyanins, flavanols, flavonols, stilbenes (resveratrol) and phenolic acids and their contents and profiles can vary significantly among different *V. vinifera* cultivars [2,35].There have been studies attempting to link phytochemical profiles with their total antioxidant activities in *V. vinifera*, but these studies were limited to just a few cultivars [4]. Here we reported phytochemical profiles of 24 *V. vinifera* cultivars and studied the relationships among total phenolics, total flavonoids, total antioxidant activities, cellular antioxidant activities, and antiproliferative activities in these cultivars. These 24 *V. vinifera* cultivars were a part of the core collection of *V. vinifera* germplasm preserved at the USDA-ARS *Vitis* clonal repository in Davis, California and they were selected for this study on the basis of their compositions and contents of polyphenolic compounds reported in a previous study [2].

The present study, using a different, more accurate detection method for the total phenolic compounds, confirmed our previous result that the 24 cultivars selected for this study had a wide range of variation in their total phenolic contents. The mean total phenolic content ranged from 95.3 (Muscat of Alexandria) to 686.5 (Royalty) mg of gallic acid equivalents/100 g FW. In contrast, Yang et al. [4] reported that the total phenolic contents of 14 wine grapes ranged from 201.1 to 424.6 mg of gallic acid equivalent/100 g of fruits. The difference in these observations, to a large extent, was certainly attributed to different cultivars used in the studies. It should be pointed out that there could be some other factors which may contribute to these differences. For example, samples collected from different vineyards or vintages could have significant influence on the total phenolic content as reported by Arnous et al [36] and Kallithraka et al. [37]. The wide range of variation of total phenolic contents among the grape cultivars in this study offered us a unique opportunity to investigate the correlative relationships of phenolic compounds with some health

Table 2. Mean contents of total phenolics, total flavonoids, ORAC and PSC values among different types of grapes (mean±SD).

	Primary use		Color type	
	Table grape	Wine grape	Colored grape	Green yellow grape
Total phenolics (GAE/100 g)	250.8±31.6[b]*	444.6±42.0[a]	399.3±37.7[a]	250.0±51.4[b]
Total flavonoids (CE/100 g)	264.8±44.0[b]	526.3±64.2[a]	462.8±57.4[a]	269.6±65.8[b]
ORAC values (TE/100 g)	830.6±122.0[b]	1769.0 ±206.1[a]	1553.0 ±186.4[a]	818.6±181.4[b]
PSC values (VE/100 g)	107.4±33.2[b]	248.7±39.1[a]	236.4±33.8[a]	56.5±13.1[b]
CAA with PBS wash (QE/100 g)	17.5±12.9[b]	31.6±29.2[a]	32.0±26.8[a]	10.3±4.8[b]
CAA with no PBS wash (QE/100 g)	27.7±18.4[b]	51.4±33.8[a]	49.3±31.8[a]	22.5±16.0[b]

Note: GAE, CE, TE and VE represent mg of gallic acid, catechin, Trolox and vitamin C equivalent respectively; QE represents μmol of quercetin equivalent.
*Values with no letters in common in each row are significantly different ($P<0.05$).

Table 3. The relationship among total phenolics, total flavonoids, ORAC, PSC and CAA values.

	Total phenolics	Total flavonoids	ORAC values	PSC values
Total phenolics	1			
Total flavonoids	0.888*			
ORAC values	0.924	0.862		
PSC values	0.757	0.737	0.853	
CAA values(no PBS wash)	0.788	0.848	0.899	0.809
CAA values(PBS wash)	0.574	0.741	0.736	0.739

*All correlation coefficients were significant at $P<0.01$.

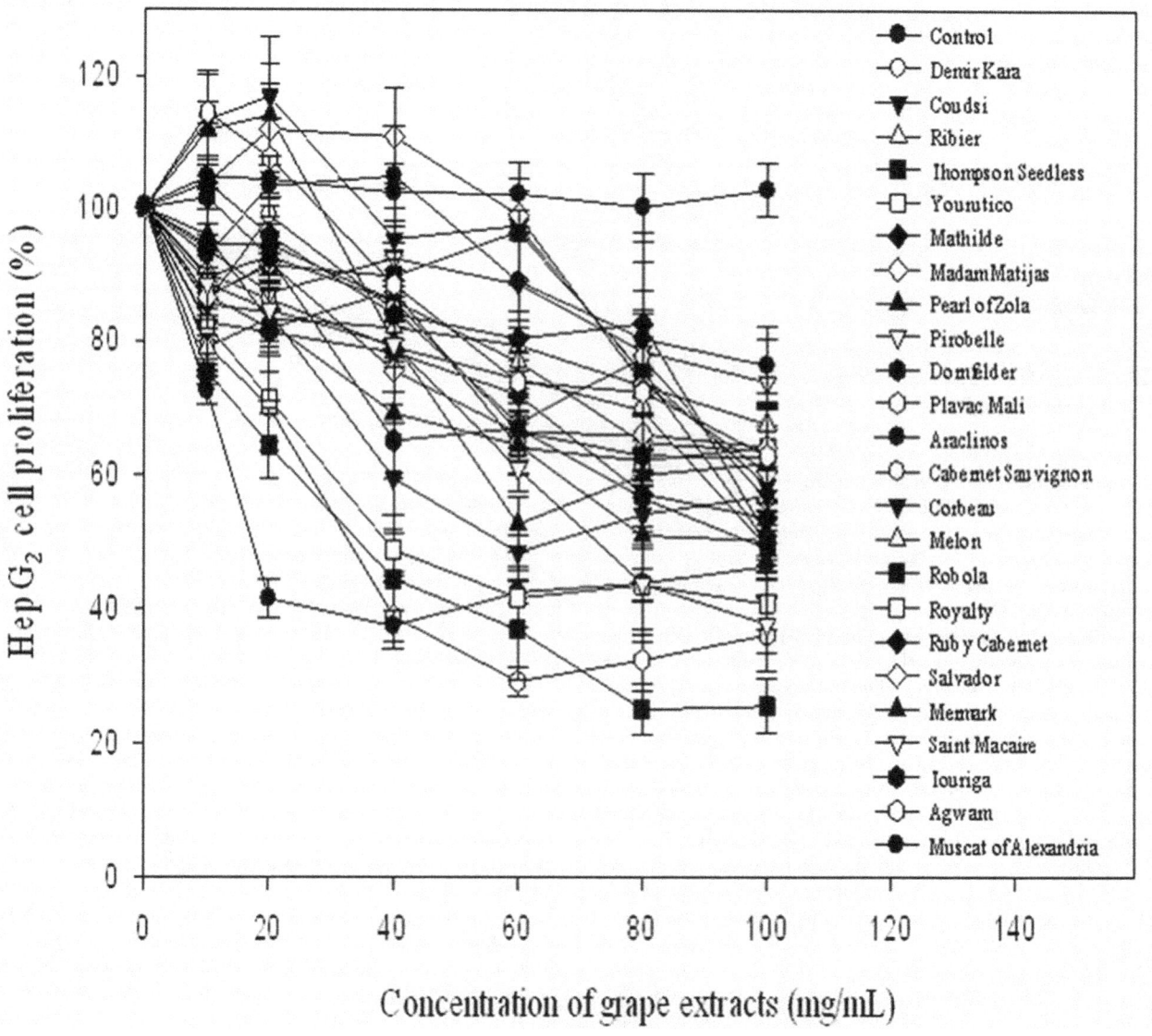

Figure 1. Inhibition of HepG2 cell proliferation by 24 grape extracts.

Table 4. Antiproliferative activities (EC_{25} and EC_{50}) and cytotoxicities (CC_{50}) of grape extracts towards human HepG2 liver cancer cells.

Cultivar	Antiproliferative Activity	
	EC_{25} (mg/mL)	EC_{50} (mg/mL)
Araclinos	1.53±0.25	ND*
Robola	10.40±1.41	33.63±4.04
Plavac Mali	11.57±2.43	38.70±11.73
Royalty	14.50±1.57	54.97±2.40
Corbeau	25.43±4.40	ND
Mathilde	30.83±11.17	ND
Saint Macaire	31.13±2.81	79.90±13.14
Mermark	39.43±2.82	ND
Pearl of Zola	39.97±2.17	ND
Demir Kara	42.40±3.76	138.23±10.17
Touriga	43.90±1.40	97.00±6.53
Salvador	48.30±8.02	ND
Yourutico	49.50±6.15	ND
Cabernet Sauvignon	55.80±4.90	ND
Ribier	66.07±6.54	ND
Dornfelder	66.50±6.99	ND
Ruby Cabernet	68.93±3.16	132.80±3.03
Agwam	71.47±2.34	ND
Melon	89.10±1.14	ND
Coudsi	90.80±4.74	129.40±2.91
Madam Matijas	95.67±2.30	ND
Thompson Seedless	101.27±3.30	ND
Pirobelle	108.00±16.81	ND
Muscat of Alexandria	112.77±6.52	ND

*ND, not detected, CC50>150 mg/mL.

benefit measurements such as antioxidant and antiproliferative activities. In consistence with our earlier study [2], the total phenolic contents of wine grapes were significantly higher than those of table grapes ($P<0.05$); and total phenolics of colored grape cultivars were higher than those of green yellow grape cultivars ($P<0.05$, Table 2).

This study was the first to determine total flavonoid content in grapes using a new method developed [38]. Determination of total flavonoids in foods is challenging. The most common methods for determining the content of total flavonoids include Aluminum Chloride ($AlCl_3$) colorimetric assay [4,38] and high performance liquid chromatography (HPLC). However, these two methods each have certain limitations. $AlCl_3$ colorimetric assay only measures partial flavonoids and therefore cannot be used accurately to determine total flavonoids [38,39]. HPLC is an excellent method to determine individual flavonoids, but cannot be used to determine the total flavonoids because the method is limited by the availability of flavonoid standards and many un-identified flavonoids are present in foods [33,38]. The SBC assay developed by our lab can detect all types of flavonoids, including flavones, flavonols, flavonones, flavononols, isoflavonoids and anthocyanins [38]. The total flavonoid contents in the 24 grape cultivars, measured by SBC assay varied from 94.7 to 1055 mg of catechin equivalents/100 g FW. Anthocyanins were the dominant in total flavonoids in colored grape cultivars according to our

previous study [2]. The total flavonoid contents of wine grapes were significantly higher than those of table grapes ($P<0.05$); and total flavonoids of colored grape cultivars were higher than those of green yellow grape cultivars ($P<0.05$, Table 2) because colored grape cultivars had high anthocyanins content and most wine grape cultivars were colored grapes. These results were consistent with those reported as the sum of individual flavonoids using HPLC [2].

The contribution of total flavonoids to total phenolics, calculated on the micromole basis, ranged from 40.3 to 91.3%. Further analysis found no significant differences in the contributions of total flavonoids to total phenolics between table and wine grapes, or between colored and green yellow grapes. These results suggested that flavonoids are one of the major phytochemicals in grapes and their relative contributions to the total phenolics are not dependent on the types of grape cultivars (i.e. wine vs. table grapes).

The mean ORAC values of wine grapes were 1769±206.1 mg of Trolox equivalents/100 g FW and significantly higher than those in table grapes ($P<0.05$, Table 2). In addition, the ORAC values of colored grapes were significantly higher than those of green yellow grapes ($P<0.05$, Table 2).

The PSC values for the 24 grape cultivars ranged from14.2 (Muscat of Alexandria) to 557 (Royalty) mg of vitamin C equivalents/100 g FW, indicating that there was a wide-range of variation across all 24 grape cultivars. Adom and Liu [24] reported the PSC value of red grapes was 371 mg of vitamin C equivalents/100 g FW, which was in the range of the results we reported here. The mean of PSC values of wine grapes was significantly higher than that of table grapes ($P<0.05$, Table 2). Similarly, the PSC values of colored grape cultivars were higher than those of green yellow grape cultivars (Table 2).

The 24 cultivars were also evaluated for their antioxidant activity in the CAA assay. The CAA values of Royalty, Salvador, Ruby Cabernet, Demir Kara, Dornfelder and Corbeauwere higher than other cultivars (Figure 1). These cultivars also had higher total phenolics and total flavonoids, and higher ORAC and PSC values. Among these cultivars, Royalty, Salvador and Corbeau are red flesh grape cultivars and tend to be rich in anthocyanins. The mean CAA values were 32.0 μmol of QE/100 g of grape for colored grape cultivars and 10.3 μmol for green yellow grape cultivars using the PBS wash protocol. But higher CAA values were observed when no PBS wash protocol was used (49.3 μmol of QE/100 g of grape for colored grape cultivars and 22.5 μmol for green yellow grape cultivars). We also observed that the CAA values in wine grapes were significantly higher than these in table grapes. The general trends of these results were in agreement with those of Wolfe et al. [29,30], but this study covered a much larger range of cultivar variation in terms of both color and end-product types.

The inhibition of HepG2 liver cancer cell proliferation by extracts from the 24 grape cultivars ranged from 25 to 82% at the concentration of 100 mg/mL extracts applied (Figure 1). Yang et al. [4] reported that the phytochemical extracts of Pinot Noir and Chardonnay varieties exhibited strong antiproliferative activity towards HepG2 cells with the lowest EC_{50}values of 17.0±0.8 mg/mL and 18.1±0.1 mg/mL ($p<0.05$), respectively. On the other hand, the phytochemical extracts of Vidal Blanc and DeChaunac showed a relatively weak antiproliferative activity with higher EC_{50}values of 52.1±2.1 and 52.2±3.0 mg/mL, respectively [4]. Our results in this study were similar to the results of Vidal Blanc and DeChaunacas reported by Yang et al. [4]. In consistent with other results, the colored grape cultivars had higher antiproliferative activities than those of green yellow grape cultivars.

Correlations among total phenolics, total flavonoids, total antioxidant activity, cell proliferation, and cellular antiocxdant activities were analyzed. There were significantly correlative relationship among total phenolics, total flavonoids and antioxidant activities, which were in agreement with what was reported by Arnous et al [36] and Kallithraka et al. [37]. No significantly correlative relationships were found between total antioxidant and antiproliferative activities against HepG2 liver cancer cells ($P>0.05$). Additionally, the total phenolic and flavonoid contents of grapes did not correlate with the antiproliferative activities. These results were consistent with what were previously reported in literature [17,39], suggesting that specific phytochemicals, rather than the total phytochemicals, might be responsible for the antiproliferative activities. Different phytochemicals might act additively and/or synergistically to contribute to the total antiproliferative activities in grapes. The CAA values for fruits were significantly positively related to total phenolic content and total flavonoids when log-transformed data were analyzed ($P<0.05$)in this study. The correlation coefficients for CAA values and total phenolics, total flavonoids, ORAC and PSC values were

higher for the no PBS wash protocol than these for the PBS wash protocol in this study. These results were consistent with what Wolfe et al. reported previously [23]. This is in contrast to a study involving vegetable extracts, in which prevention of DCFH oxidation in HepG2 cells by vegetable extracts was not correlated to ORAC or total phenolics [40]. Our current study suggested that total phenolic content might be a better predictor for the cellular antioxidant activity of grapes than the ORAC value, despite the commonality of measuring peroxyl radical scavenging abilities in both of the antioxidant activity assays.

Among the 24 cultivars evaluated, Royalty grape showed highest total phenolic content, total flavonoid content, ORAC, PSC and CAA values. It was the one which had consistent and high correlations among these measurements. Several other cultivars such as Corbeau, Salvador and Mermark also had high levels of phytochemicals, and antioxidant activities. On the other hand, Cabernet Sauvignon, a well-known wine grape cultivar, had relative high levels of total phenolic and flavonoid contents, but its ORAC and PSC values were about average. Some of the differences are certainly due to varietal genetic background. For example, Royalty is a red flesh cultivar with abundant anthocyanins produced in whole fruits, while many other cultivars including Cabernet Sauvignon are not.

In summary, grapes are rich in phenolics and flavonoids. Although viticulture practices, environmental conditions, and post-harvest processing can affect the content of total phenolics, total flavonoids, or individual bioactive compounds in grapes and grape products, varietal or genetic difference is one of the most important factors. Here we observed tremendous variation in total phenolic and total flavonoid contents, and ORAC, PSC and CAA values among the24 grape cultivars tested. The total antioxidant activities of grapes were significantly correlated with the total phenolics and flavonoids. However, no significant correlations were found between total phenolic content, total flavonoid content, ORAC, PSC values, or CAA values with the antiproliferative activities in this study. Lack of such correlations provides both challenges and opportunities in future studies. Total phenolics, total flavonoids, ORAC, PSC and CAA values of wine grapes and colored grapes were significantly higher than those of table grapes and green-yellow grapes. Another important finding from this study was the identification of several germplasm accessions with much high contents of phenolics and flavonoids, and total antioxidant activity. This information is important for breeding grape cultivars with better nutritional qualities of wine and table grapes in the future.

Acknowledgments

We are grateful to Mr. Bernard Prins of the USDA-ARS National Clonal Germplasm Repository for Fruit and Nut Crops in Davis, California and Dr. Christopher Owens of the USDA-ARS Grape Genetics Research Unit in Geneva, New York for providing V. vinifera berry samples. USDA-ARS is an equal opportunity provider and employer.

Author Contributions

Conceived and designed the experiments: GYZ RHL. Performed the experiments: ZCL. Analyzed the data: ZCL LLC RHL GYZ. Contributed reagents/materials/analysis tools: RHL GYZ. Wrote the paper: ZCL LLC GYZ RHL. Obtained permission for use of cell line: RHL.

References

1. USDA, Center for Nutrition Policy and Promotion. Dietary Guidelines for Americans (2010) Available: http://www.cnpp.usda.gov/DGAs2010-PolicyDocument.htm. Accessed 2014 July 25.

2. Liang ZC, Christopher OL, Zhong GY, Cheng LL (2011) Polyphenolic profiles detected in the ripe berries of Vitis vinifera germplasm. Food Chem 129: 940–950.

3. Xia EQ, Deng GF, Guo YJ, Li HB (2010) Biological activities of polyphenols from grapes. *Int J Mol Sci* 11: 622–646.

4. Yang J, Martinson TE, Liu RH (2009) Phytochemical profiles and antioxidant activities of wine grapes. *Food Chem* 116: 332–339.

5. Bitsch R, Netzel M, Frank T, Strass G, Bitsch I (2004) Bioavailability and biokinetics of anthocyanins from red grape juice and red wine. *J Biomed Biotechnol* 2004: 293–298.

6. Bub A, Watzl B, Blockhaus M, Briviba K, Liegibel U, et al. (2003) Fruit juice consumption modulates antioxidative status, immune status and DNA damage. *J Nutr Biochem* 14: 90–98.

7. Garcia-Alonso M, Minihane AM, Rimbach G, Rivas-Gonzalo JC, de Pascual-Teresa S (2009) Red wine anthocyanins are rapidly absorbed in humans and affect monocyte chemoattractant protein 1 levels and antioxidant capacity of plasma. *J Nutr Biochem* 20: 521–529.

8. Jang MS, Cai EN, Udeani GO, Slowing KV, Thomas CF, et al. (1997) Cancer chemopreventive activity of resveratrol, a natural product derived from grapes. *Science* 275: 218–220.

9. Del Caro A, Cacciorro AF, Fenu PAM, Piga A (2010) Polyphenols, colour and antioxidant activity changes in four Italian red wines during storage. *Acta Aliment Hung* 39: 192–210.

10. Liang ZC, Wu BH, Fan PG, Yang CX, Duan W, et al. (2008) Anthocyanin composition and content in grape berry skin in *Vitis* germplasm. *Food Chem* 111: 837–844.

11. Nikfardjam MSP, Mark L, Avar P, Figler M, Ohmacht R (2006) Polyphenols, anthocyanins, and trans-resveratrol in red wines from the Hungarian Villany region. *Food Chem* 98: 453–462.

12. Pirniyazov AZ, Abdulladzhanova NG, Mavlyanov SM, Kamaev FG, Dalimov DN (2003) Polyphenols from *Vitis vinifera* seeds. *Chem Nat Comp* 39: 349–354.

13. Tagliazucchi D, Verzelloni E, Bertolini D, Conte A (2010) In vitro bio-accessibility and antioxidant activity of grape polyphenols. *Food Chem* 120: 599–606.

14. Valdez LB, Alvarez S, Zaobornyj T, Boveris A (2004) Polyphenols and red wine as antioxidants against peroxynitrite and other oxidants. *Bio Res* 37: 279–286.

15. Vita JA (2005) Polyphenols and cardiovascular disease: effects on endothelial and platelet function. *Am J Clin Nutr* 81: 292s–297s.

16. USDA-ARS, GRIN. Available: http://www.ars-grin.gov. Accessed 2014 July 25.

17. Sun J, Chu YF, Wu XZ, Liu RH (2002) Antioxidant and anti proliferative activities of common fruits. *J Agr Food Chem* 50: 7449–7454.

18. Singleton VL, Orthofer R, Lamuela-Raventos RM (1999) Analysis of total phenols and other oxidation substrates and antioxidants by means of Folin-Ciocalteu reagent. In *Oxidants and Antioxidants, Pt A*, 299: 152–178.

19. Okarter N, Liu CS, Sorrells ME, Liu RH (2010) Phytochemical content and antioxidant activity of six diverse varieties of whole wheat. *Food Chem* 119: 249–257.

20. He XJ, Liu D, Liu RH (2008) Sodium borohydride/chloranil-based assay for quantifying total flavonoids. *J Agr Food Chem* 56: 9337–9344.

21. Zhang MW, Zhang RF, Zhang FX, Liu RH (2010) Phenolic profiles and antioxidant activity of black rice bran of different commercially available varieties. *J Agr Food Chem* 58: 7580–7587.

22. Ou BX, Hampsch-Woodill M, Prior RL (2001) Development and validation of an improved oxygen radical absorbance capacity assay using fluorescein as the fluorescent probe. *J Agricul Food Chem* 49: 4619–4626.

23. Wolfe KL, Liu RH (2008) Structure-activity relationships of flavonoids in the cellular antioxidant activity assay. *J Agr Food Chem* 56: 8404–8411.

24. Adom KK, Liu RH (2005) Rapid peroxyl radical scavenging capacity (PSC) assay for assessing both hydrophilic and lipophilic antioxidants. *J Agr Food Chem* 53: 6572–6580.

25. Shin Y, Ryu JA, Liu RH, Nock JF, Watkins CB (2008) Harvest maturity, storage temperature and relative humidity affect fruit quality, antioxidant contents and activity, and inhibition of cell proliferation of strawberry fruit. *Postharvest Biol Technol* 49: 201–209.

26. Liu RH, Jacob JR, Tennant BC, Hotchkiss JH (1992) Nitrite and nitrosamine synthesis by hepatocytes isolated from normal woodchucks (*Marmotamonax*) and woodchucks chronically infected with woodchuck hepatitis virus. *Cancer Res* 52: 4139–4143.

27. Liu RH, Jacob JR, Hotchkiss JH, Cote PJ, Gerin JL, et al. (1994) Woodchuck hepatitis virus surface antigen induces nitric oxide synthesis in hepatocytes: possible role in hepatocarcinogenesis. *Carcinogenesis* 15: 2875–2877.

28. Yoon H, Liu RH (2008) Effect of 2 alpha-hydroxyursolic acid on NF-kappa B activation induced by TNF-alpha in human breast cancer MCF-7 cells. *J Agr Food Chem* 56: 8412–8417.

29. Wolfe KL, Liu RH (2007) Cellular antioxidant activity (CAA) assay for assessing antioxidants, foods, and dietary supplements. *J Agr Food Chem* 55: 8896–8907.

30. Wolfe KL, Kang XM, He XJ, Dong M, Zhang QY, et al. (2008) Cellular antioxidant activity of common fruits. *J Agr Food Chem* 56: 8418–8426.

31. Felice DL, Sun J, Liu RH (2009) A modified methylene blue assay for accurate cell counting. *J Funct Foods* 1: 109–118.

32. Yoon H, Liu RH (2007) Effect of selected phytochemicals and apple extracts on NF-kappa B activation in human breast cancer MCF-7 cells. *J Agr Food Chem* 55: 3167–3173.

33. Liu RH (2004) Potential synergy of phytochemicals in cancer prevention: Mechanism of action. *J Nutr* 134: 3479S–3485S.

34. Liu RH (2003) Health benefits of fruit and vegetables are from additive and synergistic combinations of phytochemicals. *Am J ClinNutr* 78: 517S–520S.

35. Jia ZS, Tang MC, WU JM (1999) The determination of flavonoid contents in mulberry and their scavenging effects on superoxide radicals. *Food Chem* 64: 555–559.

36. Arnous A, Makris DP, Kefalas P (2002) Correlation of pigment and flavanol content with antioxidant properties in selected aged regional wines from Greece. *J Food Comp Analysis* 15: 655–665.

37. Kallithraka S, Aliaj L, Makris DP, Kefalas P (2009) Anthocyanin profiles of major red grape (*Vitis vinifera* L.) varieties cultivated in Greece and their relationship with in vitro antioxidant characteristics. *Inter J Food Sci Technol* 44:2385–2393.

38. He X, Liu RH (2008) Phytochemicals of apple peels: isolation, structure elucidation, and their antiproliferative and antioxidant activities. *J Agr Food Chem* 56: 9905–9910.

39. Chu YF, Sun J, Wu XZ, Liu RH (2002) Antioxidant and anti proliferative activities of common vegetables. *J Agr Food Chem* 50: 6910–6916.

40. Eberhardt MV, Kobira K, Keck AS, Juvik JA, Jeffery EH (2005) Correlation analyses of phytochemical composition, chemical, and cellular measures of antioxidant activity of broccoli (*Brassica oleracea* L. var. italica). *J Agr Food Chem* 53: 7421–7431.

23

Hemoglobin–Albumin Cluster Incorporating a Pt Nanoparticle: Artificial O$_2$ Carrier with Antioxidant Activities

Hitomi Hosaka[1], Risa Haruki[1], Kana Yamada[1], Christoph Böttcher[2], Teruyuki Komatsu[1]*

1 Department of Applied Chemistry, Faculty of Science and Engineering, Chuo University Tokyo, Japan, **2** Research Center of Electron Microscopy, Institute of Chemistry and Biochemistry Freie Universität Berlin, Berlin, Germany

Abstract

A covalent core–shell structured protein cluster composed of hemoglobin (Hb) at the center and human serum albumins (HSA) at the periphery, Hb-HSA$_m$, is an artificial O$_2$ carrier that can function as a red blood cell substitute. Here we described the preparation of a novel Hb-HSA$_3$ cluster with antioxidant activities and its O$_2$ complex stable in aqueous H$_2$O$_2$ solution. We used an approach of incorporating a Pt nanoparticle (PtNP) into the exterior HSA unit of the cluster. A citrate reduced PtNP (1.8 nm diameter) was bound tightly within the cleft of free HSA with a binding constant (K) of 1.1×10^7 M^{-1}, generating a stable HSA-PtNP complex. This platinated protein showed high catalytic activities for dismutations of superoxide radical anions (O$_2^{\bullet-}$) and hydrogen peroxide (H$_2$O$_2$), i.e., superoxide dismutase and catalase activities. Also, Hb-HSA$_3$ captured PtNP into the external albumin unit ($K = 1.1 \times 10^7$ M^{-1}), yielding an Hb-HSA$_3$(PtNP) cluster. The association of PtNP caused no alteration of the protein surface net charge and O$_2$ binding affinity. The peripheral HSA-PtNP shell prevents oxidation of the core Hb, which enables the formation of an extremely stable O$_2$ complex, even in H$_2$O$_2$ solution.

Editor: Eugene A. Permyakov, Russian Academy of Sciences, Institute for Biological Instrumentation, Russian Federation

Funding: This work was supported by a Grant-in-Aid for Scientific Research on Innovative Area ("Coordination Programming" Area 2107, No. 21108013) from MEXT (Ministry of Education, Culture, Sports, Science and Technology) Japan, Chuo University Grant for Special Research, and Joint Research Grant from the Institute of Science and Engineering, Chuo University. The funders had no role in study design, data collection and analysis, decision to publish, or preparation of the manuscript.

Competing Interests: The authors have declared that no competing interests exist.

* Email: komatsu@kc.chuo-u.ac.jp

Introduction

Hemoglobin (Hb)-based O$_2$ carriers (HBOCs) have been studied extensively as a substitute for red blood cells (RBCs) in transfusion medicine and as O$_2$ therapeutic reagents [1–5]. Nevertheless, none satisfies all requirements for use in clinical situations [6,7]. A common side-effect is mild hypertension resulting from nitric oxide (NO) depletion by Hb diffused into the extravascular space [8,9]. Actually, NO is an endothelial-derived relaxing factor. Moreover, HBOCs show faster autoxidation of Hb to the ferric heme form (metHb) than the native Hb shows [10–12]. Autoxidation of Hb produces a superoxide radical anion (O$_2^{\bullet-}$), which is disproportionated to hydrogen peroxide (H$_2$O$_2$) [13]. These reactive oxygen species (ROS) promote the oxidation of Hb. In RBC, antioxidant systems include superoxide dismutase (SOD) and catalase, which catalytically scavenge O$_2^{\bullet-}$ and H$_2$O$_2$, and thereby protect the Hb function. In ischemia-reperfusion when the ischemic tissue is reperfused with O$_2$, xanthine oxidase converts xanthine and hypoxanthine into O$_2^{\bullet-}$ [14–16]. Overproduction of O$_2^{\bullet-}$ and subsequently H$_2$O$_2$ causes not only tissue injury, but also further oxidation of Hb. Consequently, in clinical situations involving ischemia-reperfusion, HBOC with antioxidant activity is expected to be tremendously useful. Chang et al. first synthesized polyHb-SOD-catalase conjugate and demonstrated the reduction of the autoxidation

rate of Hb [17]. Kluger et al. reported that the metHb formation was inhibited in structurally defined Hb-SOD dimer [18]. Silaghi-Dumitrescu et al. prepared Hb copolymer with rubrerythrin, non heme iron enzyme [19]. These Hb-(antioxidant enzyme) conjugates displayed both O$_2$ carrying and antioxidant properties. However, a specific enzyme is necessary to scavenge the individual ROS, and it denatures gradually.

More recently, we synthesized a covalent core–shell structured protein cluster comprising Hb at the center and human serum albumins (HSA) at the periphery, Hb-HSA$_m$ (m = 2, 3, 4), which acts as a unique HBOC (Figure 1) [20]. Since HSA contains only one cysteinyl thiol at position 34, we exploited a heterobifunctional crosslinker, N-succinimidyl 4-(N-maleimidomethy) cyclohexane-1-carboxylate (SMCC), as a connector between the Cys-34 residue of HSA and the surface lysyl ε-amino groups of Hb. The major product is the Hb-HSA$_3$ heterotetramer in triangular form with an HSA-binding number (m) of three. HSA, the most prominent plasma protein, demonstrates low permeability in the vasculature walls because of the electrostatic repulsion between the negatively charged albumin surface [isoelectric point (pI): 5.0] and glomerular basement membrane around the endothelial cells [21]. From this physiological perspective, the surface net charge of the Hb-HSA$_m$ cluster is satisfactorily negative (pI: 5.1–5.2) [20]. Intravenous transfusion of the Hb-HSA$_m$ cluster is expected to enable

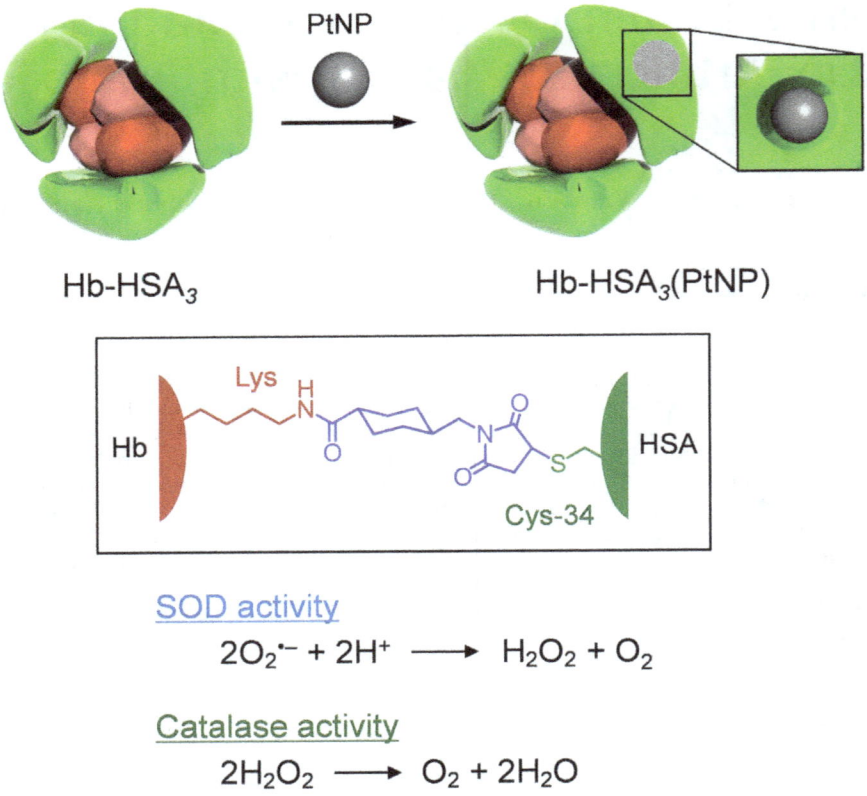

SOD activity

$$2O_2^{\bullet -} + 2H^+ \longrightarrow H_2O_2 + O_2$$

Catalase activity

$$2H_2O_2 \longrightarrow O_2 + 2H_2O$$

Figure 1. Schematic illustrations of Hb-HSA$_3$(PtNP) cluster. The Cys-34 of HSA and the surface Lys group of Hb were connected covalently with a crosslinking agent (SMCC). A PtNP was bound within the cleft of the exterior HSA unit and performed SOD and catalase activities.

long-term circulation without extravasation. Moreover, it might not elicit an unfavorable increase in blood pressure.

If one were able to confer antioxidant properties to the external HSA unit of Hb-HSA$_m$, then this construct would become a promising O$_2$ carrier with high resistance towards oxidation reactions. In this context, we chose Pt nanoparticle (PtNP) as a potential candidate. PtNPs have been widely investigated for a variety of applications, such as fine chemical synthesis, fuel cell fabrications, and biomedical treatments. It was reported that PtNP is an extremely effective catalysis for both O$_2^{\bullet -}$ and H$_2$O$_2$ dismutations (Figure 1) [22–24]. (i) The high ROS scavenging activities of PtNP depend on greater surface area per mass relative to large particle [22,23]. (ii) Almost no cytotoxicity was observed even after adherent cells were exposed to PtNPs [23]. We have found that small PtNP (1.8 nm diameter) is incorporated into HSA, and the obtained HSA-PtNP complex showed SOD and catalase activities with high efficiency. The Hb-HSA$_3$ also possesses the capability of binding PtNP into the exterior HSA shell. The resultant Hb-HSA$_3$(PtNP) cluster forms a very stable O$_2$ complex, even in aqueous H$_2$O$_2$ solution (Figure 1). This artificial O$_2$ carrier, having triple functionalities (O$_2$ transport, O$_2^{\bullet -}$ dismutation, H$_2$O$_2$ dismutation) might be useful in clinical conditions with ischemia-reperfusion. The Hb-HSA$_3$(PtNP) cluster would deliver O$_2$ to the ischemic tissue, and simultaneously protect Hb and tissues from damaging effects of reperfusion injury.

Materials and Methods

Materials and apparatus

Human serum albumin (HSA) was purchased from Japan Blood Products Organization. Pure bovine Hb was purified from bovine blood purchased from Tokyo Shibaura Zouki Co., Ltd. [20]. Hydrogen hexachloroplatinate(IV) hexahydrate (H$_2$PtCl$_2$•6H$_2$O), xanthine, and catalase (from bovine liver) were purchased from Wako Pure Chemical Industries Ltd. Ferricytochrome c (Cyt. c, from bovine heart) was purchased from Sigma-Aldrich Co. Xanthine oxidase (XOD, from butter milk) was purchased from Oriental Yeast Co., Ltd. Mn(III)-terakis(*N*-methylpyridinium) porphyrin (Mn-TMPyP) was purchased from Frontier Scientific Corp. Other chemicals of special grades were used without further purification. The water was deionized (18.2 MΩcm) using water purification systems (Elix UV and Milli Q Reference; Millipore Corp.). Isoelectric focusing (IEF) was performed using an electrophoresis power supply (EPS 601; GE Healthcare UK Ltd.) with an IEF gel (Novex pH 3–10; Invitrogen Corp.). The protein marker used was an IEF calibration kit Broad pI (pH 3–10; GE Healthcare UK Ltd.).

Synthesis of PtNP

The citrate-reduced PtNP was prepared as described in a report of a study by Bond et al. [25]. To the refluxed aqueous H$_2$PtCl$_2$•6H$_2$O solution (271 µM, 85.5 mL), 1 wt% trisodium citrate dihydrate in water (4.5 mL) was added and then refluxed continuously for 1 h with stirring. The solution changed to dark brown. After cooling slowly to 25°C, the obtained PtNP solution was washed with water using an ultrafilter (Q0100, 10 kDa

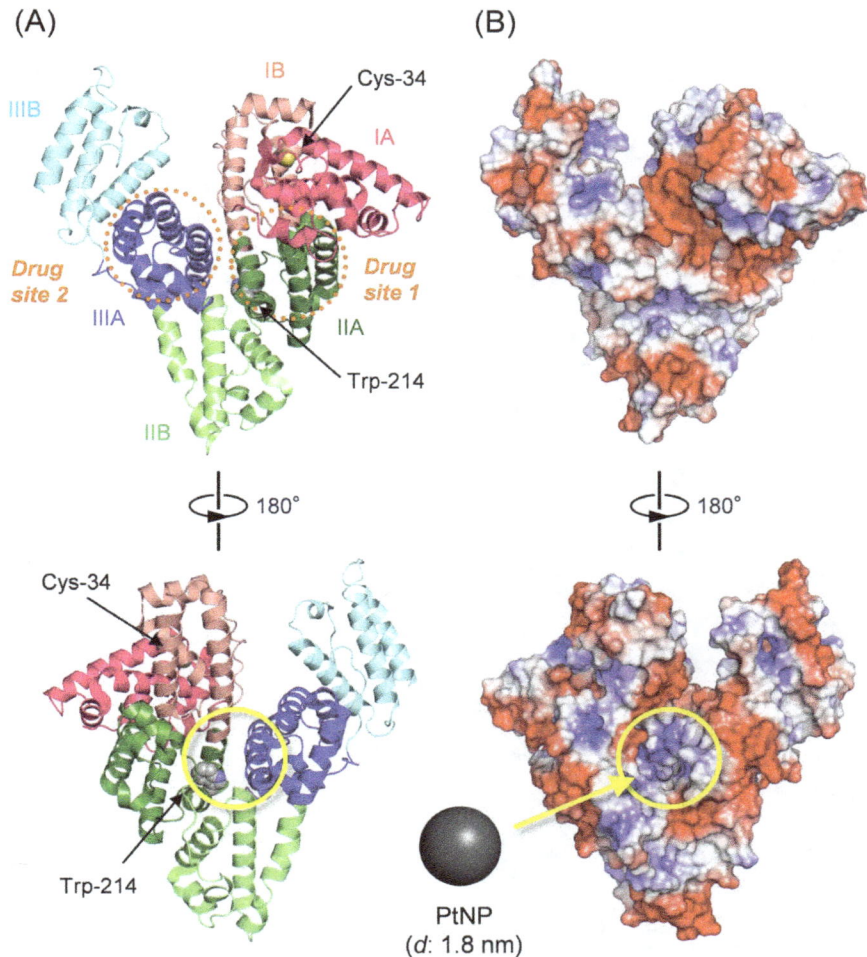

Figure 2. Crystal structure of HSA (PDB 1E78, ref. 31) and the PtNP binding site. (A) HSA structure involving the positions of drug site 1 (subdomain IIA, dark green), drug site 2 (subdomain IIIA, dark blue), Cys-34, and Trp-214. Cys-34 and Trp-214 are depicted in space-filling representation. The upper image and lower images respectively show the "front side" and "back side". (B) Surface electrostatic potential representations of HSA in the same orientations illustrated in (A). Blue and red respectively represent positive charge and negative charge density. Possible binding site of PtNP in the positively charged cleft between subdomain IIA and IIIA is indicated by a yellow circle. These images were produced based on crystal structure coordinates using PyMOL (Schrödinger K. K., CA, USA).

MWCO; Advantec Toyo Kaisha Ltd.) in an UHP-76K ultra-holder. Finally, the medium was concentrated up to 50 μM as PtNP using the UHP-76K ultraholder. The resultant PtNP colloid solution was stored in a refrigerator at 4°C.

Preparation of Hb-HSA₃ cluster

The Hb-HSA₃ cluster was prepared according to our previously reported procedure with some modifications [20]. Typically, a DMSO solution of heterobifunctional crosslinker, N-succinimidyl 4-(N-maleimidomethy)cyclohexane-1-carboxylate (SMCC; Tokyo Chemical Industry Co., Ltd.) (20 mM, 4 mL) was added dropwise into phosphate buffered saline (PBS) solution (pH 7.4) of carbonyl Hb (0.1 mM, 40 mL), and the mixture was stirred for 3 h in the dark at 4°C. After removing unreacted crosslinker by gel filtration chromatography (GFC) with a Sephadex G25 (superfine) column, the obtained SMCC-bound Hb (maleimide activated Hb) was concentrated to 40 mL ([Hb] = 0.1 mM) using a centrifugal concentrator (Vivaspin 20 ultrafilter, 10 kDa MWCO; GE Healthcare UK Ltd.). Then this solution was added slowly into the PBS solution of HSA (1 mM, 40 mL) with subsequent stirring

under dark conditions for 14 h at 4°C. A part of reaction mixture was applied to size-exclusion chromatography (SEC) on an HPLC system (LaChrom Elite; Hitachi High-Technologies Corp.) with a Shodex Protein KW-803 column (Showa Denko K.K.) using phosphate buffer (PB, pH 7.4, 50 mM) as the mobile phase. The elution curve exhibited new multiple peaks at the high molecular weight region. The three major components were identified as Hb-HSA₄ heteropentamer (minor), Hb-HSA₃ heterotetramer, and Hb-HSA₂ heterotrimer [20]. Then the resultant solution was subjected to GFC with a Superdex 200 pg in XK50/60 column (GE Healthcare UK Ltd.) using PBS (pH 7.4) as the running buffer. We collected all major fractions before the HSA peak. The unreacted free HSA was excluded completely. By Hb and total protein assays [20], the average HSA/Hb ratio of the harvested Hb-HSA$_m$ cluster was found to be 2.8–3.2, which is indicated as Hb-HSA₃. Finally, the obtained Hb-HSA₃ solution was condensed ([Hb] = 5 g/dL) using a Vivaspin 20 ultrafilter (30 kDa MWCO) and stored in a refrigerator at 4°C.

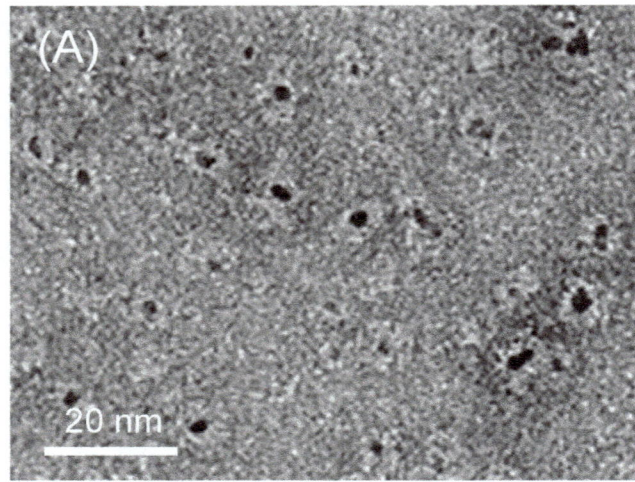

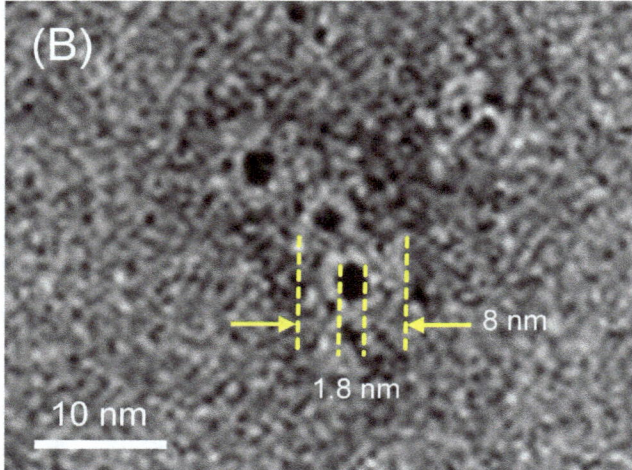

Figure 3. TEM images of HSA-PtNP complexes. The sample was negatively stained with 1% uranyl acetate.

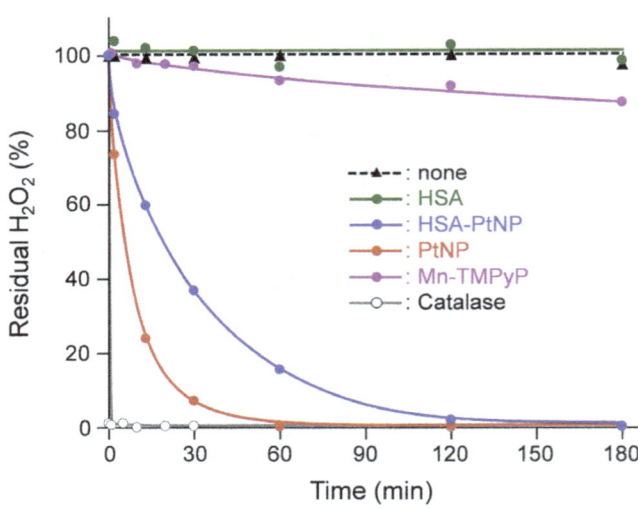

Figure 4. Time course of residual H_2O_2 percentage in 0.1 mM H_2O_2 solution with HSA-PtNP complex. [Sample] = 1 µM at 25°C.

Preparation of HSA-PtNP complex and Hb-HSA₃(PtNP) cluster

The medium of PtNP solution was exchanged to PBS (pH 7.4) using a Vivaspin 20 ultrafilter (10 kDa MWCO). A PBS solution of HSA (0.51 mM, 0.1 mL) was added slowly to the PtNP solution (10.2 µM, 5 mL, PBS), and the mixture was incubated for 1 h with gentle stirring in the dark at 25°C, yielding HSA-PtNP complex (PtNP/HSA = 1/1). Similarly, the Hb-HSA₃ solution (0.51 mM, 0.2 mL, PBS) was added to the PtNP solution (10.2 µM, 10 mL, PBS). Then the mixture was incubated for 1 h with gentle stirring in the dark at 25°C, affording Hb-HSA₃(PtNP) cluster (PtNP/Hb-HSA₃ = 1/1).

Determination of binding constants of PtNP for HSA and cluster

Binding constants (K) of PtNP for HSA and Hb-HSA₃ cluster were determined using fluorescence quenching measurements of albumin by PtNP titration according to the literature [26]. Fluorescence of the HSA or Hb-HSA₃ ([HSA unit] = 10 µM) (E_m: 340 nm) solution (PBS, pH 7.4) was quenched upon binding of PtNP (0–0.3 µM). The plots of $log(F_o–F)/F$ vs. log[PtNP] were produced from the data to obtain the K values and binding number.

CD measurements

Circular dichroism (CD) spectra were obtained using a spectropolarimeter (J-820; Jasco Corp.). The sample concentration was 0.2 µM in PBS. Quartz cuvettes with 10-mm thickness were used for measurements of 200–250 nm.

Table 1. $O_2^{•-}$ scavenging activity (IC_{50}) and H_2O_2 scavenging activity (T_{50}) of HSA-PtNP complex at 25°C.

Enzyme mimics	IC_{50} (µM)[a]	T_{50} (min)[b]
HSA	N.D.	N.D.
PtNP	0.12	6
HSA-PtNP	0.16	19
Mn-TMPyP	0.8[c]	N.D.
Cu,Zn-SOD	0.03[d]	–
Catalase	–	≈0.1

[a]In PB solution (pH 7.8, 50 mM).
[b]In PBS solution (pH 7.4), [H_2O_2] = 0.1 mM.
[c]Ref. 29.
[d]Ref. 33. In PB solution (pH 7.8, 45 mM).

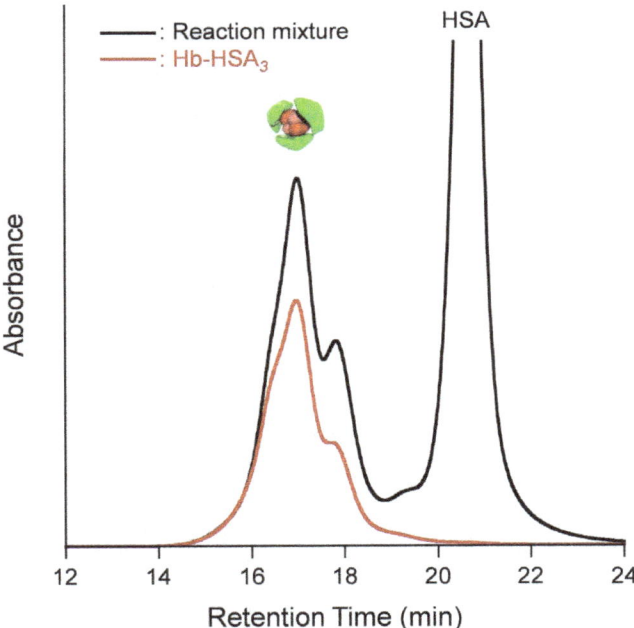

Figure 5. SEC profile of Hb-HSA₃ cluster. Black line: reaction mixture of SMCC-bound Hb and HSA, red line: separated Hb-HSA₃.

TEM measurement

Droplets of HSA-PtNP ([protein] = 0.35 mg/mL) were applied to amorphous carbon film covered 200-mesh grids (Quantifoil R1/4 with a hole diameter of approximately 1 μm; Quantifoil Micro Tools GmbH, Jena, Germany), which had been hydrophilized before use by plasma treatment (8 W, 60 s) in a Baltec Med 020 device (Leica Microsystems). After the supernatant fluid was blotted with a filter paper, an aqueous uranyl acetate (1 w/v %) was applied for another 45 s and the grids were eventually left to air-dry after blotting. Then the grids were transferred into a transmission electron microscope (Tecnai F20 microscope equipped with field emission gun operated at a 160 kV accelerating voltage; FEI Co.). Images were recorded using a CCD camera (Eagle 4k-CCD device; FEI Co.) operated at a binning factor of 2 (2,048×2,048 pixel).

O₂•⁻ scavenging activity (xanthine–XOD–Cyt. c assay)

O₂•⁻ scavenging activity (SOD activity) of the HSA-PtNP complex was determined using the Cyt. c reduction technique, in which O₂•⁻ was produced in situ by a xanthine–XOD reaction [27,28]. The experiments were performed according to our previously reported procedure [29]. To the PB solution (pH 7.8, 50 mM, 3.0 mL) containing Cyt. c (10 μM), xanthine (50 μM), and catalase (500 U/mL) in a 10-mm path length optical quartz cuvette, an amount of XOD sufficient to give an initial rate of $\Delta A_{550} = 0.025$ min⁻¹ (without HSA-PtNP complex) (approximately 2.0 mU/mL) was injected at 25°C. After the addition of XOD, increases in the absorption at 550 nm based on the reduced-form Cyt. c was monitored at 25°C. From the absorbance increase, the initial rate constant (v_i) was determined at various concentration of HSA-PtNP complex. The IC₅₀ value is defined as the 50% inhibition concentration of Cyt. c reduction. The same experiments were also conducted for PtNP and HSA.

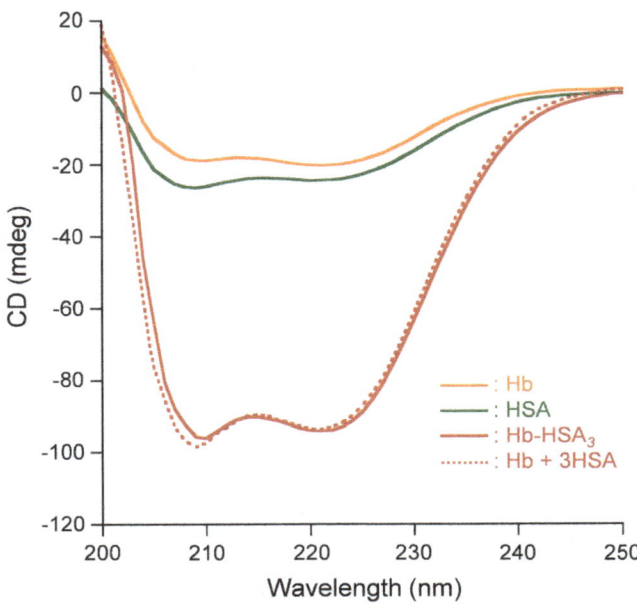

Figure 6. CD spectra of Hb, HSA, and Hb-HSA₃. [Sample] = 0.2 μM in PBS solution (pH 7.4) at 25°C.

H₂O₂ scavenging activity (quantitative peroxide assay)

H₂O₂ scavenging activity (catalase activity) of the HSA-PtNP complex was evaluated by measuring the concentration of residual H₂O₂ using the Pierce Quantitative Peroxide Assay Kits (Thermo Fisher Scientific Inc.). The HSA-PtNP solution (50 μM, 41 μL) was added to the aqueous solution of H₂O₂ (102 μM, 2.0 mL) in a vial bottle. Then the mixture was incubated with gentle stirring at 25°C. The 50 μL sample was pipetted out regularly from the reaction mixture and HSA-PtNP was removed using a centrifugal

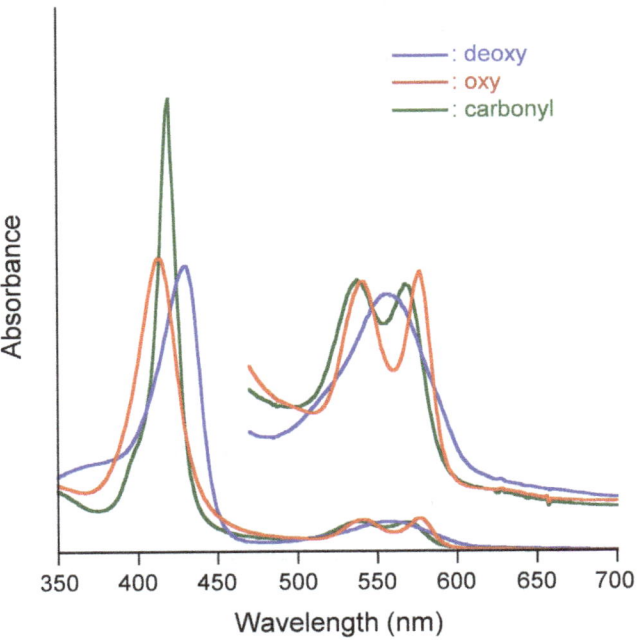

Figure 7. Visible absorption spectral changes of Hb-HSA₃ cluster. In PBS solution (pH 7.4) at 25°C.

Table 2. Visible absorption spectral data of Hb-HSA$_3$ and Hb-HSA$_3$(PtNP) clusters in PBS solution (pH 7.4) at 25°C.

| Hemoproteins | λ_{max} (nm) | | |
	deoxy	oxy	carbonyl
Hb-HSA$_3$	430, 556	413, 541, 577	420, 538, 569
Hb-HSA$_3$(PtNP)	430, 554	413, 541, 576	419, 536, 568
Hb[a]	430, 555	414, 541, 577	420, 538, 569
HbA[b]	430, 555	415, 541, 577	419, 540, 569

[a]From ref. 20.
[b]HbA (human adult Hb), from ref. 36.

filter device (Microcon Ultracel YM-30; Millipore Corp.). Then 20 µL of the filtrate was mixed with the working reagent (200 µL) in a hole of a 96-well cell culture plate. The absorbance at 555 nm based on the (xylenol orange)-Fe(III) complex was measured using a Microplate Reader (iMark; Bio-Rad Laboratories, Inc.). From absorption at 550 nm, the concentration of residual H_2O_2 in the sample was determined using the calibration line ([H_2O_2] = 0–100 µM) prepared in advance. The T_{50} value is defined as time required for quenching half of H_2O_2. The same experiments were also conducted for PtNP, HSA, catalase, and Mn-TMPyP.

O$_2$ binding property

The visible absorption spectra of deoxy (under N_2), oxy (under O_2), and carbonyl (under CO) forms of the Hb-HSA$_3$ and Hb-HSA$_3$(PtNP) clusters ([Hb]: 10 µM, PBS, pH 7.4) were obtained in accordance with our previously reported procedures using a UV–Visible spectrophotometer (8543; Agilent Technologies Inc.) equipped with a temperature control unit (89090A; Agilent Technologies Inc.) [20]. The O$_2$ affinity (P_{50}: O$_2$-partial pressure where Hb is half-saturated with O_2) and Hill coefficient (n) were determined using an automatic recording system for O$_2$-equilibrium curve (Hemox Analyzer; TCS Scientific Corp.) using PBS (pH 7.4) at 37°C. The sample was oxygenated by an increasing O$_2$-partial pressure and deoxygenated by flushing with N_2.

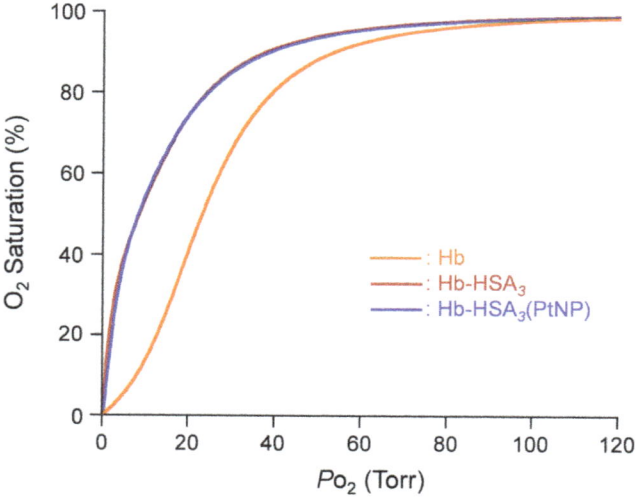

Figure 8. O$_2$ equilibrium curves of Hb-HSA$_3$ and Hb-HSA$_3$(PtNP) clusters. In PBS solution (pH 7.4) at 37°C.

O$_2$ complex stability

The O$_2$ complex stability of the Hb-HSA$_3$ cluster was evaluated using the first-order autoxidation rate constant (k_{ox}) of the central Hb. The PBS solution (pH 7.4) of oxyHb-HSA$_3$ cluster ([Hb] = 10 µM, 2 mL) was put into a 10-mm-path length optical quartz cuvette. The top of the cuvette was sealed with a gas permeation film (AeraSeal Film MAF710; Gel Co.), which allows air exchange and which prevents water evaporation. The absorption intensity at 630 nm (A_t) based on metHb formation was monitored under aerobic conditions at 37°C. After the measurement, the entirely oxidized metHb-HSA$_3$ cluster was prepared by addition of slightly excess K_3[Fe(CN)$_3$], and its absorption intensity (A_{100}) was observed. From the absorbance increase, the k_{ox} value was ascertained using nonlinear least-squares curve fitting techniques. The same experiments were conducted for native Hb and Hb-HSA$_3$(PtNP) cluster.

The O$_2$ complex stability of the cluster in 20 µM H_2O_2 solution was evaluated by the time course of metHb formation level because the mechanism of the Hb oxidation was complicated. To the PBS solution (pH 7.4) of oxyHb-HSA$_3$ cluster ([Hb] = 10 µM, 2 mL) in a 10-mm-path length quartz cuvette, aqueous H_2O_2 (2 mM, 20 µL) was added, and the absorption intensity at 630 nm (A_t) was measured under aerobic conditions with gentle stirring for 180 min at 25°C. The top of the cuvette was sealed with a gas permeation film. After the measurement, a slightly excess K_3[Fe(CN)$_3$] was added to determine the absorption intensity of the entirely oxidized metHb form (A_{100}). From the absorbance increase, the metHb level [$(A_t–A_0)/(A_{100}–A_0)\times100$ (%)] (A_0: absorption intensity at 630 nm before H_2O_2 injection) was ascertained. The same experiments were carried out for native Hb, Hb-HSA$_3$(PtNP) cluster, and simple mixture of Hb/HSA-PtNP/HSA (1/1/2, molar ratio).

Results and Discussion

Synthesis and structure of HSA–PtNP complex

Enzymatic activities of PtNP have attracted considerable attention because of their potential applications for medical use [22–24]. Shirahata et al. reported high $O_2^{\bullet-}$ and H_2O_2 dismutation activities of PtNPs and the highest enzyme reactivity at a particle size of about 2.0 nm [23]. In the circulatory system, the small PtNP (ca. 2 nm diameter) might be captured by HSA. However, the enzymatic properties of such postulated HSA-PtNP complex have not been reported in the relevant literature. We have now prepared the HSA-PtNP complex and have examined its $O_2^{\bullet-}$ and H_2O_2 dismutation activities.

HSA is a heart-shaped monomeric protein (66.5 kDa) consisting of three homologous domains (I–III), each of which contains two subdomains: A and B (Figure 2A) [30,31]. Many water insoluble

Table 3. O_2 binding parameters of Hb-HSA$_3$ and Hb-HSA$_3$(PtNP) clusters in PBS solution (pH 7.4) at 37°C.

Hemoproteins	P_{50} (Torr)	n (–)	k_{ox} (h^{-1})
Hb	23	2.6	0.037
Hb-HSA$_3$	9	1.5	0.035
Hb-HSA$_3$(PtNP)	9	1.5	0.039

metabolites (fatty acids, bilirubin, thyroxin, etc.) and commonly used drugs (warfarin, diazepam, ibuprofen, etc.) bind to the principle ligand binding sites in subdomain IIA and IIIA of HSA: so-called drug sites 1 and 2 [32]. To embed a PtNP into this protein interior, we prepared citrate-reduced PtNP with a diameter of 1.5–2.0 nm [25]. TEM images clearly showed the formation of uniform PtNPs with diameter (d) of 1.8 nm. The PtNP concentration was calculated as 1.25 μM based on the Pt^{2+} concentration and particle size. The resultant aqueous PtNP solution was concentrated up to 50 μM using an ultrafiltration device. The medium was exchanged to phosphate buffered saline (PBS, pH 7.4). No precipitation was found for over one year at 4°C.

The complexation of PtNP to HSA was conducted by adding HSA to the PtNP solution (PtNP/HSA = 1/1). Gel permeation chromatography (Sephadex G25) of the obtained protein displayed a single elution peak. Notably, TEM images demonstrated the formation of equivalent complex of HSA and PtNP (Figure 3A). Close inspections of TEM micrographs revealed that each PtNP is accommodated in the center of the protein (Figure 3B). One feasible binding mode is a covalent linkage between the thiol residue (Cys-34) of HSA and the PtNP surface. Nevertheless, nonmercapt HSA, in which Cys-34 is oxidized, also formed a similar HSA-PtNP complex, indicating that the covalent S-Pt bond is unlikely. Another possible binding force is electrostatic attraction between the negatively charged surface of PtNP and a positively charged region of the protein. Based on the electrostatic potential representation of HSA, we found a positively charged cleft between subdomain IIA and IIIA (Figure 2B). In

fact, the fluorescence emission intensity of the HSA solution (λ_{em}: 340 nm) was quenched by addition of PtNP. It is caused primarily by an energy transfer from the tryptophan (Trp)-214 residue in subdomain IIA (Figure 2A) to the bound PtNP. From titration measurements [26], the binding constant (K) and binding number of PtNP with HSA were calculated respectively as 1.1×10^7 M^{-1} and 1.1. We reasoned that one PtNP binds to the positively charged cleft of HSA on the back side, yielding a 1:1 HSA-PtNP complex. The obtained dark-brown protein solution was stable over one year at 4°C.

Antioxidant activities of HSA–PtNP complex

The SOD activity of the HSA-PtNP complex was evaluated in phosphate buffered (PB) solution using the xanthine–(xanthine oxidase)–ferricytochrome c (Cyt. c) assay [27–29]. In the presence of the HSA-PtNP complex, the Cyt. c reduction by $O_2^{\bullet-}$ was inhibited significantly. The IC$_{50}$ value (the concentration of enzyme necessary to attain 50% inhibition of the Cyt. c reduction) of the HSA-PtNP complex was determined to be 0.16 μM (Table 1). Under our experimental conditions, the reduction of Cyt. c was not suppressed by HSA alone. For that reason, SOD activity of the albumin protein is excluded. The IC$_{50}$ of HSA-PtNP complex is smaller than that of the best synthetic SOD model Mn(III)-tetrakis(N-methylpyridinium)porphyrin (Mn-TMPyP) [29] and resembled the value of native Cu,Zn-SOD [33]. We infer that the HSA-PtNP complex possesses a strong capability to catalyze the dismutation of $O_2^{\bullet-}$.

Next, the catalase activity of the HSA-PtNP complex was examined by measuring the H_2O_2 decomposition. In the presence of HSA-PtNP, the H_2O_2 concentration declined considerably and reached zero after 180 min (Figure 4). The T_{50} value (time required for quenching half of H_2O_2) of HSA-PtNP was 19 min (Table 1). On the one hand, with the coexistence of HSA alone, the concentration of H_2O_2 was not changed. These results imply that the catalase activity of HSA-PtNP complex was based on the PtNP in the protein. While the T_{50} value is at least two order of magnitude larger than that of native catalase, this platinated protein showed much higher H_2O_2 dismutation activity than Mn-TMPyP [34]. Overall, we concluded that the HSA-PtNP complex shows strong abilities to catalyze the dismutation of both $O_2^{\bullet-}$ and H_2O_2.

Synthesis and structure of Hb-HSA$_3$(PtNP) cluster

The Hb-HSA$_3$ cluster with the average HSA/Hb ratio of 3.0 was synthesized according to our previously reported procedure with some modifications (See Materials and Methods). Size exclusion chromatography (SEC) of the reaction mixture of SMCC-bound Hb and HSA exhibited new peaks of Hb-HSA$_4$ heteropentamer (shoulder), Hb-HSA$_3$ heterotetramer, and Hb-HSA$_2$ heterotrimer (Figure 5); the major product was Hb-HSA$_3$ (42%). By gel filtration chromatography (GFC), all the cluster fractions were harvested together (yield: 80% based on Hb). Unreacted free HSA was removed completely (Figure 5). The

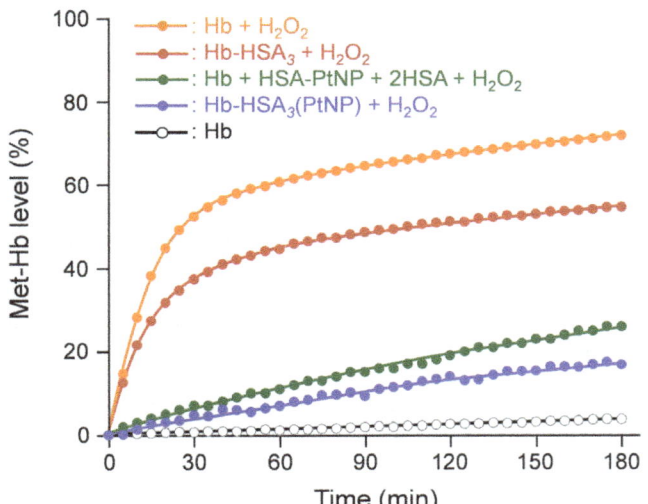

Figure 9. Time course of metHb level of Hb-HSA$_3$ and Hb-HSA$_3$(PtNP) clusters. [Hb] = 10 μM in 20 μM H_2O_2 solution at 25°C.

average HSA/Hb ratio was determined to be 2.8–3.2 using Hb and total protein assays. This protein cluster is shown as Hb-HSA$_3$. The CD spectral pattern and intensity of the Hb-HSA$_3$ cluster agreed well with the sum of the Hb spectrum and a three-fold-enlarged HSA spectrum (Figure 6). This observation also supports the average HSA/Hb as 3 (mol/mol).

Then the Hb-HSA$_3$ solution was added slowly to the PBS solution of PtNP, yielding Hb-HSA$_3$(PtNP) hybrid cluster (PtNP/Hb-HSA$_3$ = 1/1). From titration measurements [26], the K value and binding number of PtNP with the exterior HSA unit were ascertained as 1.1×10^7 M^{-1} and 1.1, which are equal to the data observed for free HSA. The affinity of PtNP with HSA moiety of the cluster is satisfactorily high. Even though, PtNP may transfer to other plasma proteins after intravenous administration. To avoid such intermolecular exchanging reaction in vivo, covalent attaching of PtNP to the HSA unit would be beneficial. The isoelectric point (pI: 5.1) of Hb-HSA$_3$ was unaltered by PtNP incorporation. HSA has a high molecular surface net charge, thereby the pI value is known to be shifted slightly by ligand binding [35]. Thus, our result suggests that the PtNP is not adhered onto the HSA surface, but that it is embedded into the HSA shell.

O$_2$ affinity and O$_2$ complex stability

The visible absorption spectral patterns of the Hb-HSA$_3$ cluster in PBS solution (pH 7.4) under N$_2$, O$_2$, and CO atmosphere (deoxy, oxy, and carbonyl forms) were fundamentally the same as those of Hb-HSA$_3$ tetramer and native Hb (Figure 7, Table 2) [20,36]. In contrast, the PBS solution of Hb-HSA$_3$(PtNP) cluster exhibited strong absorbance over the entire visible range. It is ascribed to the superposing of the PtNP absorption onto the Hb-HSA$_3$ spectrum. Nevertheless, the absorption maxima of the Hb-HSA$_3$ and Hb-HSA$_3$(PtNP) clusters showed good mutual agreement, indicating that PtNP caused no alternation of electronic states of the hemes in Hb (Table 2).

The P_{50} (O$_2$-partial pressure where Hb is half-saturated with O$_2$) and cooperativity coefficient (Hill coefficient, n) of Hb-HSA$_3$ cluster (Figure 8, Table 3) were identical to the values of isolated Hb-HSA$_3$ tetramer [20]. Moderate O$_2$ affinity of Hb-HSA$_3$ cluster than native Hb might be attributable to the fact that the Cys-93(β) residue in Hb was blocked by the crosslinking agent SMCC and that Lys-82(β) was exploited as a binding partner of Cys-34 of HSA [20]. Nonetheless, the high O$_2$ affinity might be favorable in application as a potential O$_2$ carrier. Winslow et al. demonstrated that HBOC with a low O$_2$ affinity engenders excessive O$_2$ release in the arterioles and thereby invokes autoregulatory vasoconstriction [37,38]. Intaglietta et al. reported that lower P_{50} (10 Torr) RBC provides improvement of microvascular function in comparison to the higher P_{50} (50 Torr) RBC in a hemorrhagic shocked hamster model [39]. In light of these investigations, the lower P_{50} might be effective to decrease arteriole O$_2$ transport, potentially eliminating undesired cardiovascular side effects.

Then the equilibrium between O$_2$ and Hb-HSA$_3$(PtNP) cluster was measured to investigate the effect of PtNP on the O$_2$ affinity. The P_{50} and n values of the Hb-HSA$_3$(PtNP) cluster were, respectively, 9 Torr and 1.5 (Figure 8, Table 3). The O$_2$ binding parameters were unaffected by the PtNP association to the HSA shell. We inferred that the Hb-HSA$_3$(PtNP) cluster retained two important benefits for RBC substitute: (i) negative surface net charge and (ii) high O$_2$ affinity.

The O$_2$ complex stability of the Hb-HSA$_3$(PtNP) cluster in PBS (pH 7.4) was evaluated using the autoxidation rate constant (k_{ox}) of the core Hb at 37°C. The k_{ox} value of native Hb was ascertained as 0.037 h^{-1}; this result is well consistent with previously reported data [10,40]. Remarkably, the Hb-HSA$_3$ cluster showed a similar k_{ox} (0.035 h^{-1}) to that of native Hb. The oxyHb nuclei maintain high stability after conjugation with HSA. This fact contrasts with the fact that other HBOCs (PEGylated Hb, polymerized Hb, cross-linked Hb) show larger k_{ox} values relative to naked Hb [10–12]. A possible explanation of the stable O$_2$ complex of our cluster is the enwrapping effect with HSA, which originally possesses a weak antioxidant property. As described earlier in this report, HSA itself showed no measurable SOD or catalase activities in our experimental conditions with a large excess amount of O$_2^{\cdot-}$ and H$_2$O$_2$ (Table 1). Actually, HSA is known to be the predominant antioxidant in plasma (in vivo). Blache et al. estimated that 70% of the free-radical trapping activity of serum is attributed to HSA [41]. Otagiri et al. found that the antioxidant capabilities of HSA are attributable to the six methionine residues and Cys-34 [42]. Therefore, we inferred that covalent enwrapping with HSAs stabilizes the core Hb structure and affords a weak antioxidant effect to the hemes in Hb.

Unexpectedly, the k_{ox} value of Hb-HSA$_3$(PtNP) cluster (0.039 h^{-1}) was almost identical to those observed for Hb-HSA$_3$ and Hb. Kim et al. synthesized various protein-coated PtNPs and analyzed their ROS scavenging activities [24]. They demonstrated that O$_2^{\cdot-}$ and H$_2$O$_2$ dismutation activities of the protein-coated PtNPs are greatly affected by the physicochemical properties and interior shape of the protein shells. In the Hb-HSA$_3$(PtNP) cluster, the PtNP is bound to the cleft on the back side of HSA (Figure 2B), whereas the Cys-34 connection site to the Hb center is located on the front side of HSA. The accessibility of O$_2^{\cdot-}$ and H$_2$O$_2$ from the Hb to PtNP in the HSA shell might be restricted because no accessible channel exists in the proteins.

Finally, we investigated the O$_2$ complex stability of Hb-HSA$_3$(PtNP) cluster in aqueous H$_2$O$_2$ solution. The H$_2$O$_2$ concentration in the human blood is assumed to be tens of micromolars ($\leq$35 μM) [43]. Therefore, the oxidation rates of Hb-HSA$_3$(PtNP), Hb-HSA$_3$, and Hb in aqueous 20 μM H$_2$O$_2$ solution were examined. The time courses of the absorbance increase at 630 nm (which is due to metHb formation) were markedly different in these protein solutions (Figure 9). Native Hb showed a biphasic autoxidation curve. Approximately 50% Hb is oxidized rapidly in the initial phase within 30 min, followed by a second slow oxidation process. The metHb formation level reached 72% after 180 min. It is accepted that the α subunits in Hb are oxidized easily with respect to the β subunits [13]. Because the heme concentration was 40 μM ([Hb] = 10 μM), the α subunit oxidation occurred first, and subsequently the β subunits were oxidized.

The rate of metHb formation, however, was somewhat low in the Hb-HSA$_3$ cluster. In the initial phase, the metHb level increased to 37% within 30 min, followed by a slow oxidation reaction. This low rate appears to be attributable to a wrapping effect of HSA shell. As expected, the Hb-HSA$_3$(PtNP) cluster was remarkably stable in H$_2$O$_2$ solution. We observed no initial oxidation process and only 17% metHb after 180 min, which is 24% of the value of native Hb. This result derives from the high antioxidant activity of the HSA-PtNP unit at the periphery. Actually the oxidation rate of Hb in the coexistence of HSA-PtNP and HSA (Hb/HSA-PtNP/HSA = 1/1/2), that are not covalently linked, was higher than that of the cluster. We can therefore conclude that the HSA-PtNP shell acts as an efficient scavenger for external H$_2$O$_2$ and achieves protection of the core Hb.

Conclusion

A citrate-reduced PtNP ($d = 1.8$ nm) binds strongly within a cleft of HSA, generating a stable HSA-PtNP complex. This platinated protein showed high $O_2^{-\bullet}$ and H_2O_2 dismutation activities. The Hb-HSA₃ cluster also captured PtNP into the external HSA unit. The obtained Hb-HSA₃(PtNP) cluster formed an extremely stable O_2 complex even in H_2O_2 solution. These results suggest that the Hb-HSA₃(PtNP) cluster with (i) negative surface net charges, (ii) high O_2 affinity, and (iii) antioxidant activities can be of tremendous medical importance as an alternative material to RBCs for transfusion in many clinical situations involving ischemia-reperfusion injury.

Author Contributions

Conceived and designed the experiments: HH TK. Performed the experiments: HH RH KY CB TK. Analyzed the data: HH RH KY CB TK. Contributed to the writing of the manuscript: CB TK.

References

1. Squires JE (2002) Artificial blood. Science 295: 1002–1005.
2. Pearce LB, Gawryl MS, Rentko VT, Moon-Massat PF, Rausch CW (2006) HBOCs-201 (Hemoglobin Glutamer-250 (Bovine), Hemopure): clinical studies. In: Winslow RM, editor. Blood substitutes. San Diego: Elsevier. 437–450.
3. Jahr JS, Sadighi A, Doherty L, Li A, Kim HW (2011) Hemoglobin-based oxygen carriers: history, limits, brief summary of the state of the art, including clinical trials. In: Bettati S, Mozzarelli A, editors. Chemistry and biochemistry of oxygen therapeutics: From transfusion to artificial blood. West Sussex: John Wiley & Sons. 301–316.
4. Kluger R, Lui FE (2013) HBOCs from chemical modification of Hb. In: Kim HW, Greenburg AG, editors. Hemoglobin-based oxygen carriers as red cell substitutes and oxygen therapeutics. Berlin Heidelberg: Springer-Verlag. 159–183.
5. Mondery-Pawlowski CL, Tian LL, Pan V, Gupta AS (2013) Synthesis approaches to RBC mimicry and oxygen carrier systems. Biomacromolecules 14: 939–948.
6. Natanson C, Kern SJ, Lurie P, Banks SM, Wolfe SM (2008) Cell-free hemoglobin-based blood substitutes and risk of myocardial infarction and death. J Am Med Assoc 299: 2304–2312.
7. Kluger R (2010) Red cell substitutes from hemoglobin –Do we start all over again? Curr Opin Chem Biol 14: 538–543.
8. Shultz SC, Grady B, Cole F, Hamilton I, Burhop K, et al. (1993) A role of endothelin and nitric oxide in the pressor response to diaspirin cross-linked hemoglobin. J Lab Clin Med 122: 301–308.
9. Doherty DH, Doyle MP, Curry SR, Vali RJ, Fattor TJ, et al. (1998) Rate of reaction with nitric oxide determines the hypertensive effects of cell-free hemoglobin. Nat Biotechnol 16: 672–676.
10. Nagababu E, Ramasamy S, Rifkind JM, Jia Y, Alayash AI (2002) Site-specific cross-linking of human and bovine hemoglobins differentially alters oxygen binding and redox side reactions producing rhombic heme and heme degradation. Biochemistry 41: 7407–7415.
11. Buehler PW, Boykins RA, Jia Y, Norris S, Freedberg DI, et al. (2005) Structural and functional characterization of glutaraldehyde-polymerized bovine hemoglobin and its isolated fractions. Anal Chem 77: 3466–3478.
12. Hu T, Li D, Manjula BN, Acharya SA (2008) Autoxidation of the site-specifically PEGylated hemoglobins: role of the PEG chains and the sites of PEGylation in the autoxidation. Biochemistry 47: 10981–10990.
13. Tsuruga M, Matsuoka M, Hachimori A, Sugawara Y, Shikama K (1998) The molecular mechanism of autoxidation for human oxyhemoglobin. J Biol Chem 273: 8607–8615.
14. Salin ML, McCord JM (1975) Free radicals and inflammation. J Clin Invest 56: 1319–1323.
15. McCord JM (1985) Oxygen-derived free radicals in postischemic tissue injury. N Eng J Med 312: 159–163.
16. McCord JM, Edeas MA (2005) SOD, oxidative stress and human pathologies: a brief history and a future vision. Biomed Phamacother 59: 139–142.
17. D'Agnilloo F, Chang TMS (1998) Polyhemoglobin-superoxide dismutase-catalase as a blood substitute with antioxidant properties. Nat Biotechnol 16: 667–671.
18. Alagic A, Koprianiuk A, Kluger R (2005) Hemoglobin-superoxide dismutase-chemical linkages that create a dual-function protein. J Am Chem Soc 127: 8036–8043.
19. Hathazi D, Mot AC, Vaida A, Scurtu F, Lupan I, et al. (2014) Oxidative protection of hemoglobin and hemerythrin by cross-linking with a nonheme iron peroxidase: potentially improved oxygen carriers for use in blood substitutes. Biomacromolecules 15: 1920–1927.
20. Tomita D, Kimura T, Hosaka H, Daijima Y, Haruki R, et al. (2013) Covalent core-shell architecture of hemoglobin and human serum albumin as an artificial O_2 carrier. Biomacromolecules 14: 1816–1825.
21. Michel CC (1996) Transport of macromolecules through microvascular walls. Cardiovasc Res 32: 644–653.
22. Kajita M, Hikosaka K, Iitsuka M, Kanayama A, Toshima N, et al. (2007) Platinum nanoparticle is a useful scavenger of superoxide anion and hydrogen peroxide. Free Radical Res 41: 615–626.
23. Hamasaki T, Kashiwagi T, Imada T, Nakamichi N, Aramaki S, et al. (2008) Kinetic Analysis of superoxide radical-scavenging and hydroxyl radical-scavenging activities of platinum nanoparticles. Langmuir 24: 7354–7364.
24. San BH, Moh SH, Kim KK (2012) The effect of protein shells on the antioxidation activity of protein-encapsulated platinum nanoparticles. J Mater Chem 22: 1774–1780.
25. Bond GC (1956) The research of ethylene with deuterium over various types of platinum catalyst. Trans Faraday Soc 52: 1235–1244.
26. Cañaveras F, Madueño R, Sevilla JM, Blázquez M, Pineda T (2012) Role of the functionalization of the gold nanoparticles surface on the formation of bioconjugates with human serum albumin. J Phys Chem C 116: 10430–10437.
27. Faulkner KM, Liochev SI, Fridovich I (1994) Stable Mn(III) porphyrins mimic superoxide dismutase in vitro and substitute for in vivo. J Biol Chem 269: 23471–23476.
28. Ohtsu H, Shimazaki Y, Odani A, Yamauchi O, Mori, et al. (2000) Synthesis and characterization of imidazolate-bridged dinuclear complexes as active site models of Cu,Zn-SOD. J Am Chem Soc 122: 5733–5741.
29. Kato R, Akiyama M, Kawakami H, Komatsu T. (2014) Superoxide dismutase activity of the naturally occurring human serum albumin-copper complex without hydroxyl radical formation. Chem Asian J 9: 83–86.
30. Curry S, Madelkow H, Brick P, Franks N (1998) Crystal structure of human serum albumin complexed with fatty acid reveals an asymmetric distribution of binding site. Nat Struct Biol 5: 827–835.
31. Bhattacharya AA, Curry S, Frank N (2000) Binding of the general anesthetics and halothane to human serum albumin. J Biol Chem 275: 38731–38738.
32. Ghuman J, Zunszain PA, Petitpas I, Bhattacharya AA, Otagiri M, et al. (2005) Structural basis of the drug-binding specificity of human serum albumin. J Mol Biol 353: 38–52.
33. Weser U, Schubotz LM (1981) Imidazole-bridged copper complexes as Cu₂Zn₂-superoxide dismutase models. J Mol Catal 13: 249–261.
34. Day BJ, Fridovich I, Crapo JD (1997) Manganic porphyrins possess catalase activity and protect endothelial cells against hydrogen peroxide-mediated injury. Arch Biochem Biophys 347: 256–262.
35. Evenson MA, Deutsch H (1978) Influence of fatty acids on the isoelectric point properties of human serum albumin. Clin Chim Acta 89: 341–354.
36. Antonini E, Brunori M (1971) Hemoglobin and myoglobin in their reactions with ligands. In: Neuberger A, Tatum EL, editors. North-Holland research monographs. Frontiers of biology, vol. 21. Amsterdam: North-Holland Publisher Co. 13–39.
37. Rohlfs RJ, Bruner E, Chiu A, Gonzales A, Gonzales ML, et al. (1998) Arterial blood pressure responses to cell-free hemoglobin solutions and the reaction with nitric oxide. J Biol Chem 273: 12128–12134.
38. Winslow RM (2003) Current status of blood substitute research: towards a new paradigm. J Intern Med 253: 508–517.
39. Intaglietta M, Johnson PC, Winslow RM (1996) Microvascular and tissue oxygen distribution. Cardiovas Res 32: 632–643.
40. Elmer J, Buehler PW, Jia Y, Wood F, Harris DR, et al. (2010) Functional comparison of hemoglobin purified by different methods and their biophysical implications. Biotechnol Bioeng 106: 76–85.
41. Bourdon E, Blache D (2001) The importance of proteins in defense against oxidation. Antioxid Redox Signal 3: 293–311.
42. Iwao Y, Ishima Y, Yamada J, Noguchi T, Kragh-Hansen U, et al. (2011) Quantitative evaluation of the role of cysteine and methionine residues in the antioxidant activity of human serum albumin using recombinant mutants. IUBMB Life 64: 450–454.
43. Halliwell B, Clement MV, Long LH (2000) Hydrogen peroxide in the human body. FEBS Lett 486: 10–13.

24

Relationships between Degree of Polymerization and Antioxidant Activities: A Study on Proanthocyanidins from the Leaves of a Medicinal Mangrove Plant *Ceriops tagal*

Hai-Chao Zhou[1,2,3]*, Nora Fung-yee Tam[2,3]*, Yi-Ming Lin[1], Zhen-Hua Ding[1], Wei-Ming Chai[1], Shu-Dong Wei[1]

1 Key Laboratory of the Ministry of Education for Coastal and Wetland Ecosystems, Xiamen University, Xiamen, China, **2** Department of Biology and Chemistry, City University of Hong Kong, Hong Kong SAR, China, **3** Futian-CityU Mangrove R&D Centre, City University of Hong Kong Shenzhen Research Institute, Shenzhen, China

Abstract

Tannins from the leaves of a medicinal mangrove plant, *Ceriops tagal*, were purified and fractionated on Sephadex LH-20 columns. ^{13}C nuclear magnetic resonance (^{13}C-NMR), reversed/normal high performance liquid chromatography electrospray ionization mass spectrometry (HPLC-ESI MS) and matrix-assisted laser desorption/ionization time-of-flight mass spectrometry (MALDT-TOF MS) analysis showed that the tannins were predominantly B-type procyanidins with minor A-type linkages, galloyl and glucosyl substitutions, and a degree of polymerization (DP) up to 33. Thirteen subfractions of the procyanidins were successfully obtained by a modified fractionation method, and their antioxidant activities were investigated using 2,2-diphenyl-1-picrylhydrazyl (DPPH) scavenging capacity and ferric reducing antioxidant power (FRAP) method. All these subfractions exhibited potent antioxidant activities, and eleven of them showed significantly different mean DP (mDP) ranging from 1.43 ± 0.04 to 31.77 ± 1.15. Regression analysis demonstrated that antioxidant activities were positively correlative with mDP when around mDP <10, while dropped and then remained at a level similar to mDP = 5 with around 95 $\mu g\ ml^{-1}$ for DPPH scavenging activity and 4 mmol AAE g^{-1} for FRAP value.

Editor: Silvia Mazzuca, Università della Calabria, Italy

Funding: This present work was supported by the National Natural Science Foundation of China (URL: http://isisn.nsfc.gov.cn/egrantweb/): no. 41306084 (Hai-Chao Zhou's funding), no. 41176090 (Zhen-Hua Ding's funding), and no. 31070522 (Yi-Ming Lin's funding). The funders had no role in study design, data collection and analysis, decision to publish, or preparation of the manuscript.

Competing Interests: The authors have declared that no competing interests exist.

* Email: zhouhc2013@gmail.com (HCZ); bhntam@cityu.edu.hk (NFYT)

Introduction

Ceriops tagal, a mangrove plant species in the Family of Rhizophoraceae, is one of the medicinal plants in East and Southeast Asia and is often used as a tradition herb to treat diseases such as hemorrhages and malignant ulcers [1,2]. The phytochemistry of *C. tagal* has been reported, however, the work mainly focused on the isolation and identification of terpenoid compounds in root, with little attention on leaf tannins [3,4]. The high level of tannins in the leaves of Rhizophoraceae is known to deter feeding by herbivores [5], but leaf tannins also show a diversity of other biological activities [6]. The unexplored tannins could be novel potential resources of bioactive compounds in mangrove plants. So far, the chemical properties of *C. tagal* tannins have not yet been determined and the structure-activity relationships of tannins are still not clear.

Tannins comprised as much as 20–40% dry weight in the leaf and bark of mangrove plants [7,8]. Compared with hydrolysable tannins, condensed tannins (proanthocyanidins) are more abundant. They are commonly found in mangrove plants [9,10,11,12,13,14] and are also the main component of the

polyphenols in our diet [15]. Because of their antioxidant activities and other potentially health-promoting qualities, proanthocyanidins have attracted more and more research interests in recent years [16,17,18,19]. Proanthocyanidins are oligomers and polymers of flavan-3-ol that are bound together with B-type and A-type linkages [15]. The chemistry and biological features of proanthocyanidins largely depend on their structure, particularly the molecular weight that is also expressed as degree of polymerization (DP) [20,21]. Ariga et al. [22] found that the ability to scavenge free radicals was proportional to DP for simple flavonoid oligomers. Hagerman et al. [23] proved that tannins with highly polymerized and many hydroxyl groups are more potent antioxidants than the simple phenolics. Recently, some findings reported that the increasing DP may enhance the antioxidant power of condensed tannins [24,25]. However, those previous works either limitedly detected the simple oligomers or directly studied the bulk mixture of polymers. The relationships between DP and antioxidant activity of proanthocyanidins, therefore, remained largely unknown.

The present study, therefore, aims to (I) achieve a complete structural characterization of tannins from the leaf of a mangrove plant (*C. tagal*) by a combination of analytical techniques, including ^{13}C-NMR, MALDI-TOF MS, thiolysis degradation, and reversed/normal-phase HPLC-ESI MS; and (II) establish an efficient fractionation method to obtain a series of proanthocyanidins with different DP; and (III) explore the possible relationships between DP and antioxidant activity of condensed tannins.

Materials and Methods

2.1. Chemicals and Materials

Water used in this experiment was purified on a Millipore Milli-Q apparatus. HPLC grade dichloromethane, acetonitrile (CH_3CN), methanol, trifluoroacetic acid (TFA), acetic acid, and all analytical grade solvents (acetone, methanol, n-Butanol etc.) were obtained from Sinopharm (Sinopharm, Shanghai, China). Sephadex LH-20, Folin-Ciocalteu reagents, acetone-d_6, deuteroxide (D_2O), Amberlite IRP-64 cation-exchange resin, cesium chloride (CsCl), 2,5-dihydroxybenzoic acid (DHB), benzylmercaptan, 2,2-diphenyl-1-picrylhydrazyl (DPPH), 2,4,6-tripyridyl-S-triazine (TPTZ), ascorbic acid (AA), and all HPLC standards were purchased from Sigma (St. Louis, MO, USA). The mature leaves (the third pair with fully expanded and dark green) of a mangrove plant *C. tagal* (Rhizophoraceae *Ceriops*), were collected from a mangrove forest in the Dongzhai harbor (19°56′N, 110°34′E), Hainan, China. The leaves were immediately freeze-dried, ground, and stored at −20°C prior to analysis.

2.2. Extraction, Purification and Fractionation of Mangrove Tannins

The extraction procedure was conducted according to the method of Zhou et al. [25] The crude tannins extract (Fc) obtained was applied to a Sephadex LH-20 column (50×1.5 cm i.d.). The procedure of purification and fractionation of F_C was shown in Figure 1. Purified tannins (Fp) and nine subfractions (F1 to F9) were evaporated under vacuum to remove organic solvents followed by freeze-dried. The last two subfractions (F8 and F9)

were re-loaded on a Sephadex LH-20 column (50×1.0 cm i.d.) to yield seven more subfractions as shown below.

2.3. Determination of Total Phenolics and Extractable Condensed Tannins

The procedure described by Zhou et al. [25] to estimate the total phenolics and extractable condensed tannins of the entire leaf of *C. tagal* was used, and the concentrations were determined by Folin-Ciocalteu [26] and butanol-HCl method [27], respectively. The content of the total phenolics (TP) and extractable condensed tannins (ECT) were calculated using the purified tannins (Fp) obtained from the purification of crude tannins extract (Fc) as the standard. Both of them were expressed as mg Fp equivalents per gram dry weight of leaves.

2.4. ^{13}C-NMR Analysis

The obtained purified tannins from *C. tagal* leaves were dissolved in acetone-d_6/D_2O. A Varian Metcury-600 spectrometer (Palo Alto, CA, USA) was employed in the 150 MHz mode [10].

2.5. Reversed-phase HPLC-ESI MS Analysis Followed by Thiolysis

The modified method described by Zhou et al. [9] was carried out to characterize Fp and obtain the mDP for each subfraction. Condensed tannins were thiolysis degraded by benzylmercaptan, and then the degradation products were analyzed on an Agilent 1200 system (Agilent, Palo Alto, CA, USA) interfaced to a QTRAP 3200 (Applied Biosystems, Foster, USA) with a 250 mm×4.6 mm i.d. 5.0 μm Hypersil ODS column (Elite, Dalian, China).

2.6. Normal-phase HPLC-ESI MS Analysis

Normal-phase HPLC-ESI-MS analysis was conducted according to the method of Hellstrom et al. [28] with minor modifications. The HPLC-MS equipment consisted of an Agilent

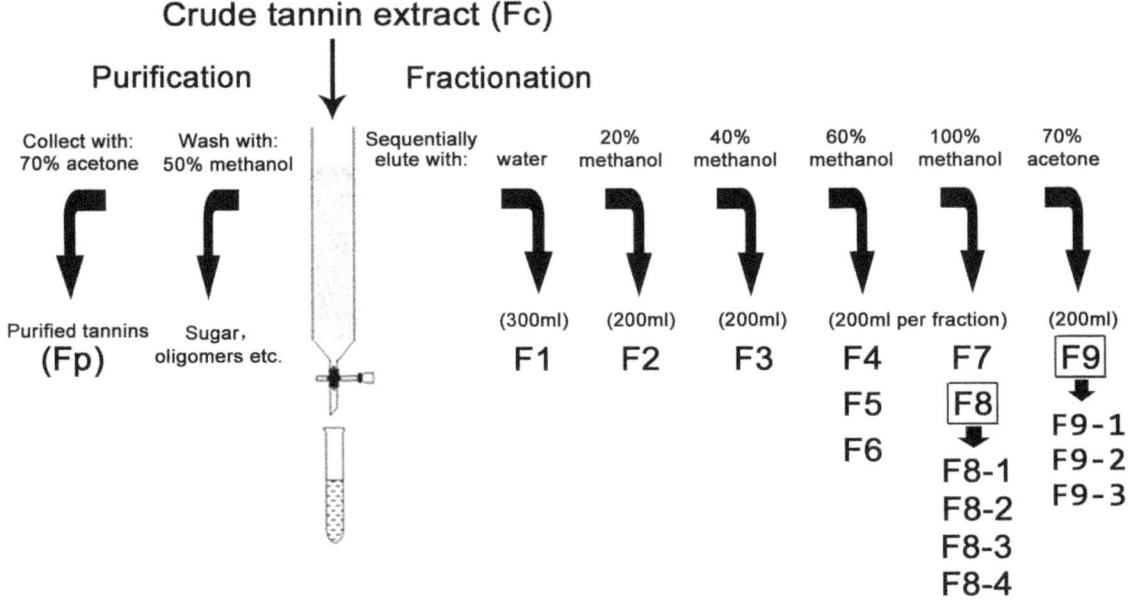

Figure 1. Sephadex LH-20 column chromatography for purification and fractionation of tannins from *C. tagal* leaves. Subfraction 8 (F8) and Subfraction 9 (F9) were further fractionated into F8-1 to F8-4 and F9-1 to F9-3, respectively.

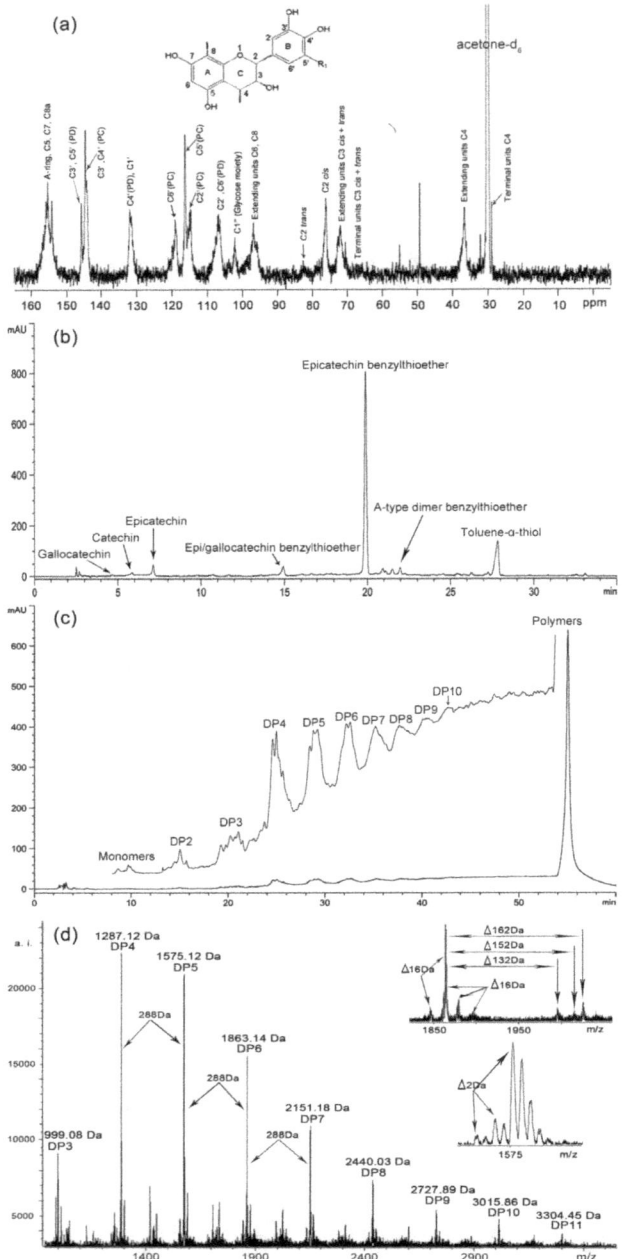

Figure 2. Structural characterization of purified tannins (Fp) from *C. tagal* leaves. Analyzed by (a) ^{13}C-NMR, (b) reversed-phase HPLC-ESI-MS, (c) normal-phase HPLC-ESI-MS and (d) MALDI-TOF MS.

1200 liquid chromatograph system as described above with a 250 mm×4.6 mm i.d. 5.0-μm Silica Luna column (Phenomenex, Darmstadt, Germany).

2.7. MALDI-TOF MS Analysis

The MALDI-TOF MS analyses were performed on a Bruker Reflex III instrument (Germany). The measurements were carried out using the conditions reported in the previous work [29]. Reflectron modes coupled with linear modes were further applied to show the different DP distribution of condensed tannins.

2.8. Determination of Antioxidant Activities

Antioxidant activities of each subfraction were evaluated by DPPH radical scavenging activity (DPPH method) [30] and ferric ion reducing antioxidant power (FRAP method) [31]. Their results were expressed in EC_{50} value (μg mL^{-1}) and mmol ascorbic acid equivalents per gram of each subfraction (mmol AAE g^{-1}), respectively. Procedures were modified according the method of Zhou et al. [32].

2.9. Statistical Analysis

Results of antioxidant activity were expressed as mean ± standard deviation of three independent determinations. A parametric one-way analysis of variance (ANOVA) was used to test any significant difference among subfractions at $P < 0.05$, and all the data without transformation fulfilled the two assumptions, normal distribution and homogeneity in variance of the parametric test. All statistical analyses were performed with SPSS 17.0 for Windows.

2.10. Ethics Statement

Field permit issued by The Mangroves of Dongzhai Gang National Reserve in Dongzhai harbor, Hainan, China, allowed us to collect the mature leaves of *C. tagal* in this site. They also confirmed that our study did not involve any endangered or protected species.

Results and Discussion

3.1. Contents of Total Phenolics and Extractable Condensed Tannins

The total phenolics and extractable condensed tannins content of *C. tagal* leaves was 27.4±2.8% and 20.1±2.0%, respectively. Phenolic compounds (including proanthocyanidins) were considered to be the major contributor to the antioxidant activity of vegetables, fruits or medicinal plants [25,33,34]. The high level of total phenolics and extractable condensed tannins in *C. tagal* leaves strongly suggested that some of their pharmacological effects could be attributed to the presence of these valuable constituents.

3.2. Structural Characterization of Purified Tannins (Fp)

Fp from *C. tagal* leaves were initially analyzed by ^{13}C-NMR spectroscopy, and the signal assignment was performed according to the method of Czochanska et al. [35] The ^{13}C-NMR spectrum (Figure 2a) showed that the characteristic ^{13}C peaks are consistent with that of condensed tannins with dominant procyanidin units and a minor amount of prodelphinidins. The signals near 145 ppm, which arise from quaternary C3′ and C5′ in prodelphinidins (PD), and C3′ and C4′ in procyanidin (PC) units were used to estimate the PD:PC ratio [35]. In this experiment, direct integration of the signals at 146.2 ppm (PD units) and 145.2 ppm (PC units) gave a PD:PC ratio of about 8:92. The signals at 115.1 ppm (C2′), 116.1 ppm (C5′) and 119.0 ppm (C6′) further supported the presence of PC, so does signals at 107–108 ppm (C2′ and C6′) and 133.0 ppm (C4′) for PD. The structural diversity of the linkage (A type and B type) and the stereoisomer of catechin and epicatechin units were apparent from the spectrum. Specially, C5, C7, and C8a carbons of procyanidins appeared at 150 to 160 ppm [36]. The region between 70 and 90 ppm was sensitive to the stereochemistry of the C-ring. The ratio of the 2,3-*cis* to 2,3-*trans* isomers could be determined through the distinct differences in their respective C2 chemical shifts. C2 gave a resonance at 76.7 ppm for the *cis* form and

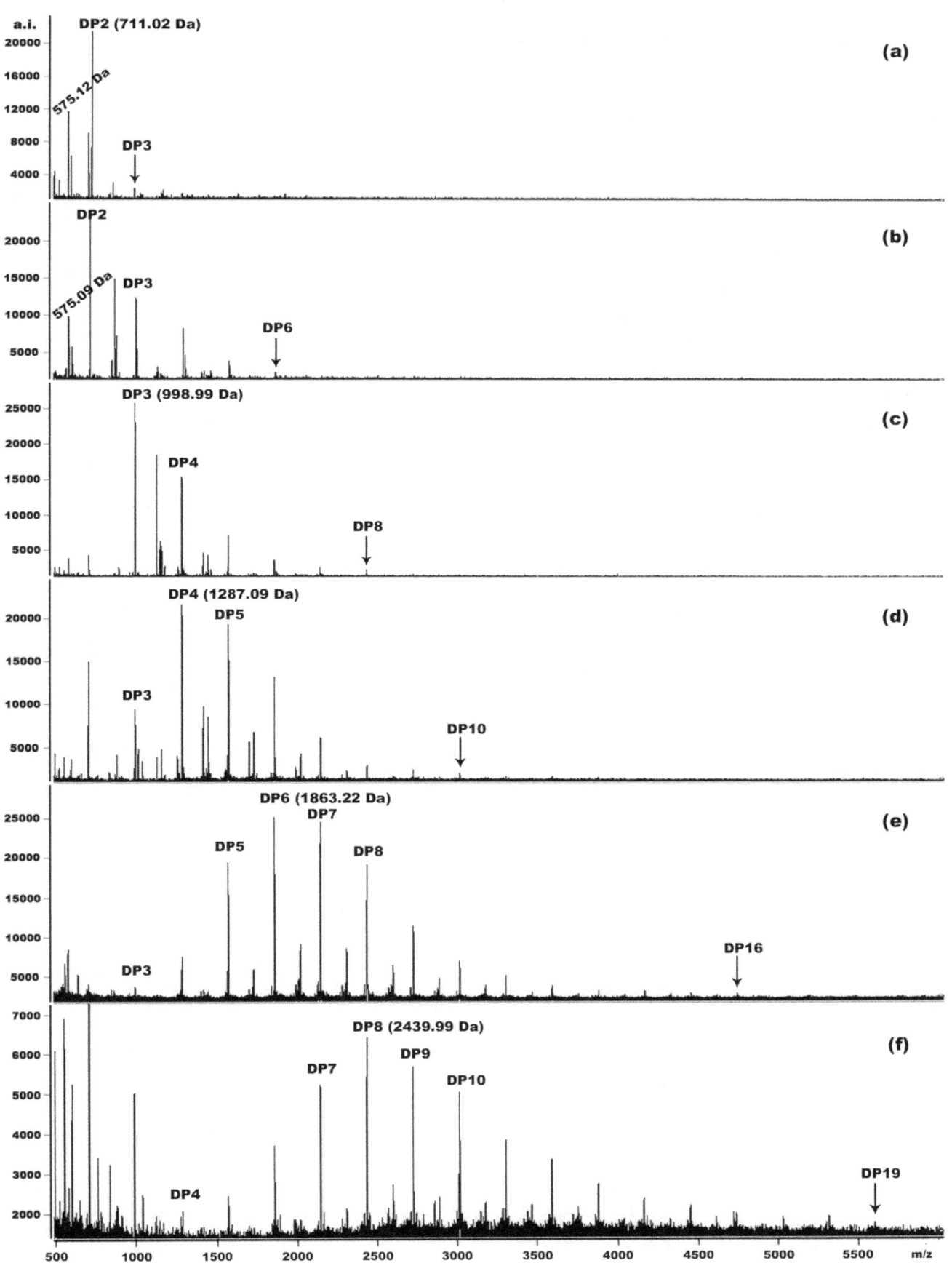

Figure 3. Reflectron mode MALDI-TOF mass spectra for tannin subfraction F4 (a) to F9 (f). Degree of polymerization (DP) of the predominant polymers and largest polymer in each mass spectrum of subfraction are labeled.

83.6 ppm for the *trans* form. From the peak areas, it was estimated that the *cis* isomer is dominant. The C3 generally have their chemical shift 66–68 ppm in terminal units, and 72–74 in extending units [37]. Theoretically, the intensity of the C3 signal in terminal units relative to that of the signal in extension units could be used for elucidating the polymer chain length [35]. However, the application of this technique for the quantification of molecular weight suffered from inaccuracy due to the low signal-to-noise ratio [24]. Similarly, the spectra obtained in the present study also showed low signal-to-noise ratio (Figure 2a). The C4 atoms of the extension units showed a broad peak at around 37 ppm, while the terminal C4 exhibits 29 ppm [37], next to the solvent peak acetone-d_6. The signal at 103.1 ppm would attribute to C1″ of the glycose moiety connected to the C3 position, while the signals of C2″ (74.0 ppm), C3″ (76.3 ppm), C4″ (71.4 ppm) and C5″ (75.7 ppm) were overlapped with the chemical shifts 70–80 ppm referred above [38]. These results thus showed that the mangrove condensed tannins of *C. tagal* leaves were predominantly constituted of epicatechin, the main constitutive monomer with some glycosides.

Although the ^{13}C-NMR spectrum revealed complex structural characteristics of the condensed tannins of *C. tagal* leaves, this analytical technique hardly allows the determination of the polymer chain length and the chemical constitution of individual chains. Moreover, the sequential succession of monomer units in individual chains could not be elucidated [17,36,39]. To overcome these problems, further characterization was continued by means of HPLC-ESI-MS and MALDI-TOF MS.

The nature of flavan-3-ol units within proanthocyanidins was analyzed by acid catalysis in the presence of benzylmercaptan as nucleophile, which had been widely used [18,21,36]. In the thiolysis reaction, terminal units of condensed tannins are released as free flavanoids, while external units are distinguished as benzylthioether adducts. The extension and terminal units of proanthocyanidins could be distinguished by reversed-phase HPLC analysis (Figure 2b). The dominant peak observed was epicatechin benzylthioether along with five other smaller peaks for epicatechin, epi/gallocatechin benzylthioether, A-type dimer benzylthioether, catechin and gallocatechin. These results suggested that epicatechin was the dominant flavan-3-ol unit, which occurred as both terminal and extension units in proanthocyanidins of *C. tagal* leaves.

Normal-phase HPLC analysis was able to further reflects the heterogeneity of the proanthocyanidin mixture [28]. Although the peak band became broader in higher DP and was hard to be resolved, this normal-phase HPLC chromatogram of Fp (Figure 2c) indicated the presence of proanthocyanidins with DP from monomers to decamers (DP10). The results suggested the heterogeneity of proanthocyanidins in their constituent units and the linkages between them.

Figure 2d showed the MALDI-TOF mass spectra of Fp, recorded as Cs$^+$ adducts in the positive ion reflectron mode, and the enlarged spectrums demonstrated the good resolution. These results showed large structural heterogeneity of condensed tannins in the mangrove plant *C. tagal*. The mass increased from DP3 (999.08 Da) to DP11 (3304.45 Da) by 288 Da, which corresponded to one epi/catechin monomer unit. For each multiple, substructures with 16 Da, both higher and lower, could be found (Figure 2d). These masses had been identified as heteropolymers of repeating flavan-3-ol units, which demonstrated the coexistence

of epi/gallocatechin and epi/afzelechin [39]. Additionally, there were three series peaks with distances of 162 Da, 152 Da, and 132 Da. The first series (162 Da) represented glucosylated heteropolyflavans containing the flavanone [39]. The second series (152 Da) was corresponding to the addition of one galloyl group at the heterocyclic C-ring [12,24]. And the 132 Da distance may have two different interpretations according to previous works. Reed et al. [39] interpreted it as the substitutions by pentoside (132 Da), while Xiang et al. [40] explained as quasimolecular ions [M+2Cs$^+$−H]$^+$ that generated by simultaneous attachment of two Cs$^+$ and loss of a proton. The distinct decrease of 2 Da illustrated the nature of interflavan bonds including A-type and B-type linkages. The data from MALDI-TOF MS well agreed with the results obtained by ^{13}C-NMR, reversed and normal-phase HPLC-ESI-MS analyses.

3.3. MALDI-TOF MS Analysis of Proanthocyanidin Subfractions

For the first step, ten subfractions (F1-F9) were obtained by fractionation on Sephadex LH-20 as described in methods section. After analysis by MALDI-TOF MS, fractions F1-F3 showed a cluster of unresolved peak lower than 500 Da (data not shown), which could be the sugars and other impurities. The spectra obtained by reflectron modes of the seven collected subfractions from fractions F4 (Figure 3a) to F9 (Figure 3f) clearly displayed the distinct predominant polymers and polymer range. The signals of lower oligomers were firstly detected in fraction F4 dominated by dimers (Figure 3a). Fractionation of the lower oligomers that could lead to the saturation of detector can significantly enhance the sensitivity of the detection of large polymers of proanthocyanidins under MALDI-TOF analysis [17]. Therefore, the last subfraction (F9) detected in the present study reached as high as DP19 polymer (Figure 3f), which demonstrated that the fractionation method could improve the detection of high DP polymers compared with the spectra of Fp (Figure 2d), even though it did not completely overcome the discrimination of high molecular weight polymers [17].

Since F8 and F9 still possessed of large DP distribution (with DP3-DP16 and DP4-DP19, respectively), a refined fractionation method was conducted to re-fractionate F8 (Figure 4) and F9 (Figure 5), which yielded other four and three subfractions, respectively, as well as with increasing trends for their predominant polymers and polymer range.

After refined fractionation for F8 and F9, as high as DP27 and DP33 polymer were detected in linear spectra of F8-4 (Figure 4) and F9-3 (Figure 5), respectively. It suggested that proanthocyanidins in *C. tagal* leaves can reach DP33, or even more. The present study clearly demonstrates a higher DP and clearer DP distribution could be obtained than many previous works reporting a high DP in various plant materials using MALDI-TOF MS [17]. In the present study, however, we showed a higher DP and clearer DP distribution than those described. Both the reflectron and linear modes of MALDI-TOF MS have been successfully applied to the analysis of proanthocyanidins. The high mass resolution power of reflectron modes allows distinguishing the mass of the isotopic peaks with enough accuracy to compare with the theoretical calculated mass, while the linear mode provides better information about the mass distribution, especially for the high DP of proanthocyanidins [17,29].

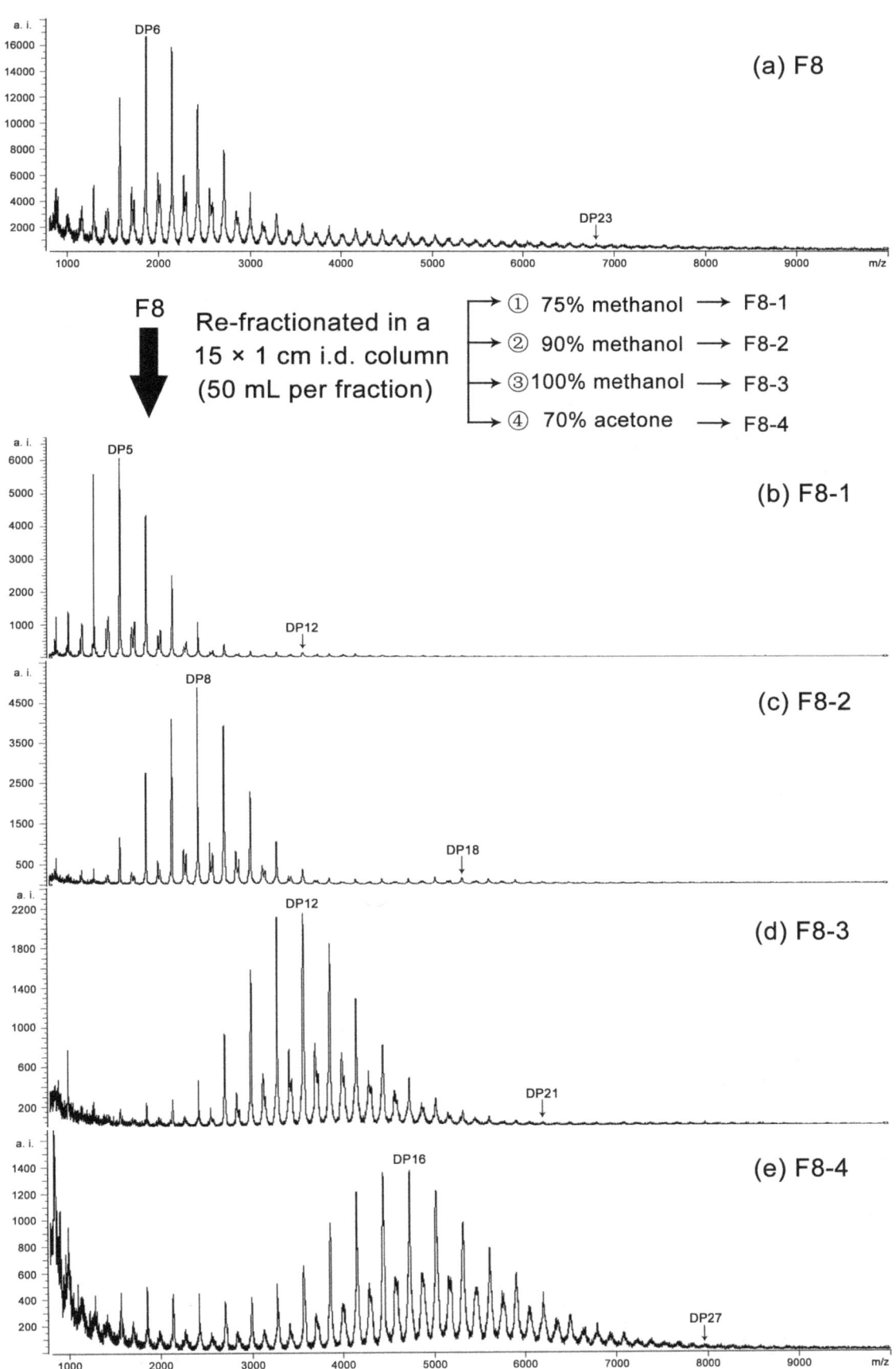

Figure 4. Linear mode MALDI-TOF mass spectra for tannin subfraction F8 (a). F8-1 (b) to F8-4 (e) were obtained by the further re-fractionation of F8. Degree of polymerization (DP) of the predominant polymers and largest polymer in each mass spectrum of subfraction are labeled.

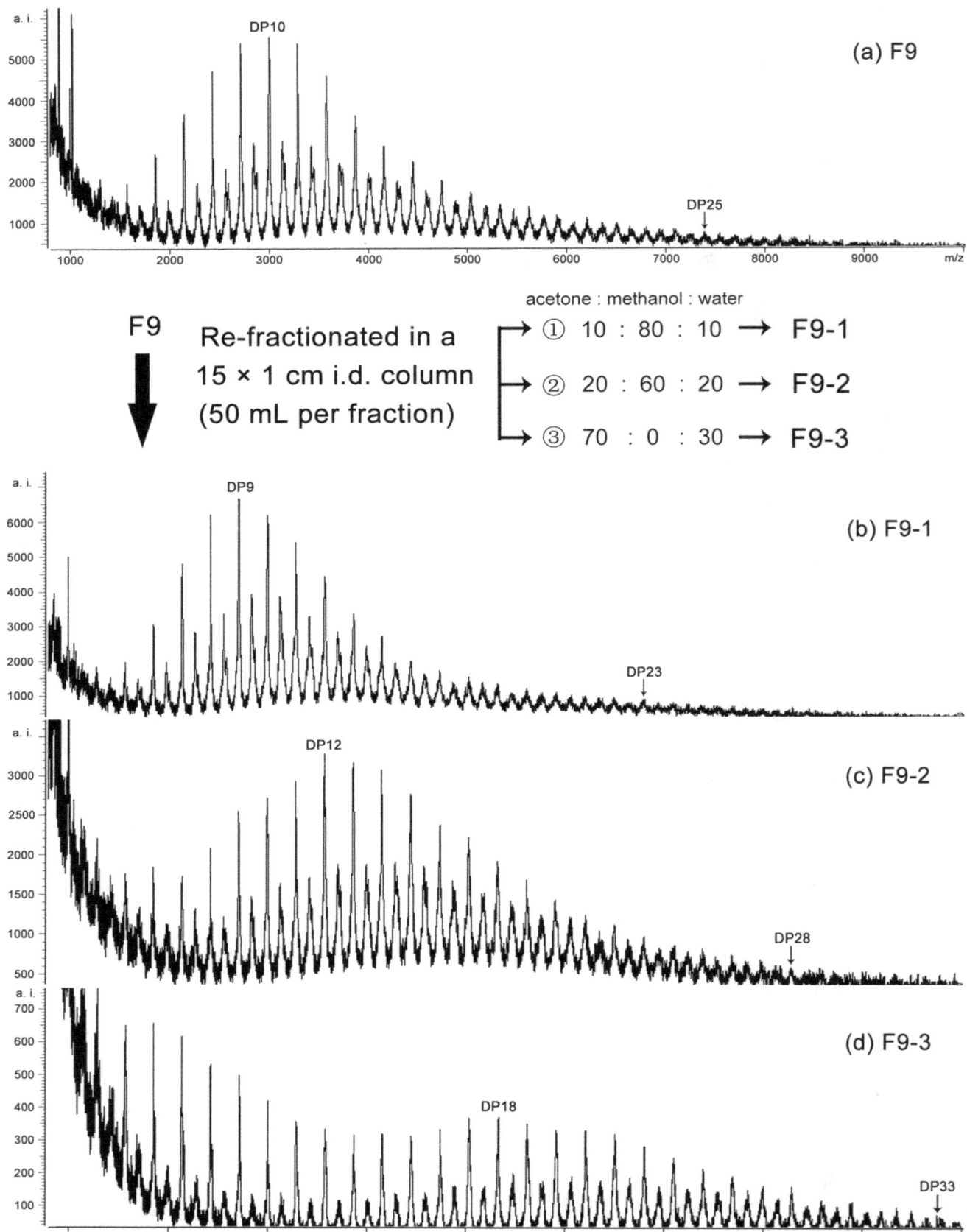

(a) F9

F9 → Re-fractionated in a 15 × 1 cm i.d. column (50 mL per fraction)

acetone : methanol : water
① 10 : 80 : 10 → F9-1
② 20 : 60 : 20 → F9-2
③ 70 : 0 : 30 → F9-3

(b) F9-1

(c) F9-2

(d) F9-3

Figure 5. Linear mode MALDI-TOF mass spectra for tannin subfraction F9 (a). F9-1 (b) to F9-3 (d) were obtained by the further re-fractionation of F9. Degree of polymerization (DP) of the predominant polymers and largest polymer in each mass spectrum of subfraction are labeled.

3.4. Mean Degree of Polymerization (mDP) Analysis of Proanthocyanidin Subfractions

Although the above MALDI-TOF mass spectra graphically showed the differences among each proanthocyanidin subfraction, it only qualitatively illustrated the changes of DP. Therefore, mDP was calculated using thiolysis degradation coupled with reversed-phase HPLC-ESI MS. Table 1 listed the mDP of 13 proanthocyanidin subfractions from *C. tagal* leaves. The mDP increased with the eluted order on Sephadex LH-20: ranging from 1.43 ± 0.04 (F4) to 16.14 ± 0.58 (F9), from 7.26 ± 0.31 (F8-1) to 16.14 ± 0.58 (F8-4) and from 14.19 ± 0.50 (F9-1) to 31.77 ± 1.15 (F9-3), which agreed with the results of MALDI-TOF analysis.

These results confirmed that the fractionation method used in this study was effective to gradually separate the plant proanthocyanidins according to DP. The occurrence of proanthocyanidins in complex mixtures makes it difficult to firmly establish the effect of molecular size (or DP) on the biologically important properties of proanthocyanidins [41]. Therefore, the distinct subfractions obtained from this fractionation method were subjected to explore the relationship between DP and antioxidant activities for proanthocyanidins.

3.5. Antioxidant Activities and Their Relationships with mDP

DPPH method is a reliable and reproducible method to determine the radical scavenging activity of proanthocyanidins extracted from plant or fruits [24,42]. FRAP method is based on the redox reaction of ferric ion in the presence of a reducer. The reduction capacity of a compound may serve as a significant indicator of its potential antioxidant activity [43]. Table 1 summarized the antioxidant activities (DPPH and FRAP method)

of proanthocyanidin subfractions from *C. tagal* leaves. The highest antioxidant activities were found in F8 with $EC_{50/DPPH}$ value 78.13 ± 1.14 µg ml^{-1} and FRAP value 5.87 ± 0.08 mmol AAE g^{-1}, while the lowest in F4 with $EC_{50/DPPH}$ value 133.45 ± 2.24 µg ml^{-1} and FRAP value 2.58 ± 0.04 mmol AAE g^{-1}. FRAP showed a similar trend to the results of DPPH assay. Although these two antioxidant assays with different mechanisms were used in this study, other antioxidant assays, such as peroxyl radical scavenging capacity or even assays for peroxidative/prooxidative enzymes should be included in further research, such as peroxyl radical scavenging capacity or even assays for peroxidative/prooxidative enzymes. Compared with the results from other plant materials, such as *Delonix regia* [18], *Litchi chinensis* [25] and *Garcinia mangostana* [32], proanthocyanidins from *C. tagal* leaves also exhibited substantial radical scavenging activity and reduction capacity.

Regression analysis was performed to establish the relationship between antioxidant activities and mDP (Figure 6). It demonstrated that antioxidant activities was positively correlative with mDP when mDP <10, while dropped and then remained at a level with around 95 µg ml^{-1} for $EC_{50/DPPH}$ value and 4 mmol AAE g^{-1} for FRAP value (similar to the antioxidant level of mDP = 5). Although some researchers attempted to study this relationship of proanthocyanidins, the work was limited to low oligomers and low DP. For instance, Gaulejac et al. [44] found an increase in the activity of procyanidins, but their work only focused on the oligomers from 1 to 4 units. Es-Safi et al. [24] simply compared the commercial standard substances (catechin and B$_3$ procyanidin dimer) with pear juice polymeric proanthocyanidins. Jerez et al. [42] showed an increase of the antiradical activity of *Pinus pinaster* and *Pinus radiata* procyanidins up to 6–7 mDP (the mDP obtained by thiolysis with cysteamine). Recently, Zhou et al. [32]

Table 1. Mean degree of polymerization (mDP) and antioxidant activities of each subfraction obtained by fractionation of proanthocyanidins from *C. tagal* leaves.

Fractions	mDP	$EC_{50/DPPH}$ value[I] (µg ml^{-1})	FRAP value[II] (mmol AAE g^{-1})
F4	1.43 ± 0.04 k	133.45 ± 2.24 a	2.58 ± 0.04 h
F5	2.49 ± 0.14 j	114.74 ± 2.24 b	3.60 ± 0.05 g
F6	3.52 ± 0.34 i	103.56 ± 0.79 c	4.05 ± 0.13 ef
F7	7.78 ± 0.29 h	83.60 ± 0.62 g	5.46 ± 0.05 b
F8	10.52 ± 0.62 g	78.13 ± 1.14 h	5.87 ± 0.08 a
F8-1	7.26 ± 0.31 h	84.11 ± 1.41 g	5.19 ± 0.05 c
F8-2	11.72 ± 0.56 f	97.04 ± 1.04 d	3.83 ± 0.07 fg
F8-3	16.84 ± 0.82 d	86.40 ± 0.24 fg	4.28 ± 0.11 e
F8-4	24.62 ± 1.15 b	98.49 ± 1.48 d	3.80 ± 0.01 fg
F9	16.14 ± 0.58 d	87.14 ± 1.68 f	4.52 ± 0.23 d
F9-1	14.19 ± 0.50 e	96.49 ± 1.55 d	3.80 ± 0.16 fg
F9-2	22.01 ± 0.81 c	93.17 ± 2.01 e	4.05 ± 0.08 ef
F9-3	31.77 ± 1.15 a	105.37 ± 1.22 c	3.66 ± 0.09 g

[I]The $EC_{50/DPPH}$ value (µg ml^{-1}) is the concentration of each subfraction at which the scavenging activity is 50%. Values with different letters in the same column are significantly different at $P<0.05$ level.
[II]The FRAP value (mmol AAE g^{-1}) is expressed in mmol ascorbic acid equivalents per gram of each subfraction. Values with different letters in the same column are significantly different at $P<0.05$ level.

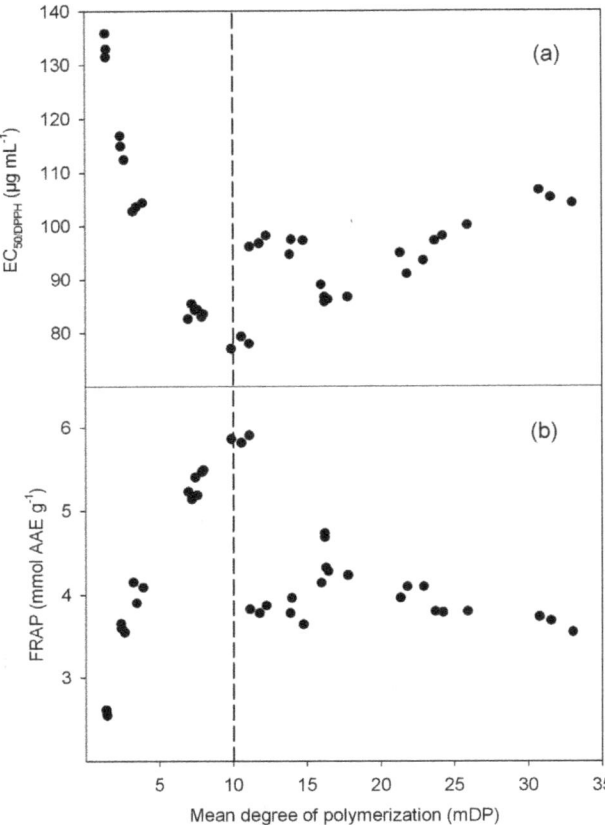

Figure 6. Regression analysis for mean degree of polymerization (mDP) and antioxidant activity. (a) $EC_{50/DPPH}$ value and (b) FRAP value correlated with mDP for each proanthocyanidin subfraction from *C. tagal* leaves.

reported an increasing antioxidant activity of mangosteen tannins with mDP between 2.71 and 9.27, but decreasing in mDP = 16.80, and proposed that 9–10 mDP could be considered as a dividing and critical point for predicting the structure-activity of mango-

steen condensed tannins, which was confirmed with the present results. However, all those previous works could not provide more information for the higher DP. In this study, a decrease and then remain at a lower antioxidant level was showed when mDP >10. However, the antioxidant activity based on the DPPH test in the present study could be influenced by the amount of total polyphenols. Zhou et al. [32] also showed linear relationships between DPPH and total polyphenol content. Similarly, antioxidant tests were positively correlated with total polyphenols [32,45,46,47]. So only the mDP data may not be sufficient to explain the differences in EC_{50} and it is better to explain the differences in EC_{50} with reference to total phenolics and ECT in each fraction. Further investigations on the relationships between antioxidant activity and polyphenols in mangrove leaves and their mechanisms are required. Further investigations on the relationships between antioxidant activity and polyphenols in mangrove leaves and their mechanisms are required.

Mangrove tannins from *C. tagal* leaves were successfully characterized by ^{13}C-NMR, MALDI-TOF MS, HPLC-ESI MS and column chromatographic fractionation. They had substantial DPPH free radical scavenging ability and ferric reducing antioxidant power, which could be used as a new source of antioxidants. The major challenge in the research on condensed tannins is probably the difficulty in obtaining them in an individual molecular form. The complete purification of a procyanidin with a DP above five is almost impossible. Therefore, for studying their structures and properties, more or less mixtures polymerized are often employed [48]. More, the synergistic effects of active mixtures make plant extracts and fractions more interesting than the pure compounds for functional food applications [42]. In the present study, an effective fractionation method was established to elucidate more about the relationships between DP and antioxidant activities. The established relationships could be used as a theoretical method for predicting the structure-activity of proanthocyanidins.

Author Contributions

Conceived and designed the experiments: HCZ NFYT. Performed the experiments: HCZ WMC SDW. Analyzed the data: HCZ YML. Contributed reagents/materials/analysis tools: HCZ YML ZHD. Wrote the paper: HCZ NFYT.

References

1. Lin P (2006) Marine higher plant ecology. Science Press, Beijing: 76–80.
2. Rastogi RP, Mehrotra BN (1991) Compendium of Indian medicinal plants. Publications & Information Directorate, New Delhi 1.
3. Chen JD, Qiu Y, Yang ZW, Lin P, Lin YM (2008) Dimeric diterpenes from the roots of the mangrove plant *Ceriops tagal*. Helvetica Chimica Acta 91: 2292–2298.
4. Chacha M, Mapitse R, Afolayan AJ, Majinda RRT (2008) Antibacterial diterpenes from the roots of *Ceriops tagal*. Natural Product Communications 3: 17–20.
5. Feller IC, Whigham DF, O'Neill JP, McKee KL (1999) Effects of nutrient enrichment on within-stand cycling in a mangrove forest. Ecology 80: 2193–2205.
6. Mainoya J, Mesaki S, Banyikwa F (1986) The distribution and socio-economic aspects of mangrove forests in Tanzania. Man in the Mangroves: the Socio-Economic Situation of Human Settlements in Mangrove Forests: 87–95.
7. Kraus TEC, Dahlgren RA, Zasoski RJ (2003) Tannins in nutrient dynamics of forest ecosystems - a review. Plant and Soil 256: 41–66.
8. Lin YM, Liu JW, Xiang P, Lin P, Ye GF, et al. (2006) Tannin dynamics of propagules and leaves of *Kandelia candel* and *Bruguiera gymnorrhiza* in the Jiulong River Estuary, Fujian, China. Biogeochemistry 78: 343–359.
9. Zhou HC, Tam NYF, Lin YM, Wei SD, Li YY (2012) Changes of condensed tannins during decomposition of leaves of *Kandelia obovata* in a subtropical mangrove swamp in China. Soil Biology and Biochemistry 44: 113–121.
10. Zhang LL, Lin YM, Zhou HC, Wei SD, Chen JH (2010) Condensed tannins from mangrove species *Kandelia candel* and *Rhizophora mangle* and their antioxidant activity. Molecules 15: 420–431.
11. Maie N, Pisani O, Jaffe R (2008) Mangrove tannins in aquatic ecosystems: Their fate and possible influence on dissolved organic carbon and nitrogen cycling. Limnology and Oceanography 53: 160–171.
12. Oo CW, Pizzi A, Pasch H, Kassim MJ (2008) Study on the structure of mangrove polyflavonoid tannins with MALDI-TOF mass spectrometry. Journal of Applied Polymer Science 109: 963–967.
13. Hernes PJ, Benner R, Cowie GL, Goni MA, Bergamaschi BA, et al. (2001) Tannin diagenesis in mangrove leaves from a tropical estuary: A novel molecular approach. Geochimica Et Cosmochimica Acta 65: 3109–3122.
14. Wang Y, Zhu H, Tam NFY (2014) Polyphenols, tannins and antioxidant activities of eight true mangrove plant species in South China. Plant and Soil 374: 549–563.
15. Khanbabaee K, van Ree T (2001) Tannins: Classification and definition. Natural Product Reports 18: 641–649.
16. Alasalvar C, Karamac M, Kosinska A, Rybarczyk A, Shahidi F, et al. (2009) Antioxidant Activity of Hazelnut Skin Phenolics. Journal of Agricultural and Food Chemistry 57: 4645–4650.
17. Monagas M, Quintanilla-Lopez JE, Gomez-Cordoves C, Bartolome B, Lebron-Aguilar R (2010) MALDI-TOF MS analysis of plant proanthocyanidins. Journal of Pharmaceutical and Biomedical Analysis 51: 358–372.
18. Chai WM, Shi Y, Feng HL, Qiu L, Zhou HC, et al. (2012) NMR, HPLC-ESI-MS, and MALDI-TOF MS analysis of condensed tannins from *Delonix regia* (Bojer ex Hook.) Raf. and their bioactivities. Journal of Agricultural and Food Chemistry 60: 5013–5022.

19. Chen XX, Shi Y, Chai WM, Feng HL, Zhuang JX, et al. (2014) Condensed tannins from *Ficus virens* as tyrosinase inhibitors: structure, inhibitory activity and molecular mechanism. Plos One 9: e91809.

20. Gu LW, Kelm M, Hammerstone JF, Beecher G, Cunningham D, et al. (2002) Fractionation of polymeric procyanidins from lowbush blueberry and quantification of procyanidins in selected foods with an optimized normal-phase HPLC-MS fluorescent detection method. Journal of Agricultural and Food Chemistry 50: 4852–4860.

21. Li CM, Leverence R, Trombley JD, Xu SF, Yang J, et al. (2010) High molecular weight persimmon (*Diospyros kaki* L.) proanthocyanidin: a highly galloylated, A-linked tannin with an unusual flavonol terminal unit, myricetin. Journal of Agricultural and Food Chemistry 58: 9033–9042.

22. Ariga T, HAMANO M (1990) Radical scavenging action and its mode in procyanidins B-1 and B-3 from azuki beans to peroxyl radicals. Agricultural and Biological Chemistry 54: 2499–2504.

23. Hagerman AE, Riedl KM, Jones GA, Sovik KN, Ritchard NT, et al. (1998) High molecular weight plant polyphenolics (tannins) as biological antioxidants. Journal of Agricultural and Food Chemistry 46: 1887–1892.

24. Es-Safi NE, Guyot S, Ducrot PH (2006) NMR, ESI/MS, and MALDI-TOF/MS analysis of pear juice polymeric proanthocyanidins with potent free radical scavenging activity. Journal of Agricultural and Food Chemistry 54: 6969–6977.

25. Zhou HC, Lin YM, Li YY, Li M, Wei SD, et al. (2011) Antioxidant properties of polymeric proanthocyanidins from fruit stones and pericarps of Litchi chinensis Sonn. Food Research International 44: 613–620.

26. Makkar HPS, Blümmel M, Borowy NK, Becker K (1993) Gravimetric determination of tannins and their correlations with chemical and protein precipitation methods. Journal of Agricultural and Food Chemistry 61: 161–165.

27. Terrill TH, Rowan AM, Douglas GD, Barry TN (1992) Determination of extractable and bound condensed tannin concentrations in forage plants, protein concentrate meals and cereal grains. Journal of the Science of Food and Agriculture 58: 321–329.

28. Hellstrom J, Sinkkonen J, Karonen M, Mattila P (2007) Isolation and structure elucidation of procyanidin oligomers from saskatoon berries (*Amelanchier alnifolia*). Journal of Agricultural and Food Chemistry 55: 157–164.

29. Zhou HC, Lin YM, Chai WM, Wei SD, Liao MM (2011) Characterization of condensed tannins from litchi seed by reflectron modes and linear modes of MALDI-TOF MS. Acta Chimica Sinica 69: 2981–2986.

30. Brand-Williams W, Cuvelier ME, Berset C (1995) Use of a free radical method to evaluate antioxidant activity. Lebensmittel Wissenschaft and Technologie 28: 25–30.

31. Benzie IFF, Strain JJ (1996) The ferric reducing ability of plasma as a measure of "antioxidant power": the FRAP assay. Analytical Biochemistry 239: 70–76.

32. Zhou HC, Lin YM, Wei SD, Tam NFY (2011) Structural diversity and antioxidant activity of condensed tannins fractionated from mangosteen pericarp. Food Chemistry 129: 1710–1720.

33. Cai YZ, Sun M, Xing J, Luo Q, Corke H (2006) Structure-radical scavenging activity relationships of phenolic compounds from traditional Chinese medicinal plants. Life Sciences 78: 2872–2888.

34. Chirinos R, Rogez H, Campos D, Pedreschi R, Larondelle Y (2007) Optimization of extraction conditions of antioxidant phenolic compounds from mashua (*Tropaeolum tuberosum* Ruiz & Pavo) tubers. Separation and Purification Technology 55: 217–225.

35. Czochanska BZ, Foo YL, Newman RH, Porter IJ (1980) Polymeric proanthocyanidin. stereochemistry, structural units, and molecular weight. Journal of the Chemical Society, Perkin Transactions 1: 2278–2286.

36. Fu C, Loo AEK, Chia FPP, Huang D (2007) Oligomeric proanthocyanidins from mangosteen pericarps. Journal of Agricultural and Food Chemistry 55: 7689–7694.

37. Kraus TEC, Yu Z, Preston CM, Dahlgren RA, Zasoski RJ (2003) Linking chemical reactivity and protein precipitation to structural characteristics of foliar tannins. Journal of Chemical Ecology 29: 703–730.

38. Castillo-Munoz N, Gomez-Alonso S, Garcia-Romero E, Gomez MV, Velders AH, et al. (2009) Flavonol 3-O-glycosides series of *Vitis vinifera* cv. *Petit Verdot* red wine grapes. Journal of Agricultural and Food Chemistry 57: 209–219.

39. Reed JD, Krueger CG, Vestling MM (2005) MALDI-TOF mass spectrometry of oligomeric food polyphenols. Philadelphia, PA. Pergamon-Elsevier Science Ltd: 2248–2263.

40. Xiang P, Lin YM, Lin P, Xiang C, Yang ZW, et al. (2007) Effect of cationization reagents on the matrix-assisted laser desorption/ionization time-of-flight mass spectrum of Chinese gallotannins. Journal of Applied Polymer Science 105: 859–864.

41. Taylor AW, Barofsky E, Kennedy JA, Deinzer ML (2003) Hop (*Humulus lupulus* L.) proanthocyanidins characterized by mass spectrometry, acid catalysis, and gel permeation chromatography. Journal of Agricultural and Food Chemistry 51: 4101–4110.

42. Jerez M, Tourino S, Sineiro J, Torres JL, Nunez MJ (2007) Procyanidins from pine bark: Relationships between structure, composition and antiradical activity. Food Chemistry 104: 518–527.

43. Meri S, Kanner J, Akiri B, Hadas SP (1995) Determination and involvement of aqueous reducing compounds in oxidative defense systems of various senescing leaves. Journal of Agricultural and Food Chemistry 43: 1813–1815.

44. de Gaulejac N, Vivas N, de Freitas V, Bourgeois G (1999) The influence of various phenolic compounds on scavenging activity assessed by an enzymatic method. Journal of the Science of Food and Agriculture 79: 1081–1090.

45. Cai YZ, Luo Q, Sun M, Corke H (2004) Antioxidant activity and phenolic compounds of 112 traditional Chinese medicinal plants associated with anticancer. Life Sciences 74: 2157–2184.

46. Silva E, Souza J, Rogez H, Rees J, Larondelle Y (2007) Antioxidant activities and polyphenolic contents of fifteen selected plant species from the Amazonian region. Food Chemistry 101: 1012–1018.

47. Lizcano LJ, Bakkali F, Ruiz-Larrea MB, Ruiz-Sanz JI (2010) Antioxidant activity and polyphenol content of aqueous extracts from Colombian Amazonian plants with medicinal use. Food Chemistry 119: 1566–1570.

48. Guyot S, Marnet N, Drilleau JF (2001) Thiolysis-HPLC characterization of apple procyanidins covering a large range of polymerization states. Journal of Agricultural and Food Chemistry 49: 14–20.

Permissions

List of Contributors

Bertrand Sagnia
Laboratory of Microbiology and Immunology, Chantal BIYA International Reference Centre for Research on Prevention and Management of HIV/AIDS (CIRCB), Yaounde, Cameroon

Donatella Fedeli and Giancarlo Falcioni
School of Pharmacy, University of Camerino, Camerino (MC), Italy

Rita Casetti
Laboratory of Cellular Immunology, National Institute for Infectious Diseases "Lazzaro Spallanzani", Rome, Italy

Vittorio Colizzi
Laboratory of Cellular Immunology, National Institute for Infectious Diseases "Lazzaro Spallanzani", Rome, Italy
Department of Biology, University of Rome Tor Vergata, Rome, Italy

Carla Montesano
Department of Biology, University of Rome Tor Vergata, Rome, Italy

Jie Zhang, Min Zhang, Shengke Tian, Lingli Lu, M. J. I. Shohag and Xiaoe Yang
MOE Key Laboratory of Environment Remediation and Ecosystem Health, College of Environmental and Resource Sciences, Zhejiang University, Hangzhou, China

Juanjuan Fu, Yongfang Sun, Xitong Chu, Yuefei Xu and Tianming Hu
Department of Grassland Science, College of Animal Science and Technology, Northwest A&F University, Yangling, Shaanxi Province, P. R. China Abstract

Sandip Pal, Subrata Kumar Dey and Chabita Saha
Department of Biotechnology, West Bengal University of Technology, Kolkata, West Bengal, India

Lizhen Zhang
College of Life Science, Shanxi University, Taiyuan, China

Ruihai Liu
Department of Food Science, Cornell University, Ithaca, New York, United States of America

Wei Niu
Shanxi Academy of Agricultural Sciences, Taiyuan, China

Poliana Cardoso-Gustavson
Programa de Pós-Graduação em Biodiversidade Vegetal e Meio Ambiente, Instituto de Botânica, São Paulo, São Paulo, Brazil

Vanessa Palermo Bolsoni, Debora Pinheiro de Oliveira, Maria Tereza Gromboni Guaratini and Silvia Ribeiro de Souza
Núcleo de Pesquisa em Ecologia, Instituto de Botânica, São Paulo, São Paulo, Brazil

Marcos Pereira Marinho Aidar and Mauro Alexandre Marabesi
Núcleo de Pesquisa em Fisiologia e Bioquímica, Instituto de Botânica, São Paulo, São Paulo, Brazil

Edenise Segala Alves
Núcleo de Pesquisa em Anatomia, Instituto de Botânica, São Paulo, São Paulo, Brazil

Tamara Sotelo, María Elena Cartea, Pablo Velasco and Pilar Soengas
Group of Genetics, Breeding and Biochemistry of Brassicas, Department of Plant Genetics, Misión Biológica de Galicia, Spanish Council for Scientific Research (MBG-CSIC), Pontevedra, Spain

Santa Cirmi, Nadia Ferlazzo and Michele Navarra
Department of Drug Sciences and Products for Health, University of Messina, Messina, Italy

Roberto Risitano, Monica Currò, Daniela Caccamo and Riccardo Ientile
Department of Biomedical Sciences and Morphofunctional Imaging, University of Messina, Messina, Italy

Pietro Campiglia
Department of Pharmaceutical and Biomedical Sciences, University of Salerno, Fisciano, Salerno, Italy

Xiaomin Li, Wang Zhu, Cheng Chen and Zhihong Sun
Key Laboratory for Bio-Feed and Molecular Nutrition, Southwest University, Chongqing, P. R. China

Zhiliang Tan and Jinghe Kang
Key Laboratory of Agro-Ecological Processes in Subtropical Region, Hunan Research Center of Livestock & Poultry Sciences, South-Central Experimental Station of Animal Nutrition and Feed Science in Ministry of Agriculture, Institute of Subtropical Agriculture, The Chinese Academy of Sciences, Changsha, Hunan, P.R. China

Tan Yang
Key Laboratory for Bio-Feed and Molecular Nutrition, Southwest University, Chongqing, P. R. China,
Key Laboratory of Agro-Ecological Processes in Subtropical Region, Hunan Research Center of Livestock & Poultry Sciences, South-Central Experimental Station of Animal Nutrition and Feed Science in Ministry of Agriculture, Institute of Subtropical Agriculture, The Chinese Academy of Sciences, Changsha, Hunan, P.R. China

Chang Li, Xiaofei Xu and Yuanjiang Pan
Department of Chemistry, Zhejiang University, Hangzhou, P. R. China

Xiu Jun Wang
Department of Pharmacology, School of Medicine, Zhejiang University, Hangzhou, P. R. China

Yongsheng Chen Hong Wang and Xinbo Guo
School of Light Industry and Food Sciences, South China University of Technology, Guangzhou, Guangdong, The People's Republic of China

Gaoyan Wang
Department of Food Science, Cornell University, Ithaca, New York, United States of America

Chaohua Cheng and Gonggu Zang
Institute of Bast Fiber Crops, Chinese Academy of Agricultural Sciences, Changsha, Hunan, The People's Republic of China

Rui Hai Liu
Department of Food Science and Institute of Comparative and Environmental Toxicology, Stocking Hall, Cornell University, Ithaca, New York, United States of America

Shuichi Shibuya, Yusuke Ozawa, Kenji Watanabe, Naotaka Izuo and Toshihiko Toda
Department of Advanced Aging Medicine, Chiba University Graduate School of Medicine, Chiba, Japan

Koutaro Yokote
Department of Clinical Cell Biology and Medicine, Chiba University Graduate School of Medicine, Chiba, Japan

Takahiko Shimizu
Department of Advanced Aging Medicine, Chiba University Graduate School of Medicine, Chiba, Japan
Molecular Gerontology, Tokyo Metropolitan Institute of Gerontology, Itabashi, Tokyo, Japan

Li Zuo and William J. Roberts
Radiologic Sciences and Respiratory Therapy Division, School of Health and Rehabilitation Sciences, Davis Heart and Lung Research Institute, The Ohio State University College of Medicine, Columbus, Ohio, United States of America
Department of Biological Sciences, Oakland University, Rochester, Michigan, United States of America

Peter D. Wagner
Department of Medicine, University of California San Diego, La Jolla, California, United States of America

Juliana Kishek
Department of Biological Sciences, Oakland University, Rochester, Michigan, United States of America

Thomas M. Best
Division of Sports Medicine, Department of Family Medicine, Sports Health and Performance Institute, The Ohio State University, Columbus, Ohio, United States of America

Michael T. Chien
Department of Biology, Kalamazoo College, Kalamazoo, Michigan, United States of America

Philip T. Diaz
Division of Pulmonary, Allergy, Critical Care & Sleep Medicine, The Ohio State University Wexner Medical Center, Columbus, Ohio, United States of America

Valdecir F. Ximenes
Department of Chemistry, Faculty of Sciences, São Paulo State University (UNESP), Bauru, São Paulo, Brazil

Luis Octavio Regasini, Dulce H. S. Silva and Vanderlan S. Bolzani
Department of Organic Chemistry, Nuclei of Bioassays, Ecophysiology and Biosynthesis of Natural Products (NuBBE), Institute of Chemistry, Saõ Paulo State University (UNESP), Araraquara, São Paulo, Brazil

Maria Luiza Zeraik
Department of Chemistry, Faculty of Sciences, São Paulo State University (UNESP), Bauru, São Paulo, Brazil
Department of Organic Chemistry, Nuclei of Bioassays, Ecophysiology and Biosynthesis of Natural Products (NuBBE), Institute of Chemistry, Saõ Paulo State University (UNESP), Araraquara, São Paulo, Brazil

Maicon S. Petrônio and Luiz Marcos da Fonseca
Department of Clinical Analysis, School of Pharmaceutical Sciences, Sa~o Paulo State University (UNESP), Araraquara, São Paulo, Brazil

Dyovani Coelho and Sergio A. S. Machado
Institute of Chemistry of São Carlos, São Paulo University (USP), São Carlos, SP, Brazil

Suraporn Matragoon and Azza B. El-Remessy
Clinical and Experimental Therapeutics, University of Georgia, Augusta, Georgia, United States of America Culver Vision Discovery Institute, Augusta, Georgia, United States of America Charlie Norwood VA Medical Center, Augusta, Georgia, United States of America

Mohammed A. Abdelsaid and Adviye Ergul
Clinical and Experimental Therapeutics, University of Georgia, Augusta, Georgia, United States of America Department of Physiology, Georgia Regents University, Augusta, Georgia, United States of America Charlie Norwood VA Medical Center, Augusta, Georgia, United States of America

Jiao Liu, Yang Bai, Jin Li, Er-wei Liu, Jun He, Xiucheng Jiao, Xiu-Mei Gao and Bo-li Zhang
Tianjin State Key Laboratory of Modern Chinese Medicine, Tianjin University of Traditional Chinese Medicine, Tianjin, China

Yan-xu Chang
Tianjin State Key Laboratory of Modern Chinese Medicine, Tianjin University of Traditional Chinese Medicine, Tianjin, China State Key Laboratory of New-tech for Chinese Medicine Pharmaceutical Process, Kanion Pharmaceutical Co., Ltd, Lianyungang, China

Zhen-zhong Wang and Wei Xiao
State Key Laboratory of New-tech for Chinese Medicine Pharmaceutical Process, Kanion Pharmaceutical Co., Ltd, Lianyungang, China

Thomas J. O'Grady, A. Gregory DiRienzo and Margaret A. Gates
University at Albany, School of Public Health, Rensselaer, New York, United States of America

Cari M. Kitahara
Division of Cancer Epidemiology and Genetics, National Cancer Institute, National Institutes of Health, Rockville, Maryland, United States of America

Samantha Serra Costa
Faculty of Pharmacy, Federal University of Bahia, Rua Barão de Geremoabo, s/n, Ondina, CEP 40171- 970, Salvador, Bahia, Brasil

Janice Izabel Druzian, Carolina Oliveira de Souza and Alaíse Gil Guimarães
Faculty of Pharmacy, Department of Bromatological Analysis, Federal University of Bahia, Rua Barão de Geremoabo, s/n, Ondina, CEP 40171-970, Salvador, Bahia, Brasil

Bruna Aparecida Souza Machado
Faculty of Technology, SENAI/CIMATEC, Serviço Nacional de Aprendizagem Industrial - SENAI, Av. Orlando Gomes, nu 1845, Piatã , CEP 41650-010, Salvador, Bahia, Brasil

Adrianna Wojtal-Frankiewicz, Piotr Frankiewicz and Tomasz Jurczak
Department of Applied Ecology, University of Lodz, Lodz, Poland

Joanna Bernasińska and Krzysztof Gwoździński
Department of Molecular Biophysics, University of Lodz, Lodz, Poland

Yong Sui, Qian Wu, Xiaopeng Li, Bijun Xie and Zhida Sun
College of Food Science and Technology, Huazhong Agricultural University, 1 Shizishan Street, Hongshan District, Wuhan 430070, Hubei, China

Juan Xiao and Mingwei Zhang
Key Laboratory of Functional Foods, Ministry of Agriculture/ Guangdong Key Laboratory of Agricultural Products Processing, Sericultural & Agri-Food Research Institute Guangdong Academy of Agricultural Sciences, 133 Yiheng Road, Dongguan Zhuang, Tianhe District, Guangzhou 510610, Guangdong, China

Shuyi Li
College of Food Science and Engineering, Wuhan Polytechnic University, Xuefu South Road, Changqing Garden, Dongxihu District, Wuhan 430023, Hubei, China

Maja D. K. Fedder
Laboratory of Reproductive Biology, Scientific Unit, Regional Hospital of Horsens, Horsens, Denmark

Henrik B. Jakobsen and Ina Giversen
Nerthus ApS, Lejre, Denmark

Lars P. Christensen
Department of Chemical Engineering, Biotechnology and Environmental Technology, Faculty of Engineering, University of Southern Denmark, Odense, Denmark

Erik T. Parner
Department of Public Health, Section for Biostatistics, University of Aarhus, Aarhus, Denmark

Jens Fedder
Laboratory of Reproductive Biology, Scientific Unit, Regional Hospital of Horsens, Horsens, Denmark
Centre of Andrology, Fertility Clinic, Department D, Odense University Hospital, Odense, Denmark

Zhenchang Liang
Institute of Botany, the Chinese Academy of Sciences, Beijing, China
Department of Food Science, Cornell University, Ithaca, New York, United States of America

Rui Hai Liu
Department of Food Science, Cornell University, Ithaca, New York, United States of America

Lailiang Cheng
Department of Horticulture, Cornell University, Ithaca, New York, United States of America

Gan-Yuan Zhong
USDA -ARS Grape Genetics Research Unit, Geneva, New York, United States of America

Hitomi Hosaka, Risa Haruki, Kana Yamada and Teruyuki Komatsu
Department of Applied Chemistry, Faculty of Science and Engineering, Chuo University Tokyo, Japan

Christoph Böttcher
Research Center of Electron Microscopy, Institute of Chemistry and Biochemistry Freie Universität Berlin, Berlin, Germany

Yi-Ming Lin, Zhen-Hua Ding, Wei-Ming Chai and Shu-Dong Wei
Key Laboratory of the Ministry of Education for Coastal and Wetland Ecosystems, Xiamen University, Xiamen, China

Hai-Chao Zhou
Key Laboratory of the Ministry of Education for Coastal and Wetland Ecosystems, Xiamen University, Xiamen, China
Department of Biology and Chemistry, City University of Hong Kong, Hong Kong SAR, China
Futian-CityU Mangrove R&D Centre, City University of Hong Kong Shenzhen Research Institute, Shenzhen, China

Nora Fung-yee Tam
Department of Biology and Chemistry, City University of Hong Kong, Hong Kong SAR, China
Futian-CityU Mangrove R&D Centre, City University of Hong Kong Shenzhen Research Institute, Shenzhen, China

Index

www.ingramcontent.com/pod-product-compliance
Lightning Source LLC
Chambersburg PA
CBHW080411190526
45161CB00003B/205